D0085150

ALLE·ZEIT·WACH
1842

M. E. Müller · M. Allgöwer
R. Schneider · H. Willenegger

Manual of INTERNAL FIXATION

Techniques Recommended by the AO Group

Second Edition, Expanded and Revised

In Collaboration with
W. Bandi · A. Boitzy · R. Ganz · U. Heim · S. M. Perren
W. W. Rittmann · Th. Rüedi · B. G. Weber · S. Weller

Translated by J. Schatzker

With 345 Color Illustrations and 2 Templates for Preoperative Planning

Springer-Verlag
Berlin Heidelberg New York 1979

Foreign language editions of the 1st edition

German: Manual der Osteosynthese
Springer-Verlag
Berlin · Heidelberg · New York 1969

French: Manuel d'Ostéosynthèse
Masson et Cie., Éditeurs
Paris 1970, 1974

Italian: Manuale della Osteosintesi
Aulo Gaggi, Editore, Bologna 1970

Japanese: 図説 骨折の手術 AO法
Igaku Shoin, Tokyo 1971

Spanish: Manual de Osteosíntesis
Editorial Científico-Médica
Barcelona 1971, 1972, 1975, 1977

Foreign language edition of the 2nd edition

German: Manual der Osteosynthese
Springer-Verlag
Berlin · Heidelberg · New York, 1977

Spanish: Manual de Osteosíntesis
Springer-Verlag
Berlin · Heidelberg · New York 1979

French: Manuel d'Ostéosynthèse
Springer-Verlag
Berlin · Heidelberg · New York 1979

Other foreign language editions in preparation

ISBN 3-540-09227-7 2. Auflage Springer-Verlag Berlin Heidelberg New York
ISBN 0-387-09227-7 2nd edition Springer-Verlag New York Heidelberg Berlin

ISBN 3-540-05219-4 1. Auflage Springer-Verlag Berlin Heidelberg New York
ISBN 0-387-05219-4 1st edition Springer-Verlag New York Heidelberg Berlin

Library of Congress Cataloging in Publication Data. Main entry under title: Manual of internal fixation. Translation of Manual der Osteosynthese.
Edition for 1970 entered under M.E. Müller.
Bibliography: p. Includes index. 1. Internal fixation in fractures. I. Müller, Maurice Edmond. D103.15M8313 1979 617'.15 78-20743

Printed in Germany.

Reproduction of the figures: Gustav Dreher GmbH, Stuttgart

Typesetting, printing, and bookbinding by Universitätsdruckerei H. Stürtz AG, Würzburg

2124/3130-543210

Preface to the Second German Edition

The goal of the Association for the Study of the Problems of Internal Fixation (AO/ASIF) is not just the propagation of internal fixation. The name of the Association implies its activity; it concerns itself with the problems of internal fixation. It has two main spheres of activity. The one deals with the question of indications for internal fixation in fracture treatment. The other deals with biomechanical improvements of internal fixation for fractures, osteotomies and non-unions.

The difficulties and failures which over the decades plagued internal fixation were, apart from wound infection, largely the result of lack of knowledge of basic scientific facts. Bone healing in the presence of internal fixation was poorly understood. The Association (AO) began its studies by focusing on this problem. These efforts led to the foundation of the Laboratory for Experimental Surgery in Davos. A research programme was developed which encompassed the fields of biology, biomechanics and metallurgy. Research in these fields was pursued at Davos as well as in the associated hospitals and institutions. New bone instruments and implants were developed and constantly modified in accordance with the new biomechanical discoveries made in the basic laboratory and associated hospitals. The evaluation of the newly developed methods was greatly enhanced by careful clinical documentation of the patients during their initial treatment and follow-up. Another important source of suggestions and advice have been the AO basic and advanced courses which have been held regularly both in Switzerland and in other countries. They represent an international effort off considerable magnitude: by the end of 1976, 55 basic courses had been held with almost 20,000 general and orthopaedic surgeons having participated.

The manual, as its name implies, is designed chiefly to convey technical details. The technical recommendations, however, are based upon fundamental research, as well as upon the vast clinical experience of over 50,000 operatively treated fractures, osteotomies and nonunions. Simply because in the manual we dwell on the operative methods of stabilization of most of the skeleton, one should not assume that we wish to minimize the issue of indications for operative intervention. Although in this second edition we discuss the indications in greater detail than we did in the first, they still remain the personal responsibility of the surgeon. Operative fracture treatment is a most worthwhile but a very difficult and demanding therapeutic method. We must reiterate what we stated in 1963 in our book *The Technique of Internal Fixation of Fractures* and in the first edition of this manual in 1969. Internal fixation should not be carried out by an inadequately trained surgeon, nor without the necessary equipment and adequate sterile operating room conditions. Advocates of internal fixation who lack self-criticism are much more dangerous than sceptics or outright opponents. We hope that our readers will understand our efforts in this direction and pass on to us any constructive criticism.

We wish to express our gratitude to all the members of the AO who took part in the revision of this manual as well as to our marvellous artist Mr. K. OBERLI and to our dear and dependable collaborator Miss E. MOOSBERGER.

Berne, August 1977

M.E. MÜLLER
M. ALLGÖWER
R. SCHNEIDER
H. WILLENEGGER

Contents

SPECIAL PART

SUPPLEMENT

List of Addresses

Authors

MÜLLER, MAURICE E., Prof. Dr. med., Direktor der Klinik für Orthopädische Chirurgie, Inselspital, CH-3010 Bern

ALLGÖWER, MARTIN, Prof. Dr. med., Vorsteher der Chirurgischen Universitätsklinik, Kantonsspital, Department of Surgery, CH-4004 Basel

SCHNEIDER, ROBERT, Prof. Dr. med., Alpenstraße 15, CH-2500 Biel

WILLENEGGER, HANS, Prof. Dr. med., President of the AO-International, Murtenstraße 35, CH-3008 Bern

Contributors

BANDI, WALTER, Prof. Dr. med., Chefarzt, Bezirksspital, CH-3800 Interlaken

BOITZY, ALEXANDRE, PD Dr. med., Ch. de Craivavers 6, CH-1012 Lausanne

GANZ, REINHOLD, Dr. med., Klinik für Orthopädie und Chirurgie des Bewegungsapparates, Inselspital, CH-3010 Bern

HEIM, URS, PD Dr. med., Chefarzt, Kreuzspital, CH-7000 Chur

PERREN, STEPHAN M., PD Dr., Leiter des Laboratoriums für Experimentelle Chirurgie, Forschungsinstitut, CH-7270 Davos-Platz

RITTMANN, WILLI WERNER, PD Dr., Chirurgische Universitätsklinik, Department of Surgery, Kantonsspital, CH-4004 Basel

RÜEDI, THOMAS, PD Dr. med., Chirurgische Universitätsklinik, Department of Surgery, Kantonsspital, CH-4004 Basel

WEBER, BERNHARD G., Prof. Dr. med., Chefarzt der Klinik für Orthopädie und Chirurgie des Bewegungsapparates, Kantonsspital, CH-9006 St. Gallen

WELLER, SIEGFRIED, Prof. Dr. med., Direktor des BG Unfallkrankenhauses, D-7400 Tübingen

Translator

JOSEPH SCHATZKER, M.D., B. Sc. (Med.), F.R.C.S. (c), Associate Professor, Division of Orthopaedic Surgery, University of Toronto, Ontario, Canada

Introduction

The first part of this manual deals with the experimental and scientific basis and the principles of the AO/ASIF method of stable internal fixation. It deals with the function and main use of the different AO implants, the use of the different AO instruments, and with the essentials of the operative technique and of postoperative care. It also discusses the handling of the most important postoperative complications.

The second part deals at length with the AO recommendations for the operative treatment of the most common closed fractures in the adult. This has been organized in anatomical sequence. The discussion of the closed fractures is followed by a discussion of open fractures in the adult, then by fractures in children and finally by pathological fractures.

The third part presents, in a condensed fashion, the application of stable internal fixation to reconstructive bone surgery.

GENERAL CONSIDERATIONS

1 Aims and Fundamental Principles of the AO Method

The Chief Aim of Fracture Treatment is the Full Recovery of the Injured Limb

In every fracture there is a combination of damage to both the soft tissues and to bone. Immediately after the fracture and during the phase of repair, we see certain local circulatory disturbances, certain manifestations of local inflammation, as well as pain and reflex splinting. These three factors, that is, circulatory disturbances, inflammation and pain, when combined with the defunctioning of bone, joints and muscle, result in the so-called *fracture disease*. Fracture disease is a clinical state which is manifested by chronic oedema, soft-tissue atrophy, osteoporosis and joint stiffness. Every type of fracture treatment must therefore encompass the treatment not only of the fracture but also of all the associated local reactions.

Life is Movement, Movement is Life

This should be the guiding principle of fracture care! Full, active, pain-free mobilization results in a rapid return to normal of the blood supply to both the bone and the soft tissues. It also enhances articular cartilage nutrition by the synovial fluid, and when combined with partial weight bearing, it greatly decreases post-traumatic osteoporosis by restoring an equilibrium between bone resorption and bone formation. *A satisfactory internal fixation is achieved only when external splinting is superfluous and when full active pain-free mobilization of muscles and joints is possible.* This thesis is the AO's main objective and is best achieved by a stable internal fixation which will last for the whole duration of the bone healing.

Stability means not only lasting adaptation of the fragments but also the prevention of any microscopic movement between the bone fragments, whether of a fracture or of an osteotomy. Over the past 20 years *inter-fragmental compression* has proved itself to be the most dependable method of immobilization. Certain fractures can also be satisfactorily immobilized by *splinting*, as for instance by means of an intra-medullary nail as originally proposed by KÜNTSCHER (see p. 104).

These methods of internal fixation have decided advantages. They not only result in a pain-free convalescence but also in a considerably shortened hospitalization time and in a shortened period of disability. Post-traumatic dystrophies, malunions and pseudarthroses are now rare complications.

In addition to the publications of LAMBOTTE, LANE and EGGERS, there have been many publications on methods of internal fixation which emphasized the importance of fragment immobilization. All these methods lacked the stabilizing influence of inter-fragmental compression. DANIS was the first to recognize the importance of inter-fragmental compression and he was the first to use it clinically. In 1949 he demonstrated that fractures of the forearm immobilized with his 'coapteurs', which exerted an axial inter-fragmental compression, could not only be moved immediately after the internal fixation but also united without any radiologically recognizable callus.

Before compression could gain widespread acceptance in clinical use, further research had to be carried out. There was a lack of basic animal experimentation and of a systematic histological analysis of callus formation in the presence and in the absence of stable internal fixation. These gaps in the knowledge were closed by the experiments carried out by the AO. Here we must mention in particular WAGNER, WILLENEGGER, SCHENK, PERREN, RAHN, GALLINARO, HUTZSCHENREUTER, J. MÜLLER, PUDDU and SCHWEIBERER, who demonstrated that in order for a method of internal fixation to be reliable and result in uncomplicated bone healing, it must be based on sound principles of mechanical stability in the presence of a preserved blood supply to bone and soft tissues.

These experiments, together with the carefully developed implants and instruments tested in the treatment of over 50,000 fractures, have come to comprise the cornerstones of the AO method.

1.1 Aims of the AO Method

Rapid Recovery of the Injured Limb

This is accomplished by:

Anatomic reduction of the fracture fragments particularly in joint fractures.

Preservation of the blood supply to the bone fragments and the soft tissue by means of atraumatic surgery.

Stable internal fixation designed to fulfill the local biomechanical demands.

Early active painfree mobilization of muscles and joints adjacent to the fracture. In this way the development of "fracture disease" is prevented.

The fulfilment of these four conditions is the prerequisite for a perfect internal fixation. Such fixation will result in the best healing not only of the bone but also of all components of the injury.

Principle of the AO

Early active pain-free mobilization of the muscles and joints of the injured limbs in patients with multiple fractures.

Fig. 1 The value of early active pain-free mobilization of all joints is particularly well demonstrated in patients with *multiple fractures*.

a A patient with an explosion fracture of the left femur (a′), a supracondylar fracture of the right femur (a″) and an open fracture of the left tibia (a‴).

b Function 4 weeks after operative treatment. Normal function of the hip, knee and foot. The axial alignment of the limbs is normal and partial weight bearing is possible.

c Four months after internal fixation with condylar plates. (A "T" plate could have been used as a buttress plate for this fracture of the proximal tibia. It would have been just as good and much simpler.)

1

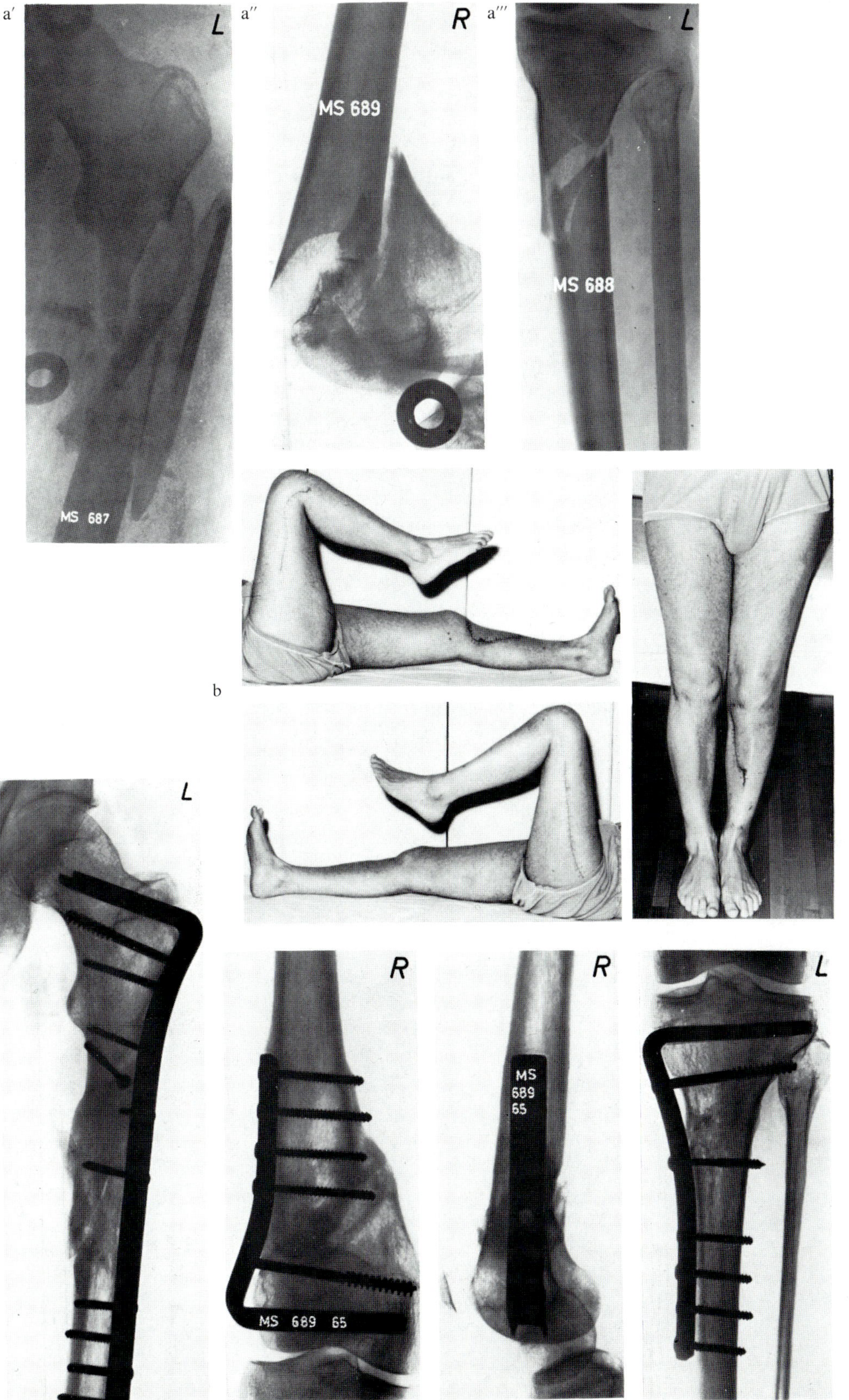

Many of the shaft fractures particularly of the humerus and tibia are best treated by non-operative methods. In intra-articular fractures, however, an absolutely perfect anatomic reduction is required *if a full functional recovery is to be achieved.*

Knowledge of the anatomy and of joint mechanics, a three-dimensional appreciation of the problem, technical ability and a careful preoperative plan of the procedure, are all basic requirements, if a perfect result is to be obtained. In intra-articular fractures, an AP and lateral radiograph are rarely enough to demonstrate all the fracture lines. We feel, therefore, that in addition, oblique projections must always be obtained. Frequently radiographs of the uninvolved limb are essential. At times it is also necessary to resort to tomography to demonstrate articular depressions. In the treatment of intra-articular fractures, we have found the following principle useful: first reconstruct the joint congruency. This is accomplished by an anatomic reduction and stable fixation of all the articular fragments. Once this is achieved, the reconstructed articular component is fixed to the shaft. It is also necessary to repair torn ligaments, tendons and joint capsules.

When considering early mobilization of joint fractures, one must take into account the damage to the associated soft tissues. Knee stiffness following surgery of the shaft of the femur and of supracondylar fractures can be prevented by postoperative positioning of the leg for 4–5 days with the knee bent to 90°. Stiffness is avoided by preventing the stripped muscles from reattaching themselves too far proximally and by preventing adhesions of the suprapatellar pouch (see Fig. 103b). When the ankle is involved in surgery it should be kept at 90° for 4–5 days. This will prevent an equinus deformity and result in more rapid healing of the soft tissues. After 4–5 days, active mobilization of both the knee and ankle should begin.

Fig. 2 In *comminuted intra-articular fractures,* rigid internal fixation is usually possible, but in order to obtain a full range of movement it is sometimes necessary to sacrifice part of a joint surface.

a A badly comminuted fracture of both the radius and ulna.

b Nine months later. The comminuted radial head was resected.

c, d Range of movement 1 month after surgery: full extension, flexion of the elbow to 150°, 90° pronation and 100° supination.

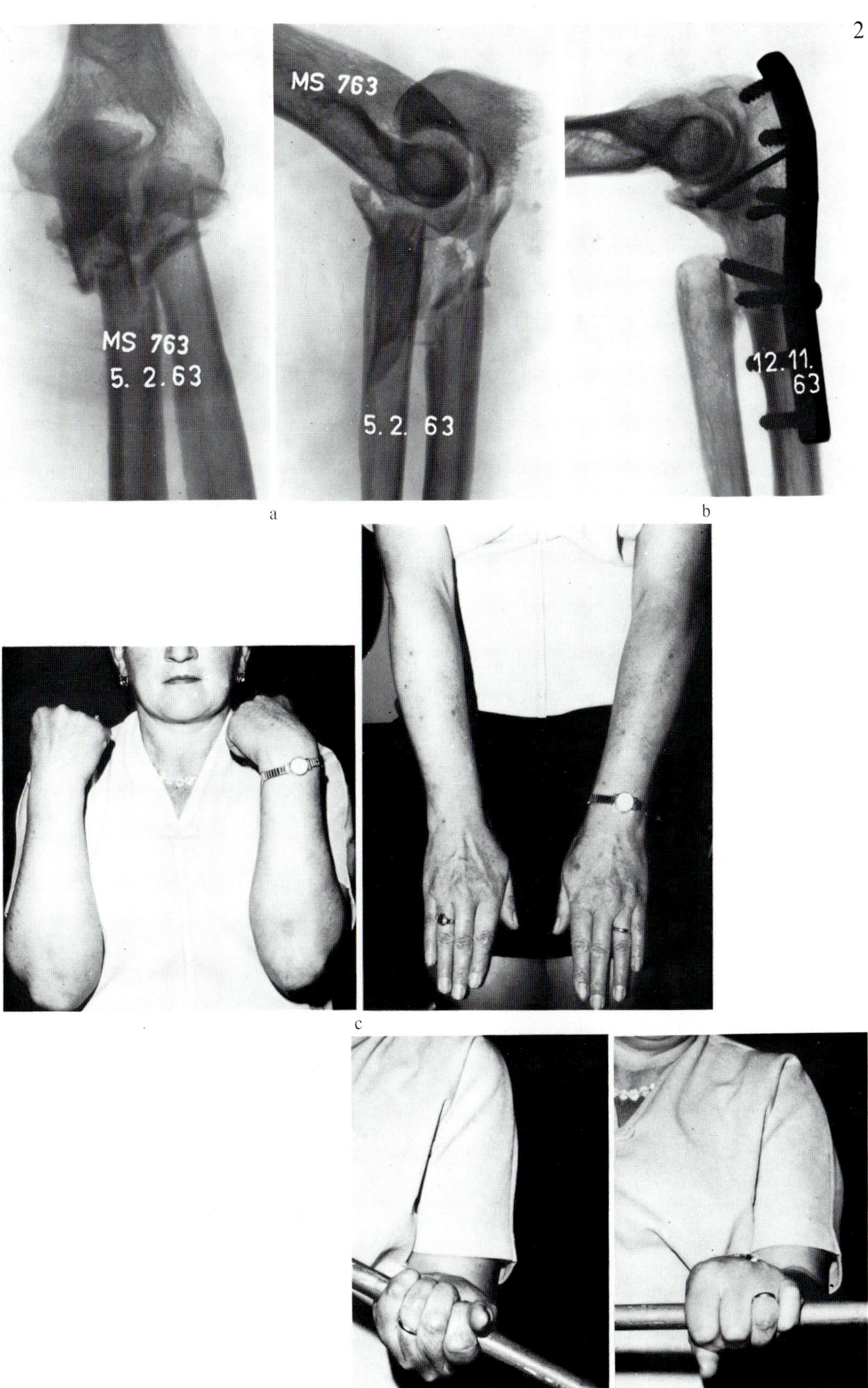

a b c d

1.2 Basic Principles of the AO Method

1.2.1 Histology of Bone Healing in the Presence of Stable Internal Fixation

The characteristic feature of union of bone without rigid fixation is the formation of a periosteal and endosteal callus. In the fracture gap there is initially the formation of connective tissue and cartilage which is then secondarily replaced by bone.

In 1935 KROMPECHER showed in his experiments on the skulls of embryonal rats that in an area free from mechanical forces, primary vascular bone formation could occur. He postulated that such *primary vascular bone formation* was also possible in fracture healing if the fragments were rigidly immobilized.

In 1963 SCHENK and WILLENEGGER were able to demonstrate histologically, first in the dog and subsequently in human material, that primary vascular bone union did occur in the healing of fractures. They also demonstrated that primary vascular bone union is the rule when the fragments are rigidly immobilized and when their blood supply is preserved. Subsequently, RAHN et al. were able to demonstrate primary bone union in many different animals. They also showed that such union would occur in the presence of full weight bearing as long as rigid fixation was maintained. Thus these investigators demonstrated that they were dealing with a universal biological principle of bone healing which occurs in the presence of absolutely rigid fixation.

The experiment of SCHENK was carried out on a transverse osteotomy of a dog radius which was then rigidly immobilized with a compression plate. The human material came from fractured bones which had been immobilized by means of lag screws and neutralization plates. In both, *gap* and *contact* healing was demonstrated (Fig. 3c and e).

Fig. 3 *Schematic representation of healing of an osteotomy of a dog's radius under compression* (drawn from histological material of SCHENK).

a The cortex adjacent to the plate is in contact. There is no ingrowth of mesenchymal cells either from the periosteum or endosteum. Note that there is a gap in the cortex opposite the plate.

b The bone ends in contact immediately adjacent to the plate show no changes within the first 3–4 weeks.

c From the fourth week onwards the bone ends in contact show an active Haversian remodelling. There is a proliferation of Haversian canals which grow across both living and dead cortex bridging the osteotomy. *This is contact healing.*

d The gap in the cortex opposite the plate is invaded by blood vessels which appear within the first 8 days. These are accompanied by osteoblasts which deposit osteoid. This gives rise to bone lamellae which are oriented at 90° to the long axis.

e From the fourth week onwards these inter-fragmentary, transversely oriented lamellae of bone are replaced by axially oriented osteons. *This is gap healing.*

f High magnification of a remodelling osteon shows resorption and bone formation adjacent to one another. At the head are osteoclasts (*a*) which give rise to a resorption canal, which is then invaded by capillary sprouts (*b*). The circumferentially oriented osteoblasts (*c*) give rise to a new osteon (*d*).

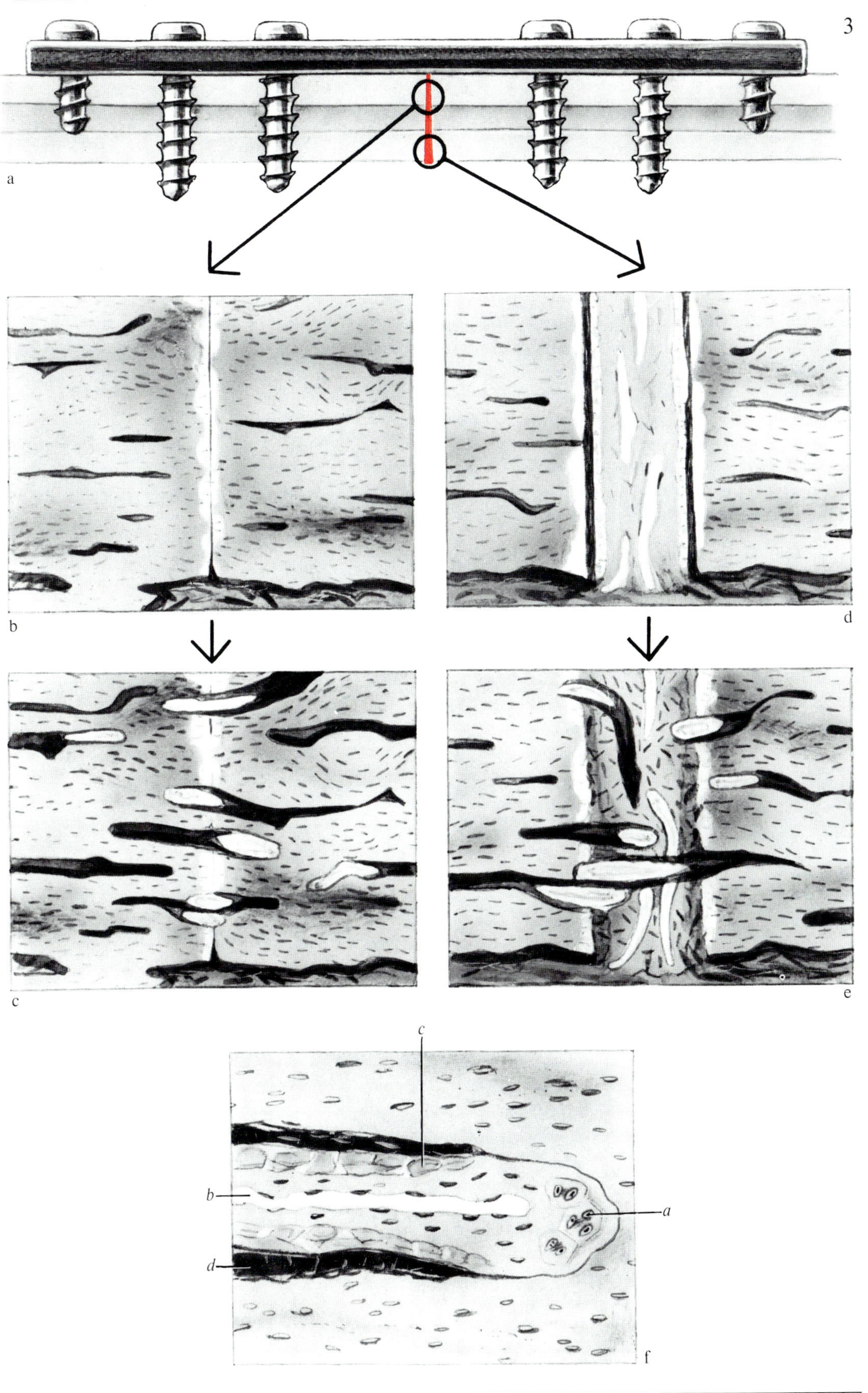
a
b
c
d
e
c
b
a
d
f

1.2.2 Bone Reaction to Compression

PERREN and his collaborators utilized inter-fragmental compression to achieve rigid immobilization of osteotomies of the tibia of sheep. The plates (compression plates) which were used to secure this axial compression were fitted with strain gauges. The amount of compression was checked daily. These investigators demonstrated that in the presence of a stable compression fixation of cortex, the compression decayed slowly. By the end of 2 months it was reduced to approximately 50%.

This same experiment but without osteotomy of the tibia showed a similar decay of the compression. In this way PERREN was able to prove that the slow decay of the compression was not due to shortening of the fragments and resorption but due to Haversian remodelling.

The initial rapid fall in the compression can be demonstrated also in fresh cadaver bone. It is therefore a purely mechanical phenomenon, the result of the visco-elasticity of bone.

PERREN et al. were also able to demonstrate that bone can tolerate very high static compression (over 300 kp/cm^2) without undergoing pressure necrosis.

Compression greatly enhances the rigidity of internal fixation. Since the compression in the presence of rigid internal fixation decays slowly while union is progressing, we can see how the mechanical advantages of compression can be utilized without any adverse biological consequences.

Fig. 4 PERREN'S *experiments.*

a A plate instrumented with strain gauges which register with an accuracy ±1.5 kp in the range of 300 kp.

b Such a plate can be brought under tension by means of a tension device (see Fig. 32). The tension of the plate corresponds to the compression of the fracture.

c The instrumentation of a tibia of a sheep with such a plate. The wire leads are brought out subcutaneously up to the trochanteric area. They serve to link the plate with a recording instrument.

d The three standard curves established by PERREN:
A The application of a compression plate to fresh cadaver bone. Tension of approximately 100 kp. Measurements continued for 3 months. Initially a rapid fall. Subsequently tension maintained without decay.
B A similarly applied plate but to an intact tibia of a sheep: Initially a rapid fall followed by a gradual decay.
C An identical experiment to *B*, with the addition of an osteotomy of the tibia. The curve is identical to *B*: initially a rapid fall followed by a gradual decay.

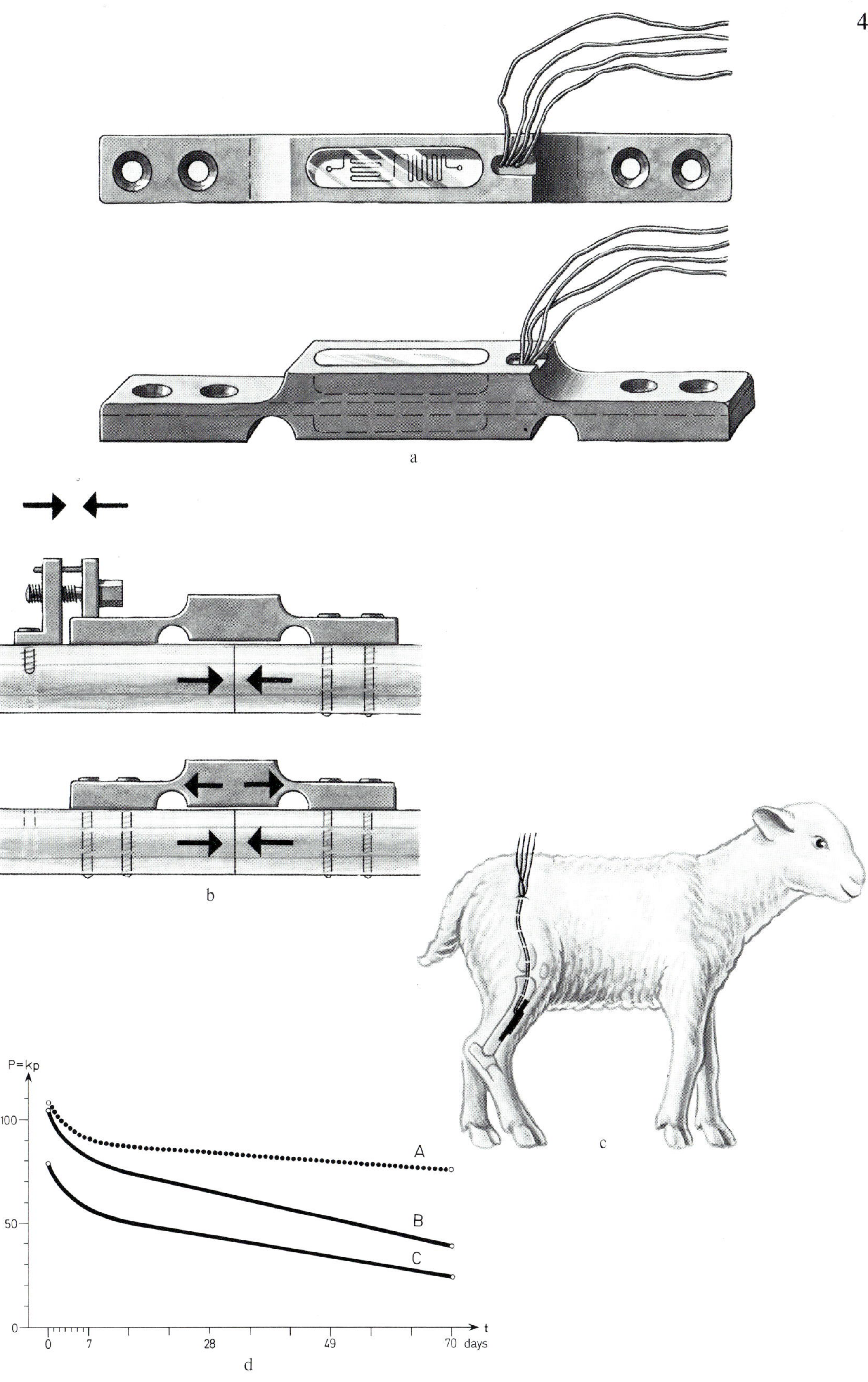
a
b
c
P=kp
100
50
0
A
B
C
t
0
7
28
49
70
days
d

1.2.3 Bone Reaction to Movement

A fracture which has been rigidly fixed by means of inter-fragmental compression (lag screw) demonstrates absolutely no movement between the fragments. In very sharp contrast to this is the fracture which has been stabilized by means of an intramedullary nail or by means of external fixators. A fracture, which has been *splinted,* will always demonstrate movement between the fragments even if only of microscopic dimension (see p. 27).

1. HUTZSCHENREUTER et al. (1969) illustrated that the amount of callus is directly proportional to the relative movement between the fragments.

2. PERREN et al. demonstrated that resorption at the interface between implant and bone is the result of inadequate pre-stress between the contact surfaces, which in turn is directly related to the pre-stress of the implant. If under functional loading a force is generated which is opposite in direction to the pre-stress but equal to or greater than the pre-stress, then the contact surface will become unloaded (zero load) and may even become loaded in the opposite direction to the original pre-stress. This signifies micro-instability and micro-movement at the contact surface which leads to resorption. The same applies to a fracture. When the fracture is stably fixed, no micro-movements occur and therefore no resorption takes place. When stable fixation is not achieved, as for instance in the closed treatment of fractures, the resorption of the fracture ends is the characteristic feature of such treatment. This was clearly demonstrated by L. BOEHLER.

3. The influence of the relationship between static and dynamic forces on the stability of internal fixation as conceived by PERREN is as follows: Movement at the interface will not occur as long as the static forces (pre-stress) are greater than the dynamic forces (functional load). As soon as the static load is insufficient (either absent or insufficient pre-stress) the functional load will lead to micro-movement at the contact interface. This mechanical micro-instability will lead to resorption and secondary loosening of the implant (Fig. 6).

Fig. 5 *The formation of callus is dependent on mechanical conditions* (HUTZSCHENREUTER). A thin, easily deformable plate (a) results in certain movement at the fracture interface. This results in a formation of a large callus (irritation callus). It is characterized by an indistinct and irregular margin and a non-homogeneous structure.
In an identical experiment where the fragments are, however, fixed with a plate of sufficient rigidity (b), the resultant callus is very much smaller and has a distinct sharp edge and is of homogeneous uniform structure.

Fig. 6 *Movement and bone resorption* (PERREN). A strain gauge instrumented plate (a) was fixed to bone in such a way that two screws on the left were pre-stressed with respect to each other. No movement took place between the plate, the screws and the bone on the left side. On the right side only one screw was inserted without pre-stress. This resulted in a repetitively changing load. It could be demonstrated that where static pre-stress was sufficient to result in stability, no resorption took place (b). Where pre-stress was insufficient and where repetitive change from tension to compression occurred, resorption did occur with secondary loosening of the implant (c).

5

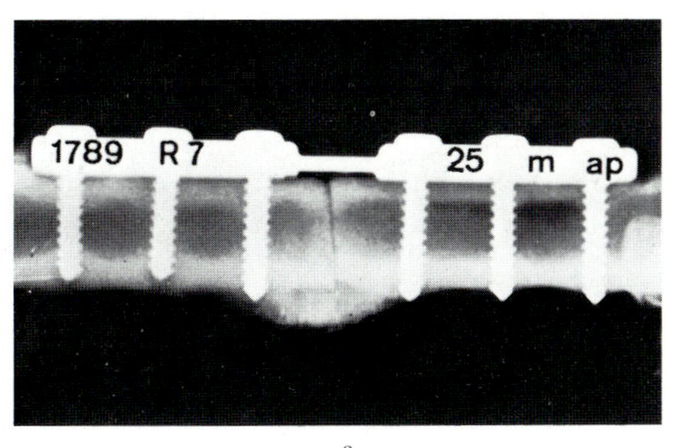

a

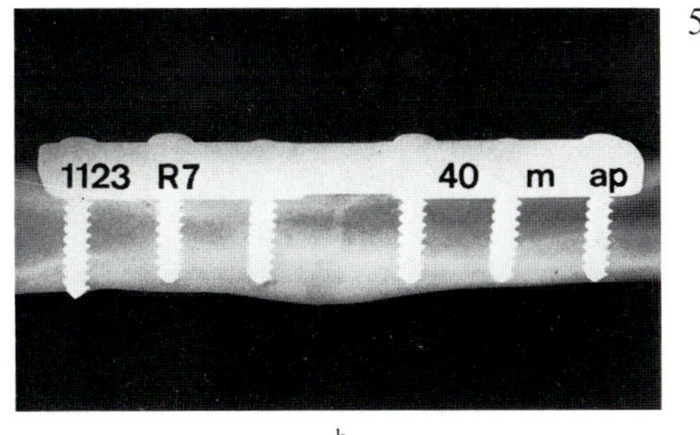

b

6

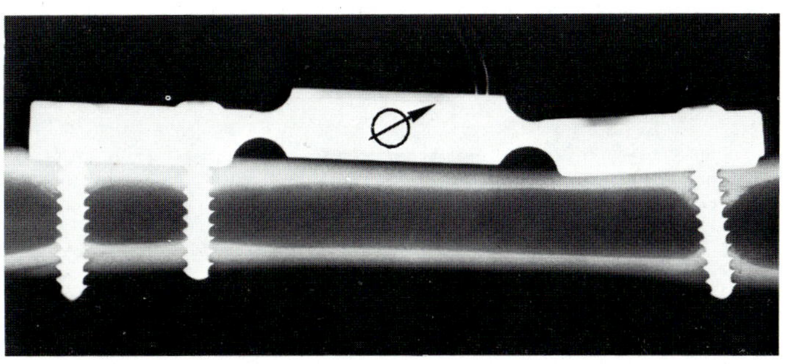
a

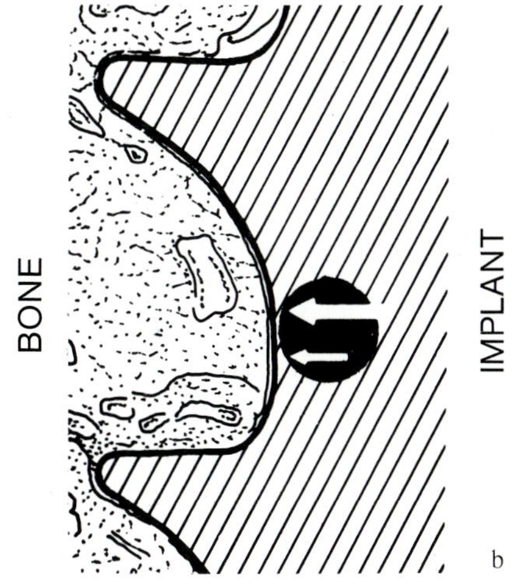

b

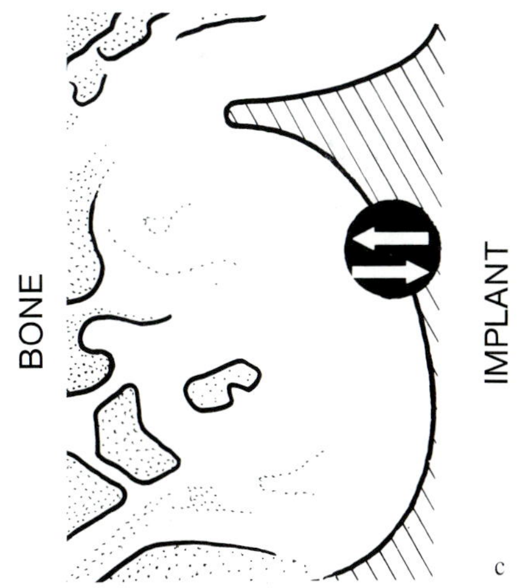

c

1.2.4 Reaction of Bone to Metallic Implants

Stable internal fixation has certain prerequisites. These are: an implant of high tissue tolerance, combined with a strong and durable fixation of this implant to bone.

Corrosion gives rise to metallic ions in bone. These can influence the tissue tolerance. Metals which conform to standards, as a rule, do not give rise to corrosion which influences bone healing. There are three types of corrosion: fretting corrosion, galvanic corrosion, and frictional corrosion.

Fretting corrosion occurs typically in stainless steel, and is influenced by inclusions. Therefore, the AO has insisted on great purity of the stainless steel* used in its implants, which exhibit fretting corrosion only in combination with frictional corrosion.

Galvanic corrosion results from the use of dissimilar metals. Therefore the use of dissimilar metals must be avoided.

Galvanic corrosion can also result from the contamination of implants by instruments (metal transfer). All stainless steel which is malleable and tolerated by tissue is relatively soft. The AO instruments and implants have been manufactured in such a way that under most circumstances the transfer occurs from the implant to the instrument and is completely harmless.

Frictional Corrosion: Metal implants are protected from corrosion by a passive layer. This passive layer regenerates very rapidly "in vivo". Scratches which occur during surgery are therefore quite harmless. Motion, as in the presence of instability, results in marked increase in frictional corrosion.

Tissue Tolerance: WAGNER demonstrated that AO stainless steel even after prolonged direct contact with bone does not give rise to any adverse histological changes. Allergic reactions are extremely rare.

Bone Remodelling: MATTER in plating intact bone was able to demonstrate that static compression does not result in bone remodelling. The temporary remodelling of bone which is seen with plating is the result of other factors such as vascularity, the rigidity of the implant, etc. Because of the temporary remodelling, implant removal must not be rushed. In double plating, the plates should be removed one at a time and at least 6 months apart (Refractures: see Sect. 3.7.4).

* The AO steel corresponds to the most important technical standards for implantation: SNV 056 506; ASTM F 55 and F 138–139; ISO: TC 150, DIS 5832; DIN 5880; BS 35311: part I.

Fig. 7 *The bone adjacent to an AO cortex screw* (WAGNER).

a The bone has grown along the surface of the screw and forms a mirror image of the screw profile. Haemopoietic tissue is separated from the metal in places by only a thin trabeculum of bone. There is no evidence of resorption or of marrow fibrosis. The screw was in position for 2 months.

b We can observe living osteocytes in lacunae in close proximity to the metal. The canaliculi reach the screw surface. There is no evidence of resorption. The screw was in place for 9 months.

c *Bone remodelling deep to the plate* (MATTER). The cortex adjacent to the plate of an intact tibia of a sheep shows marked bone remodelling. This remodelling is so marked that after 1–2 months, we find marked osteoporosis, which disappears again after a prolonged period.

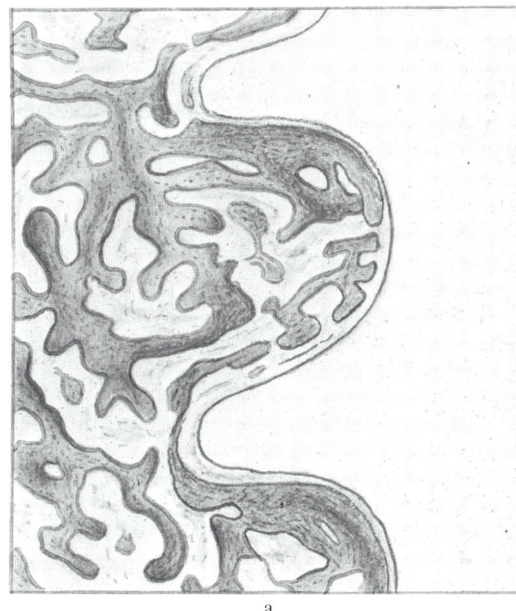

a

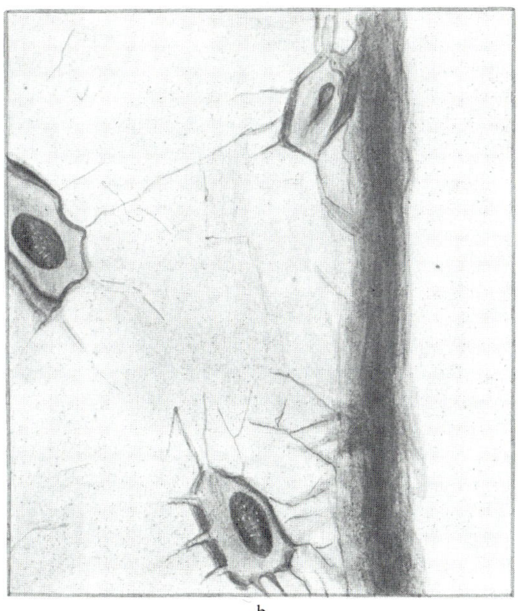

b

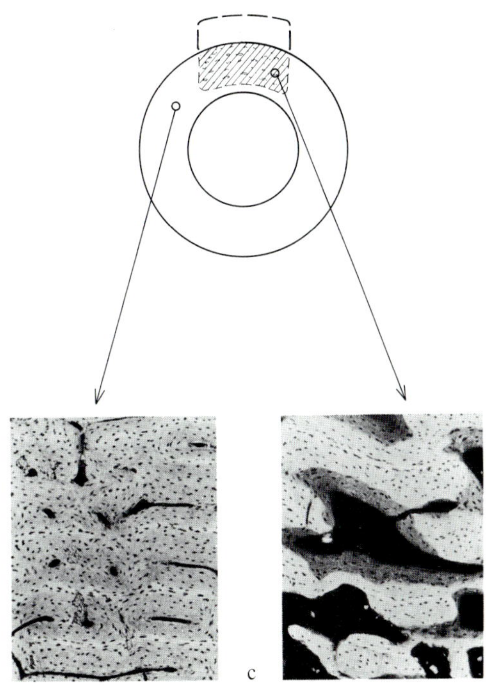

c

1.2.5 Documentation

The backbone of the AO method is the careful documentation of over 30,000 operative cases, most of which have been carefully followed up. Thousands of code cards were completed and fed into computers for analysis. Over 250,000 radiographs including those taken before and after operation, and at 4 and 12 months after surgery, were microfilmed, printed and fixed on to punch cards. It is only through this detailed, painstaking documentation that the members of the AO were able to assess the potential as well as the results of the methods, and to eliminate certain mistakes. This kind of follow-up required a great deal of patience and excellent organization. The arduous task turned out, in the long run, however, to be most worthwhile because today we are able to prove or disprove any hypothesis with fact. The study of failures was particularly valuable. We established the rule that a single failure resulting from the correct application of the method and not from its misuse required re-evaluation and re-design of the technique. Over the years, many of the procedures have been gradually changed and improved, expanded in scope, and newly standardized. Our results and experiences are discussed annually at the Davos Instructional Courses on the operative method of fracture care.

The systematic analysis of failures has shown that most of these were due to a disregard of biomechanical principles. This means that most failures are preventable. Whenever a surgical procedure has been developed to such a high standard that success or failure can be predicted, and that success can almost always be achieved if sufficient care is taken, it can be considered most dependable. This does not mean, however, that these procedures are easy and that everyone is capable of carrying them out. On the contrary, years of practice and experience are required in order to master all the intricacies of the AO method.

Everyone who uses the AO method should carry out a follow-up at 4 and 12 months of all the patients so treated. This is the only way that a surgeon can come to have a realistic assessment of his own skill and of the potential and limitation of the method in his own hands. He will also come to recognize complications and will learn to treat them promptly.

Note: Comments on documentation and on the radiological signs of fracture healing are to be found in Sect. 3.5.6, page 146.

Fig. 8 *AO documentation.*

a Punch cards showing the most important information on one side with the microfilms of the radiographs pasted on the back.

b The three AO code cards – those for the initial hospitalization, those for the 4-month follow-up and those for the final review at 1 year.

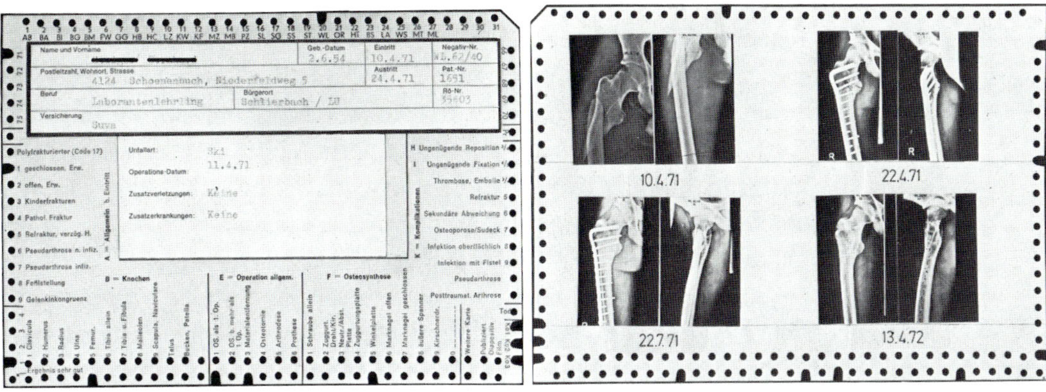

a

1. IDENTIFIKATION
AO-DOKUMENTATION
TRAUMATOLOGIE
BLATT A
2. KLINIK-NR. 3. EINTRITTSDATUM 4. AUSTRITTSDATUM
5.-7. ANZAHL FRAKTUREN UND OPERATIONEN
8. VERSICHERUNG
9. ALLGEMEINE ZUSATZVERLETZUNGEN
10. ALLGEMEINE ZUSATZERKRANKUNGEN
11. VERSORGUNG BEI ENTLASSUNG
12. SCHWELLUNG BEI ENTLASSUNG
13. HAUTINZISION BEI ENTLASSUNG

AO-DOKUMENTATION
TRAUMATOLOGIE
BLATT B
20. UNFALLDATUM 21. DIAG.-NR.

AO-DOKUMENTATION
TRAUMATOLOGIE
BLATT C
46. NK-DATUM 47. DIAG.-NR.

b

1.2.6 Surgical Bone Instruments

The surgical bone instruments developed by the AO comprise an integrated uniform system of sufficient scope to permit the fixation of the most complex fracture, pseudarthrosis, osteotomy or arthrodesis.

The technical commissions of the AO have over the years developed and carefully tested new implants and instruments which for the most part were designed to achieve pre-stress and compression fixation of bone. In addition certain existing implants were adopted and modified in order to make the procedure simpler, safer and the fixation more rigid. Good examples of these are the modified external compression clamps of CHARNLEY and the modified intra-medullary nails of KUENTSCHER and HERZOG. The development of such instruments as the bone-holding clamps, the bone spreaders, the bone retractors and air-driven motors have made it possible to carry out an internal fixation in accordance with modern mechanical principles.

These developments have been made possible by the AO surgeons, who since 1958 have subjected the instruments and implants to the most critical trials. Furthermore, these developments could not have taken place without the personal commitment and the wealth of ideas of R. MATHYS and the metallurgical knowledge of F. STRAUMANN.

The AO instruments have been copied by almost all the large surgical instrument manufacturers. These copies have played a major role in the dissemination of the AO method. The quality and the design of the original AO implants, which are distributed by Synthes, have been most thoroughly tested by the technical commissions of the AO. The mixing of implants of different manufacture is most inadvisable because of the increased danger of corrosion.

We have divided our instrumentation into standard implants and standard instruments which suffice for the fixation of most fractures and which were accepted by all members of the AO group. In addition we have instruments and implants which are obtainable "upon request". These are used for the fixation of certain special fractures or are the favourite instruments or implants of only some surgeons. In the first part of this manual we shall discuss only the standard AO-instrumentation.

Note: The holding cassettes are perforated and can be sterilized. We recommend suitable wrapping of these cassettes and their sterilization in an autoclave at 135° C and 2 atm. We do not recommend hot air sterilization at 180° C because the heat results in dulling of instruments and damage to the air motors and their hoses.

Fig. 9 *Standard instrumentation of the AO.*

I *Cassettes for compression fixation*

a Instruments for screwing and plating (basic instrumentation).

b Screw cassette.

c Plate cassette: for round hole or dynamic compression plates (DCP).

d Special hip instruments such as the AO angled plates for the fixation of peri-articular fractures of the femur.

II *Instrumentation for intra-medullary nailing*

a The cassette which contains instruments for intra-medullary reaming as well as instruments for insertion and in extraction.

b Intra-medullary nail for the tibia and femur and the reaming guide and the nail guide.

9/I

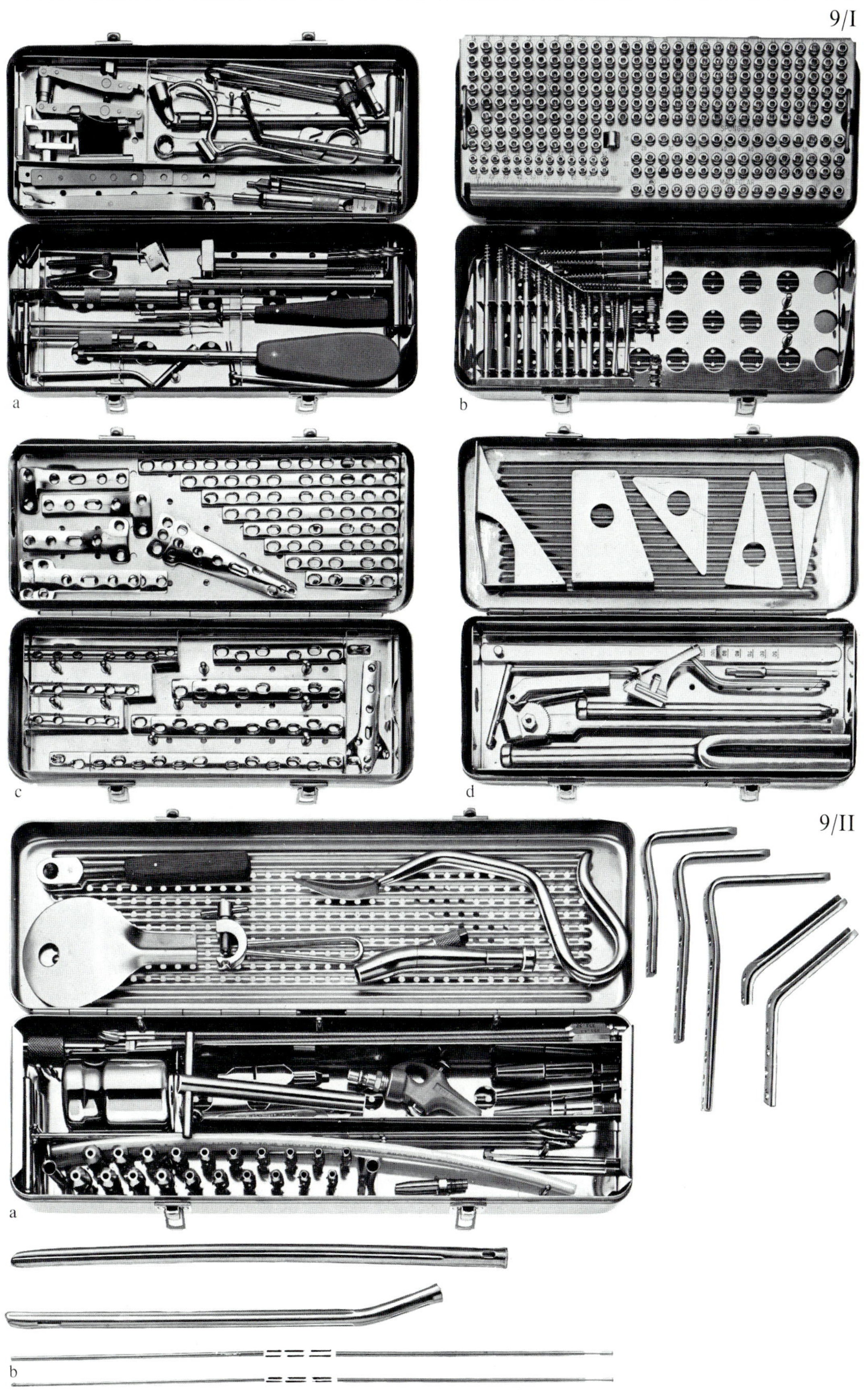

9/II

Small-Fragment Instrumentation and the External Fixators

Initially we developed only standard implants and instruments for the fixation of fractures of long bones. Subsequently we developed small-fragment instrumentation for the fixation of hand fractures and foot fractures. At the same time we developed the stable external fixators. Instruments for tension band wiring as well as general instruments have been standardized and are now available in special cassettes.

Note: The metal used in the AO implants is under constant scrutiny and assay by experts of the STRAUMANN metallurgical research institute in Waldenburg. Despite very high specifications, approximately 30% of the stainless steel is rejected.

Fig. 9 (continued).

III *Small-fragment cassettes*

a Instrumentation.

b The cassette for the implants. Note the empty holes in the screw holder. These have been designed for the 2.0-mm and the 1.5-mm mini-cortex-screws which are available for certain hand procedures.

IV *External fixators:* Tubes, Steinmann pins, Schanz screws and various clamps, enough for some 4–6 cases. If desired, suspension rolls and spreaders are also obtainable for suspension of extremities.

V *Instrumentation for tension band wiring:* Various pliers and tension band wiring instruments.

9/III

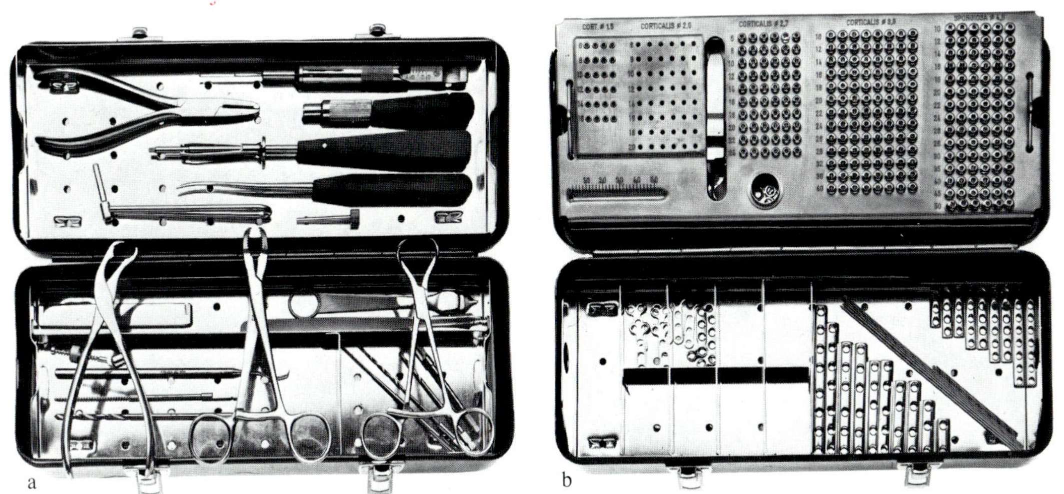

a b

9/IV

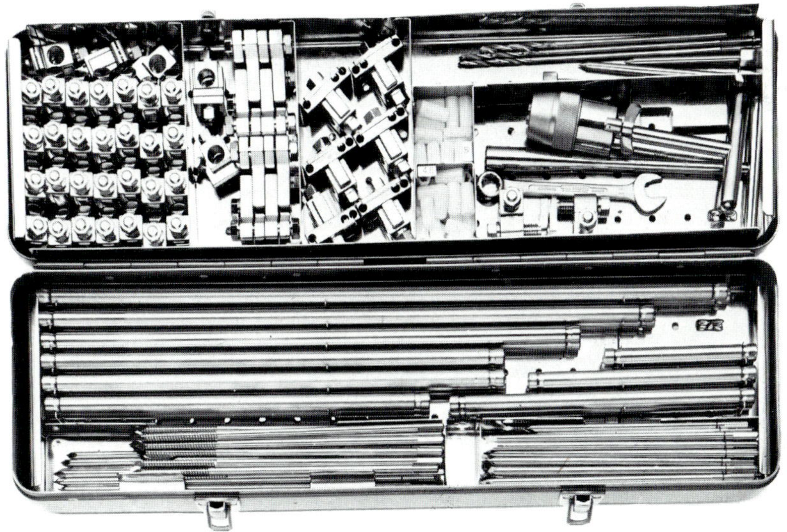

9/V

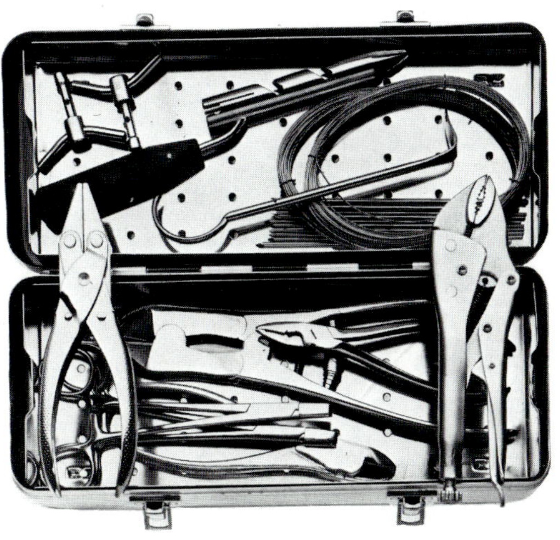

Ordinary Standard Instruments and Motors of the AO

The multitude of available instruments designed for bone surgery makes it very difficult to determine which are necessary and useful and which are not. To facilitate matters, the instruments which we recommend and which we have described in this manual have been put together in cassettes.

We cannot conceive how a plating in accordance with our principles could be undertaken without the necessary bending press and bending irons. More than 50% of the straight plates must be contoured – which involves bending and at times also twisting. These instruments therefore form an integral part of the AO instrumentation.

Note:

1. In addition to the standard implants and instruments, and the aforementioned instruments available on "special request" we have a whole range of instruments designed for veterinary surgeons (both small and large animals), and also for mandibular surgery.

2. Mini-instruments and mini-implants are available on request. They can be stored in the cassettes for small instruments and implants where space has been set aside for them.

Fig. 10 *Ordinary standard instruments and motors of the AO.*

I *Ordinary instruments (two cassettes)*

a Self-centring bone-holding forceps (three sizes).

b Reduction forceps (160 and 220 mm).

c Reduction forceps with points (130 mm).

d Reduction forceps with points (200 mm).

e Hohmann retractors (bone retractors with a short tip 8 and 18 mm wide and a long tip 18 mm wide).

f Hohmann retractor with a broad tip (24 mm).

g Periosteal elevators: curved 12 mm wide and straight 6 mm wide.

h Chisel with exchangeable blade 16 mm wide.

i Gouge 10 mm wide.

k Hammer 500 g.

II *Instruments for contouring (individually available)*

a Bending press.

b Bending pliers.

c Bending irons.

III *AO compressed-air-driven power tools (individually available)*

a Small air drill with forward and reverse gears.

b Medullary reamer.

c Oscillating bone saw.

10/I

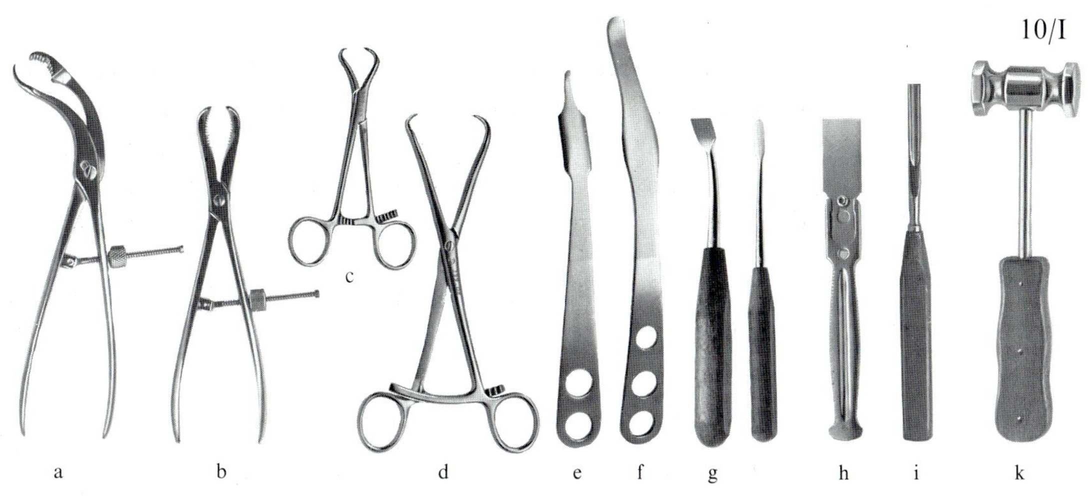

10/II

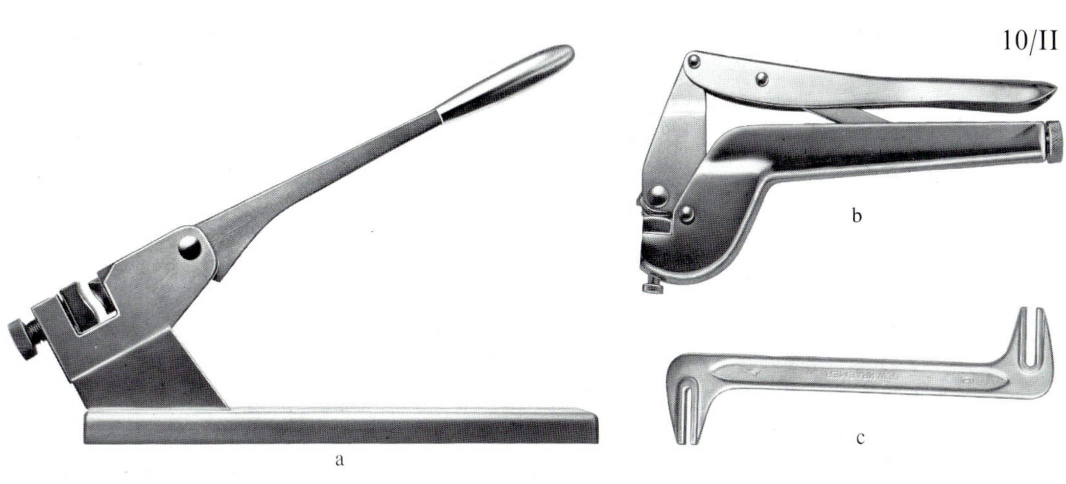

10/III

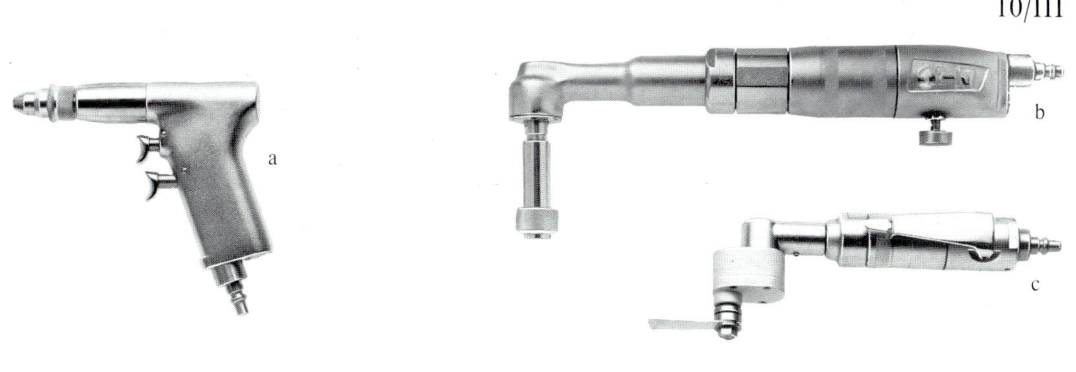

1.3 Principles of the AO Method

Within the scheme of the AO method, there are two principal methods of achieving internal fixation: the one is *inter-fragmental compression* and the other is *splinting*.

Each principal method has its indications and contra-indications as well as its methodology. For each method there is 'the best way' of achieving it. This is governed in part by the instrumentation. In certain cases it is possible to combine the two methods.

Both basic methods, inter-fragmental compression and splinting, have similar aims: (1) Anatomic reduction and healing of the fracture and (2) functional after-care.

1.3.1 Inter-fragmental Compression

Inter-fragmental compression increases the friction between the fragments and in this way the stability of the internal fixation. It neutralizes the torsional shearing and bending forces and increases the load tolerance of the internal fixation. The fixation must be so stable that movement of the limbs is possible, while the compressed bone fragments which have a preserved blood supply go on to primary bone union.

In order to achieve lasting inter-fragmental compression, the implants must be pre-stressed, that is, placed under tension, and the bone surfaces which are compressed must be relatively large. Whenever the implants are subjected to a load which repetitively changes its direction, the metal becomes subjected to cyclic stresses. Furthermore movement takes place between the bone fragments. The result of this is local resorption of bone, the formation of an irritation callus, delayed bone healing or non-union and implant failure (see p. 154).

Interfragmental compression may be *static* or *dynamic*:

1. In static compression, the tension applied to the implant results in compression at the fracture interface. A lag screw is a perfect example of static compression. Other good examples are the single or double compression plates and axial compression exerted through an external fixator.

2. In dynamic compression, the fragments are not only compressed by the pre-stress of the implant (lag screw, compression plate) but also are subjected to additional compression which results from harnessing forces generated at the level of the fracture when the skeleton comes under normal physiological load. An excellent example of dynamic compression is the transverse fracture of the patella. If a wire, passed over the front of the patella and around the quadriceps and infra-patellar tendon, is tightened, it will result in compression of the cortex adjacent to the wire. The opposite cortex will gape slightly, but will, however, come under compression when the knee is flexed and the quadriceps is contracted. A plate applied on the tension side of the bone will give rise to dynamic compression in the same way. During active movement and loading, there will be a rise in the inter-fragmental compression. This mode of increasing the stability of the fixation by means of dynamic compression is called *tension band fixation*.

1.3.2 Splinting

Splinting, in sharp contrast to inter-fragmental compression, almost never results in rigid fixation of the fragments. The result of this is that the cortex heals by means of a callus which contains intermediary phase cartilage and connective tissue. The callus, as seen radiologically, may be small or large.
We should distinguish between internal and external splinting.

1. Good examples of *internal splinting* of a fracture are: the intra-medullary nail in diaphyseal fractures, the 130° plate in fractures of the femoral neck and inter-trochanteric fractures, and Kirschner wire fixation of certain metaphyseal fractures, in children.

2. Good examples of *external splinting* of a fracture are the external fixators which can be applied as a frame around the bone or on one side as in leg lengthening.

Note: Simple straight plates are not a reliable means of splinting.

1.3.3 Combination of Inter-fragmental Compression and Splinting

If a lag screw is used to stabilize a fracture under inter-fragmental compression, it must be protected by a plate or a nail. Thus lag screws when used for fixation of shaft fragments are protected by *'neutralization plates'* and when used to fix metaphyseal fragments are protected by *'buttress plates'*. The *Kirschner* wires which are used in combination with tension band wiring, serve to neutralize the shearing and torsional stresses.

2 Means by Which Stable Internal Fixation is Achieved

Stable internal fixation can be obtained by means of screws, wires, plates, intramedullary nails, external fixators and by composite internal fixation.

We shall now discuss the standard of the AO implants, the way the implants and instruments are used, as well as the chief indications for the use of the different implants.

2.1 Lag Screws

A lag screw is the best and most important way of achieving static inter-fragmental compression. A screw will act as a lag screw when its purchase is only in the far cortex and when its thread passes freely through the cortex adjacent to the screw head.

In epiphyseal and metaphyseal areas we use classic lag screws, the so-called *cancellous screws*. They have a relatively thin core and a wide thread. If such screws are inserted into cortex they cannot be removed. With healing cortical bone grows right up to the shaft. These screws cannot cut a thread through the cortex in reverse. If one attempts to remove them even after a few months, excessive torque is generated and often the screw will break. A *cortical screw* is fully threaded. It will act as a lag screw only when its purchase is confined to the far cortex and when it glides through the near cortex. For the screw to glide through the cortex the diameter of the hole must be at least equal to the outer diameter of the screw thread.

Fig. 11 *The three standard AO cancellous screws.*

I *6.5-mm cancellous screw with spherical head*

a With a 32-mm thread and a 4.5-mm shaft.

b With a 16-mm thread and a 4.5-mm shaft.

c Head with a hexagonal recess 3.5 mm wide and a hemispherical profile of the screw head.

d Screw thread 6.5 mm, core 3 mm and pitch 2.75 mm.

e Corresponding drill bit 3.2 mm with an 80° tip, hollow ground, to be used in cancellous bone. If the shaft passes through thick cortex, the cortex must be perforated with a 4.5-mm drill bit.

II *4.0-mm cancellous screw with spherical head*

a Shaft 2.3 mm.

b Head profile hemispherical, hexagonal recess 2.5 mm wide.

c 4-mm thread diameter, core 1.9 mm, pitch 1.75 mm.

d Corresponding drill bit 2.0 mm.

III *Malleolar screw*

a Shaft 3.0 mm.

b Screw head same as 6.5-mm cancellous screw.

c Screw thread 4.5-mm diameter, same profile as the 4.5-mm cortex screw, core 3 mm.

d Corresponding drill bit 3.2 mm.

IV *Washers for cancellous screws*

a 13-mm washer for the 6.5-mm-core cancellous and malleolar screws.

b 7-mm washer for the 4-mm cancellous screw.

c 11-mm and 8-mm serrated washers for ligamentous avulsions.

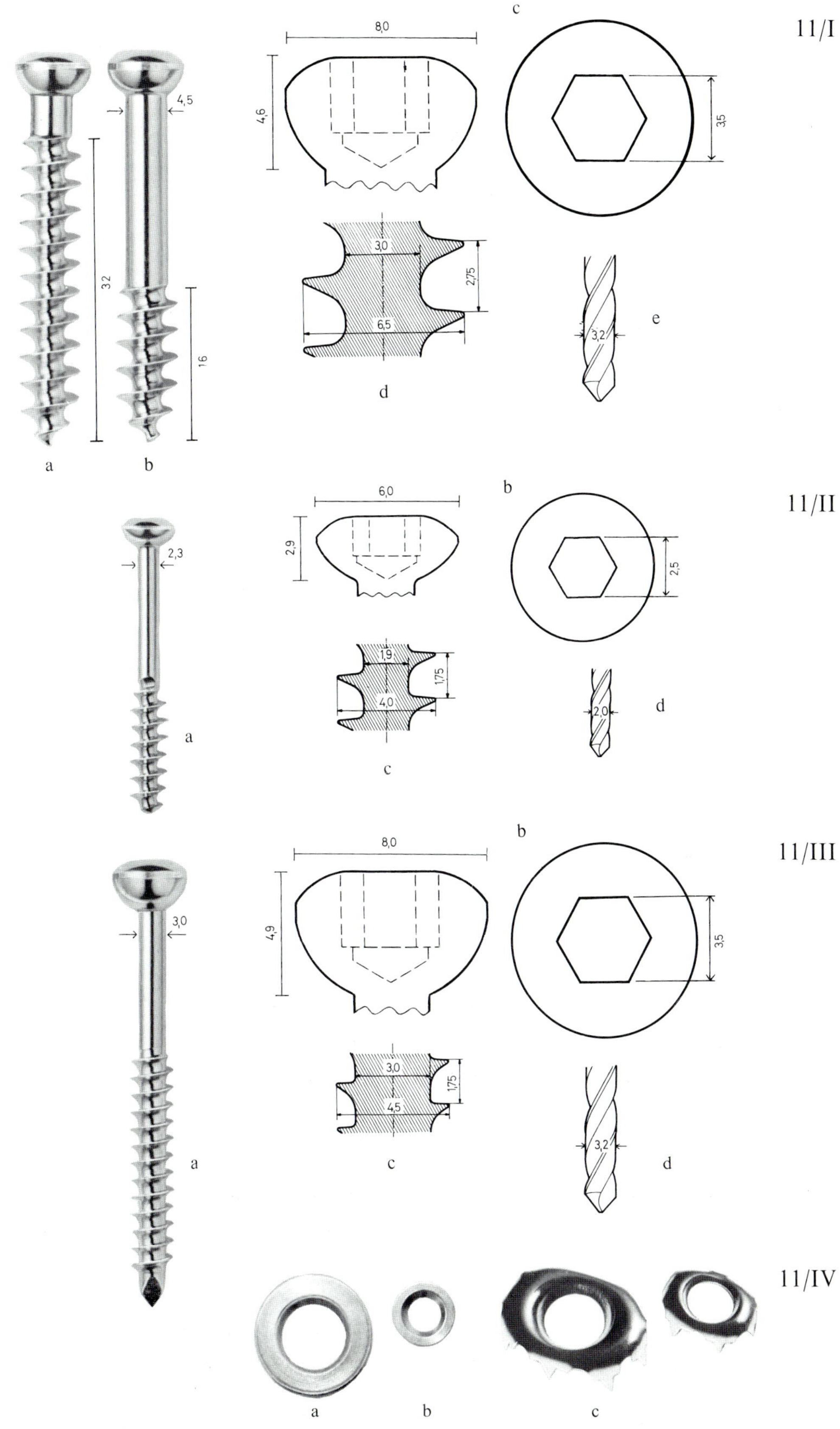
11/I
a
b
4,5
32
16
c
8,0
4,6
3,5
3,0
2,75
6,5
d
3,2
e
11/II
2,3
a
b
6,0
2,9
2,5
1,9
1,75
4,0
c
2,0
d
11/III
3,0
a
b
8,0
4,9
3,5
3,0
1,75
4,5
c
3,2
d
11/IV
a
b
c

2.1.1 Cancellous Lag Screws

In order to achieve inter-fragmental compression between two fragments of epiphysis or metaphysis, it is important to make certain that the thread of the cancellous screw does not cross the fracture line. After the hole is drilled, cut only the first centimetre or so of the thread with the tap before screwing in the cancellous screw. Unlike the cortex screw, the cancellous screw does not require tapping of its whole thread. In fact, when a cancellous screw is screwed into cancellous bone without tapping, it tends to compress the trabeculae together. This gives it a better hold than when the thread is tapped completely.

Malleolar screws can be inserted into the medial malleolus without pre-drilling a hole and without cutting a thread. Malleolar screws are so designed that in pure cancellous bone they can cut a thread for themselves.

> *The principle of the lag screw:*
> *The thread must not cross the fracture line.*

Note: Screw fixation of cancellous bone is useful only if the two cancellous surfaces can come in contact or better still when they can be impacted. A gap which cannot be obliterated always becomes filled with connective tissue. Therefore we recommend that whenever a defect is present it should be filled with autologous cancellous bone (see Fig. 98).

Fig. 12 *Classic indications for the different cancellous lag screws.*

a 6.5-mm cancellous screw with a 32-mm thread: the fixation of a lateral condyle in a young adult. The washers are usually necessary.

b 6.5-mm cancellous screw with the 16-mm thread: the fixation of a Volkmann's triangle (posterior lip) with the screw passing from front to back just above the ankle joint.

c 4.0-mm cancellous screws used for fixation of the medial malleolus.

d 4.0-mm cancellous screws used for fixation of the avulsed anterior syndesmotic ligament from the tibia with its tubercle of bone, the tubercle of Chaput.

e 4.0-mm cancellous screw used for fixation of epiphyseal fracture of the distal tibia (type B2, see Fig. 273).

f Malleolar screws used for fixation of an oblique fracture of the medial malleolus.

g A malleolar screw used for the fixation of a short oblique fracture of the lateral malleolus. Note that the screw traverses the bone obliquely in two planes and its tip protrudes through the cortex (see Fig. 246b).

h A malleolar screw used for fixation of the vertical component of a Y supracondylar fracture of the distal humerus. This is the first step in the internal fixation of such a fracture (see also Fig. 137).

a

b

c

d

e

f

g

h

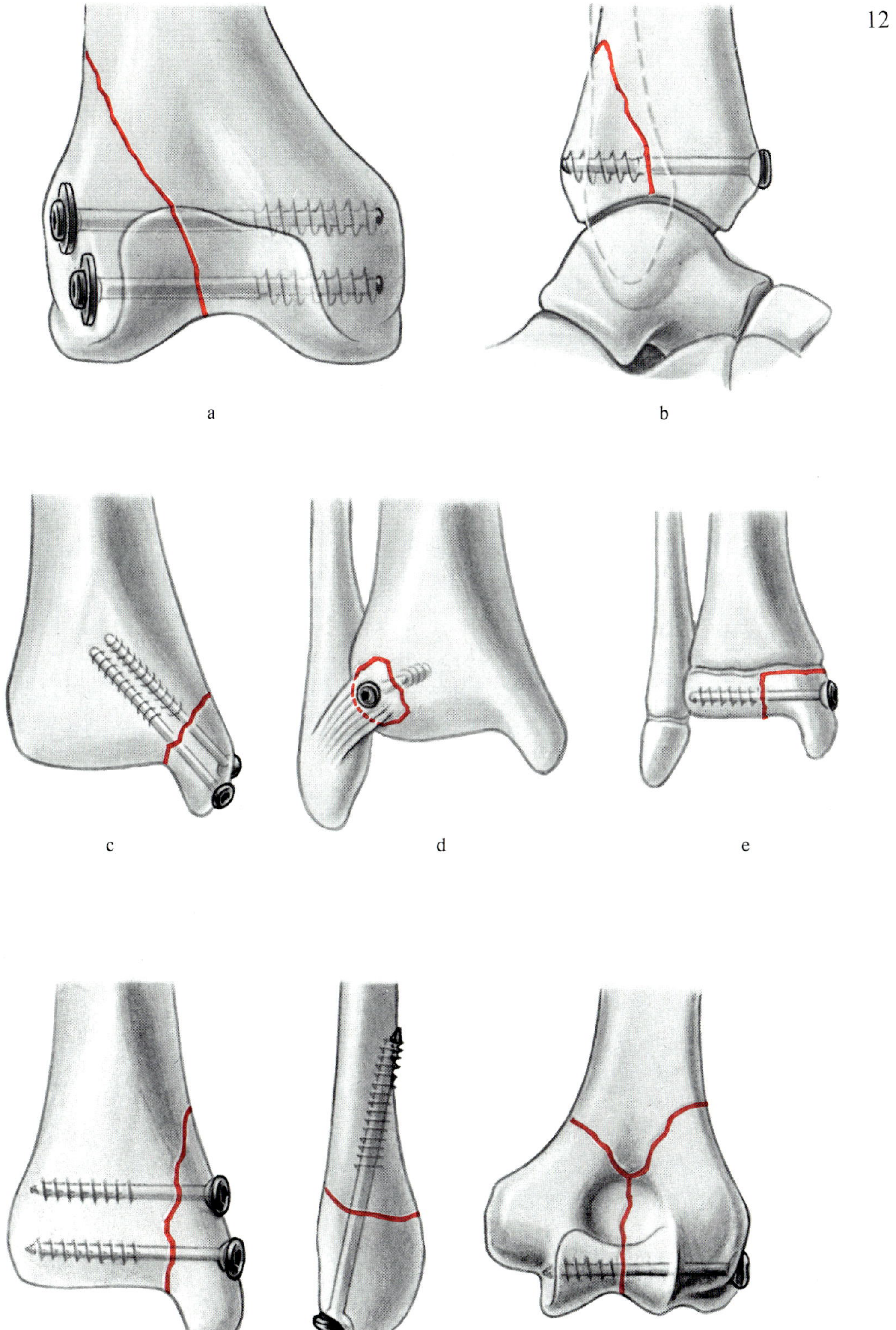

2.1.2 Cortex Screws

Since the cortex screw is fully threaded, it can act as a lag screw only when the cortex adjacent to the screw head is over-drilled so that the screw thread gains purchase only in the far cortex. The over-drilled hole adjacent to the screw head is called the *gliding hole* and the one in the far cortex, the *thread hole* (see Fig. 17a). Cortical screws are also used for fixation of plates to bone. Any screw used for the fixation of a plate, if it should cross a fracture line, should be inserted as a lag screw. The remaining screws should get purchase on both cortices.

Fig. 13 *Standard cortex screws.*

I *4.5-mm cortex screw*

a Fully threaded with a thread of 4.5-mm diameter.

b Screw head 8 mm wide and 4.6 mm high with a 3.5-mm hexagonal recess. The underside of the head is hemispherical in cross-section.

c The core is 3 mm with a 1.75-mm pitch.

d The drill bit or the thread hole is 3.2 mm and for the gliding hole, 4.5 mm.

II *3.5-mm cortex screw*

a Fully threaded with a thread of 3.5-mm diameter.

b Screw head 6 mm wide, 2.6 mm high with a 2.5-mm hexagonal recess.

c Thread with a 1.9-mm core and 1.75-mm pitch.

d 2.0-mm drill bit for the thread hole and 3.5-mm drill bit for the gliding hole.

III *2.7-mm cortex screw*

a Fully threaded with a 2.7-mm thread.

b Screw head 5 mm wide, 2.3 mm high with a 2.5-mm hexagonal recess.

c Thread with a 1.9-mm core and 1.0-mm pitch.

d 2.0-mm drill bit for the thread hole and 2.7-mm drill for the gliding hole.

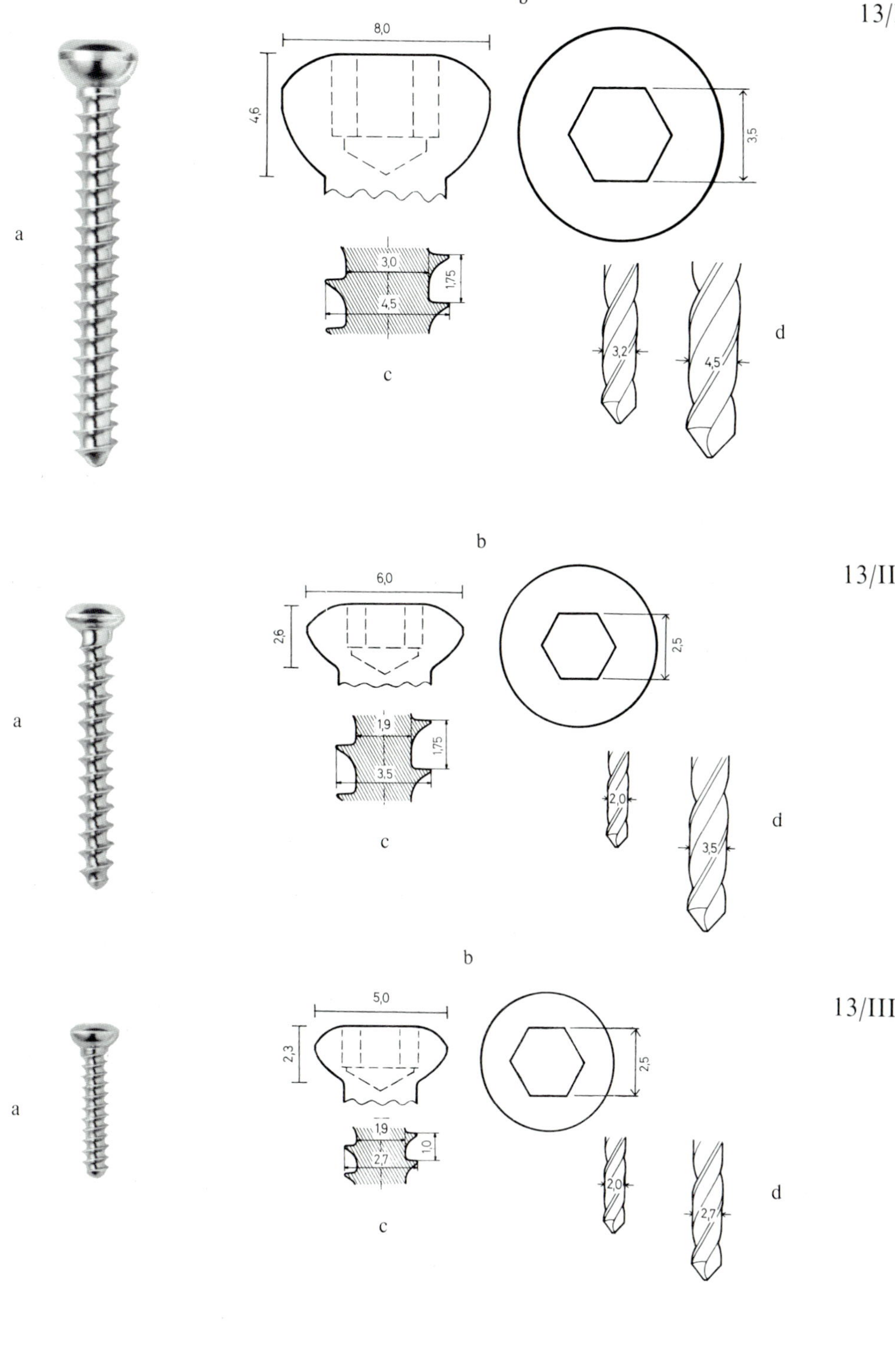
13/I
a
b
8,0
4,6
3,5
3,0
1,75
4,5
c
3,2
4,5
d
13/II
a
b
6,0
2,6
2,5
1,9
1,75
3,5
c
2,0
3,5
d
13/III
a
b
5,0
2,3
2,5
1,9
1,0
2,7
c
2,0
2,7
d

A screw can cut its own thread in cortex only when the hole drilled for it is almost the same diameter as the outer diameter of the screw thread. In sharp contrast to this is the thread cut with a tap for the corresponding AO cortex screw. These threads are deeper and wider and thus have a broad, flat, pressure-bearing surface which is almost at 90° to the axis of the screw (Fig. 14.1).

Each AO cortex screw of a given diameter has its corresponding tap. A tap is absolutely necessary when inserting a cortex screw. In cancellous bone it is not necessary to cut the thread for the screw because the cancellous trabeculae compressed by a screw inserted without pre-tapping have better holding power than those cut with the sharp tap. The 6.5-mm tap should be used only for the tapping of hard cortex.

The 4.5-mm cortex and the 6.5-mm cancellous screw have had from the very beginning (1958) a hexagonal recess in their heads. Since 1976 the small screws, the 4.0-mm, 3.5-mm and the 2.7-mm screw have been manufactured with the 2.5-mm hexagonal recess. Thus, with only two screwdrivers one can insert and remove all of the standard AO screws. A short screwdriver can also be coupled to the small air drill which has forward and reverse settings. It can be used for the insertion and removal of the large screws.

It should be noted that the 2-mm and the 1.5-mm mini-screws which are obtainable on special request have a cruciate slot in the head.

Fig. 14 *Thread profile of an AO screw and a commercially available bone screw.*

1 The thread of the AO screw is of the same diameter throughout from the screw head to the very tip so that even the last thread will afford a perfect hold. The thread is rounded, saw-blade-like in profile with flat, rounded, pressure-bearing surfaces at right angles to the axis of the screw. The tip of the screw is round without a flute. This means that the thread must be pre-cut with a sharp tap. The hole drilled for the screw (*a*) is only 0.2 mm greater than the core of the screw. The thread is cut throughout the whole length. The debris falls into the flutes of the tap.

2 In a self-tapping screw, the rate that the flute fills with bone is directly proportional to the size of the hole and the smaller the hole, the more rapid the filling of the flute. For this reason the drill hole (*b*) must be relatively large. This means that in a self-tapping screw only the tips (*c*) of the V-shaped narrow threads gain a purchase in bone.

Fig. 15

a Tap for the 6.5-mm cancellous screw for the tapping of the screw thread.

b The 4.5-mm tap with the short and long thread with rapid coupling.

c Tap holder for the 4.5-mm and the 3.5-mm taps.

d 3.5-mm tap with rapid coupling.

e The handle for the 3.5-mm and 2.7-mm taps.

f 2.7-mm tap with rapid coupling.

Fig. 16

a The 3.5-mm hexagonal screwdriver for the 6.5-mm cancellous, 4.5-mm cortex as well as the malleolar screws.

b The 2.5-mm hexagonal screwdriver for the 4.0-mm cancellous, the 3.5-mm cortex and 2.7-mm cortex screws.

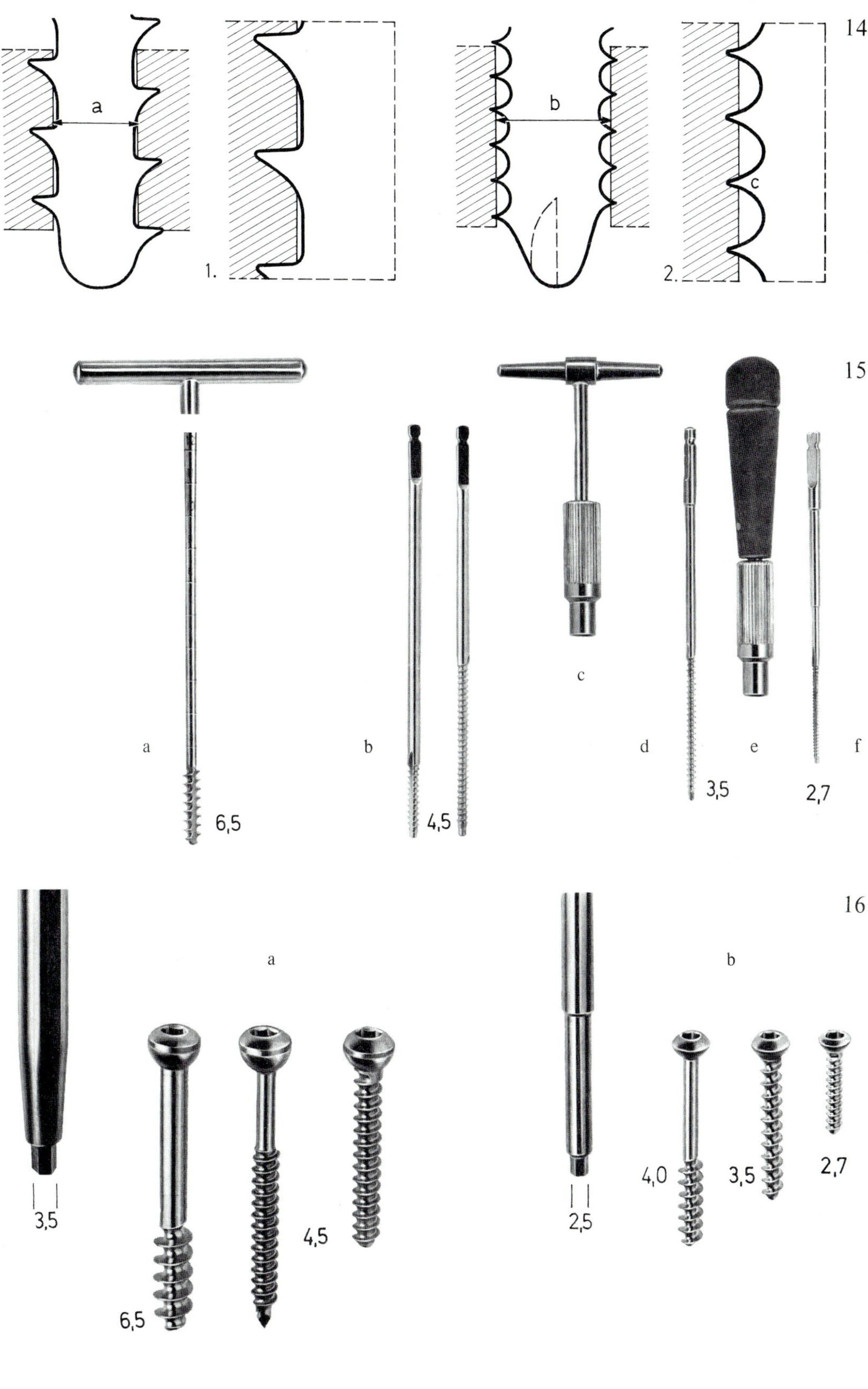
14
a
1.
b
c
2.
15
6,5
4,5
c
a
b
d
e
f
3,5
2,7
16
a
b
3,5
6,5
4,5
2,5
4,0
3,5
2,7

2.1.3 Technique of Screw Fixation

Whenever lag screws are used to secure inter-fragmental compression between two fragments, the procedure can be carried out in two ways:

1. One can carry out an accurate reduction, maintain it provisionally with reduction clamps and drill a *gliding hole* in the near cortex (Fig. 18), or

2. the *gliding hole* or the *thread hole* can be pre-drilled prior to reduction (Figs. 19 and 20).

Fig. 17 *Principle of obtaining inter-fragmental compression with a lag screw.*

a Fragments of bone when transfixed with a screw can be compressed only when the hole in the cortex adjacent to the screw head is at least as large as the outer diameter of the thread. This hole is called the gliding hole. The thread glides in this hole and gains purchase only in the opposite cortex in the thread hole. As soon as the screw head abuts on the cortex, inter-fragmental compression is generated.

b The screw can never compress if both cortices are drilled and tapped, because the cortices can never come together. Either the bone or the screw will break before there is the slightest approximation of the fragments.

Fig. 18 *Technique of lag screw fixation after reduction of the fragments.*

a With the aid of the 4.5-mm tap sleeve used here as a drill guide, drill the gliding hole with the 4.5-mm drill bit.

b Insert into the hole the 58-mm-long drill sleeve which has a 4.5-mm outer diameter and a 3.2-mm inner diameter. Push it in till its serrated end abuts on the opposite cortex. This drill sleeve ensures absolutely perfect centring of the drill bit even when oblique holes are being drilled.

c Drill the thread hole with the 3.2-mm drill bit.

d Cut a recess for the screw head with the counter-sink.

e Measure carefully the screw length with the large-fragment depth gauge.

f Cut the thread in the thread hole with the short 4.5-mm tap.

g Insert the selected 4.5-mm cortex screw either with the hand screwdriver or with the small air drill. The screw is only lightly tightened. Once all screws are inserted, they are completely tightened one after the other.

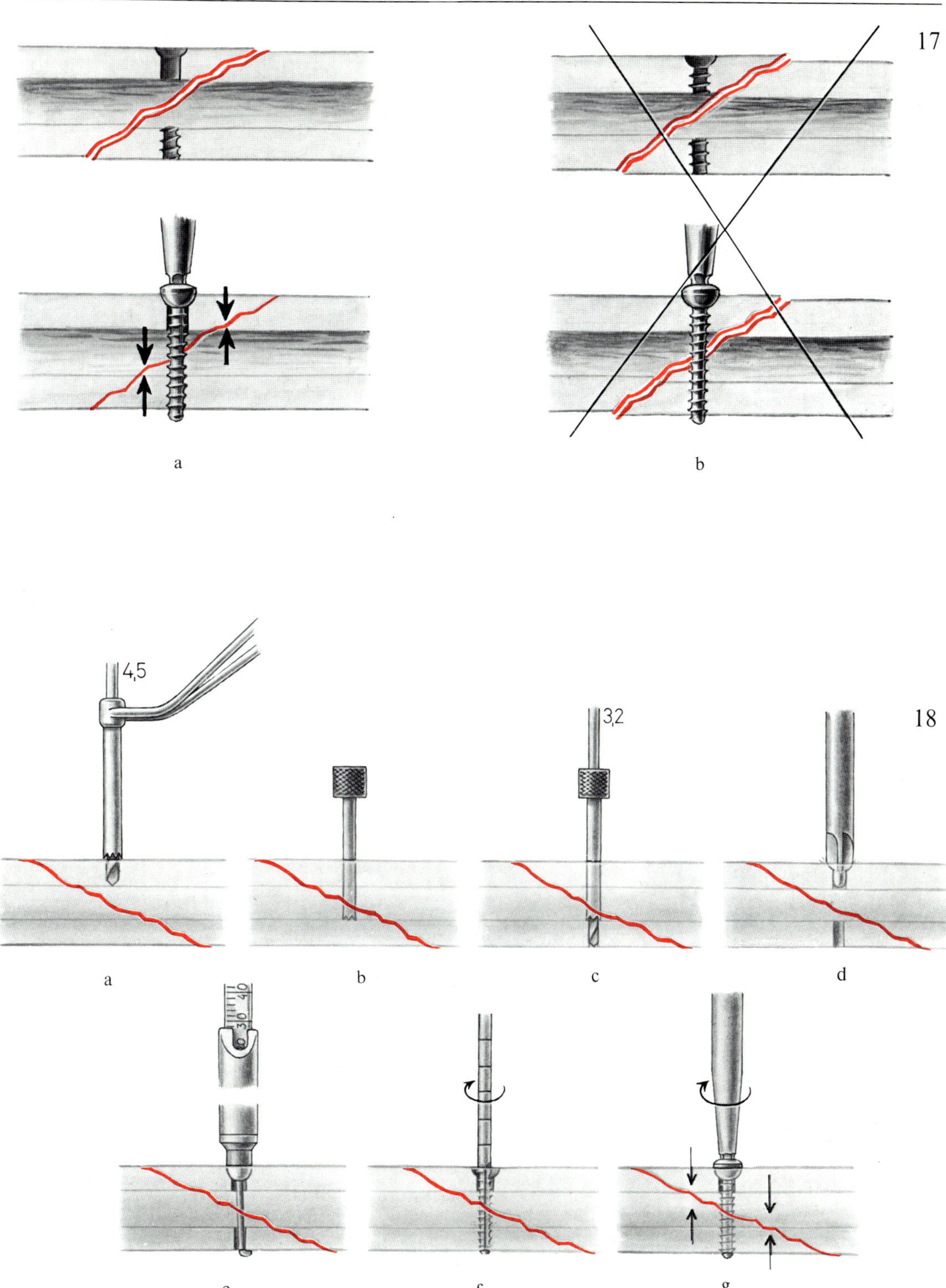
17
a
b
4,5
3,2
18
a
b
c
d
e
f
g

Technique of Lag Screw Fixation Without Prior Reduction of the Fragments

A screw has the best hold in bone when it is inserted exactly in the middle of both cortices. The opposite cortex, however, is often obscured from view once the bone is reduced (Fig. 21). Thus whenever the fragments are reduced first, it becomes extremely difficult to aim exactly for the middle of the opposite cortex with a drill bit. Therefore we have found that in practice it is far better to drill either the gliding hole or the thread hole prior to the reduction of the fragments. This technique results in the minimum stripping of the bone fragments.

The gliding hole can be drilled with the 4.5-mm drill bit, either from the periosteal or from the medullary aspect. The thread hole can also be pre-drilled with the 3.2-mm drill bit. The drilling is done from the medullary canal to the outside.

Fig. 19 *Pre-drilling of the gliding hole.*

a Use the 4.5-mm tap sleeve as a drill guide and aim for the middle of the medullary canal and for the middle of the opposite fragment and drill the cortex with the 4.5-mm drill bit or

b insert the 4.5-mm tap sleeve into the medullary canal and drill the gliding hole by drilling from inside to the outside with the 4.5-mm drill bit.

c After reduction and temporary fixation of the bone fragments with the reduction clamp, insert the 58-mm drill sleeve into the pre-drilled gliding hole and drill the second cortex with the 3.2-mm drill bit. Complete the screw fixation as in Fig. 18d–g.

Fig. 20 *Pre-drilling of the thread hole.*

a With the 3.2-mm drill bit, drill the thread hole through the middle of the bony spike.

b Insert the pointed drill guide into the thread hole and reduce the fracture.

c After temporary fixation of the fragments with a reduction clamp drill the gliding hole with the 4.5-mm drill bit. The 4.5-mm tap sleeve is inserted into the sleeve of the pointed drill guide and serves as the drill guide for the 4.5-mm drill bit. This ensures centring and co-axial drilling of the gliding hole. Complete the fixation as in Fig. 18b–g.

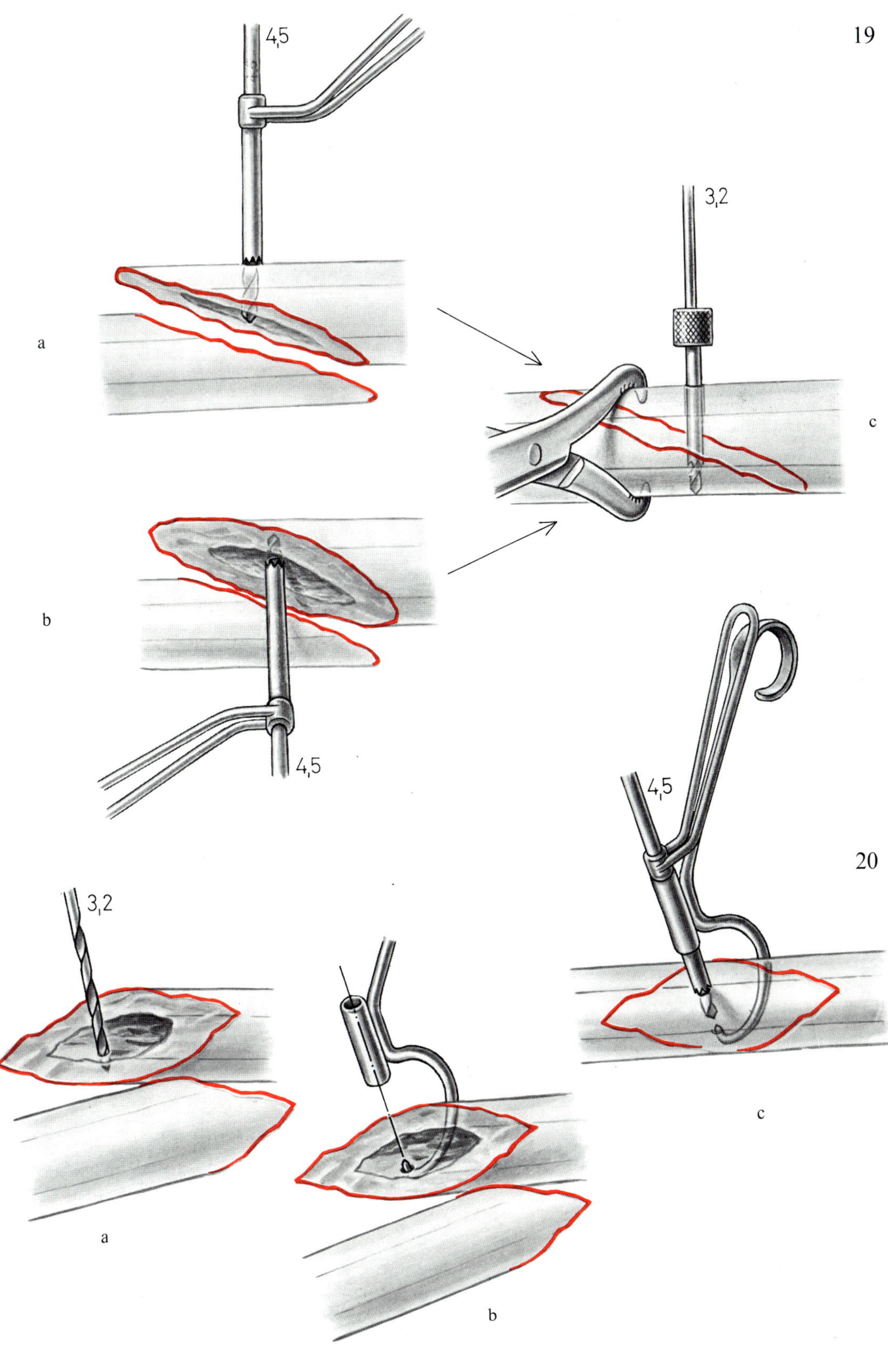
19
4,5
a
3,2
c
b
4,5
4,5
20
3,2
a
b
c

2.1.4 Orientation of Cortical Screws

In order to obtain an evenly distributed inter-fragmental compression it is necessary to centre the screw in the middle of the opposite fragment. The best hold is obtained when the screw is inserted at right angles to the long axis of the bone. When a butterfly fragment is present, the screw should be inserted in such a way that it will bisect the angle subtended between a perpendicular dropped to the long axis of the bone and one dropped to the fracture plane.

Indications for Lag Screw Fixation of the Tibia. Screws alone can be used for internal fixation whenever the fracture line is at least twice as long as the diameter of the bone at the level of the fracture. In all other cases the lag screw fixation is protected with a neutralization plate.

Fig. 21 *Screw orientation in a simple spiral fracture.* The screws must follow the spiral of the fracture if their tips are to go exactly through the middle of the opposite fragment. The screws are inserted more or less at right angles to the long axis.

a A staggering of the screw as seen on the surface.

b The cross-section: Ideally the screw tips pass exactly through the middle of the opposite cortex.

c If the thread hole is not exactly in the middle of the opposite fragment, when the screw is tightened, the fragments tend to become displaced.

d A screw inserted at right angles to the fracture plane gives rise to the best inter-fragmental compression, but provides no stability under axial loading. Under an axial load, the fragments glide upon one another with consequent loss of reduction.

e If the lag screw is inserted at right angles to the long axis, a loss of reduction on axial loading can occur only when the head sinks through the cortex or when the thread rips out of the opposite cortex. Either event is most unlikely.

Fig. 22 *A simple butterfly.* A screw (*a*) links the two main fragments. The other screws (*b*) have been inserted in such a way that they bisect the angle subtended by the perpendiculars dropped to the fracture plane and to the long axis. Final tightening of the screws should be done at the end when all have been inserted.

21

a

b

c

d

e

22

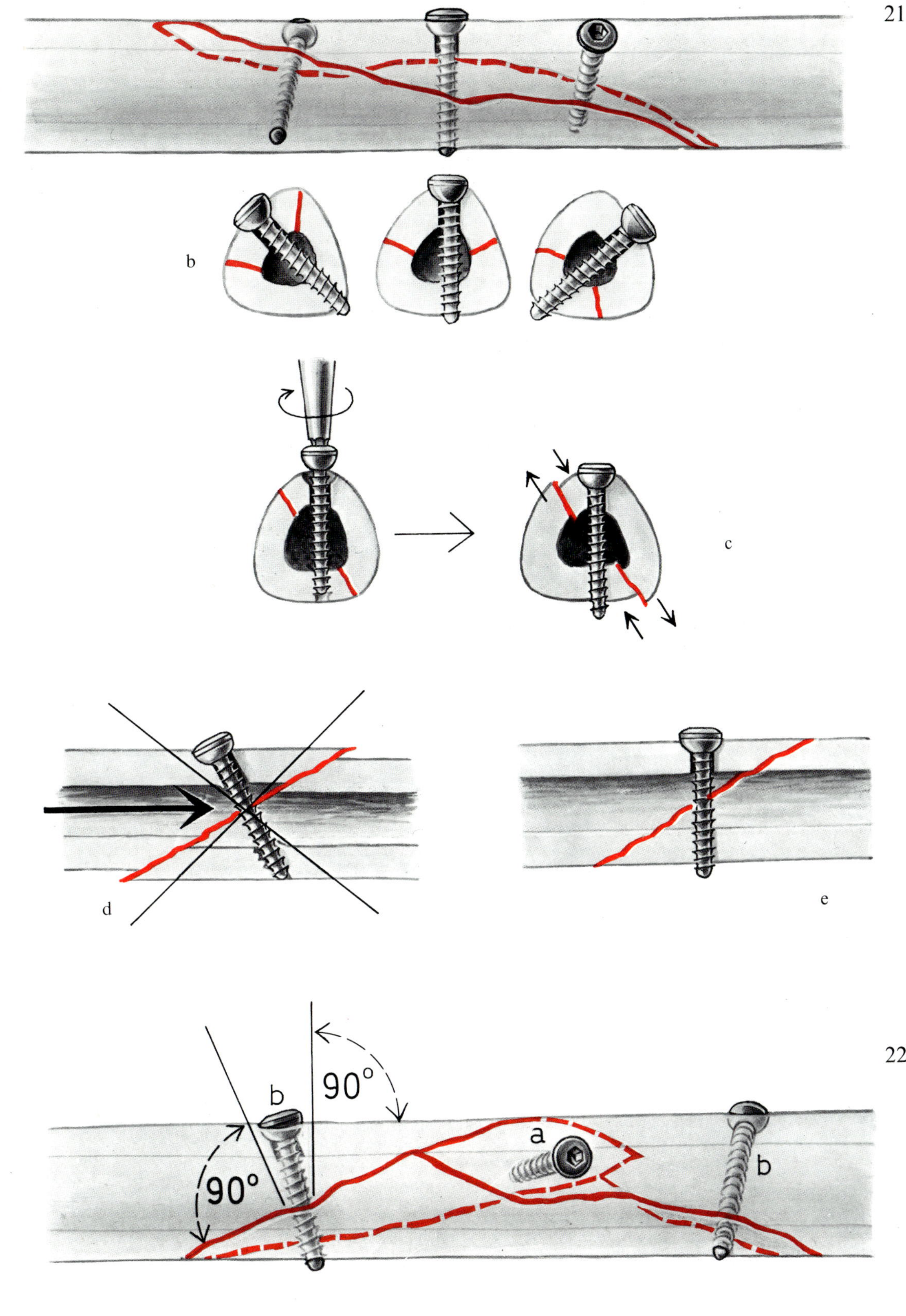

2.2 Dynamic Compression with the Tension Band

PAUWELS borrowed from mechanics the principle of tension band fixation and demonstrated its application in internal fixation of bone. Every eccentrically loaded bone is subjected to bending stresses. This results in a typical distribution of stresses with tension on the convex and compression on the concave side of bone. This is also why when such a bone fractures it displaces with a gap on the tension side. In order to restore the load-bearing capacity of an eccentrically loaded fractured bone, the tensile forces have to be absorbed by a tension band (wire, plate) and the bone itself has to be able to withstand axial compression. The pre-stressing of the plate (wire) in tension results in axial inter-fragmental compression. Loading results in a dynamic increase of this axial inter-fragmental compression.

If the medial buttress is deficient, the requirements of the tension band are not met because the bone cannot absorb compression. Under load the plate is subjected to bending stresses and soon thereafter suffers a fatigue failure and breaks.

> *Principle of the tension band*
>
> The implant absorbs the tension and the bone the compression.

2.2.1 Tension Band Wires

PAUWELS was the first to demonstrate the success of tension band wiring in a large series of fractures, osteotomies and pseudarthroses. He used a 1.2-mm wire as tension band.

The tension band wire results in *dynamic compression*. It can be employed in those cases where it can absorb all the tensile forces and where the bending and shearing forces are overcome by the friction and impaction of the fragments or by a supplementary Kirschner wire fixation which serves as internal splinting.

Fig. 23 *Schematic drawing after* PAUWELS *which illustrates the differences between load and stress and which demonstrates the principle of tension band fixation.*

a If a column with a surface area of 10 cm^2 is loaded axially with 100 kp, it is subjected to pure axial compression $D=10$ kp/cm^2.

b If the column is now subjected to eccentric loading, we have not only the axial compressive stresses but also additional bending stresses which give rise to further compressive stresses and tensile stresses. In our example the resultant compressive stresses on the medial side equal 110 kp/cm^2 ($D=110$ kp/cm^2) and on the lateral side the tensile stresses equal 90 kp/cm^2 ($Z=90$ kp/cm^2).

c, d These bending stresses can be neutralized by a chain (or a wire) which represents a tension band (c). The resultant compression corresponds to the pressure exerted by a second weight which is placed on the opposite side and equidistant from the centre of the column (d). Although the load has increased (200 kp), the total stress is reduced to a fifth ($D=20$ kp/cm^2), because the bending stresses have been completely neutralized.
If we wish to use the tension band principle in order to achieve an increase in inter-fragmental compression we must place the implant (wire or plate, see also Fig. 301) wherever we have maximal tensile forces, i.e. furthest from the load axis.

Fig. 24 *Use of the tension band in internal fixation of a fracture of the patella.* If the tension band wire is placed around the middle of the patella (a) as soon as the knee is bent the fragments come apart anteriorly. (b). If the wire is inserted over the front of the patella and through the quadriceps and patellar tendons very close to the bone in the region of Sharpey's fibres (c), all the tensile forces are absorbed and the bone comes under pure compressive stresses (d). One can then confidently begin early active mobilization.

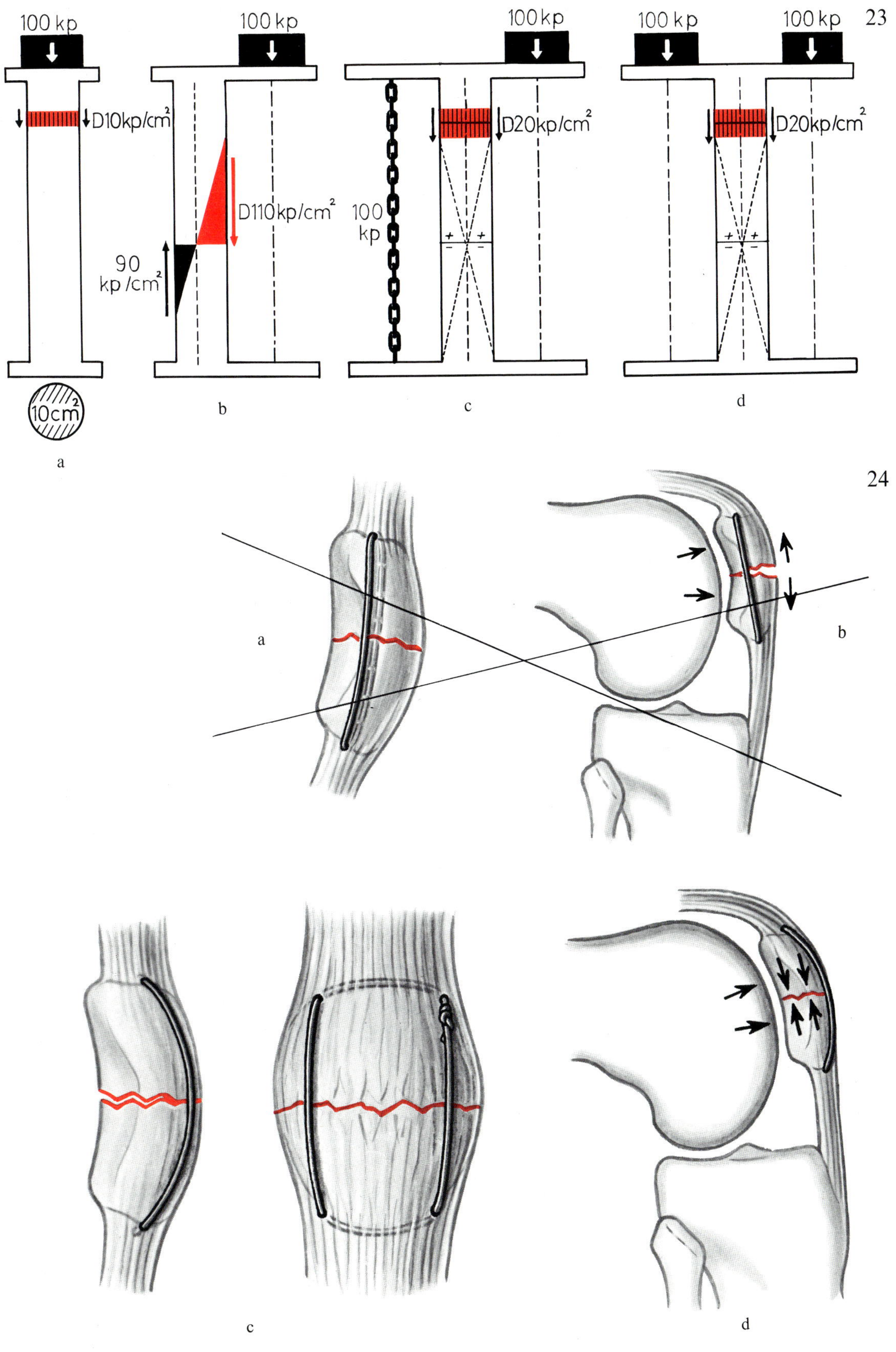
23
100 kp
D10kp/cm²
10cm²
a
100 kp
D110kp/cm²
90 kp/cm²
b
100 kp
100 kp
D20kp/cm²
c
100 kp
100 kp
D20kp/cm²
d
24
a
b
c
d

2.2.2 AO Wire Tightener

The AO has adapted and improved upon an old well-known industrial method of wire-tightening.

Fig. 25 *The technique.*

a Pass the wire around the bone, then through the eye of the wire, then through the oval hole of the tightener and finally through the hole in the little crank.

b Slowly tighten the wire by turning the crank in the slit.

c As soon as the wire begins to stretch, tip the tightener through a 90° angle which bends the wire and release the tension by reversing the turns of the crank.

d As soon as about 1 cm of wire is showing, cut the wire with a wire cutter or break the wire by bending it back and forth.

e At the end, with the aid of needle nose pliers compress the bend of the wire and if possible tuck the free end under the wire.

Fig. 26 *An example of tension band wiring of the olecranon.*

25

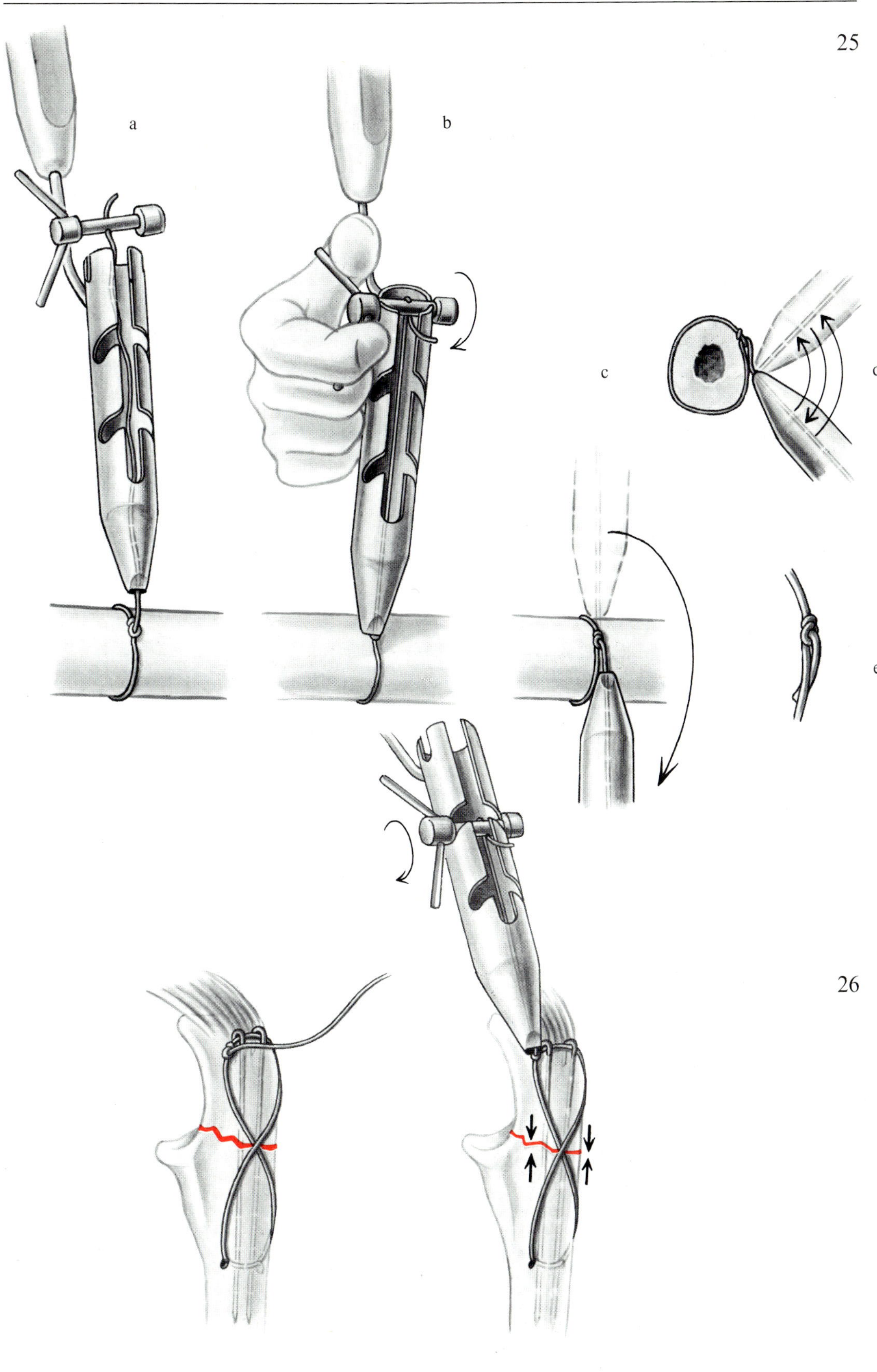

26

2.2.3 Combination of Tension Band Wire and Kirschner Wires

The Principle: The Kirschner wires add to the rotational stability and give further anchorage for the wire. The tension band wire is passed around the Kirschner wires and does not have to be passed so carefully through the tendon insertions. By twisting the wire on both sides, one can achieve a more equally distributed compression.

Note: If the wire is passed only through tendon tissue, when it is tightened, there is some danger of causing a shortening of the infra-patellar tendon which will give rise to a low levelled patella. Therefore we have come to employ more and more the combination of tension band wiring with Kirschner wires so that the tension band wire can be anchored around the Kirschner wires.

The Kirschner wires must be inserted parallel to one another. If they are crossed they not only keep the fragments apart and interfere with the inter-fragmental compression but also provide much less rotational stability.

Fig. 27 *The combination of tension band wiring with Kirschner wires.*

a A lateral and AP view of the patella (for the technique, see Fig. 207). To facilitate the eventual removal of the internal fixation bend over only one end of the Kirschner wires.

b In the fixation of an olecranon fracture, the Kirschner wires are inserted into the bones so that they come to lie in the medullary canal. The ends are then bent over. The wires are just outside the insertion of the triceps tendon.

c In fixation of the greater trochanter, particularly when comminuted, the tension band wire is not passed through the tendinous insertion, but around the ends of the Kirschner wires. The Kirschner wires add to the rotational stability of the reduced fragments.

d In avulsion fractures of the malleoli, where the fragment is either too small for screw fixation or too osteoporotic, a tension band wire is passed around the Kirschner wires.
The disadvantage of this fixation is that at the time of metal removal considerable exposure is necessary. Generally we prefer, even in small avulsion fragments, to attempt fixation of the medial malleolus with one or two 4-mm cancellous screws.

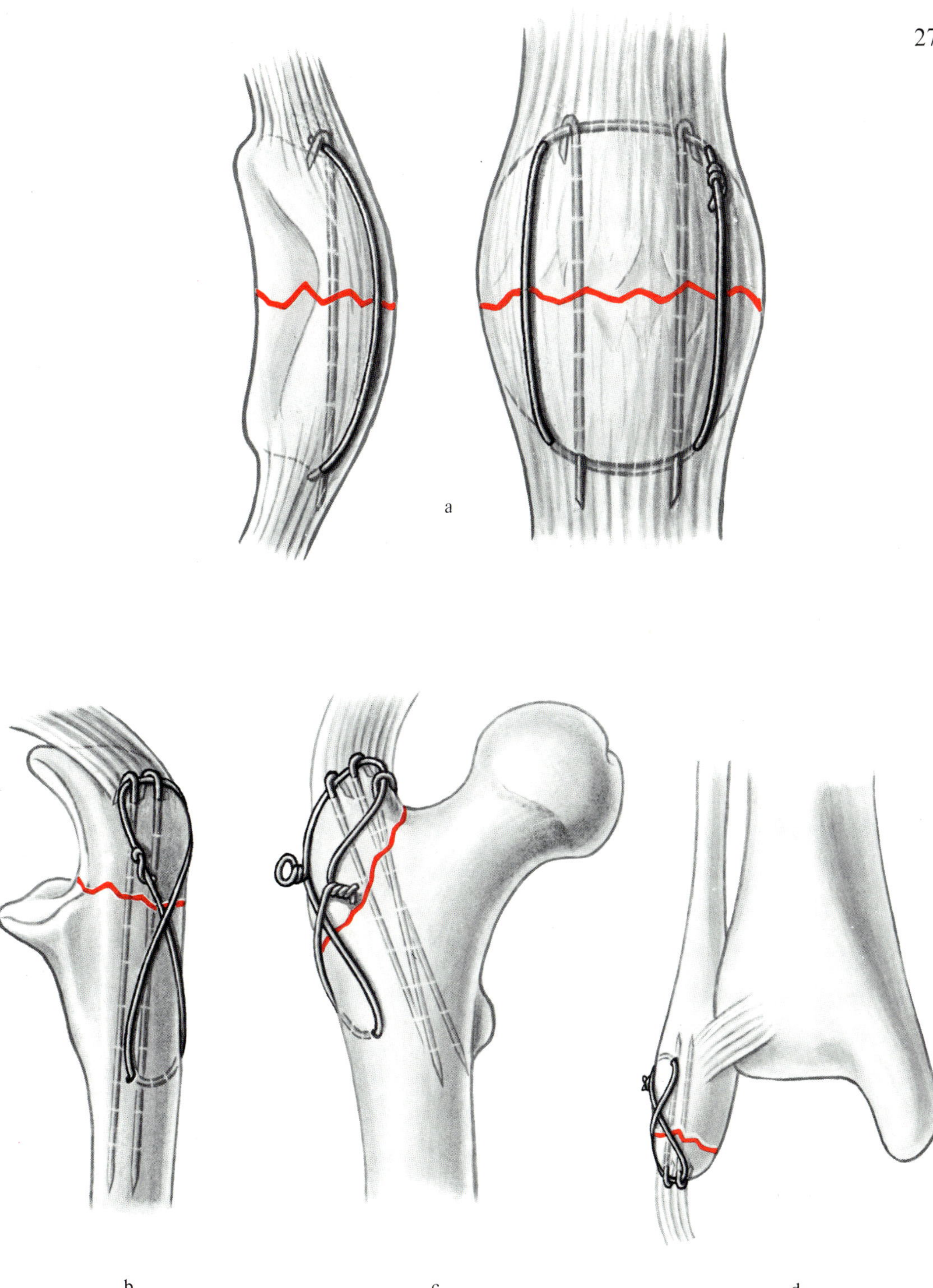
a
b
c
d

2.3 Standard AO Plates Grouped According to Shape

In general there are three groups of plates: the straight plates, the special plates and the angled plates. Of these, the straight plates are meant for diaphyses, the special plates for epiphyses and metaphyses and the angled plates, also called blade plates, for the proximal and distal femur.

Note: In the early days of the AO, in the years 1958 and 1959, of the straight plates only the wide and the narrow round hole plates were available, of the special plates only the T-plates and of the angled plates only the 90° osteotomy plate and the condylar plate could be obtained. Over the years, new methods of rigid fixation were developed and new plates had to be developed to meet the new requirements.

The semi-tubular and the one-third or small semi-tubular plate was developed in 1960. Its holes were oval (MÜLLER et al., 1963). Basic research of ALLGÖWER, PERREN, RUSSENBERGER and others led in 1965 to the development of the self-compressing plate with the semi-cylindrical load screw holes known as the dynamic compression plate (DCP) (see Fig. 45a). There are a number of advantages of the DCP. First, compression can be achieved by the eccentric placement of the screws, which means that smaller exposure is necessary. This is of particular advantage in such areas as the forearm. Furthermore, the screws no longer have to be inserted at right angles to the axis of the plate, but can be inserted obliquely. Even cancellous screws can be inserted with some tilt through all the screw holes. Experience has shown that eccentric placement of a screw in the round hole plate can destroy all the compression generated with the tension device. In the AO DC plate, the design of the screw holes has completely eliminated this danger. The plates also lead to a better stress distribution in bone as long as the screws have not been inserted in the eccentric buttress position near the fracture.

Fig. 28 *Standard AO plates.*

1 *Straight plate:* Round hole plate wide (a) and narrow (b) for the 4.5-mm screws. Tubular plates with oval holes: a semi-tubular plate for the 4.5-mm screws (c), one-third tubular plate for the 3.5-mm screws (d), the one-quarter tubular plate for the 2.7-mm screw (e). Self-compressing plates (DCP) for the 4.5-mm screws, wide (f) and narrow (g). The small DCPs for the 3.5- and 2.7-mm screws (h, i).

2 *Special plates for metaphyses:* T plates, standard (a) and small (b). The tiny T plates (c) and the right and left L plates for phalanges (d). The special buttress plates for tibial plateaus: The T plates (e) and the right and left L plates (f). The spoon plates (g) and the cloverleaf plate (h) for the distal tibia and the cobra plate for hip fusions (i).

3 *Angled plates:* The condylar plate (a) and the 130° angled plate (b), the 120° angled plate for re-positioning osteotomies (c). The 90° angled plate for inter-trochanteric osteotomies: For adults (d), teenagers (e), children (f) and small children (g) (see also p. 368).

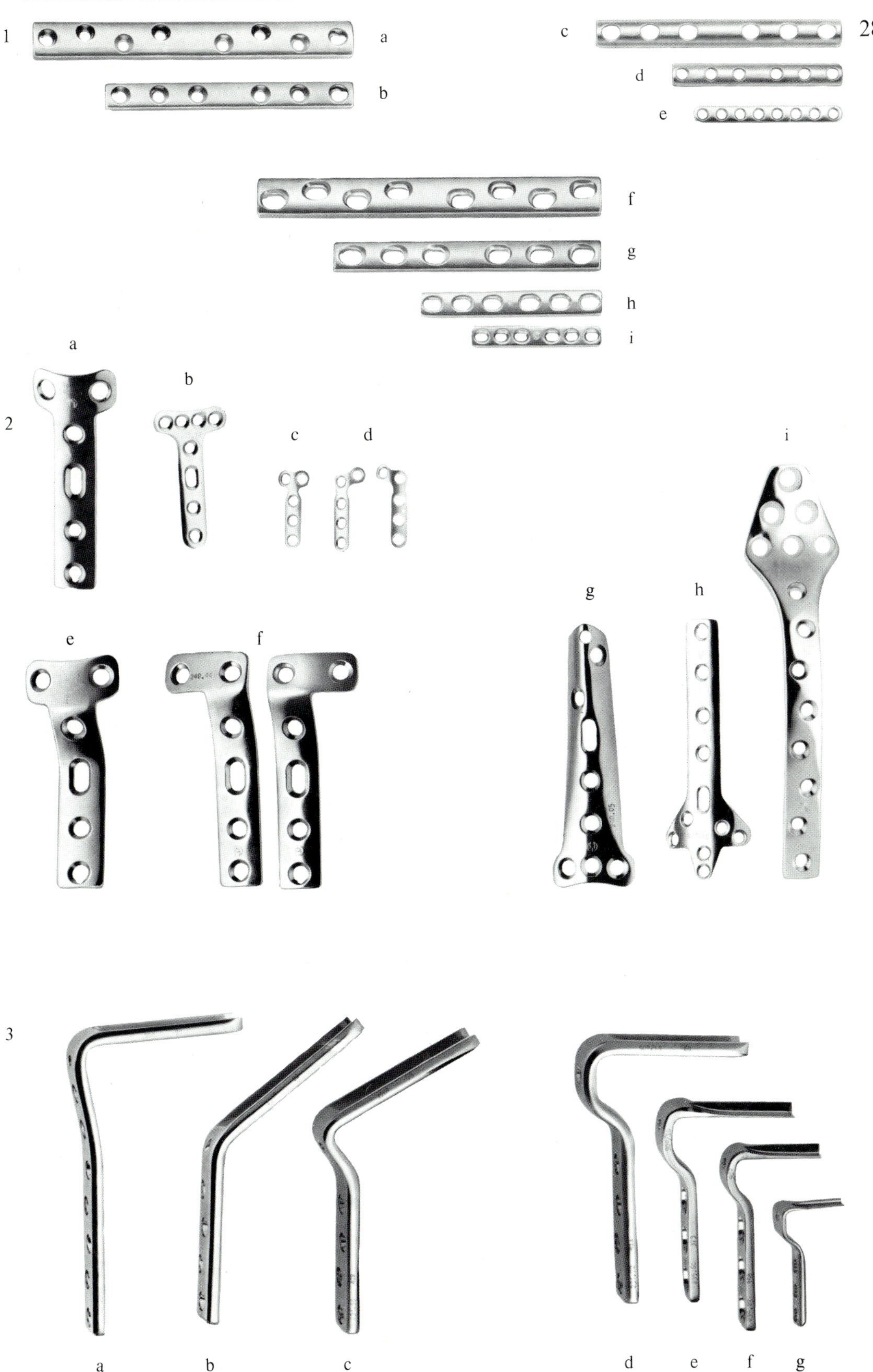
1
a
b
c
d
e
f
g
h
i
2
a
b
c
d
e
f
g
h
i
3
a
b
c
d
e
f
g

Standard AO Plates Grouped According to Function

A plate can have one or more of the following four functions. Its function depends on the manner in which it has been employed. The four functions are: static and dynamic compression, neutralization and buttressing.

1. *Static Compression:* The tensile pre-stress of the plate results in axial compression of the fracture. Such compression plating is indicated chiefly in fractures of the arm (see p. 56). The transverse boot top fracture is an excellent indication for two semi-tubular plates used as compression plates. A single compression plate without a lag screw inserted obliquely across the fracture must never be used for fractures of the tibia or femur.

2. *Dynamic Compression (Tension Band Plate):* A plate so employed takes up all the tensile stresses, and gives rise to pure compressive stresses at the level of either a pseudarthrosis, osteotomy or arthrodesis. Tension band plating should not be attempted in fresh fractures with the exception of certain subtrochanteric fractures and certain comminuted fractures of the olecranon. In fresh fractures one cannot tell most of the time the direction of the tensile stresses.

3. *Neutralization:* This is by far the most common function of the plates. Static inter-fragmental compression is achieved by means of lag screws which are inserted either alone or through the plate. Once the lag screw fixation has been achieved, the carefully contoured neutralization plate (protection plate) is fixed to the bone. Such a plate protects the lag screw fixation by neutralizing most of the torsional, shear and bending forces. The chief indications for a neutralization plate are fractures of the tibia with the exception of transverse fractures. Whenever a short oblique fracture is fixed by means of a plate, the fracture line must be bridged by a lag screw. This is a prime pre-requisite for stable fixation.

4. *Buttressing:* A plate so employed protects the cortex or a cancellous bone graft from collapsing. A buttress plate can also be used to bridge a large diaphyseal defect filled with bone graft while the bone graft is being incorporated.

Fig. 29 *Classic examples of plate function.*

a *Static compression:* Transverse fracture of the humerus. The plate must be contoured (see Fig. 33) in order to increase the stability and the amount of the fracture surface which comes under compression. A broad plate should be used because the staggered screws decrease the likelihood of longitudinal fissuring of the bone.

b *Dynamic compression (tension band plating):* A pseudarthrosis of the tibia with varus deformity. The narrow plate is applied laterally (along the tension side) and is placed under tension. This results in neutralization of all the tensile forces.

c *Neutralization:* A tibial shaft fracture with a butterfly. The butterfly fragment is first fixed with lag screws. Subsequently a narrow plate is applied on the medial side as a neutralization plate. The plate conducts all torsional, shear and bending stresses from the one to the other main fragment.

d *Buttressing:* Compression fracture of the lateral tibial plateau. The fracture is elevated and the defect bone grafted. Thereafter the fracture is buttressed with a T-buttress plate. This plate is not placed under tension but under compression.

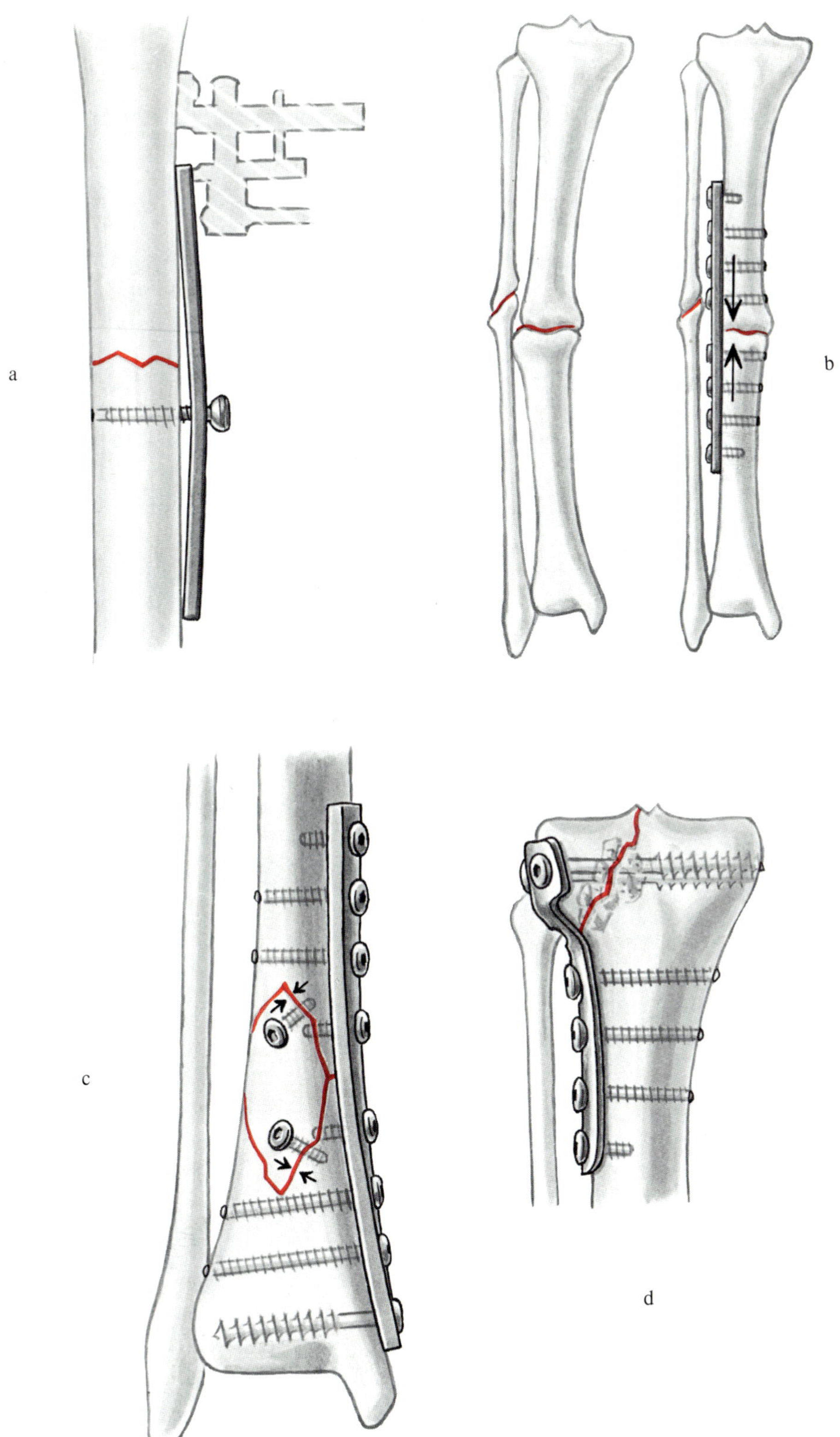
a
b
c
d

2.3.1 Straight Plates

2.3.1.1 Round Hole Plates

For 15 years, lag screws, the round hole plates, and the tension device were the most important implants of the AO used to secure rigid internal fixation with compression. Then we introduced the spherical screw head and the self-compressing plate (DCP). In 1977 we modified the screw holes in the small round hole plate. The profile of the screw hole was changed in such a way that cortical screws could be slightly angled and cancellous screws could be inserted through all the holes.

The tension device was modified and improved. The amount of tension can now be read off directly. The hook of the device has been made to swivel so that it can change direction; thus it can be used not only for compression but also for distraction. This feature is most useful when one is reducing a fracture which has shortened and which has to be lengthened.

Fig. 30 *Round hole plates.*

a Broad round hole plate, convex in profile.

b Narrow round hole plate, convex in profile.

c Profile of a screw hole which has a 5.5-mm conical recess on the underside which allows for angling of the screws. Since 1977 the diameter of the hole has been 4.4 mm.

d A 4.5-mm cortex screw can be angled through 20°.

e The cancellous screw can be inserted through all the holes of the new narrow round hole plate.

30

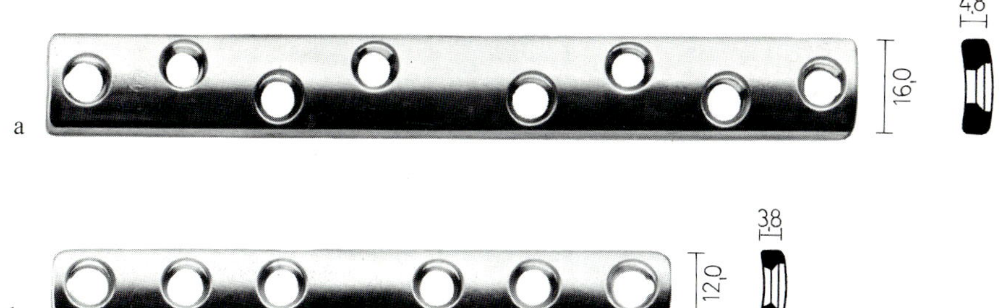

a

b

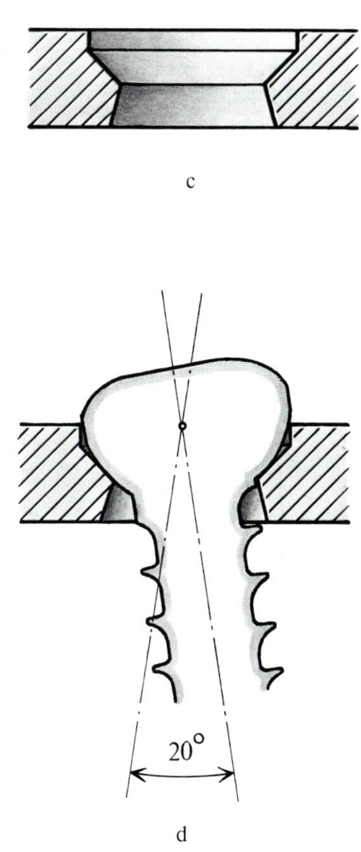

c

d

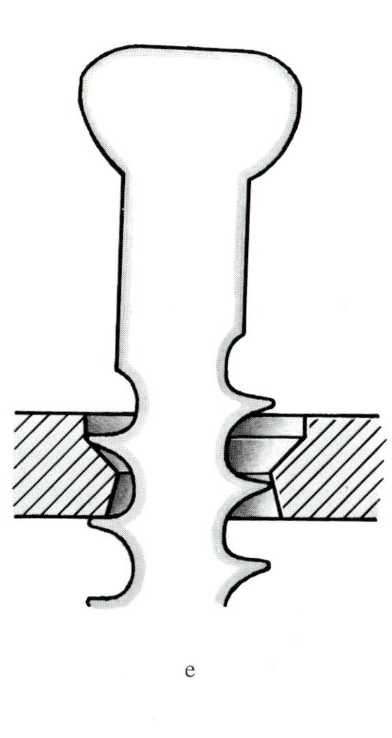

e

Pre-stressing of the Plates with the Tension Device

Most of the AO plates can be pre-stressed with the tension device. The broad and narrow round hole plates, the DC plates as well as the angled plates and cobra plates all have a recess cut into the end hole which will receive the hook of the tension device. Use of the tension device will be demonstrated with the narrow round hole plate. The tension device is not necessary for compression with the self-compressing plates such as the DC or the tubular plates. However, it should be used whenever the gap to be closed exceeds 2 mm or whenever the desired compression, as for instance in a femur, should exceed 100 kp.

Note: In compressing oblique fractures one must consider which fragment the tension plate is attached to. It should be attached in such a way that when compression is generated the plate will press the two fragments together and assist in achieving axial compression and impaction of the fragments. As a rule the tip of the opposite cortex points to the fragment to which the tension device should be fixed (see Fig. 34).

Fig. 31

a Tension device with an 8-mm excursion.

b Articulated tension device with a pressure gauge and an excursion of 20 mm. The green ring indicates a compression of 50 kp and the red ring a compression of 120 kp. In order to achieve distraction simply turn over the hook.

c Drill guide for the tension devices.

Fig. 32 *Use of the tension device.*

a Drill a 3.2-mm drill hole 1 cm from the fracture. Measure the depth. Tap the thread and reduce the fracture. Screw the plate down with a pre-selected cortical screw. Hold the reduction with a self-centring bone-holding clamp. With the drill guide for the tension device with 8-mm excursion, drill the 3.2-mm hole, which will be used to fix the tension device to bone. The tension device is fixed either to one or two cortices. This depends on the thickness and hardness of the bone.

b Fix the tension device to bone after hooking it into the end hole of the plate. With the socket wrench (Kardan key) slightly tighten the tension device, and make sure that the reduction is maintained.

c Insert the rest of the screws into the first fragment. Use the 40-mm-long 3.2-mm drill guide in drilling the screw holes to make certain that they are perfectly centred in the holes. When tapping always use the tap sleeve. This prevents the tap from catching and damaging the soft tissues.

d When the screw of the tension device is turned the fragments are brought under considerable inter-fragmental compression. With the socket wrench a compression of approximately 40–45 kp can be achieved and with the open spanner, a pressure of 120 kp and greater.

e The compression results in rigid fixation of the fragments. Once compression is generated the reduction should be checked once again. If all is well the remaining screws should be inserted.

f At the end, the tension device is removed, and a screw is inserted through the end hole of the plate. The screw is usually inserted through only one cortex. This results in a more gradual transition of forces under load between the plated segment and the rest of the bone, and the elasticity of the bone is not interrupted so suddenly.

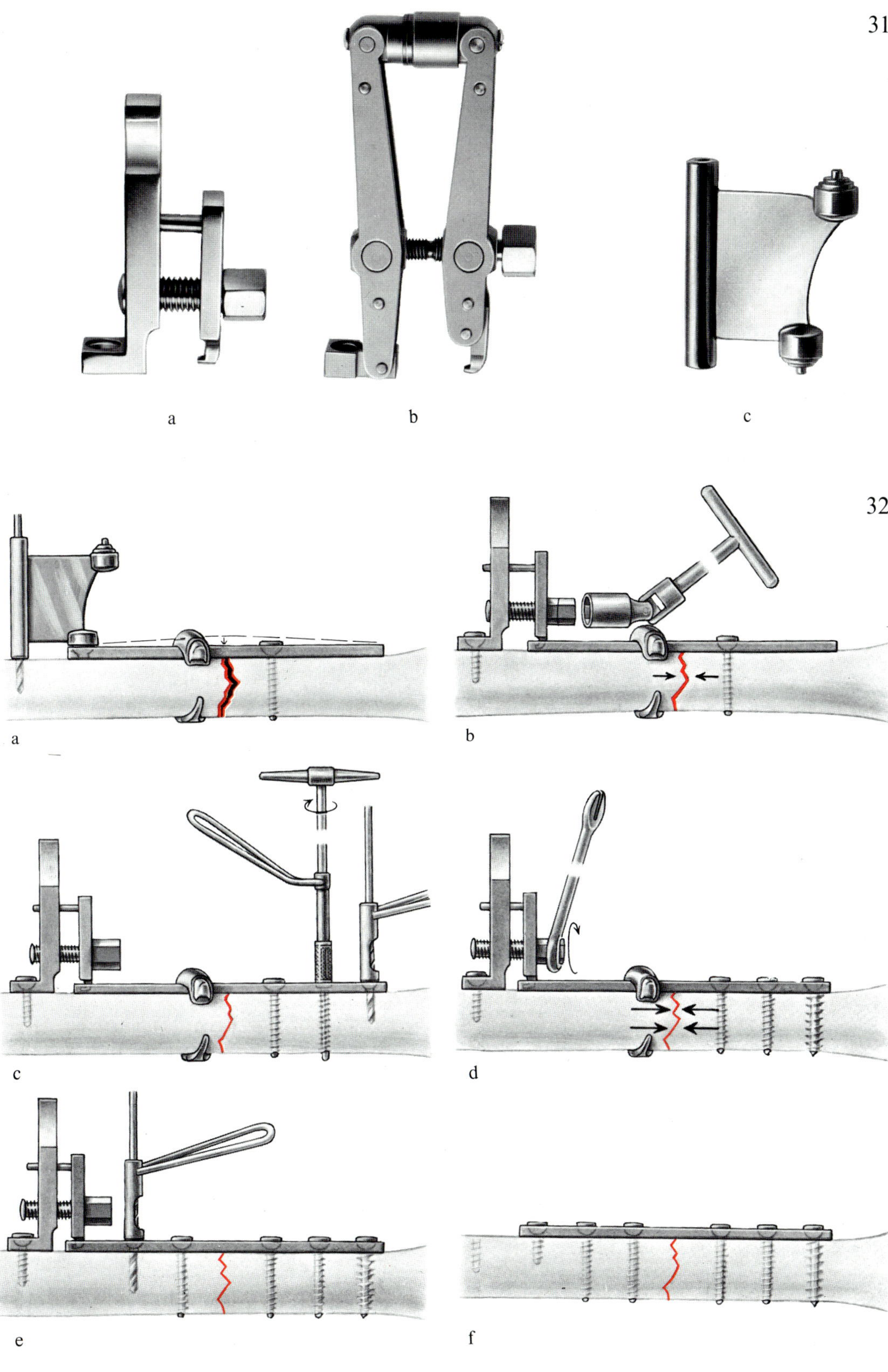
31
a
b
c
32
a
b
c
d
e
f

We shall demonstrate, with the round hole plate, the four functions of a plate as well as the four different ways of pre-stressing a plate.

The Round Hole Plate as a Static Compression Plate

Through slight over-bending of a plate through its middle, one can achieve compression of the opposite cortex and in this way increase the amount of the fracture surface which will come under compression. The increase in compression results in an increase in friction and impaction which greatly enhances the potential of the plate to neutralize all the torsional and bending forces.

This over-bending of the plate should be done through its middle, through the solid portion of the plate and not through a screw hole. It is best done in a bending press or with the bending pliers. The axial inter-fragmental compression is achieved with the aid of the tension device. Whenever possible the fracture should be crossed with an oblique lag screw. This greatly increases the stability of the internal fixation.

The over-contouring of the plate should be used only when fixing transverse or short oblique fractures of the humerus, radius and ulna. It is rarely, if ever, used in comminuted fractures. A compression plate should never be used as the sole means of fixation for fractures of the leg because it does not provide sufficient stability. It must be supplemented by compressing the fracture with a lag screw which should so be inserted through the plate if at all possible.

Fig. 33 *A transverse fracture of the humerus.*

a If a straight plate is placed under tension, only the cortex immediately adjacent to the plate comes under compression.
This will result in the plate becoming subjected under load to some bending stresses.

b After slight over-bending of the plate in its middle, the plate is fixed to the bone with one screw approximately 1 cm away from the fracture line. The hook of the tension device is then hooked through the end hole of the plate and the tension device is fixed to the bone. This results in a slight gaping of the fracture.

c As soon as the screw of the tension device is turned and tension generated, the fracture surface begins to come under compression. As further tension is generated the plate straightens out and both the plate and the bone become straight. The whole fracture surface is under compression and there is no longer any gaping at the fracture.

Fig. 34 *An oblique fracture of the radius.*

a Fix the slightly over-contoured plate to that fragment whose spike is opposite the cortex which will be adjacent to the plate.

b The tightening of the screw of the tension device will result first in a reduction and subsequently in compression.

c If the fracture is to be bridged with a 3.5-mm cortex lag screw, the cortex adjacent to the plate is perforated with a 3.5-mm drill bit. The 3.5-mm tap sleeve is used here as a drill guide. The hole is made obliquely so that the screw will cross the fracture as much at right angles as possible.

d The 2-mm drill sleeve which has a 3.5-mm outer diameter is now pushed through the gliding hole until it abuts on the opposite cortex. The thread hole is now drilled with the 2-mm drill bit and the thread is then cut with a 3.5-mm tap.

e The fracture plane is thus under both axial and inter-fragmental compression. This combination results in far greater stability than axial compression alone.

Note: If because of bone size one chooses the 2.7-mm cortex screw then the screw head will not engage in the screw hole of the plate because it is too small.

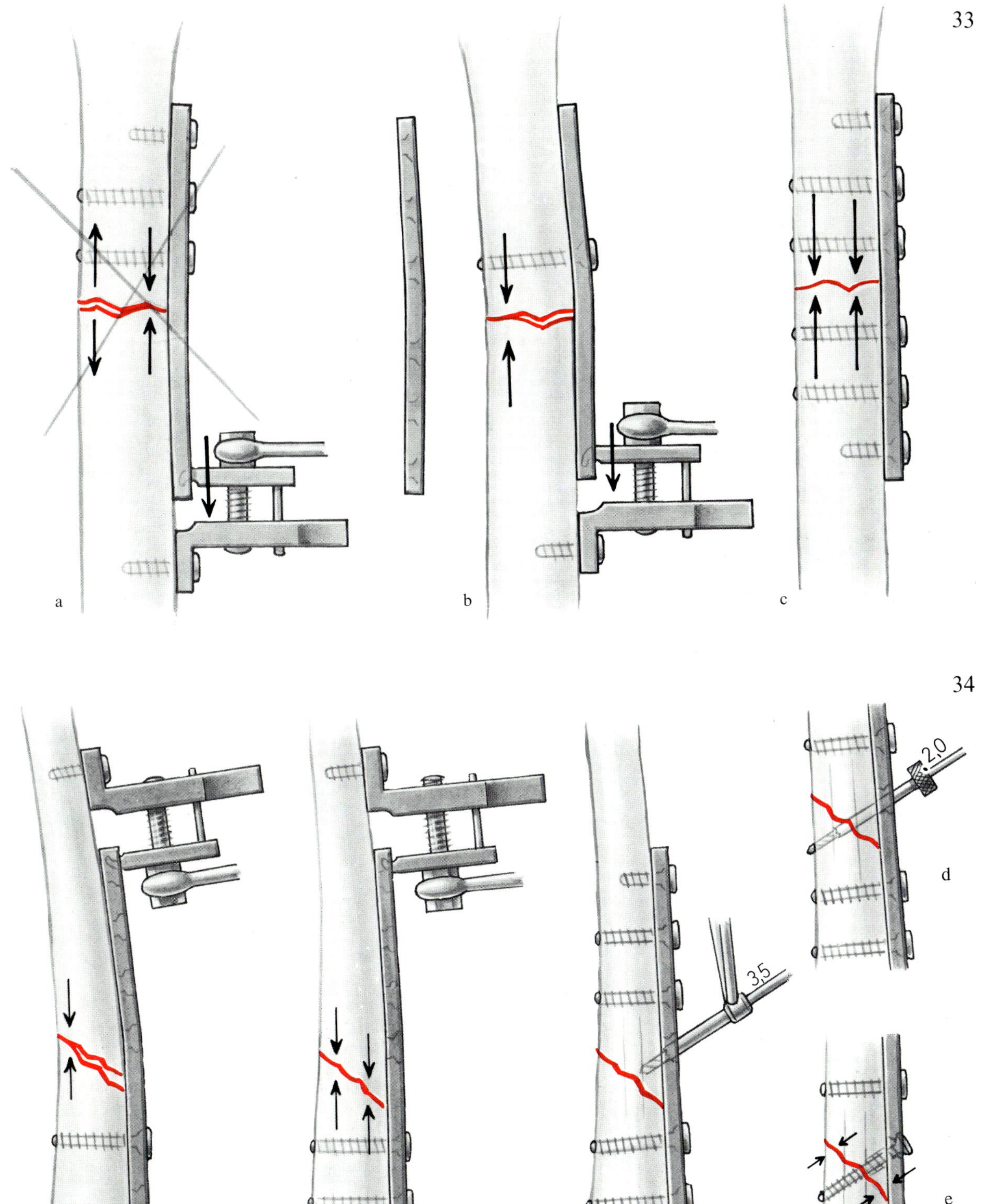
33
a
b
c
34
2,0
3,5
d
e
a
b
c

The Round Hole Plate as a Tension Band Plate (Dynamic Compression)

A tension band plate neutralizes all the tensile forces which may act on fractures, pseudarthroses or osteotomies. As already described under the tension band wiring of the patella, active loading results in an increase of the inter-fragmental compression.

Theoretically, a plate fixed to the bone without pre-stress, but on its tension side, could act as a tension band because it would come under tension as soon as muscles contracted or one loaded the limb. This tension in the plate would result in axial compression. In order, however, to have axial compression present during periods of rest as well as activity, it is necessary to pre-bend the plate and pre-stress it by applying tension.

The principle of the tension band permits correction of axial deformities with the help of a straight plate and a tension device.

Fig. 35 *Pseudarthrosis of the tibia with a varus deformity.*

a The plate is applied to the lateral surface of the tibia. The tension device is fixed to the tibia with a long screw.

b As the screw fixing the tension device to the bone is tightened, the plate adapts to the shape of the bone and bends.

c As the tension device is tightened, the plate comes under more and more tension. This results in straightening of both the plate and the underlying bone, and the pseudarthrosis comes under considerable axial compression.

Fig. 36 *Pseudarthrosis of the tibia with a posterior bowing.* The plate is applied on the posterior surface of the tibia. As in Fig. 35, the deformity is corrected with the tension device. Loading results in an increase in the axial compression.

35

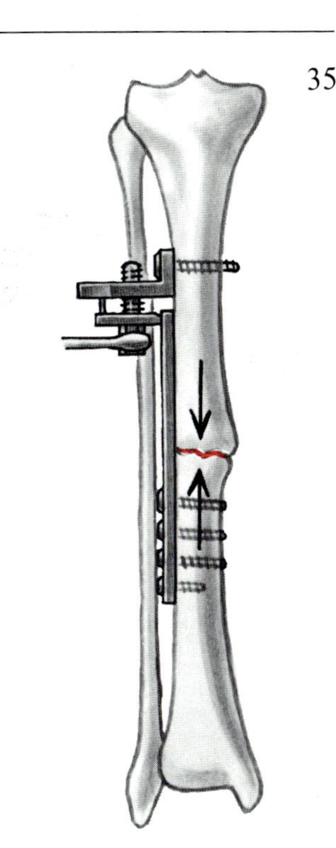

a b c

36

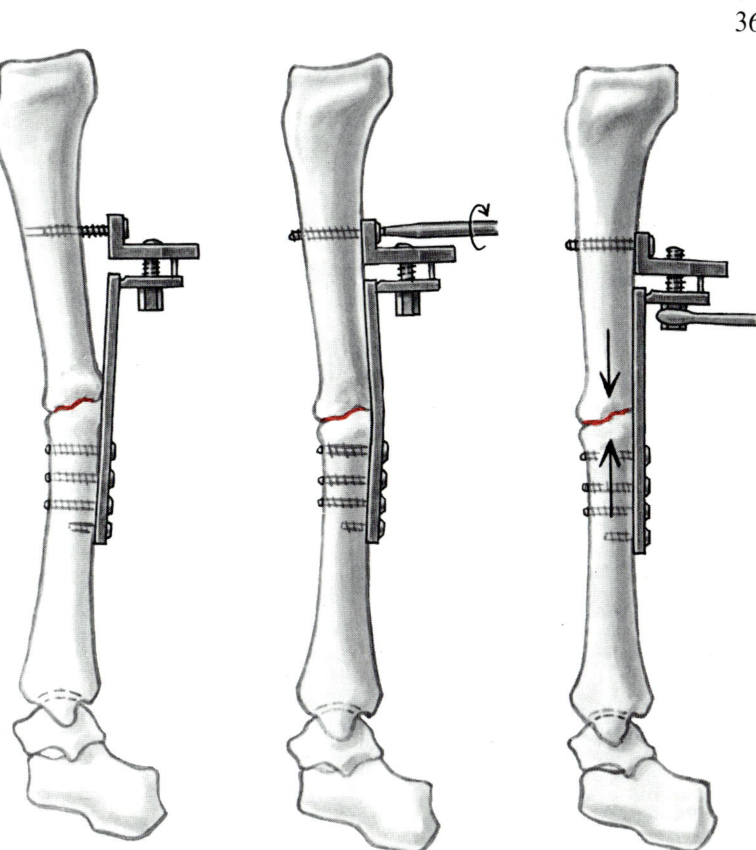

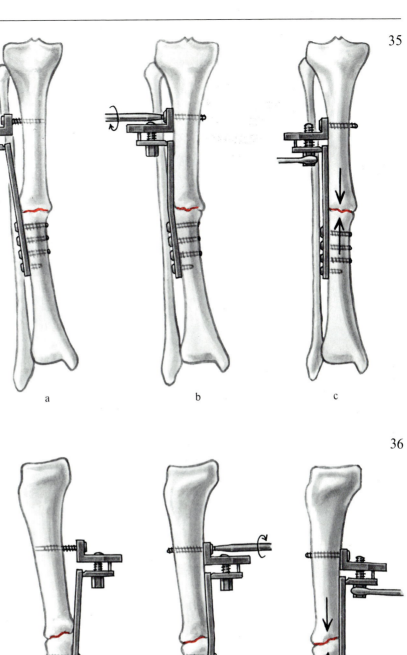

The Round Hole Plate as a Neutralization Plate

Lag screw fixation alone of shaft fractures is advisable only for the long spiral fractures of the tibia or phalanges. Lag screw fixation alone is not able to withstand much loading. In order to permit the so-called functional after-care with early movement and loading, it is necessary to bridge or protect most fracture zones with a plate. Such a plate neutralizes all the torsional, shearing and bending forces. A plate which neutralizes these forces is referred to as a *neutralization plate*.

A comminuted fracture is first reduced and fixed with lag screws (see Figs. 19 and 20). A carefully contoured plate is then fixed to the two main fragments with two to four screws. A lag screw can also be inserted through the plate as demonstrated in Fig. 38. The combination of lag screws and neutralization plating is the commonest type of internal fixation of the tibia. Whenever possible such a plate is also placed under slight pre-stress in order to add some axial compression.

Fig. 37 *Inter-fragmental compression by means of lag screws in combination with a neutralization plate.*

a Drill the gliding holes with a 4.5-mm drill bit prior to the reduction.

b Carry out the reduction, drill the thread holes and fix the butterfly with lag screws under maximal inter-fragmental compression.

c The neutralization plate bridges and protects the fracture.

Fig. 38 *The insertion of a lag screw through a plate.*

a Drill the gliding hole, reduce the fracture, drill the thread hole and tap the thread.

b A plate is then applied to bridge the fracture and the lag screw is inserted through the plate into the opposite cortex and tightened.

Note: It is advisable most of the time to reduce the fracture and insert the lag screws. Once the fracture is reduced and fixed one should contour the plate to the bone. Any lag screw which comes to lie directly under the plate is removed and is inserted through the plate.

37

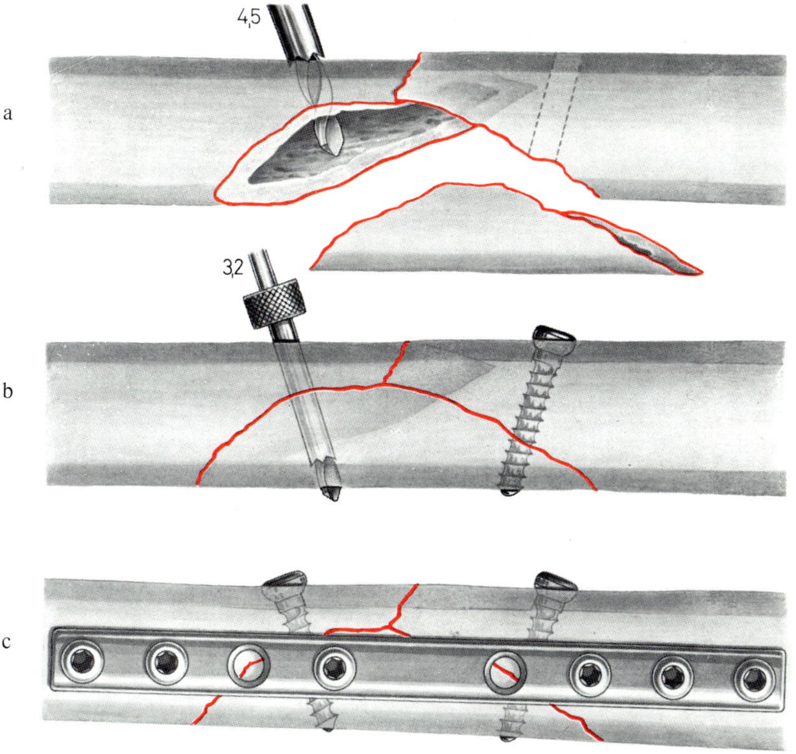

38

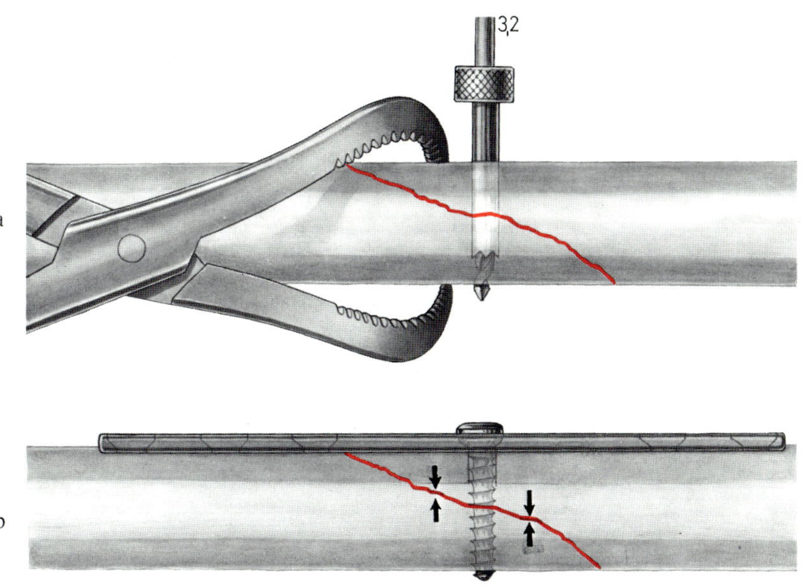

Contouring (Bending and Twisting) of the Neutralization Plate in the Distal Third of the Tibia

If a plate is to be applied to the medial surface of the distal third of the tibia, it must not only be bent but also twisted in order to lie flat on the bone. The contouring of such a plate is greatly facilitated by the malleable templates which are easily bent and twisted with the fingers. *A malleable plate,* once bent to the shape of the bone, is then used as a template for the contouring of the plate which is carried out with the bending press or the bending pliers, together with the twisting irons.

Note: The AO stainless steel permits plastic deformation without any appreciable loss in strength. Repetitive bending, however, should be avoided.

Fig. 39 *Bending and twisting the plate to make it fit the distal tibia.*

a The coloured, malleable templates are made of aluminium and have the end screw hole punched out. The other screw holes are just marked.

b After reduction of the fracture and fixation of the butterfly by means of lag screws, the template is placed on the bone and carefully contoured.

c This malleable plate will now serve as a template for the contouring of a round hole or of a DCP.

d The bending pliers.

e The bending press and the twisting irons.

f The plate is first bent with the bending press.

g The plate is held in the bending press and twisted with the twisting iron.

Note: The narrow plates can be bent by means of the bending pliers. If a plate is inserted obliquely into the press both the bend and the twist can be achieved simultaneously.

39

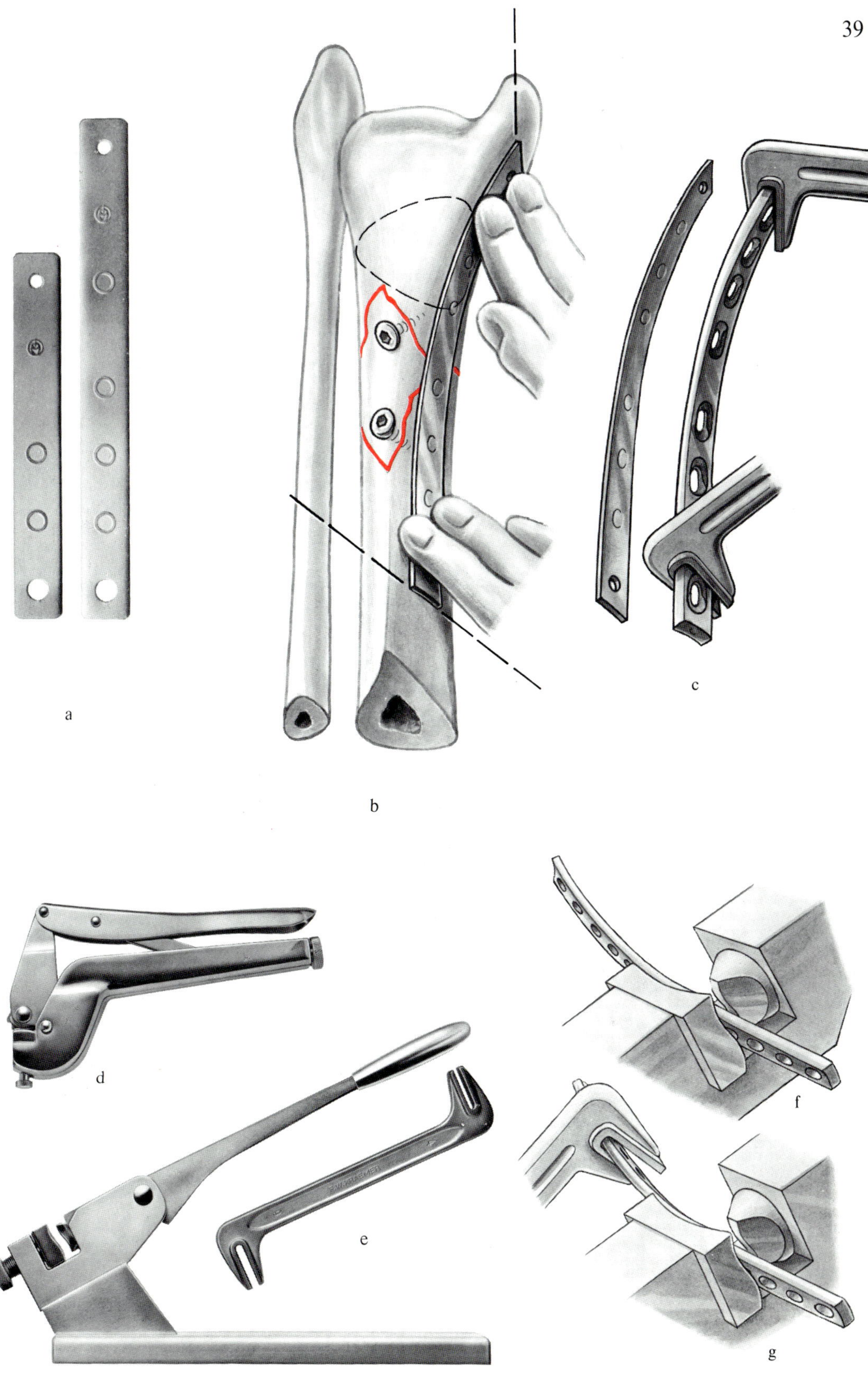

Pre-stressing of a Plate and Axial Compression of the Bone Achieved by Mismatching of Plate and Bone

This principle must be carefully observed because the wrong mismatch between plate and bone can actually result in distraction rather than in axial compression.

Note: When the double-angled blade-plate designed for repositioning osteotomies is used (see Fig. 316.3), the obliquity of the fracture or osteotomy surface provides one further possibility in the way in which inter-fragmental compression can be achieved.

Fig. 40 *The principle guiding the contouring of plates.*

a *Contouring, when the plate is to be applied to a concave portion of bone.* If at the centre the plate is 1–2 mm away from the bone, then it is shorter than the segment of bone it covers. If the plate is now fixed to the bone first through the end holes and then in a progressive manner through subsequent screws, closer and closer to the centre, the plate will be pre-stressed in tension, and the bone will be placed under axial compression. In carrying out this fixation, one should insert one screw at a time and drill the next screw hole only after the preceding screw has been tightened.

b *Distraction through the contouring of a plate when applied to a straight segment of bone.* If the middle of the plate is in contact with the bone at the fracture and the plate has been contoured in such a way that the ends stand away, then as the plate is screwed to the bone, there will be a tendency to distract the fracture through the opposite cortex.

c Examples of pre-stressing of a plate through careful contouring as applied to the distal tibia.

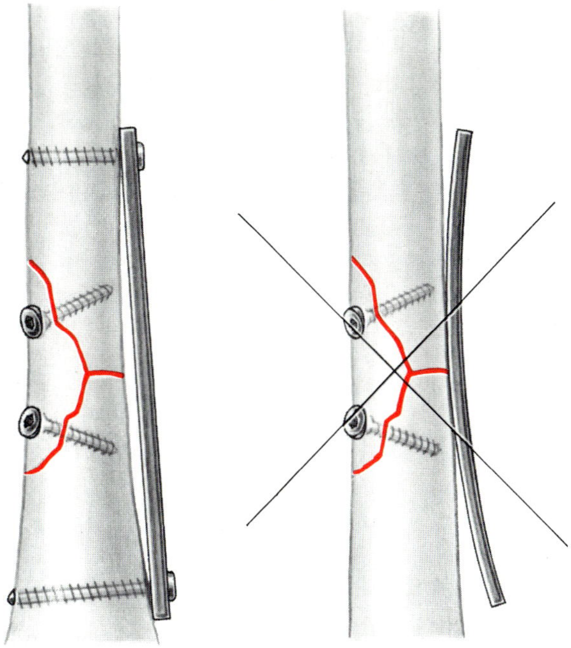

a

b

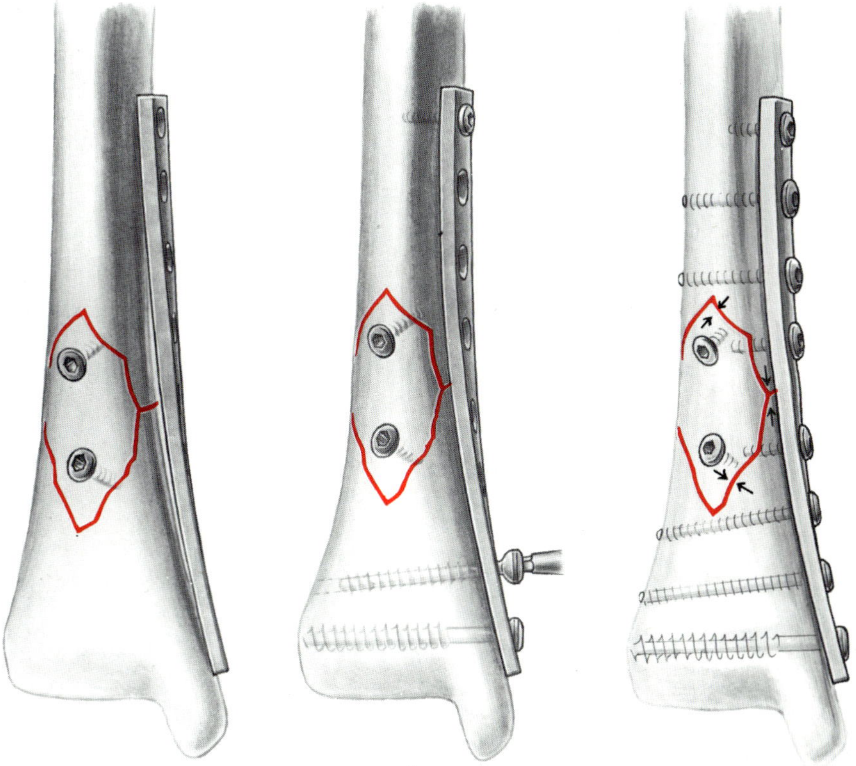

c

2.3.1.2 Tubular Plates

Originally the tubular plates were developed to be used on the anterior crest and medial edge of the tibia. They are self-compressing, only 1 mm thick, and easily deformable. Because they are so easily deformable, they should be used ideally only as tension bands where the only deforming forces are those of tension. Their great advantage is their ability to create rotational stability by their close adaptation to bone and by their edges digging into bone. Their great disadvantage lies in the fact that the screw heads protrude deeply through the screw holes which may result in the shattering of the underlying cortex. The stability of tubular plates is considerably greater than that of straight plates of equal thickness. This is the reason why the quarter-circle profile was chosen for the mini-plates.

The oval holes and the eccentric placement of screws, when properly combined, account for the self-compressing property of these plates. The fracture must be anatomically reduced. The screws are then inserted eccentrically through the oval hole as far from the fracture as possible. This results in axial compression of the fracture. The semi-tubular plate is fixed to bone with the 4.5-mm cortical screws, the one-third tubular plate with the 3.5-mm and the one-quarter tubular plate with the 2.7-mm cortical screws.

Tubular plates can also be used as buttress plates. A good example of this is their use on the distal tibia to bridge a bone defect which has been filled with a bone graft. One plate is placed on the anterior edge and one on the medial edge of the tibia.

Note: If one wishes to achieve compression with the one-third tubular plate or with the one-quarter tubular or with the mini-T-plates, their corresponding screws (i.e. the 3.5-mm cortex screw or the 2.7-mm cortex screw respectively) must be inserted eccentrically in the same manner as the 4.5-mm cortex screws for the semi-tubular plates.

Fig. 41 *Tubular plates: a segment of a tube of 6-mm radius and 1-mm thickness.*

a Semi-tubular plates with 4.5-mm cortical screws.

b One-third tubular plates with 3.5-mm cortical screws.

c One-quarter tubular plates with 2.7-mm cortical screws.

Fig. 42 *Compression with a semi-tubular plate.*

a Drill the first hole 1 cm from the fracture, tap the thread and screw the plate to the bone but do not tighten the screw. Reduce the fracture accurately and with a hook apply traction to the plate. This will result in eccentric placement of the first screw in its screw hole. While the pull is maintained, drill the second hole eccentrically as far from the fracture as possible. Use the 3.2-mm drill sleeve as a drill guide to protect the drill bit. Viewed from above and from the side.

b Tap the thread and insert the second screw, which is tightened. Tighten the first screw. As the screws are tightened, by virtue of the hemi-spherical profile of their head and their eccentric placement in the oval screw holes, they come together 1–1.5-mm. Because they are firmly anchored in bone, the bone fragments impact and come under correspondingly high axial compression.

c The remaining screws are then inserted centrally or slightly eccentrically in the screw holes.

41

a 

b 

c 1/4

42

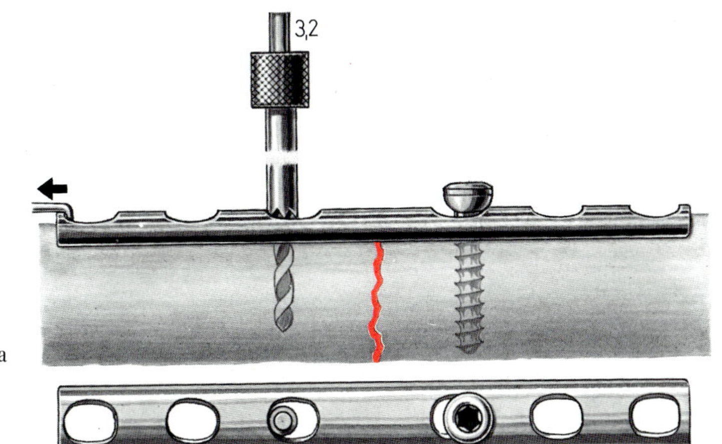

a

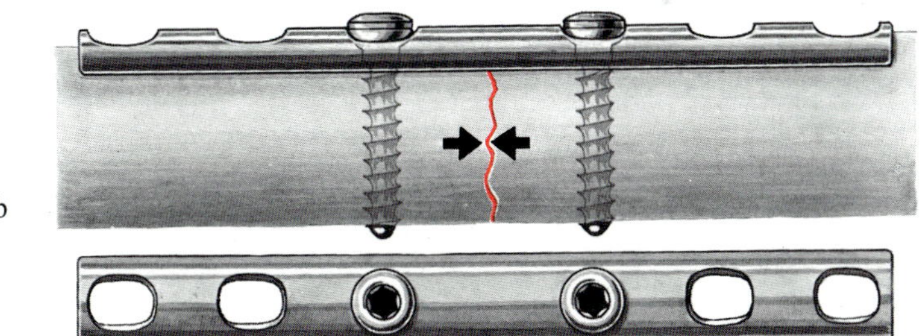

b

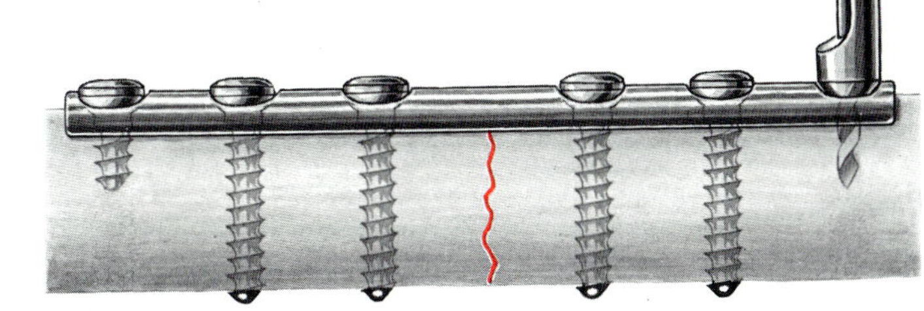

c

The chief indications for the use of the *semi-tubular plate* are the fractures of the radius, the fractures of the proximal ulna, particularly the comminuted fractures of the olecranon, and the fractures of the distal tibia, the so-called boot top fractures which are best handled with two semi-tubular plates.

The chief indication for the *one-third tubular plate* (small semi-tubular plate) is the comminuted fracture of the lateral malleolus and the transverse fracture of the fibula. This plate makes it possible to preserve the length of the lateral malleolus, even when very severe comminution is present. We have also found this plate useful in metacarpal and metatarsal fractures.

The *quarter tubular plates* are used mostly for metacarpal fractures, which can also be stabilized with the small L plates with oval holes. In most of these examples the tubular plate is used as a static compression plate. When used in comminuted fractures to bridge a defect as for instance in the boot top fracture, the tubular plates are used as buttress plates.

Fig. 43 *Examples of uses of tubular plates.*

a Fracture of the shaft of the radius, a six-hole semi-tubular plate (static compression).

b Comminuted fracture of the olecranon stabilized under axial compression with a semi-tubular plate (tension band plate).

c The use of two semi-tubular plates for the fixation of a boot top fracture with comminution of the cortex. Each plate is fixed to only one cortex (buttress plate).

d Comminuted fracture of the lateral malleolus stabilized with a one-third tubular plate (buttress plate).

e Fracture of the 5th metatarsal fixed with a one-third tubular plate (static compression).

f Fracture of the proximal phalanx fixed with one-quarter tubular plate (static compression).

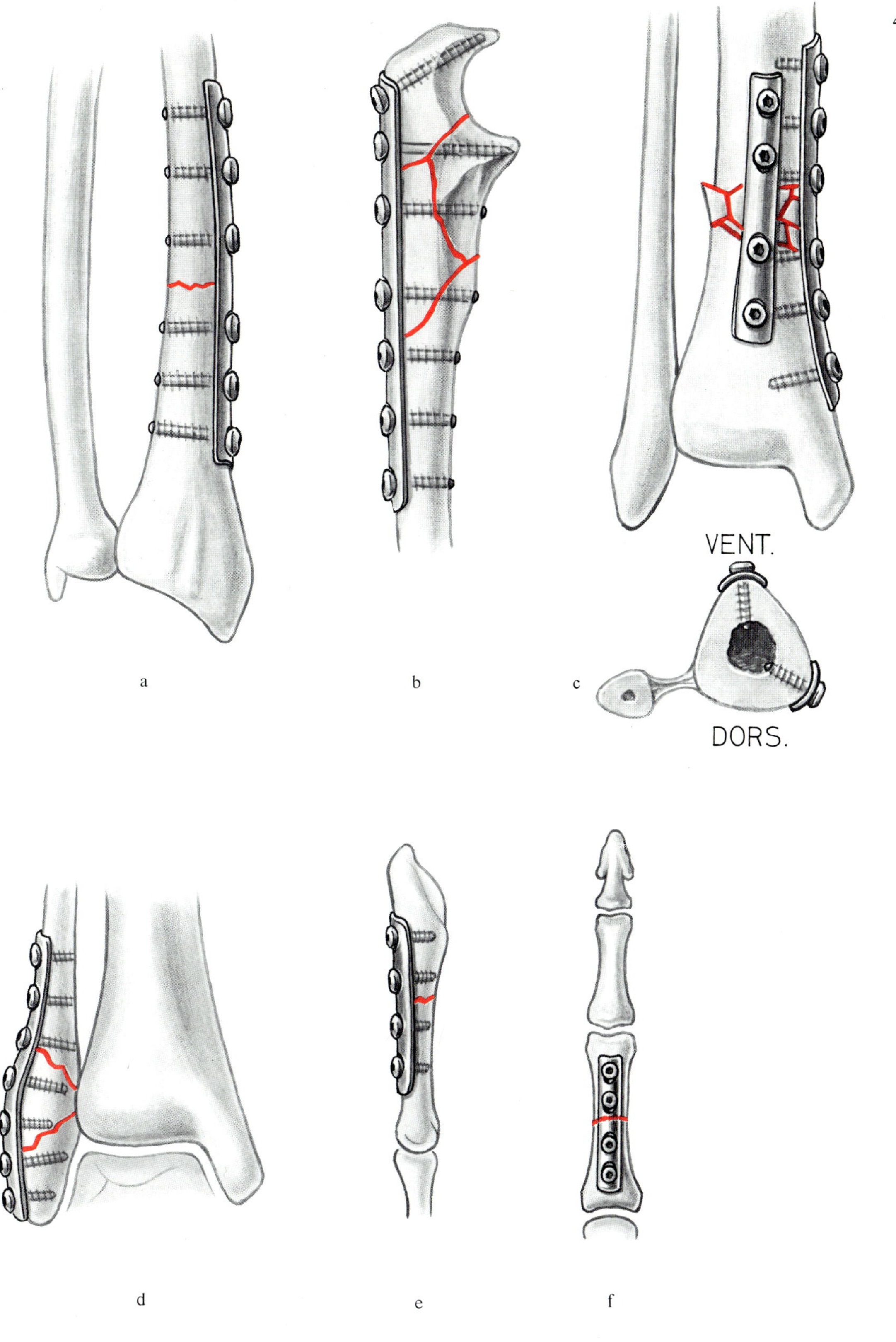
a
b
c
VENT.
DORS.
d
e
f

2.3.1.3 Dynamic Compression Plates (DCP)

The DCP represents an improvement on the traditional round hole plate. The special geometry of the screw holes has increased the potential uses of the plate. It is possible to achieve axial compression without the use of the tension device. It is also possible to angle the screws in any direction. This has made the plate more adaptable to different situations of internal fixation. The name "dynamic compression plate" reflects its function, namely its ability to cause the fragments to approximate one another and in this way achieve compression. In view of the remarks which we have made on page 50 regarding static and dynamic compression, it would be better, in order to avoid a misunderstanding, to refer to the plate in the future as a "self-compressing plate", rather than a "dynamic compression plate". The plate can also be used to achieve other plate functions. It can be used as a static compression plate, as a dynamic compression plate with and without the tension device, as a neutralization plate and also as a buttress plate. The narrow as well as the broad DCPs are represented in Fig. 44. The spherical, sliding principle is the specific characteristic of the sliding hole of the self-compressing plate (DCP) and is described in Figs. 45, 46 and 49.

Fig. 44

a Broad DCP (plate and its profile) to be used on the humerus and femur.

b The small DCP (plate and its profile) to be used on the bones of the forearm, on the tibia and on the pelvis.

c The narrow DCP to be used with 3.5-mm cortex screws (plate and its profile). It should be used when a smaller plate is required on the radius and ulna.

d The DCP to be used with 2.7-mm cortex screws (plate and its profile). It is particularly useful in mandibular surgery.

Fig. 45 *The screw hole and the spherical sliding principle.*

a A sphere (the screw head) slides down an inclined cylinder (the screw hole). This combination of downward and horizontal movement of the screw causes a horizontal movement of the underlying bone with respect to the stationary plate. A sideward movement is impossible. One aims to position the screw head at the intersection of the inclined and horizontal cylinder. At this point, the screw head has a spherical contact in the screw hole, which results in maximum stability without blocking the horizontal movement of the screw.

b The screw hole is manufactured in such a way that in shape it corresponds to the scheme of the two half-cylinders described in (a).

c Here we see the exact representation of the shape of the screw hole. The inclined half-cylinder is for self-compression or self-loading and the horizontal cylinder is to prevent blocking or obstructing the screw in its horizontal path. This guarantees that only compression will be exerted at the fracture.

d Schematic representation of the *load sliding hole* and its corresponding spherical screw head. We see in the left portion of the screw hole the inclined load plane and on the right, the horizontal sliding plane.

44

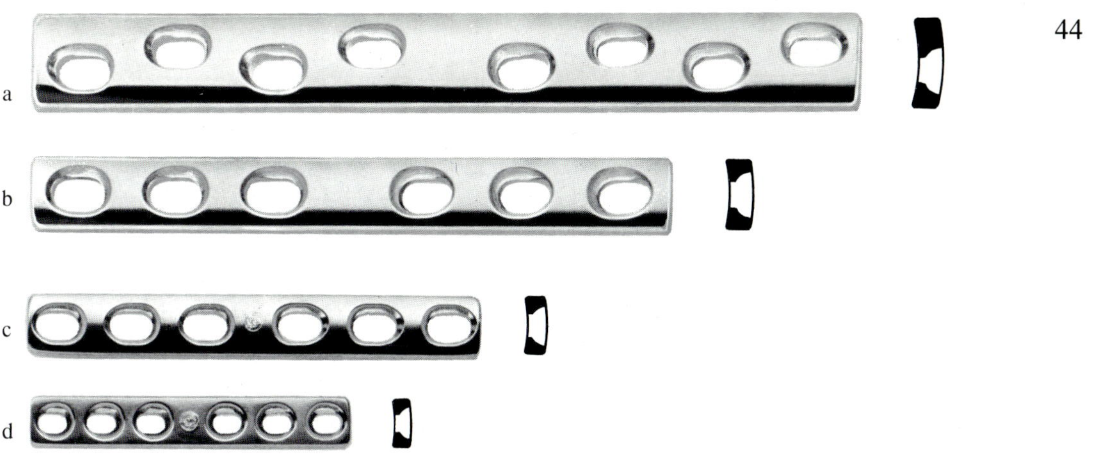

45

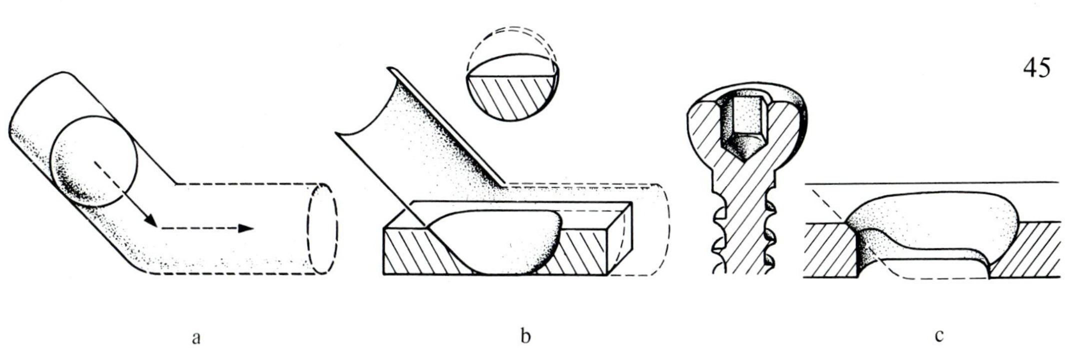

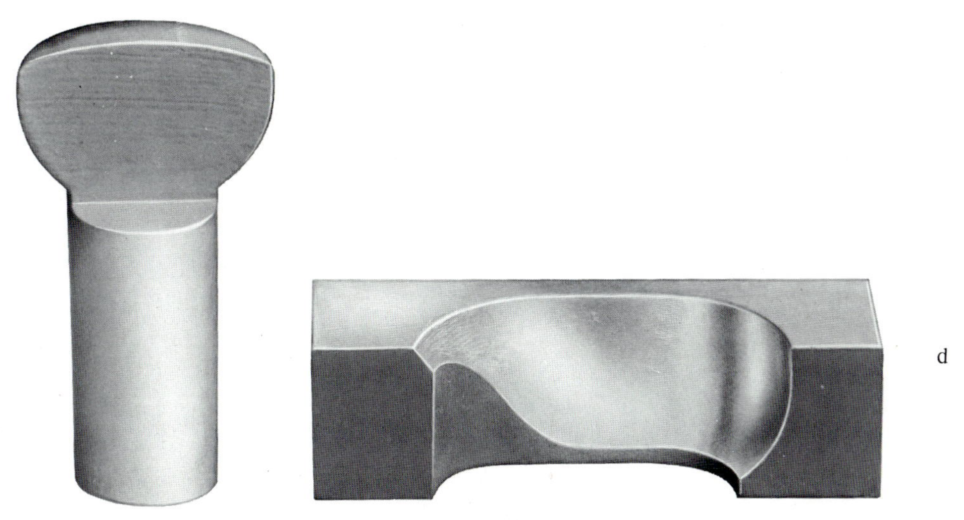

The DCP requires two drill guides. The neutral drill guide (Fig. 46a) is the one most commonly used. It has a central hole for the drill bit. It allows the screw to be inserted in its neutral position, that is, at the intersection of the two half-cylinders which make up the screw hole. (In reality even the neutral guide results in a 0.1-mm loading so that even the screws which are inserted in a neutral position when fully tightened cause a slight degree of axial compression.)

The load drill guide (Fig. 46b) has an eccentric hole for the drill bit which must be directed away from the fracture. This results in the screw being inserted initially 1 mm from the neutral position in a load position along the inclined half-cylinder. When the screw is tightened, the plate and correspondingly, the bone are moved 1 mm. This is enough, with the accuracy of reduction usually obtained under clinical circumstances, to achieve with only one screw an axial compression of 50–80 kp. In order to increase the load, that is, increase the impaction of the fragments, other screws can be inserted in the load position. This, however, is rarely necessary.

We no longer use the buttress drill guide. If we wish to use the DCP as a buttress plate, we use the 3.2-mm drill sleeve as a drill guide to drill the required screw holes (Fig. 46c). The outer diameter of this drill sleeve is 4.5 mm. When the hole is drilled with the sleeve in the screw hole and pressed against the plate towards the fracture, the resultant screw hole is drilled in such a position that, when the screw is tightened, its head is pressed against the plate in the buttress position (see p. 78).

Fig. 46 *Drill guides.*

a The neutral drill guide. This is the drill guide which is used most often with the DCP. Compression is achieved either with one or two screws inserted as load screws or with the use of a tension device. Screws which are inserted with the help of the neutral guide are inserted 0.1 mm eccentrically along the load plane. This results, when the screw is tightened, in an additional axial compression of 0.1 mm. The neutral drill guide is identified by its green colour.

b The load guide. A screw inserted with the aid of the load drill guide meets the inclined load plane 1 mm from its end. If we begin with an accurately reduced fracture, the horizontal displacement with a single load screw will result in 60–80 kp of axial inter-fragmental compression. A load drill guide is identified by its eccentric hole and the yellow colour which also warns that it must be used with caution.

c The 58-mm long, 3.2-mm drill sleeve is used now instead of the previously recommended red buttress guide. If this drill sleeve is inserted towards the fracture at the end of the horizontal half-cylinder, the screw will be automatically brought in a buttress position. If the DCP is used as a buttress plate, we recommend that all screws be inserted in their buttress position. Where a screw crosses a fracture line it should be inserted as a lag screw (see also p. 38).

Fig. 47 *Drill jigs for the narrow and broad DCPs.* These drill jigs allow us to drill the first two screw holes in such a way that the screws are automatically loaded as long as the fracture is accurately reduced. The load distance corresponds to the eccentric load drill guide of 1 mm. The advantage of the drill jig is that screw holes can be drilled without losing sight of the fracture. Loss of reduction can occur if the fracture is hidden under the plate. In this way one can be certain that reduction is maintained. The second screw position is further apart and permits the insertion of an oblique lag screw through the screw hole closest to the fracture (see Fig. 48).

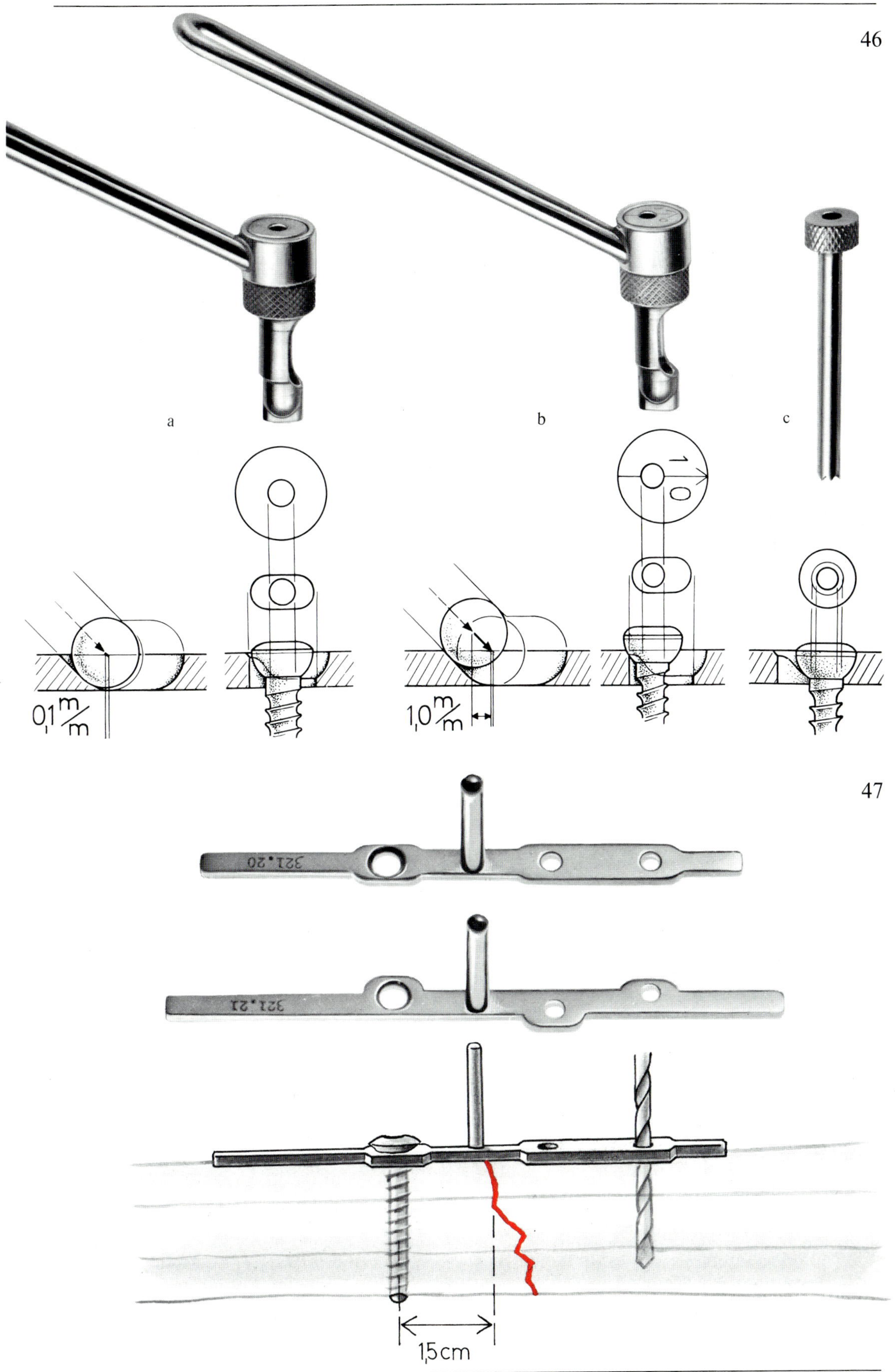
46
a
b
c
0,1 m/m
1,0 m/m
47
321.20
321.21
1,5 cm

Axial Compression with the DCP

Compression plating of transverse fractures should be carried out only in the upper extremity, that is, in fractures of the humerus, radius and ulna. We recommend that even in transverse fractures, whenever possible, one should aim to insert one oblique lag screw through the plate. This becomes essential when plating fractures of the tibia and of the femur. This is why in Fig. 48 the load screw is not the screw next to the fracture but the next one over. This makes it possible to insert an oblique lag screw through the plate and across the fracture. The drill jig illustrated in Fig. 47 has two holes for the load screws, one close to the fracture and one further away. This facilitates the insertion of the lag screw through the plate and across the fracture whenever possible.

Note: The insertion of the oblique lag screw as demonstrated in Fig. 48 is not always simple. If necessary in accordance with the principle demonstrated in Fig. 19a and b we can drill the gliding hole prior to the reduction. The drill sleeve is then inserted through the plate into the gliding hole and the plate is placed on the bone. As shown in Fig. 48b and c the plate is then placed under tension by means of the two screws. The thread hole is then drilled through the drill sleeve, the tap is used to cut the thread, and the cortical screw is inserted as a lag screw.

Fig. 48 *The combination of axial and interfragmentary compression by means of the DCP.* In the case of a short oblique fracture the rigidity of a plate fixation may be considerably increased by introducing a lag screw through the plate.

a Shows the "gliding-hole-first procedure", where the gliding hole (4.5-mm drill bit) is placed prior to reduction in the anterior fragment: (a′) from the outside in direction of the medullary cavity, (a″) from the inside of the medullary cavity to the outside. (a‴) shows the "threaded-hole-first procedure", where the threaded hole (3.2-mm drill bit) is placed first in the posterior fragment. By using the pointed drill guide it will be possible to place the gliding hole after reduction of the fragments into the proper position and direction.

b The 3.2-mm drill sleeve is inserted to prevent the plate from slipping, when the next hole is drilled on the opposite side of the fracture with the neutral drill guide. The screw is introduced after measuring and tapping.

c The drill sleeve is now removed and using the load drill guide (yellow) the next hole is placed.

d By tightening the second screw axial compression is obtained, resulting in a close adaptation of the fracture below the plate: the gap in the cortex opposite to the plate will, however, slightly open up.
We therefore now insert the drill sleeve and drill the threaded hole of the oblique lag screw.

e By driving home the lag screw we will add inter-fragmentary compression and thereby close the opened fracture line.

f The remaining screws are all introduced with the neutral drill guide.

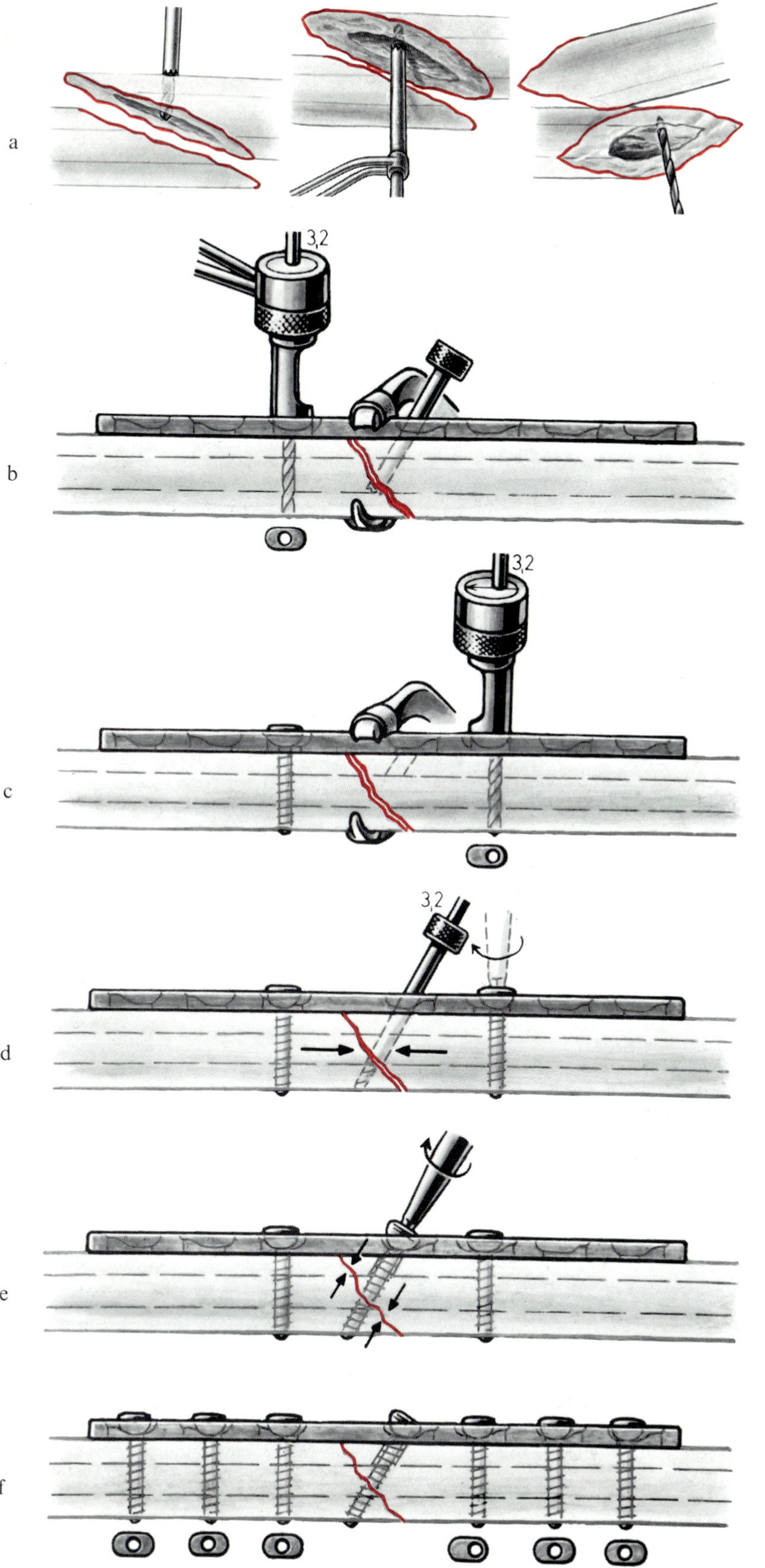
a
b
3,2
c
3,2
d
3,2
e
f

Screw Direction Through the DCP

The shape of the screw head and of the screw hole permits the angling of screws in all directions. This gives the surgeon far greater scope in adapting the screw insertion to the particular fracture pattern he is dealing with. We must stress again and again that whenever possible one should enhance the stability of the internal fixation by the insertion of an oblique lag screw through the plate across the main fracture line. This applies if the plate is used as a neutralization plate and even more so if it is used as a compression plate without other lag screws adding to the stability of the fixation. It is easy to observe this principle if one is dealing with oblique fractures (see Fig. 49d) but we recommend it also for the fixation of transverse fractures (see Fig. 51).

DCP in Conjunction with the Tension Device

If the increase in the surgical exposure does not present any problems and if the gap in the fracture line is considerable, then we would certainly recommend the use of the tension device together with the DCP. This applies above all to the plating of the femur. Furthermore, whenever a femur is plated we strongly recommend that a *cancellous bone graft be placed on the fracture opposite the plate.*

Surgeons who have become used to the round hole plate do not have to change their technique. They should carry out the internal fixation in their accustomed manner with the tension device. For the insertion of the screws they should use the neutral guide. This has the advantage of ensuring maintenance of the compression generated with the tension device because slight eccentric insertion of the screw through the screw hole of the DCP will not lead to the dissipation of compression or even to distraction, which is possible with the round hole plate.

Fig. 49 *Direction of the screws through the DCP.* The spherical geometry of the screw hole and of the screw head permit the angling of the screw in all directions. This allows one to choose the best direction for the screw and adapt it to a particular fracture situation. It facilitates, in particular, the insertion of the oblique lag screw through the plate and across the fracture.

a The drill guides can be angled to the side.

b An obliquely inserted lag screw which not only misses the area of comminution but also places the fracture planes under inter-fragmental compression.

c It is also possible to angle the drill guide in the long axis.

d The insertion of the lag screw through the screw hole of the DCP.

Fig. 50 *Schematic representation of the best biomechanical configuration of the plating of a fracture (radius, ulna, tibia and femur).* The axial compression is achieved either by inserting screws in the load position or with the tension device. The inter-fragmental compression is achieved by means of the oblique lag screw.

Fig. 51 *Use of the DCP with the tension device.* Drill a 3.2-mm drill hole approximately 1 cm from the fracture. Examine the fracture carefully and make sure that the first hole is drilled in that fragment, which under axial compression will cause the other fragment to impact against the plate and not cause the fragment to displace away from the plate. Cut the thread and lightly tighten the screw fixing the plate to the bone. Fix the tension device to the bone and begin applying axial compression. Constantly check your reduction. If necessary loosen the compression and improve on the reduction. Once maximal axial compression has been achieved, insert the remaining screws in the neutral position with the corresponding neutral drill guide. All screws which cross the fracture should be inserted as lag screws. To insert a lag screw, drill both cortices with the 3.2-mm drill bit and the neutral drill guide. Thereafter, with the 4.5-mm drill bit overdrill the cortex adjacent to the plate.

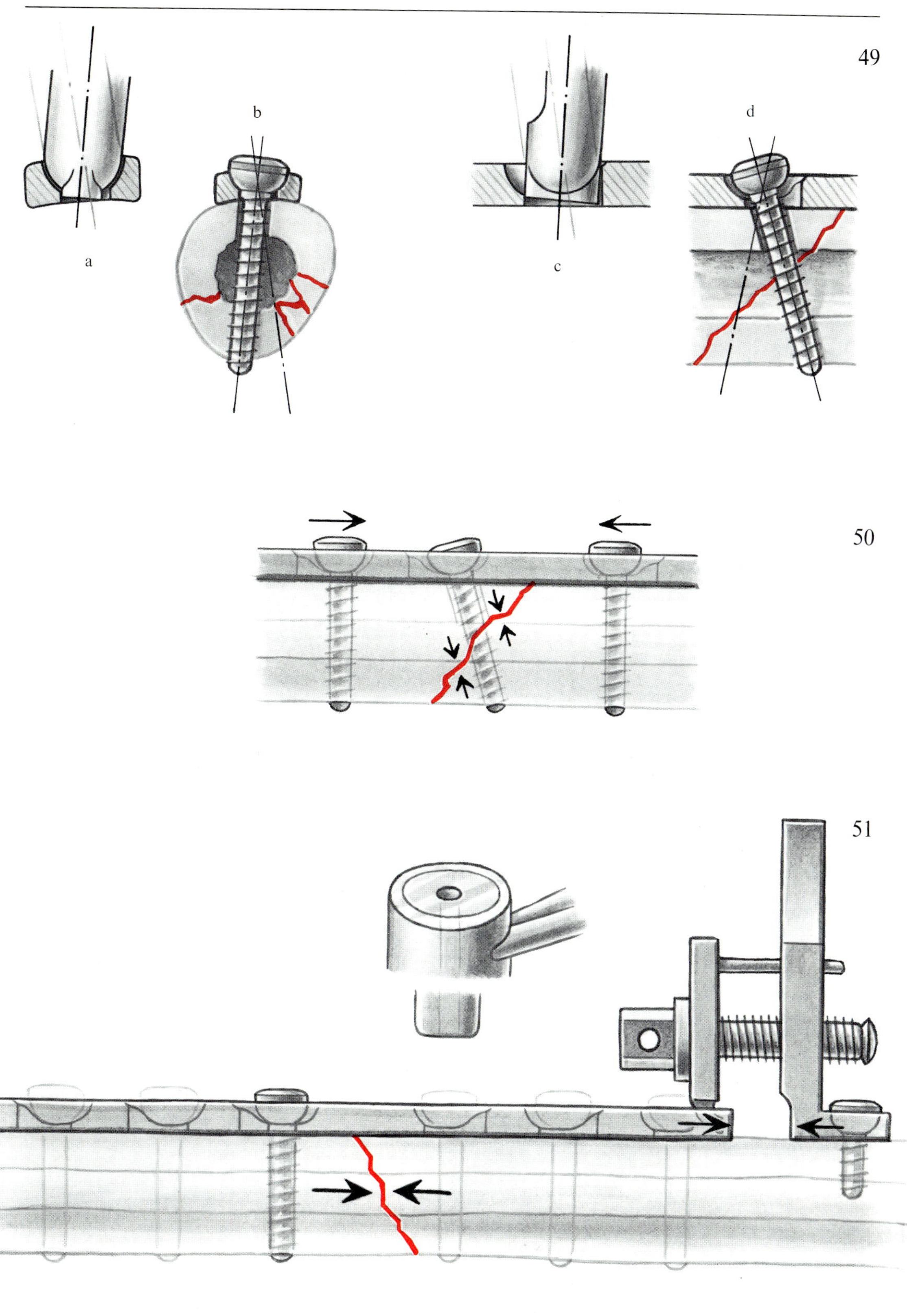
49
a
b
c
d
50
51

The DCP as Buttress Plate

If one inserts a screw through the hole of a DCP so that it will slide away from the fracture, the screw will buttress the bone. The 3.2-mm drill sleeve with its 4.5-mm outer diameter when inserted in the screw hole close to the fracture, will accurately centre the screw in its buttress position.

Generally, the buttressing of metaphyses is best carried out with the special plates such as the T plates, the L plates and the condylar plates. In many situations, however, the buttressing effect of the DCP is enough. If used as such it can also be used to fix the remaining fracture line, should this reach obliquely into the diaphysis.

Fig. 52 The DCP as buttress plate. After careful contouring of the plate to make it conform to the shape of the underlying bone, drill the required screw holes with the 3.2-mm drill sleeve. The 3.2-mm drill sleeve is inserted through the screw hole and slid towards the fracture. If the screw should cross a fracture, then the cortex adjacent to the plate is overdrilled with a 4.5-mm drill bit in order to convert the screw to a lag screw. All screws are inserted in the buttress position.

Examples:

a The proximal tibia.

b The distal tibia.

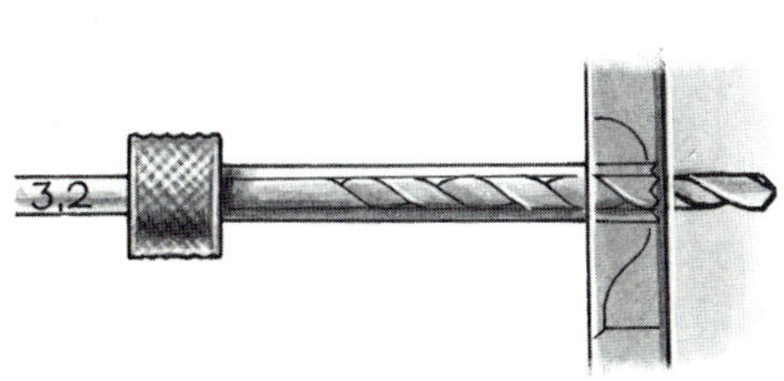

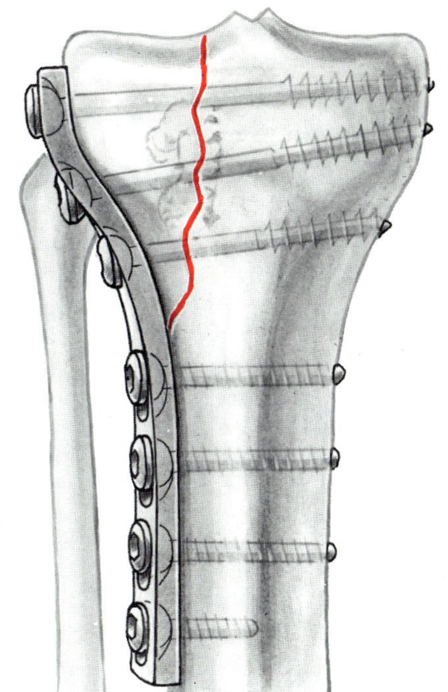

a

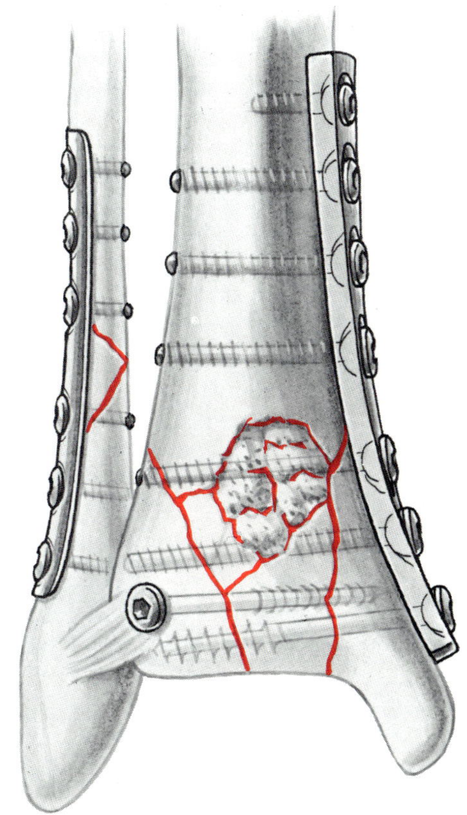

b

2.3.2 Special Plates

The special plates are "T" shaped and have been developed for the fixation of epiphyseal and metaphyseal bone. They are used chiefly as buttress plates. As such they protect a thin cortex or prevent a defect in cancellous bone from collapsing. If, in addition, there is an avulsion or a shear fragment it should be fixed under inter-fragmental compression. This is accomplished with a long threaded cancellous screw.

The special T plates have a double bend to fit the lateral plateau. The medial plateau will usually take the ordinary T plate without much contouring. The oval hole in the T plate has been designed to permit temporary fixation of the plate to the bone with some up-and-down movement of the plate still possible. Once plating is completed, one can remove this screw and insert through the oval hole an oblique lag screw to fix either the fracture or an osteotomy under inter-fragmental compression. The L plates have been developed to fit certain fracture patterns of the lateral plateau.

Fig. 53 *Buttress plates for plateau fractures.*

a A four-hole T plate.

b T buttress plate.

c, d Buttress plates.

e In profile the tibial plateau buttress plate shows its double bend.

Fig. 54 *Examples of their use.*

a The buttressing of the lateral plateau impression fracture with an L plate and cancellous bone graft. Front view and side view.

b The buttressing of a fracture of the medial tibial plateau with a T plate.

53

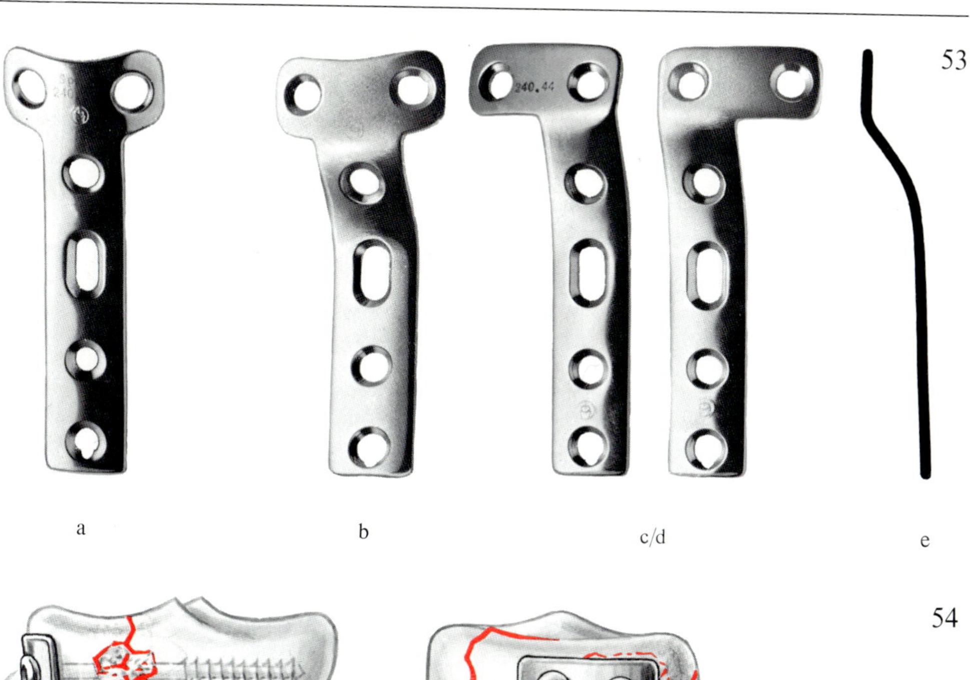

a b c/d e

54

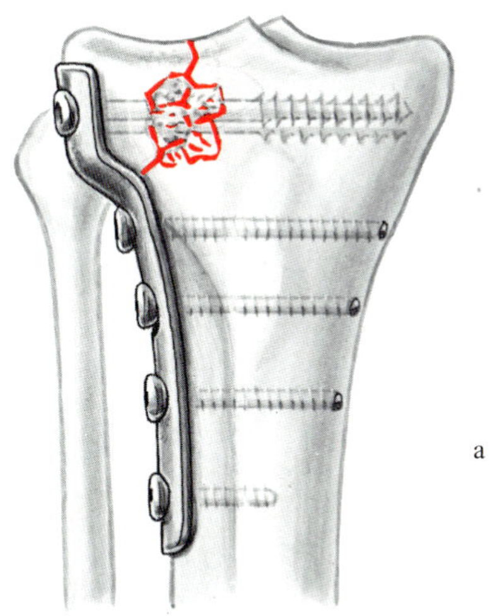

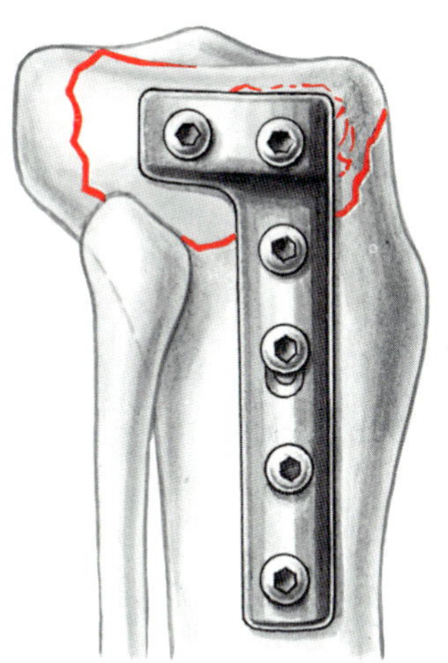

a

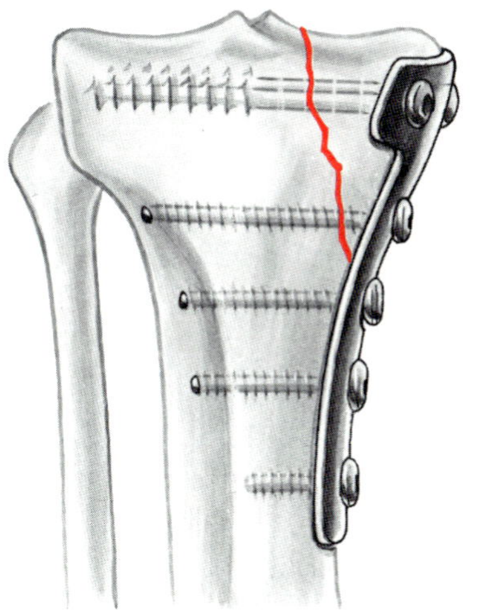

b

Buttress Plating of the Distal Tibia

We have found three plates useful for buttressing the distal tibia. These are: the spoon plate, the T plate and the cloverleaf plate. The chief indications for the spoon plate are the intra-articular fractures with a large posterior fragment and anterior comminution. The comminuted fragments are compressed between the intact posterior fragment and the anterior plate.

The cloverleaf plate is the universal plate for all other fractures of the distal tibia. The plate will prevent varus deformity. There is some danger in its use, since problems with wound healing have been encountered. We strongly recommend, therefore, that when one is buttress plating the distal tibia, the incision be made so that it does not cross over the buttress plate.

Fig. 55

a The spoon plate.

b Seen from in front, that is, from the anterior aspect when used for the fixation of the fracture of the distal tibia.

c As seen from the side. The little anterior fragments are sandwiched between the plate and the large posterior fragment. The lower edge of the plate reaches almost to the edge of the joint. The cancellous screws are inserted parallel to the joint.

Fig. 56

a The Klee or cloverleaf plate.

b The cloverleaf plate was designed to be used on the medial side. Distally it should be fixed to the bone with the 3.5-mm cortex and 4.0-mm cancellous screws. Its distal cloverleaf-shaped portion can be easily trimmed with a wire cutter should this prove necessary.

55

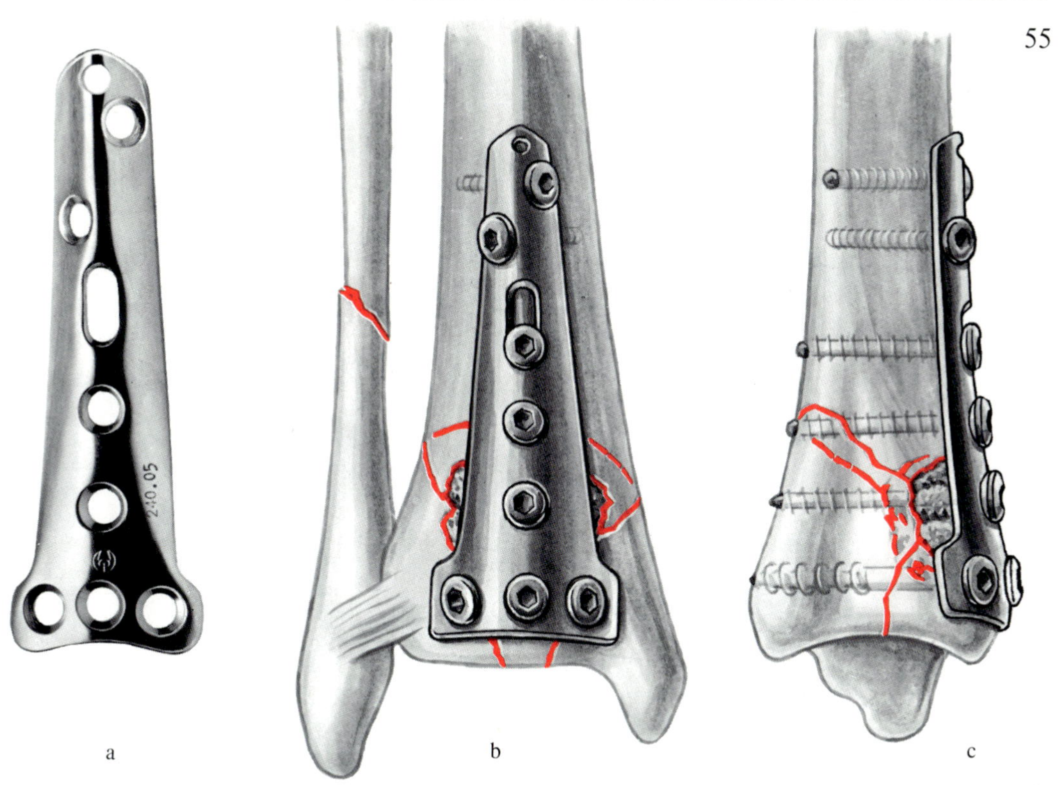

56

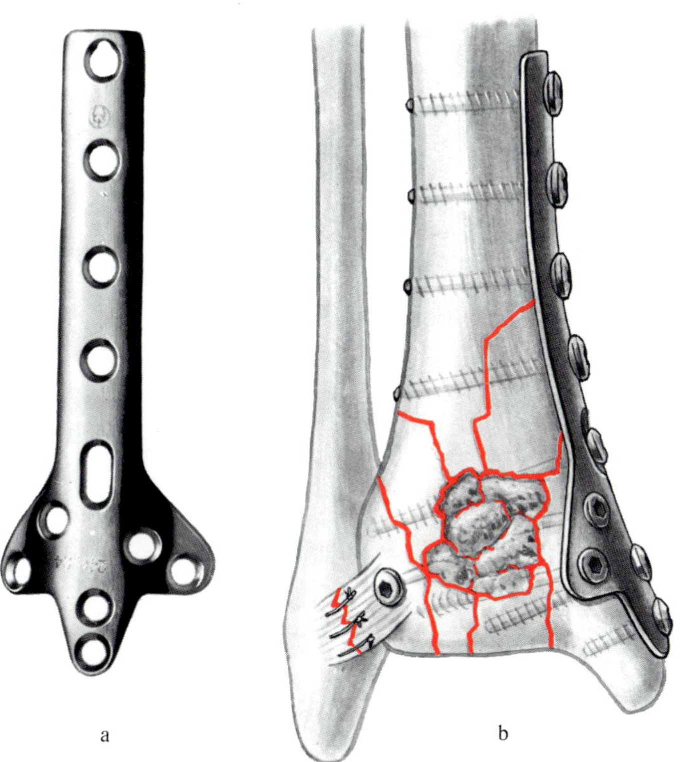

Special Plates in Other Areas

The main indications for the use of the T plate are fractures of proximal humerus such as irreducible fracture dislocations, irreducible displaced fractures of the head, subcapital pseudarthroses and proximal osteotomies. Previously we used to bridge the bicipital groove, but with time we have found that it is best to insert both screws lateral to the bicipital groove (for technique see Fig. 126).

The small T plates with three or four holes for the 3.5-mm cortex and the 4.0-mm cancellous screws are designed to be used on the distal radius as for example in the fixation of a Smith-Goyrand fracture (see Fig. 156). The small T and L plates have been developed for fixation of the hand and the foot. They are useful for fixation of corrective osteotomies of the hallux, of metatarsals, of metacarpals, and for the fixation of difficult fractures (see p. 201).

The cobra plate is used almost exclusively for hip arthrodeses (see Fig. 339). Patients with such an arthrodesis can usually be treated without a hip spica.

2.3.3 Angled Plates

In 1959 we developed the angled plates. For the blade we chose a U profile. We decided upon a single-blade unit with a fixed angle between the blade and plate portion. These angled plates are used as tension band plates or as neutralization plates for splinting of the proximal or distal femur. They have made it possible to fix the simplest as well as the most difficult fracture of the proximal and distal femur.

The advantage of the fixed angle is the increased strength and an increased corrosion resistance of the implant. The disadvantage is the technical difficulty of insertion, because the blade has to be inserted not only in the middle of the femoral neck (neck axis), but also at a pre-determined inclination to the shaft axis. Furthermore, the plate portion of the angled blade has to line up with the axis of the shaft at the end of the procedure. A pre-operative work drawing is essential. The operation must follow the work drawing accurately step by step if at the end everything is to fit together accurately.

Planning

In every peri-articular fracture of the femur, we recommend an X-ray of the contralateral uninvolved joint. The X-ray of the normal hip should be taken in internal rotation (correction of antiversion). For the distal femur one must have accurate antero-posterior and lateral X-rays centred on the joint. The outlines of the normal, proximal or distal femur are then drawn in, as are the fracture lines. The fracture pattern determines which plate is used (see Fig. 60 and page 98). The selected plate is then drawn in with the help of the template (enlargement 1.15–1.0). The guide wires should also be drawn in (see Fig. 62).

These work drawings are imperative before carrying out any corrective osteotomies of the proximal or distal femur. It is the only way the surgeon can check preoperatively the result of the osteotomy as well as his three-dimensional concept of the femur.

The instruments which we have developed will with practice, greatly facilitate the exact execution of the procedure in accordance with the work drawing. Although we do not object to the use of the image intensifier or X-rays during the procedure, we have found that in the hands of the experienced surgeon, who has mastered the use of the angled plates and their instrumentation and who has learned the atraumatic exposure of the proximal and distal femur, the use of the image intensifier or X-rays during the procedure is rarely necessary.

The use of the AO angled plates: The proximal and distal third of the femur

In the following part we shall deal only with the use of the angled plates for the treatment of fractures of the femur. Osteotomies will be dealt with in the third part, the section on reconstructive surgery.

The AO Angled Plates for the Proximal and Distal Femur

For the treatment of fractures of the proximal and distal femur we have developed the so-called condylar plate as well as the 130° angled plate. Both have a blade with a U profile.

Initially the condylar plate was used only for fractures of the distal femur. Over the years we have found the condylar plate to be most useful also in the treatment of most inter-trochanteric fractures, where it has proven itself actually superior to the 130° angled plate in the treatment of comminuted inter-trochanteric fractures. One must make certain, however, that the condylar plate is used, only for those fractures of the proximal femur which have an intact calcar, that is an intact medial buttress, or at least when the condylar plate is used, that the calcar or medial buttress are reconstructed (see p. 138 regarding the place of bone grafting).

The condylar plate has a fixed angle of 95° between its blade and plate portion. Most commonly we use the five-hole plate. For fractures which extend into the diaphysis, we have condylar plates with 7, 9 and 12 holes. The blades come in lengths from 50 to 80 mm. The condylar plate guide is a mirror image of the plate (Fig. 59g). The triangular angle guide with 50° is shown in Fig. 59h and the triple drill guide in Fig. 59i.

Subcapital fractures are fixed with the 130° angled plate with one or four holes, depending on the stability of the reduction (Fig. 58). For inter-trochanteric fractures we use the 130° angle plate with four to six holes. The 130° angled plate with nine holes is used only for the fixation of fractures with a long distal extension into the shaft.

All AO angled plates come now with dynamic compression holes.

Fig. 57 *The condylar plates.*

a The most commonly used condylar plate has five holes. The two holes next to the blade take cancellous screws; the other three take cortical screws. The blades come in varying lengths of 50, 60, 70 and 80 mm. The 70-mm blade is the one most commonly used in the proximal femur and the 50-mm blade the one most commonly used in the distal femur.

b The U profile of the blade.

c The 7-, 9- and 12-hole condylar plates for the fixation of longer fractures.

Fig. 58 *The 130° angle plates.*

a The standard four-hole 130° blade plate is used for the fixation of subcapital fractures and stable inter-trochanteric fractures. The blades come in 50-, 60-, 70-, 80-, 90-, 100- and 110-mm lengths. In-between sizes can be obtained on special request. The one most commonly used in the proximal femur is the 90-mm blade length.

b U profile of the blade.

c The 130° angled plate with one hole.

d The longer 130° angled plates with six and nine holes (these are rarely used).

57

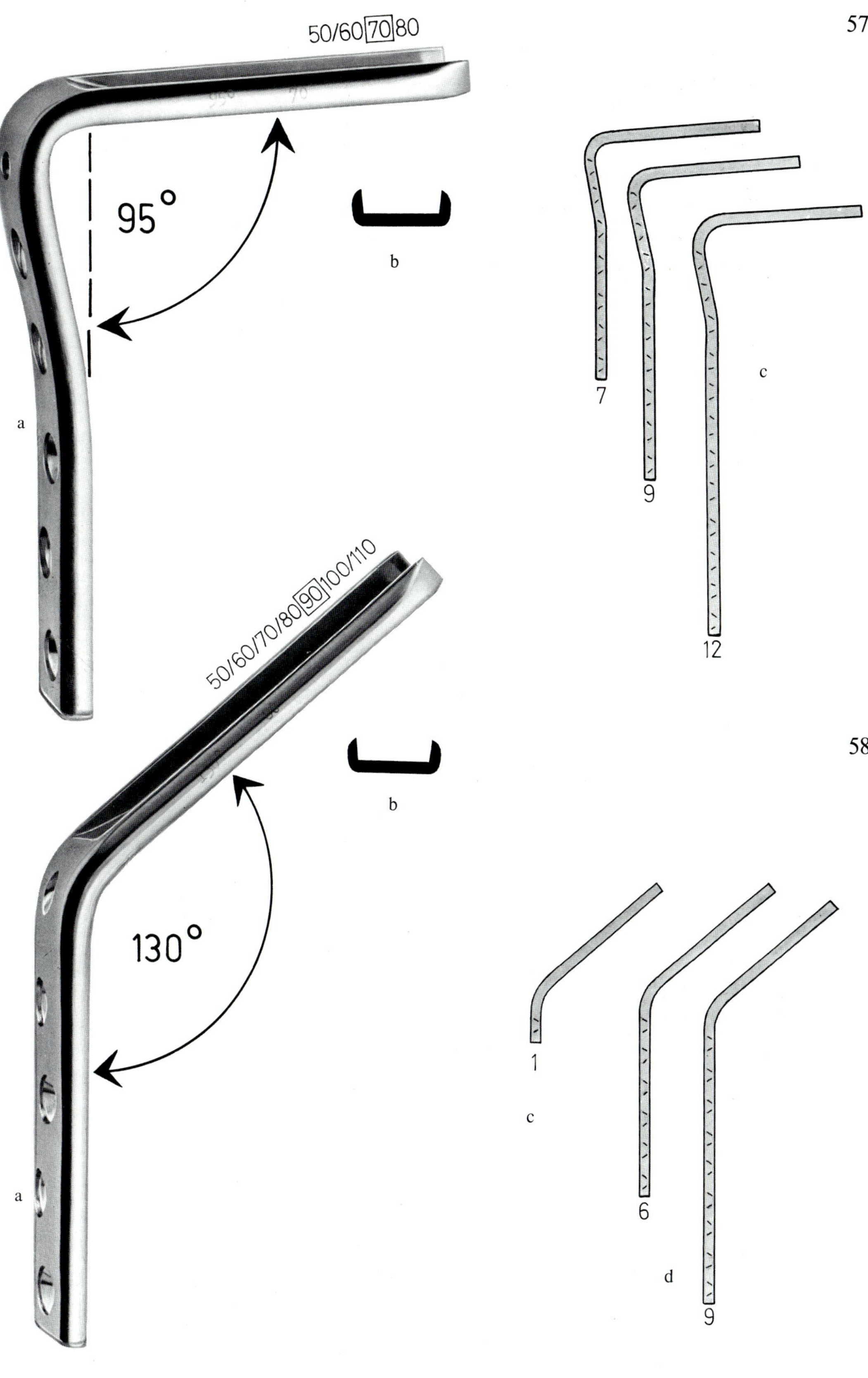

Instrumentation Required for the Insertion of AO Angled Plates

The characteristic feature of the AO technique of internal fixation with the angled plates is the need to pre-cut the channel for the blade with the seating chisel. The seating chisel has the same U profile as the blade portion of the angled plate. The open portion of the U profile is always directed towards the adjacent joint.

The fixed angle between the blade and the plate will cause some difficulty until one masters the techniques of inserting the seating chisel correctly with the aid of Kirschner wires and the available aiming devices. Thereafter, intraoperative check X-rays will rarely be necessary.

Note: The sides of the seating chisel have been ground to converge slightly and the tip has been ground in such a way that, when the seating chisel is inserted obliquely into the neck, the tip will not bite into the calcar. The obliquity of the sides enhances the centring of the seating chisel within the femoral neck.

Fig. 59 *Instrumentation.*

a The *seating chisel* used for cutting the seat for the blade in the proximal and distal femur.

b The *U profile* which corresponds to the profile of the blade of the angled plates.

c The *seating chisel guide* is used to determine the rotation of the seating chisel about its long axis. It is slid over the seating chisel. In internal fixation of the proximal and distal femur, the flap of the seating chisel guide must be in line with the long axis of the femur. The angle between the flap and the body of the seating chisel guide is set with the aid of the condylar blade guide or the triangular angle guide and the set is maintained by tightening the screw with a screwdriver.

d During insertion the rotation of the seating chisel is controlled with the *slotted hammer*. The slotted hammer serves also for removal of the seating chisel and for the hammering out of the blade holder with its corresponding plate.

e The *plate holder* is used for insertion and removal of plates. The plate should be so fastened in the plate holder that the long handle is always in line with the blade of the angled-plate.

f The *impactor* is used to drive in the last 5 mm of the blade into the bone.

g *Condylar blade guide* subtends an angle of 85° (for its use see Fig. 62b).

h The *triangular angle guide* with a 50° angle. When applied to the lateral cortex of the femur, its lateral edge, when projected, will subtend an angle of 130° with the femoral shaft.

i The *triple drill guide* has a fixed angle of 130° and can greatly simplify the insertion of the 130° angled plates. It can also be slid over the bottom of the condylar guide. Its removable guide is slid over either a 3.2-mm drill guide or a 3-mm Kirschner wire which is previously inserted in line with the neck axis and serves as the directional guide for the triple drill guide (for use see Fig. 66).

k The *router* used for conversion of a 4.5-mm drill hole to a slit.

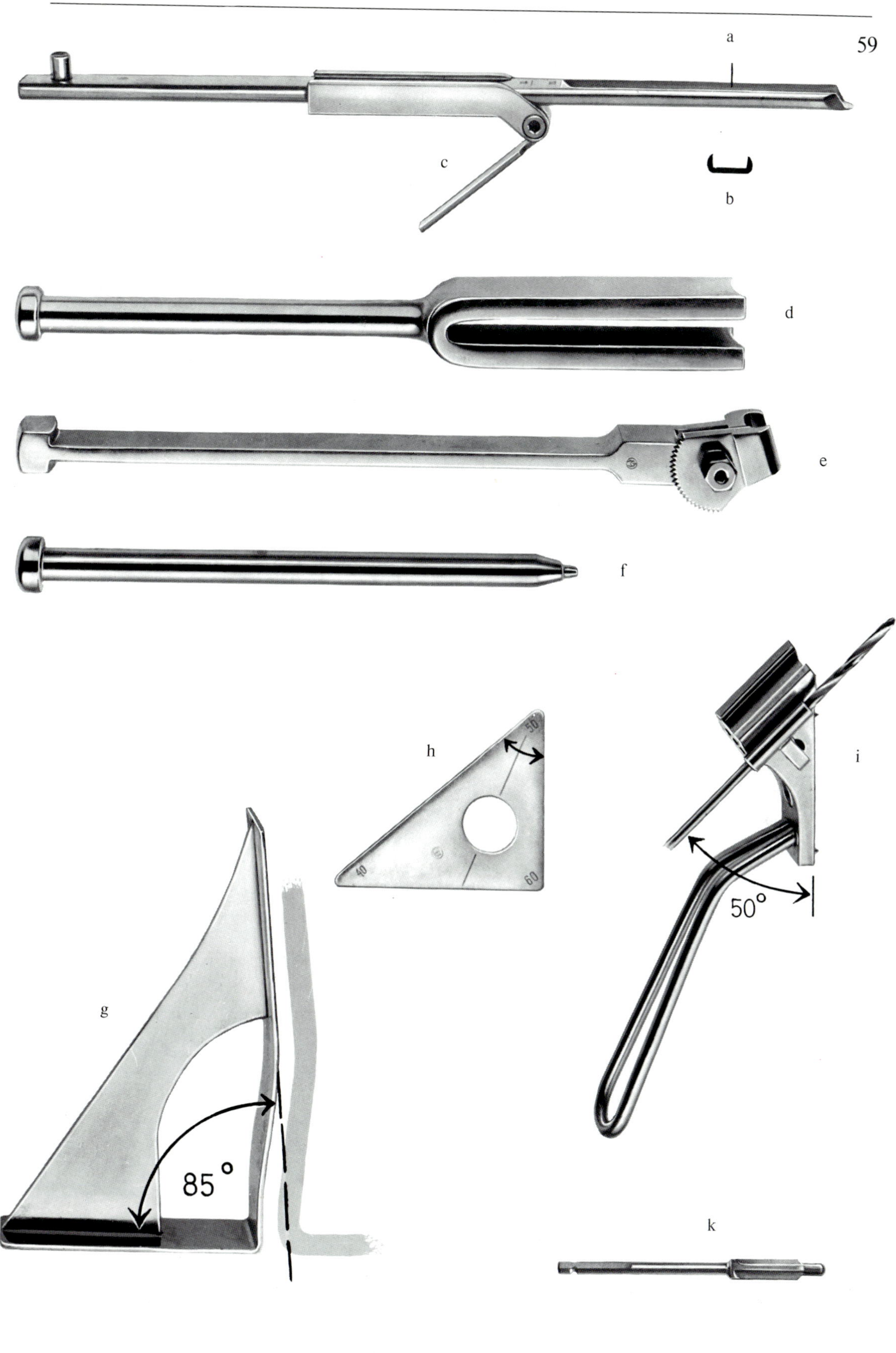
a
b
c
d
e
f
g
85 °
h
50
40
60
i
50°
k

Position of the Blades Within the Proximal Femur

In considering which is the optimum position for the blade within the proximal femur, we should note that the only area filled with bone and the only one to offer some resistance is the area of intersection between the compression and tension trabecular systems. We recommend therefore that the tip of both the condylar and the 130° angled plates be inserted into the bottom half of the femoral head, below and as close as possible to the intersection of the trabecular systems. This area of the femoral head provides the best hold for the blades.

When correctly inserted the blade of the condylar plate passes below and tangentially to the superior cortex of the neck. The blade of the 130° angle plate passes 6–8 mm above the calcar (see Figs. 60 and 61).

Note: We are opposed to the use of angled plates in children, in teenagers and in adults below age 40 (see p. 326). Their fractures should be fixed with lag screws. When the condylar plate is inserted its hold in the proximal fragment must be supplemented with either one or two lag screws (triangulation). One should reduce both subcapital and inter-trochanteric fractures in some degree of valgus (see Fig. 66/3).

The fracture should be reduced and then temporarily stabilized with Kirschner wires. This makes it possible to check the reduction of the calcar under direct vision. To do this, flex the hip to 90° and then externally rotate it.

Fig. 60 *Ideal position of the condylar plate in the proximal femur.*

Note the tip of the blade in the lower half of the femoral head. The blade passes below the superior cortex of the neck. A cortical screw has been used to fix the plate to the calcar.

Fig. 61 Note the tip of the blade of the 130° angled plate in the lower half of the femoral head and note that the blade passes 6–8 mm above the calcar. The blade enters through the lateral cortex at a point approximately 3 cm distal to the rough line of the greater trochanter.

60

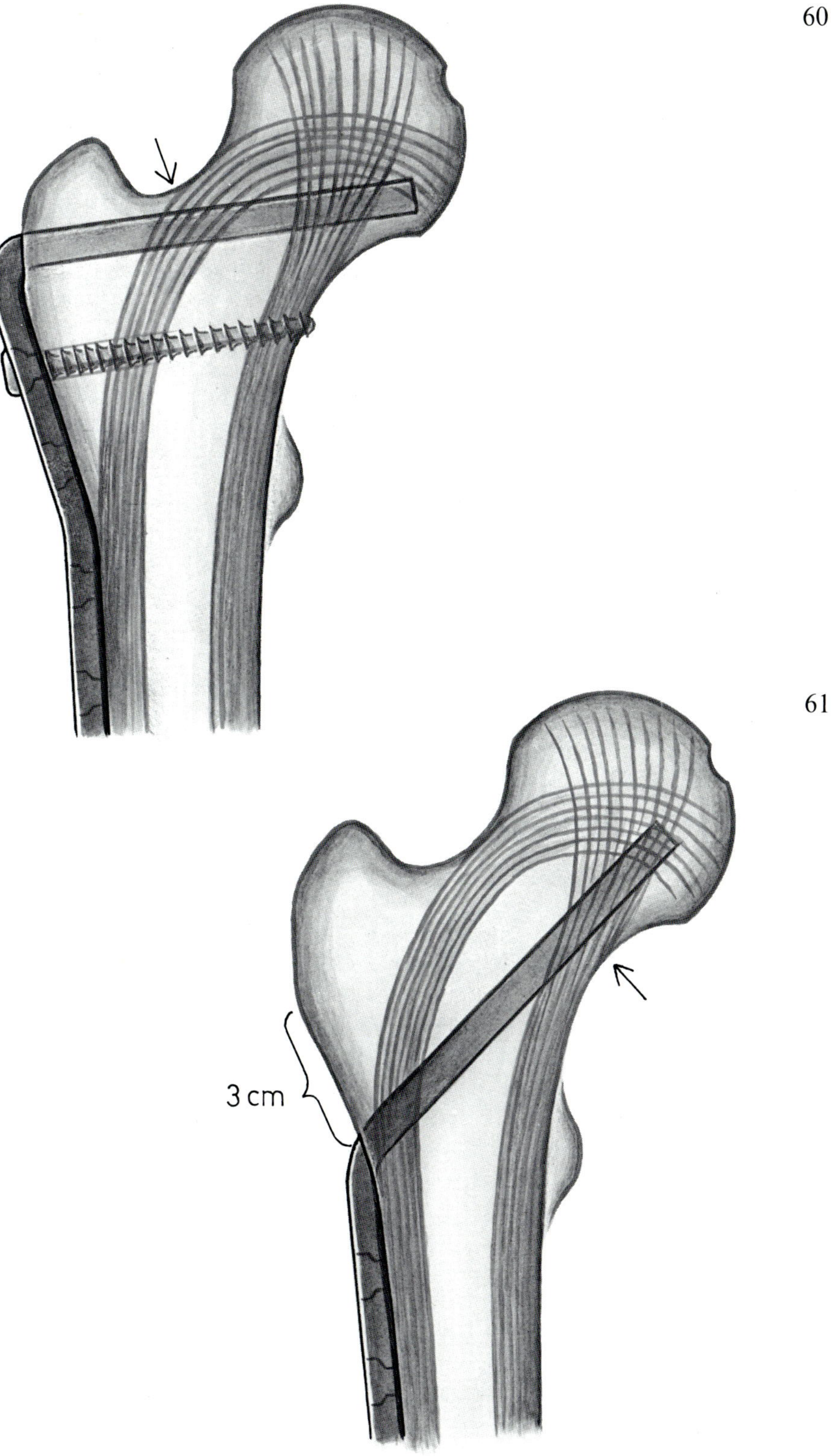

61

Proximal Femur: Preparation of the Seat for the Blade

Once the position of the blade has been determined on the preoperative work drawing, it is pre-cut in bone with the seating chisel. This can be carried out without radiological control. There are four things one must pay attention to: the axis of the femoral neck, the angle between the blade and the femoral shaft axis, the point of entry of the blade and the rotation of the seating chisel about its long axis.

a) The Neck Axis. The guide to the neck axis is the Kirschner wire which is inserted anteriorly in line with the axis of the neck. This will ensure the centring of the seating chisel in line with the neck axis.

b) The Angle Between the Blade and the Femoral Shaft Axis. When the condylar plate is to be inserted, the condylar plate guide is brought up against the lateral surface of the femur and a Kirschner wire is driven into the greater trochanter parallel to the upper edge of the condylar guide. This Kirschner wire should also be inserted parallel to the first Kirschner wire, which is the guide to the anteversion. The second Kirschner wire becomes the definitive guide wire for the insertion of the seating chisel. For the insertion of the 130° angled plate we proceed with the triangular angle guide in exactly the same way as with the condylar guide and repeat the steps (see Fig. 62b).

c) The Point of Entry. The point of entry is determined on the pre-operative drawing. A corresponding point on the femur is now opened with a chisel. Because the greater trochanter flares posteriorly some 30°–40°, the more proximal the point of entry, the more anterior it must be in order to come into line with the axis of the neck (see Fig. 62c and c′).

d) The Rotation of the Blade About Its Long Axis. In reducing a fracture, we aim to reconstruct the normal anatomy and return the bone to its pre-fracture configuration. This means that the plate must come to lie in line with the shaft axis. This alignment is determined with the seating chisel guide (see Fig. 62d).

Fig. 62 *How to determine the position of the blade in the proximal femur.*

a *The neck axis.* A Kirschner wire is inserted low down along the anterior aspect of the neck and is driven into the head.

b *The angle between the blade and the shaft axis.* For the insertion of the condylar plate, the condylar plate guide is placed along the lateral cortex. A second Kirschner wire is inserted parallel to the first Kirschner wire and parallel with the upper edge of the condylar plate guide. It is driven into the greater trochanter above the planned point of entry.

b′ When inserting the *130° angled plate,* use the triangular angle guide with the 50° angle, place it along the lateral cortex of the femur and drive a Kirschner wire into the greater trochanter which is parallel to the first Kirschner wire and parallel to the upper edge of the triangular guide.

c *Point of entry* in the greater trochanter is in the anterior half of its lateral bulge.

c′ *Point of entry* for the 130° angled plate is directly in the middle of the lateral cortex.

d, d′ *The rotation of the blade about its long axis:* Insert the seating chisel guide over the seating chisel. Set the angle between the flap of the guide and the seating chisel (85° for condylar plates and 50° for the 130° angled plates). With the slotted hammer rotate the seating chisel about its long axis till the flap of the seating chisel guide comes to lie in line with the long axis of the femur. Hammer in the seating chisel parallel to its guide wire.

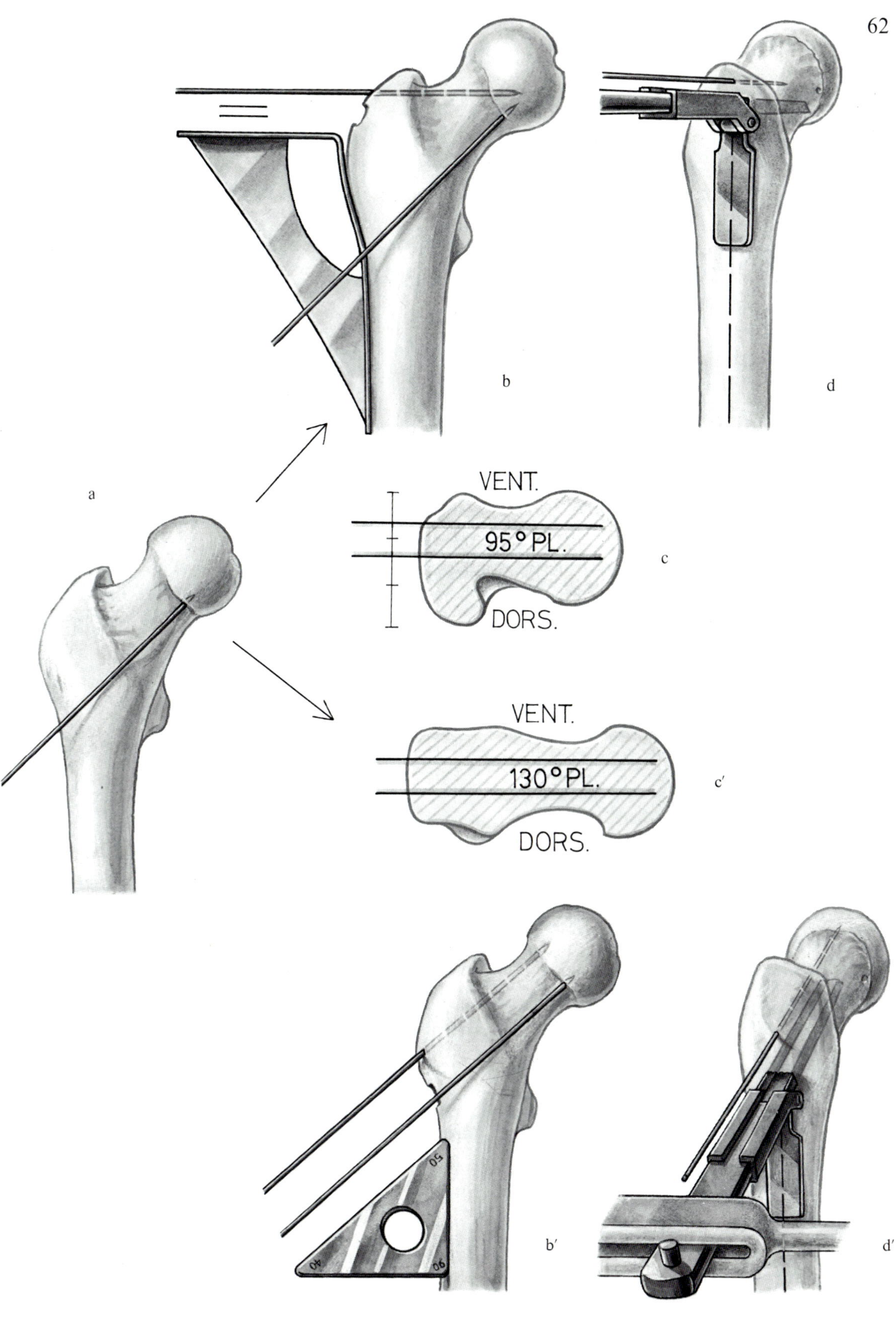
b
d
a
VENT.
95° PL.
DORS.
c
VENT.
130° PL.
DORS.
c′
b′
d′

After the insertion of the seating chisel, measure the distance from the point of entry to the middle of the femoral head and select the condylar plate of corresponding blade length. Insert this condylar plate into the plate holder. Remove the seating chisel with the slotted hammer and push the blade of your angled plate into the pre-cut channel. The blade should encounter little resistance and should require no more than light hammer blows for its insertion. When the blade strays from its pre-cut channel a resistance becomes readily apparent. Stronger blows of the hammer become necessary in order to advance the blade into the bone. Should the blade stray, remove it and prepare its seat once again with the seating chisel. When the plate is 5 mm from the lateral cortex, remove the plate holder and complete the insertion with the blade impactor.

It is important to stabilize the condylar plate in the proximal fragment with one or two screws which are inserted through the plate into the calcar. This triangulates the fixation. We use the 3.2-mm drill sleeve to guide the drill bit into the calcar and prevent its deviation (for the technique see Fig. 63).

The postero-medial buttress must be reconstructed. If there is a large fragment which consists of the lesser trochanter and adjacent bone, it must be reduced and fixed with an obliquely inserted malleolar or cancellous screw.

In reducing unstable inter-trochanteric fractures or comminuted fractures, expose the superior cortex of the neck and the ventral half of the femoral head. It is then possible to insert the seating chisel and the blade of the angled plate into the proximal fragment prior to the reduction of the fracture.

Fig. 63 *Insertion of the condylar plate.*

a In order to *triangulate the fixation* of the condylar plate in the proximal femur, insert the cortical screw as a lag screw into the calcar. Drill first with a 4.5-mm drill bit to a depth of about 4 cm. Aim at right angles to the plate and slightly distally away from the blade.

b Insert the 3.2-mm drill sleeve. This will guide the 3.2-mm drill bit and will prevent it from gliding along the calcar. This will also prevent the breaking of the drill bit. Drill the calcar, determine the screw length, tap the thread and insert the corresponding cortical screw.

c Fix the plate. Oftentimes a second lag screw can be inserted into the proximal fragment.

Fig. 64 *Fixation of the lesser trochanter.* In order to fix the fragment of the lesser trochanter, a hole is drilled through the cortex from in front of the plate in a postero-medial direction. The lesser trochanter is then fixed with a malleolar screw. The angle between this screw and those fixing the plate usually subtends 35°–45°.

Fig. 65 When you are dealing with a *comminuted fracture*, determine first the point of entry on the pre-operative drawing, then insert the seating chisel through this predetermined point. Insert the condylar plate and fix it to the proximal fragment with cortical screws. One can then proceed with the reduction and fixation of the fracture. We can use a shorter blade here because the fixation in the proximal fragment is enhanced by the cortical screws.

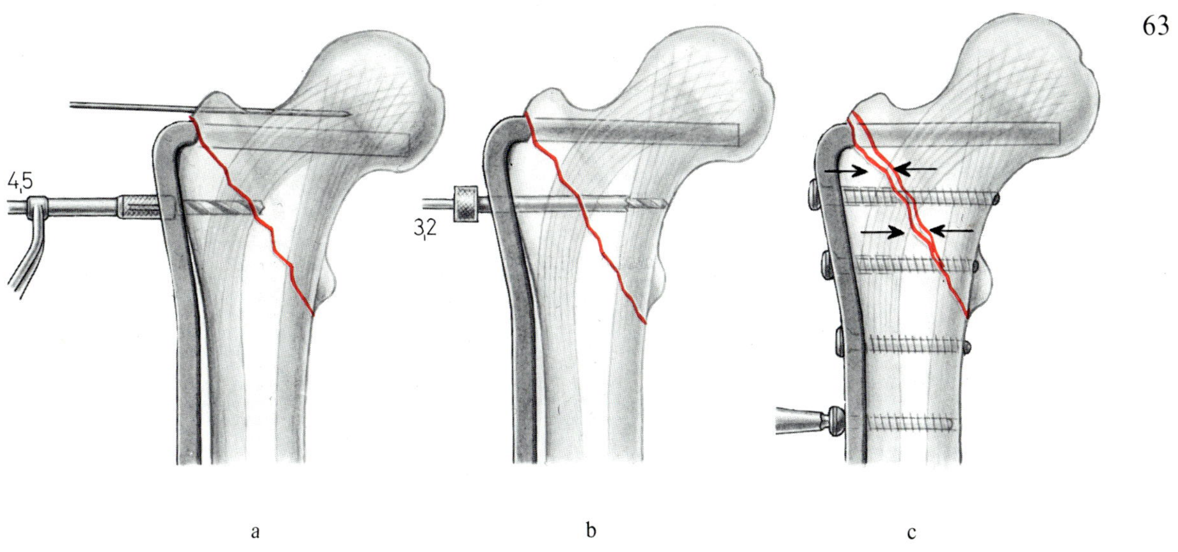

a b c

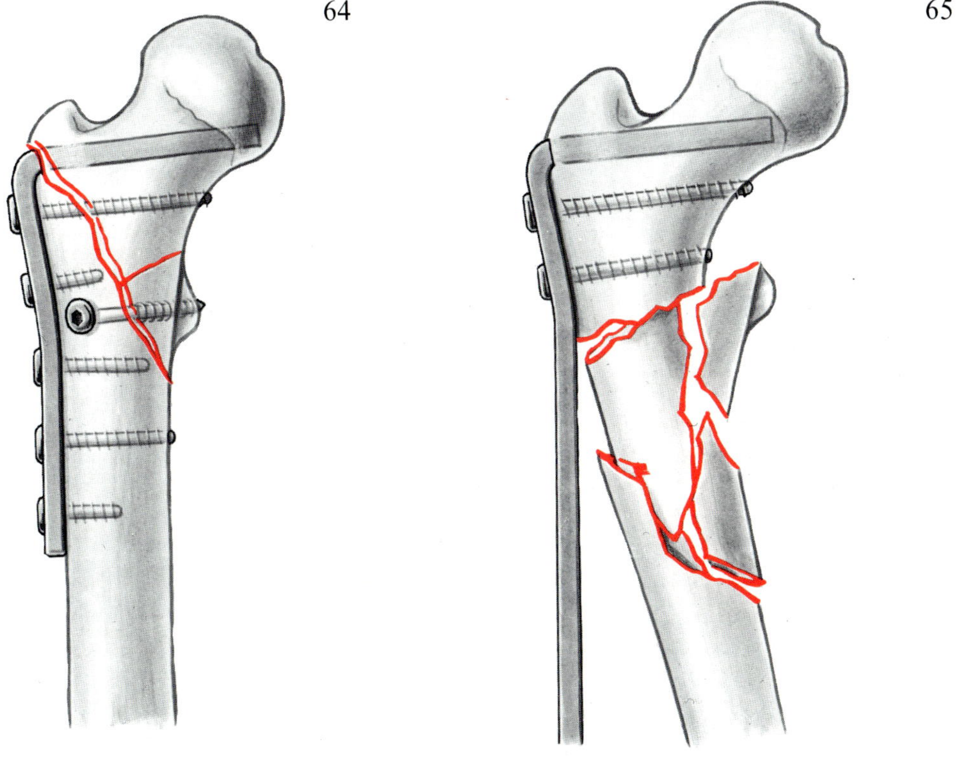

Proximal Femur: Insertion of the 130° Angled Plate

Because the lateral cortex of the proximal femur is very hard, we recommend the pre-drilling of three holes with the 4.5-mm drill bit and the use of the router. The 16-mm chisel is used to complete the hole. The distal cortex should be bevelled to receive the shoulder of the angled plate.

Note: When dealing with a subcapital fracture, insert the seating chisel first, but do not push it across the fracture (see p. 218). The guide Kirschner wire used should be 2.5 mm thick because it will be used for temporary fixation after reduction. It is inserted 1 cm above the point of entry of the seating chisel. One can pre-drill its hole with the 2.0-mm drill bit.

Fig. 66 *Preparation of the seat for the 130° angled plate.*

1 The preparation of the hole for the insertion has been made easy by the development of the triple-angle guide (*a*). Insert through the removable guide (*b*) either a 3.2-mm drill bit (*c*) or a 3.0-mm Kirschner wire. As soon as this drill bit or Kirschner wire comes to lie 8–10 mm above the calcar and in line with the neck axis one can drill the first anterior 4.5-mm drill hole to a depth of 4–6 cm. Leave the drill bit in the bone (*d*) and use another 4.5-mm drill bit to drill the remaining two holes. Remove the triple-angle guide and the drill bit and enlarge the drill holes (*e*) with the router (*f*) in order to convert the three holes to a slit (*g*). Bevel the hole distally (*h*) for a distance of a few millimetres to receive the shoulder of the angled plate. This will prevent the shattering of the lateral cortex.

2 Insert with light hammer blows the seating chisel (*i*) with its seating chisel guide (*k*) and its corresponding guide wire (*l*) until the tip of the seating chisel comes up against the fracture.

3 Reduce the fracture and check the accuracy of the reduction by visualization of the calcar. To do this flex and externally rotate the hip. If the reduction is satisfactory drive the seating chisel into the head, remove it, and replace it with the 130° angled plate. The length of the blade is determined by the depth to which the seating chisel has been driven into bone. To determine this simply, check the gradations marked on the seating chisel. One should also compare it with the predetermined length from the pre-operative drawing.

Fig. 67 *Repositioning or valgus medial displacement osteotomy and its fixation with the 130° angled plate* (for technique see Fig. 183).

66

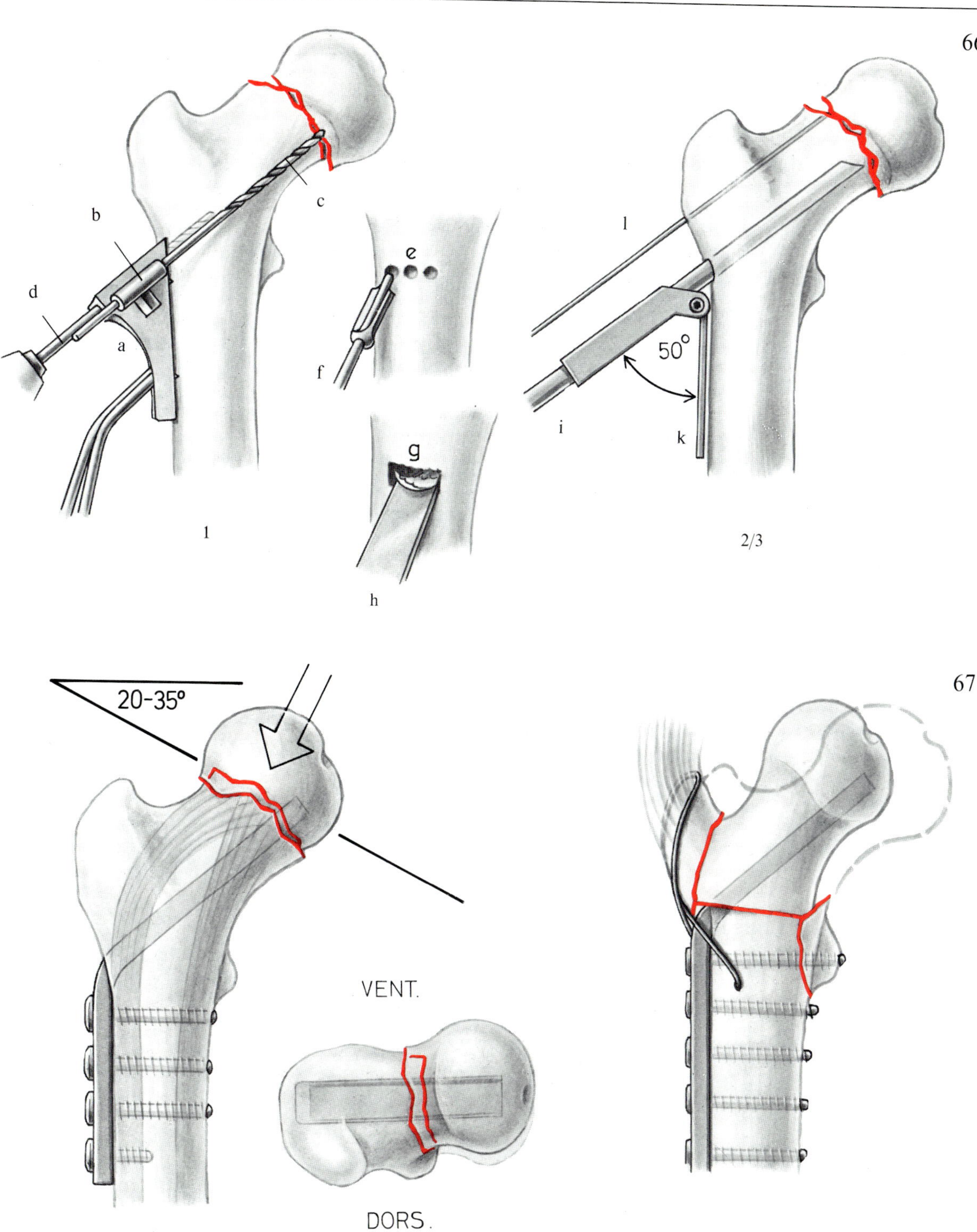

Distal Femur: Insertion of the Condylar Plate

When the condylar plate is inserted into the distal femur, one must make certain that the physiological angle of 99° between the shaft axis and the joint axis is maintained. The distal fragment should not be medialized, and the medial and lateral cortex, when projected distally, should flare out like a trumpet. In the normal femur, when the blade is parallel to the joint axis, the plate is parallel to the shaft axis. This is why the angle between the blade and plate subtends an angle of 95°. The point of entry should be 1.5 cm from the tibio-femoral joint and be in line with the middle of the femoral shaft or slightly anterior to it. This will guarantee that the plate will come to lie in line with the lateral cortex.

Distally, the seat is prepared in very much the same way as in the proximal femur.

Note: Usually in dealing with the distal femur the intra-articular intra-condylar fracture is reduced first and fixed with lag screws. The blade is then inserted and the plate is fixed to the distal fragment with the first screw. Finally the supracondylar component of the fracture is reduced and is placed under axial compression.

Positioning of the patient: We prefer to flex the knee to 90° over a sterile bolster. We can apply traction by pushing on the bolster and by increasing the flexion of the knee joint.

Fig. 68 *Position of the condylar plate in the distal femur.*

a In the frontal plane, the blade lies parallel to the knee joint axis and the plate lies along the lateral cortex of the shaft.

b In the sagittal plane, the point of entry is ventral or anterior and in line with the shaft axis and 1.5 cm from the knee joint.

Fig. 69 *Preparation of the seat for the blade with the seating chisel.*

a Bend the knee to 90° and insert the first Kirschner wire which marks the knee joint axis. The second Kirschner wire is inserted anteriorly over the lateral and medial condyle. It is the guide to the inclination of the patello-femoral joint. These two Kirschner wires indicate the desired direction for the blade.

b A third Kirschner wire is now driven 1 cm away from the joint in line with the long axis and parallel to the first two Kirschner wires. It will serve as the definitive guide for the seating chisel.

c The point of entry is prepared by drilling first three holes with the 4.5-mm drill bits and then enlarging the holes with the router. The seating chisel is then driven in. The assistant must apply firm counter-pressure against the medial condyle while the seating chisel is being driven in. The seating chisel guide must be in line with the shaft axis.

68

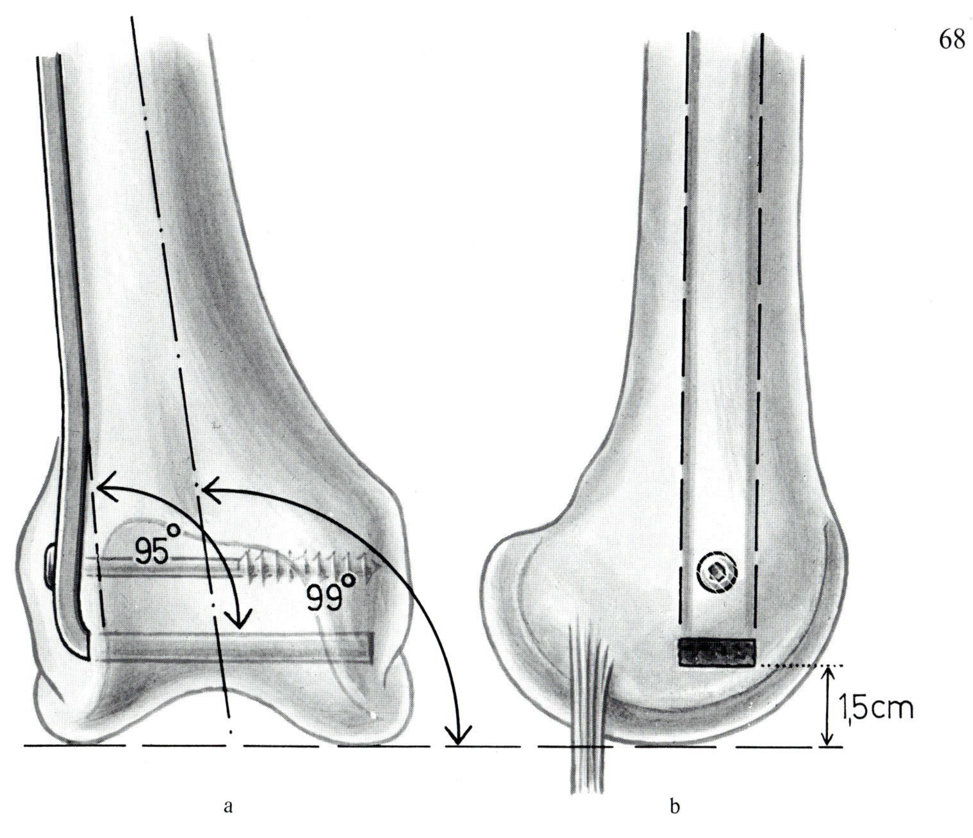

a b

69

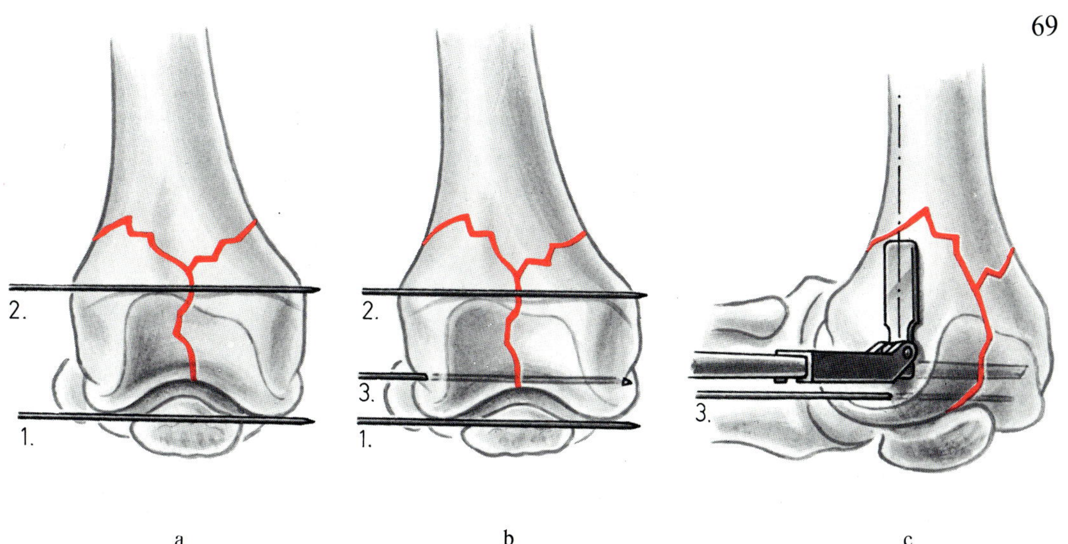

a b c

Distal Femur: Position of the Screws and the Length of the Blade

Note: In fixing fractures of the distal femur, one must pay great attention to the medial buttress. If the medial buttress is defective, then autologous cancellous bone grafting must be carried out.

Fig. 70 *Distal femur: the position of the screws and the length of the blade.*

a When the knee is bent to 90°, one can easily see where the first two guide wires should be placed, where the point of entry should be, and where the two cancellous screws should be inserted.

b The two cancellous screws which are used to fix the vertical inter-condylar component of the fracture are usually inserted slightly proximally to the point of entry of the blade. They do not have to be parallel. We recommend that washers be used to prevent the heads of the screws from sinking through the thin lateral cortex.
When determining the length of the plate, one must keep in mind that the distal femur is a rhomboid and that the medial wall is inclined at a 25° angle. Thus, a blade which on an AP X-ray would appear to be just the right length, would, in point of fact, be too long. This is why in most small patients we use only a 50-mm blade.

c In order to achieve axial compression we use the tension device even if the condylar blade has the dynamic compression holes, because the impaction usually measures at least a few millimetres.

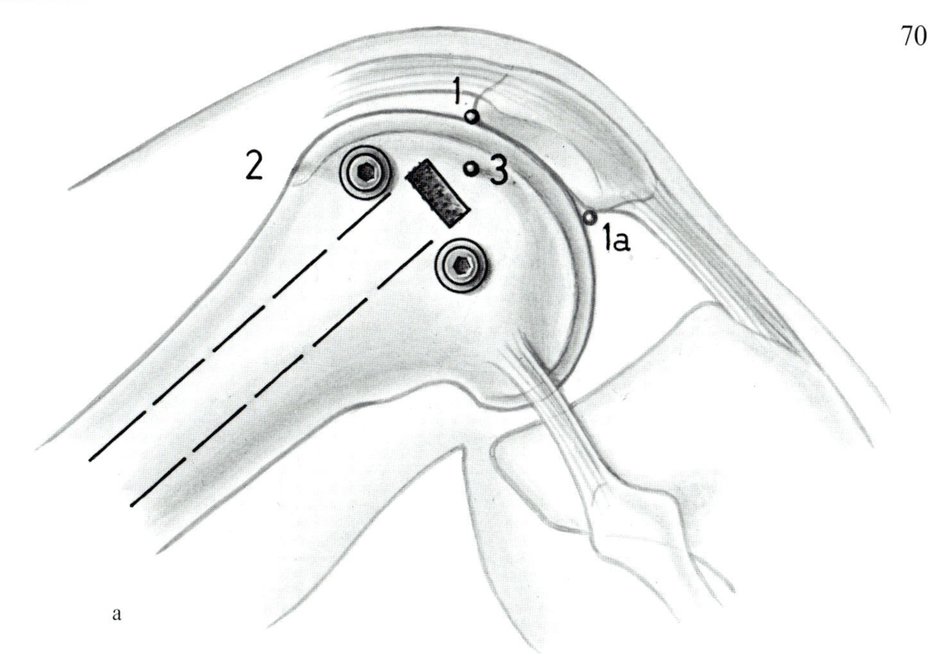

a

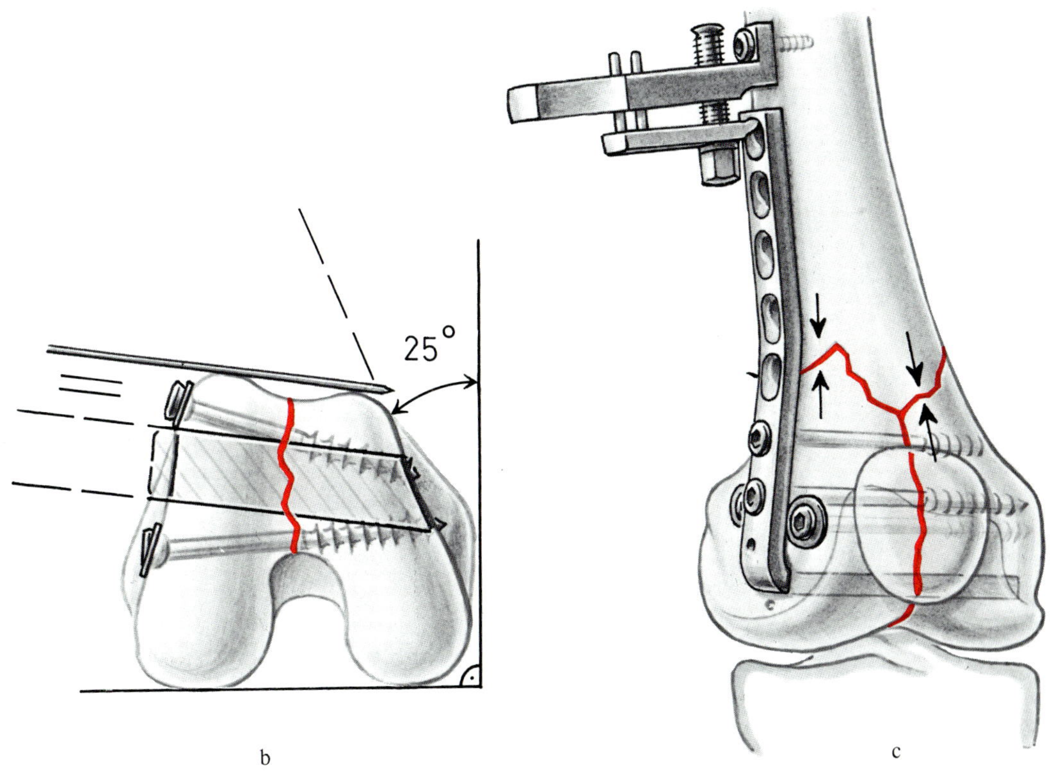

b

c

2.3.4 Mini-Implants and Mini-Instruments

Small screws and plates are often necessary for the fixation of the small bones of the hand or foot, for the fixation of avulsion of joint capsules and similar injuries. The AO has developed the "mini" instruments and implants for the surgeons who deal with these injuries. They serve together with the mini-motor with its accessories as a supplement to the small-fragment instrumentation (Fig. 71).

Fig. 71 *Mini-implants and instruments.*

a Mini-cortex-screw of 2.0 mm and 1.5 mm with a cruciate slot in the head.

b Mini-plates with oval holes: straight three-, four- and five-hole plates, T plates and L plates.

c Drill guide for the 1.1- and 1.5-mm drill bit.

d The 2-mm, 1.5- and 1.1-mm drill bit with rapid coupling.

e Mini-depth-gauge with its protective cap.

f Hand grip with rapid dental lock for

g the cruciate screwdriver,

h the 2-mm and 1.5-mm tap and

i the countersink.

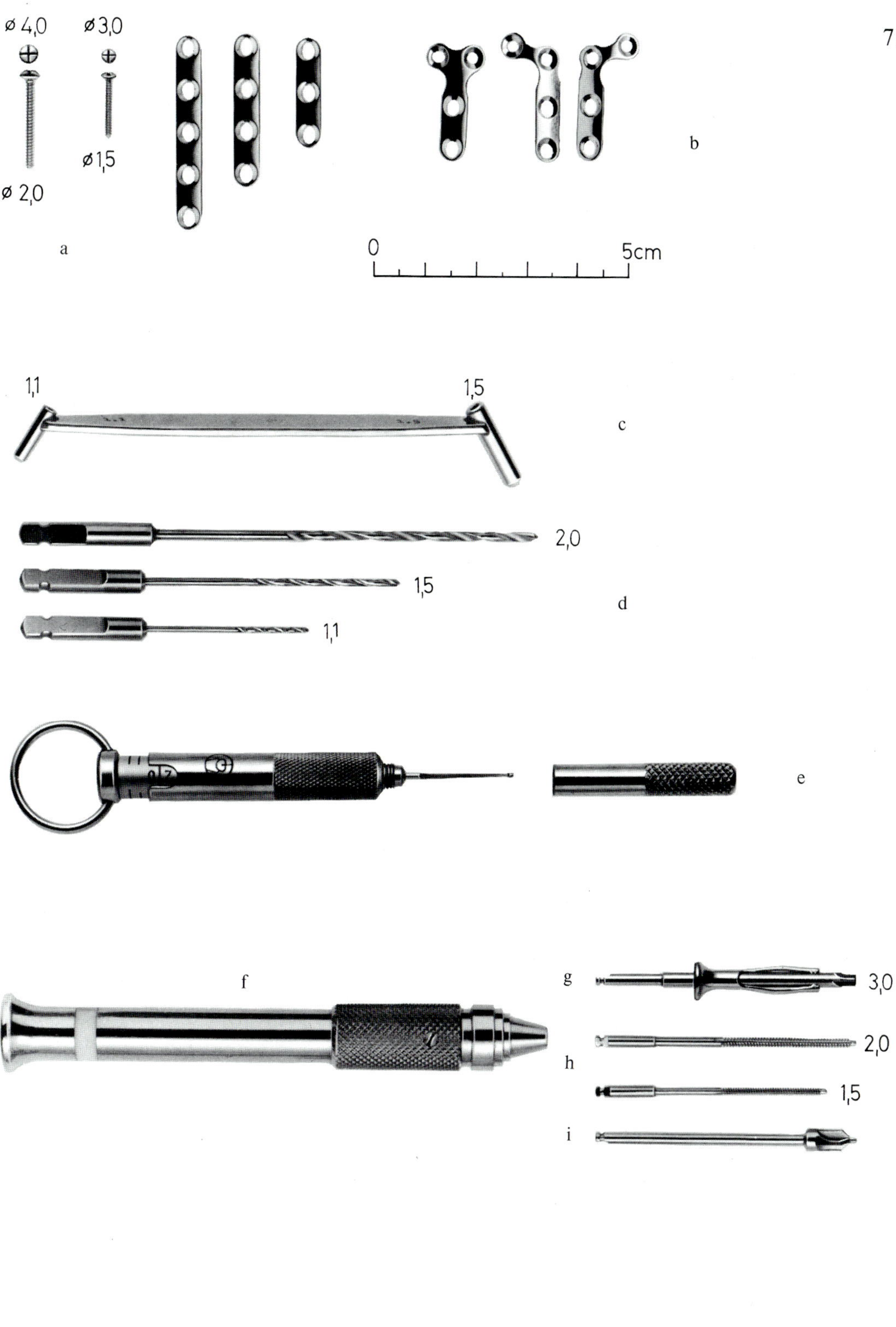
71
ø 4,0
ø 3,0
ø 1,5
ø 2,0
a
b
0
5cm
1,1
1,5
c
2,0
1,5
d
1,1
e
f
g
3,0
2,0
h
1,5
i

2.4 Intra-medullary Nailing

In 1940, KÜNTSCHER introduced his method of intra-medullary nailing. In 1960 the AO took over this method, made improvements upon it and developed it further. It has gained such importance that we feel that every bone surgeon must master it. The stability of the nail fixation is the result of the pressure generated between the elastic recoil of the deformable nail and the rigid non-deformable bone, as well as the result of intra-medullary splinting of the bone by the nail. Therefore the stability of the internal fixation depends upon the enlargement of the medullary canal to permit the insertion of an intra-medullary nail of sufficient size. The fracture is stabilized by an interference fit. It does not depend on any additional inter-fragmental compression, although axial loading does result in some inter-fragmental compression, which adds to the stability of the fixation. In general we feel that an intra-medullary nail is indicated in those circumstances where it can provide sufficient stability to permit early active use of the extremity.

Open and Closed Intra-medullary Nailing

In open intra-medullary nailing we expose the fracture and carry out an open reduction. In closed intra-medullary nailing the reduction, intra-medullary nailing and reaming are all carried out closed with the patient on a special fracture table, under radiological control, usually in the form of an image intensifier.

Open nailing permits not only an absolutely accurate reduction with accurate reduction of rotation, but also the irrigation of the medullary canal and removal of all reaming debris. One does not require a special table, X-ray or image intensifier and one is not exposed to the danger of radiation. However, for cases with a poor soft-tissue envelope, such as stasis ulceration, we prefer closed nailing.

From the case material which the AO has in its documentation there appears to be no difference in the incidence of sepsis between closed and open nailing. Rotational malreduction, however, was very much more common in the closed nailing.

Indication for Intra-medullary Nailing with Reaming of the Medullary Canal

We consider it as the method of choice for fractures of the middle third of the femur and for most of the transverse and short oblique fractures of the middle one-third of the tibia. Comminuted fractures of the middle one-third of the femur can usually be dealt with by the combination of an intra-medullary nail with screws, cerclage wiring, or plating (see Fig. 87). The intra-medullary nail has also proven itself to be an almost spectacular method for the treatment of delayed healing or nonunion of the femur and tibia.

Fractures at the junction of the proximal and middle third and middle third and distal third of the tibia are considered only a relative indication for intra-medullary nailing because they demand considerable experience. In order to stabilize these fractures adequately, that is those at the junction of the proximal and middle one-third, one often requires a supplemental screw which goes through bone and through the two oval holes in the proximal part of the nail. Those at the junction of the mid- and distal third require ledge wires (see Fig. 220c/d) for their stabilization.

We feel that intra-medullary nailing is not indicated for fractures of the upper extremity. Nailing of fractures of the humerus often resulted in damage to the shoulder joint and nailing of fractures of the radius in damage to the wrist joint. Furthermore, for fractures of the forearm, the intra-medullary nails do not provide sufficient rotational stability. They also lead to straightening out of the physiological curvature of the radius which results in the loss of pronation and supination. Peri-articular fractures are not suitable for intra-medullary nailing. The trans-articular nailing recommended by KÜNTSCHER goes against our basic concept of early movement of joints near a fracture.

Indication of Intra-medullary Nailing of Femur and Tibia Without Reaming

We recommend intra-medullary nailing without reaming for open fractures and for double fractures with an intact but displaced intervening segment. These can be nailed with a thin nail which preserves axial alignment. Most of these, however, require supplemental plaster cast support. If in 2–3 months delayed healing becomes evident then these can be re-nailed with a larger nail without much difficulty.

Note: Intra-medullary wire splinting gives no more than adaptation of the fragments and we of the AO reject it as a method.

A few of the AO members have gone over to the use of the *elastic intra-medullary nails* of SIMON-WEIDNER and of ENDER in the treatment of per-trochanteric fractures in the elderly. This procedure is rapid, relatively simple, but we have found that in our experience it does not always result in a stable fixation which would allow the patients early mobility and ambulation. Furthermore, we have found that external rotational deformities in the unstable fractures are almost the rule.

The *stack nailing* of HACKETHAL is still used in some of the AO clinics as the method of choice for transverse fractures of the humerus.

2.4.1 AO Intra-medullary Nails

The AO intra-medullary nails are particularly light and elastic. They are made out of thin-walled tubes slit for $^4/_5$ of their lengths. The closed, proximal end has a threaded inner surface, which has vastly simplified nail removal. It also allows for more accurate force transmission at the time of nail insertion. We have preserved the cloverleaf profile introduced by KÜNTSCHER. It gives the best interference fit in the areas of contact while allowing at the same time a rapid ingrowth of blood vessels into the clear spaces. This leads to a rapid re-establishment of the intra-medullary blood supply (SCHWEIBERER).

The Diameter of the Intra-medullary Nail. Until 1976, the diameter marked on the AO intra-medullary nails corresponded to the diameter of the closed tube or to the size determined when the dimension of the nail was measured with a caliper (see Fig. 74A, B and C). The cloverleaf profile, because of shaping, results in an increase of the real diameter of the nail (line D) of about 1 mm for the femoral nail and 0.5 mm for the tibial nail. The size corresponds to the size determined by means of a special template.

Since 1977, all AO intra-medullary nails have had the catalogue number stamped on them. *Under most circumstances the size of the last reamer head should correspond to the size of the intra-medullary nail.*

Ledge Wires of HERZOG: The distal end of the AO tibial nail has two oval slits which permit the protrusion of the ledge wires. The ledge wires give additional stability when the distal fragment is short.

Fig. 72 *Intra-medullary nail for the tibia.*

a This is made from a tube which has a curved proximal end (the Herzog curve). The tibial nail is inserted from the front and must be introduced close to the anterior cortex in order to protect the posterior wall of the tibia. For this same reason the slit is posterior and the tapered end of the nail also points to the back.

b The cloverleaf profile.

c In cases where the intra-medullary nail is used for the stabilization of a fracture with a short proximal fragment, further stability is achieved by inserting a cortex screw from front to back and passing it through the oval holes of the nail. The markings used since 1977 can be seen. The first number engraved on the nail corresponds to the diameter of the nail as determined by the template. The medullary canal must be reamed to the same size.

Fig. 73 *Intra-medullary nail for the femur.* Because of the anterior curvature of the femur, the intra-medullary nail must be inserted along the posterior wall in order to protect the anterior wall. Its slit is therefore anterior as is its tapered end (a/b).
The closed tube portion (c) is threaded on its inner aspect. This permits the insertion of a conical bolt for the transmission of force. This protects the nail from damage at the time of insertion and removal. The same applies for the tibial nail.

Fig. 74 *The tube profile (a) with the diameter A.*

The cloverleaf profile corresponds to the larger circles (b). The diameters *B* and *C*, which can be determined by means of a caliper, correspond to the old diameter of the tube but not to the real diameter of the nail, which is given by the circle (*c*) which circumscribes the nail. There is a three-point contact between the nail and this circle. The diameter *D* of this circle is greater than the diameters *A*, *B* or *C*, which are given by caliper measurements.

72

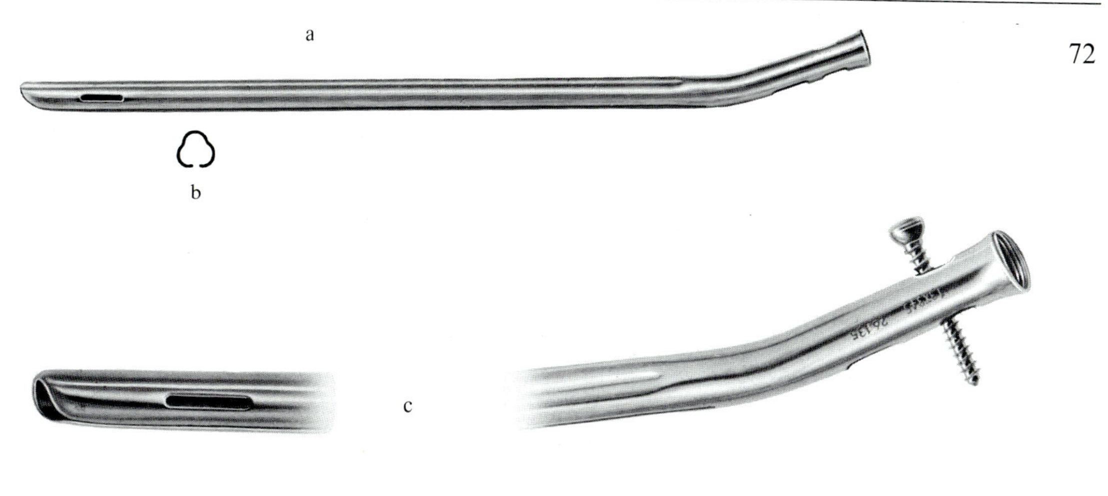

73

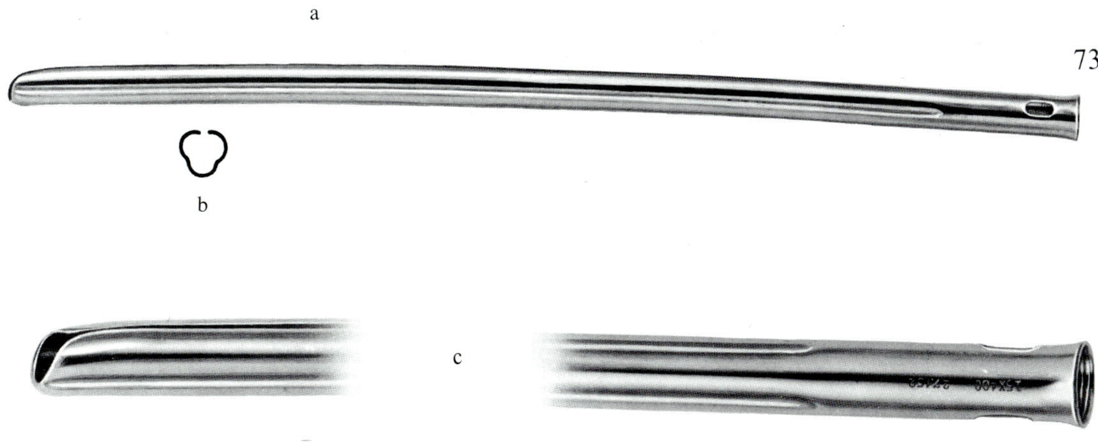

74

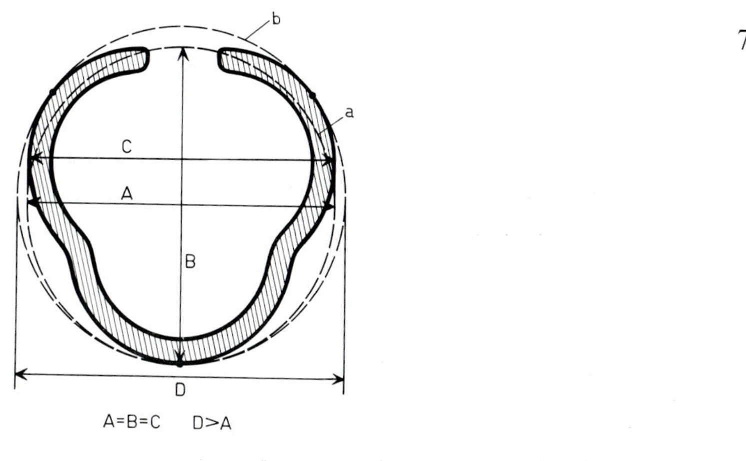

2.4.2 AO Intra-medullary Reamers

Reaming has made it possible to introduce an intra-medullary nail of sufficient diameter to provide stable internal fixation. At the end of intra-medullary reaming the canal proximally and distally to the fracture should correspond to the outer diameter of the cloverleaf. The interference fit between the nail and the medullary canal of equal dimension is the result of the fact that the nail is relatively straight and the profile of the reamed endosteal surface, because of the short reamer heads, wavy, rough and uneven.

The AO reamer shafts are elastic and the reamer heads side cutting. The compressed-air reamer drive (approx. 350 rpm) has proven itself most useful.

Note: One should not over-ream. In a short fragment a few centimetres of contact between cortex and nail are enough. The metaphysis should be reamed only with the front-cutting 9-mm fixed reamer head.

A counter-clockwise direction of the reamer drive will ruin the flexible reamer shafts.

The tip of the reamer guide can be easily bent. This greatly facilitates the passage of the reamer guide along the posterior wall of the tibia. In closed nailing the curve facilitates the centring of the reamer guide in the distal fragment. The front-cutting, fixed reamer head must not be passed over the curved portion of the reamer guide. It will bind on the curvature and become damaged.

Fig. 75 *Instruments for intra-medullary nailing.*

a The awl.

b The 6–9 mm hand reamers, necessary at times for the insertion of the reamer guide.

c, d 3-mm reamer guide with flared tip, 820 mm long with its holder (d). Its tip is slightly curved.

e The tissue guard of BÖHLER design.

Fig. 76 *Instrument set for reaming of the intra-medullary canal.*

a The reamer drive.

b The 3-mm reamer guide.

c The flexible reamer shaft with the fixed 9-mm front-cutting reamer head.

d The 8-mm flexible reamer shaft for exchangeable reamer heads 9.5–12.5 mm.

e The flexible 10-mm reamer shaft for exchangeable reamer heads 13–19 mm.

f The plastic medullary tube used to exchange the 3-mm reamer guide for the 4-mm nail guide. It can also be used for the irrigation of the medullary canal.

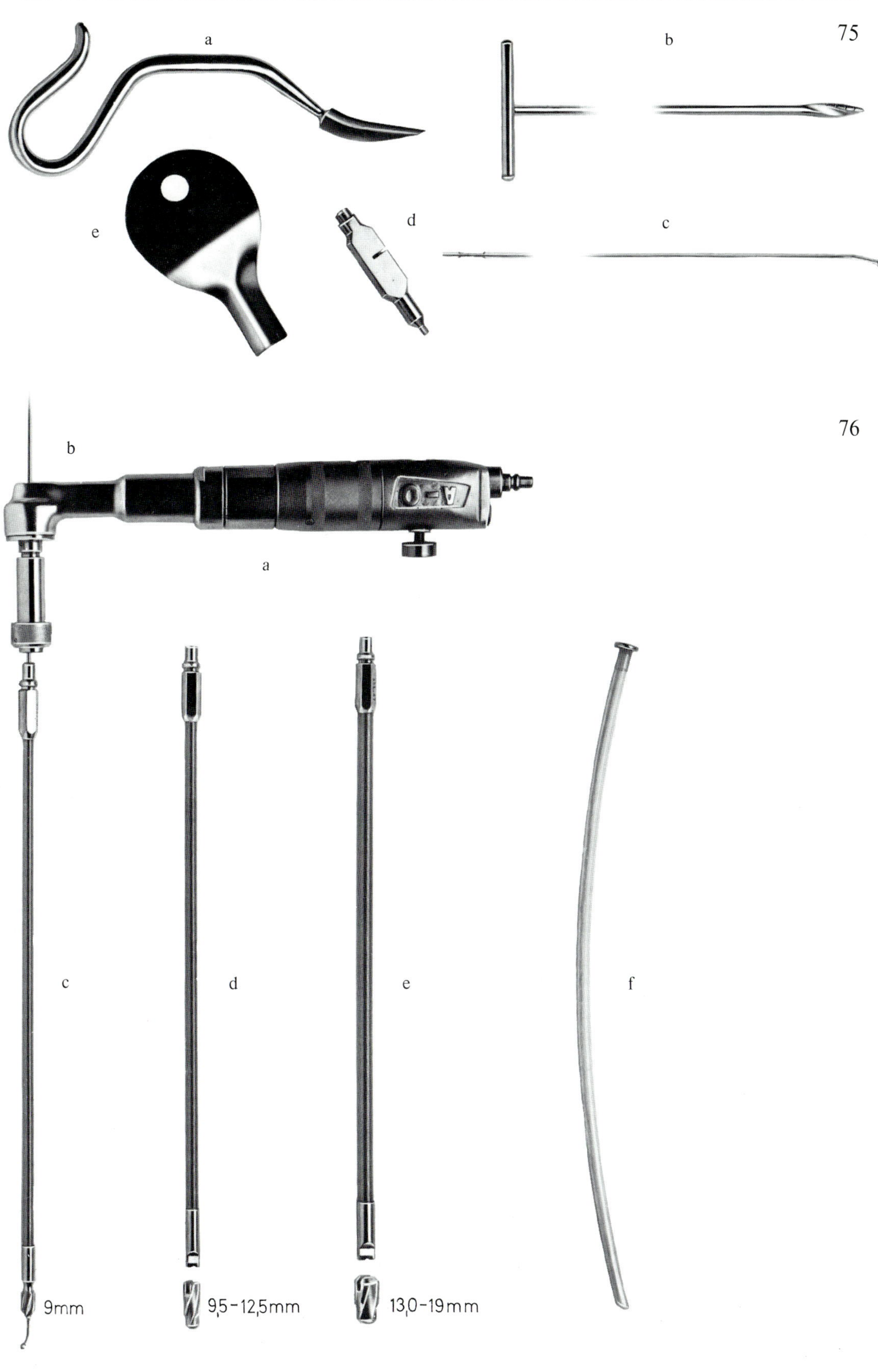
75
a
b
e
d
c
76
b
a
c
d
e
f
9mm
9,5-12,5mm
13,0-19mm

2.4.3 Technical Complications of Intra-medullary Reaming

The reaming should be done progressively, exchanging the reamer heads in 0.5-mm succession. The reaming has to be done carefully, feeling one's way. Once the reamer head begins to ream after an advance of 1–2 cm, it should be slightly withdrawn before it is advanced again. This is repeated again and again. If reaming is carried out in this way, using successive sizes, gentleness and care, the reamer should rarely, if ever, get stuck in the medullary canal. If it should get stuck, the flared end of the reamer guide allows for easy removal of the assembly from the medullary canal.

The two flexible reamer shafts of 8 and 10 mm respectively should never be confused. The thinner one is for exchangeable reamer heads from 9.5 to 12.5 mm and the thicker one for reamer heads of 13 mm and upwards.

Fig. 77 If the wrong reamer shaft is used, reaming is either impossible to carry out or the reamer head becomes easily stuck in the medullary canal because it tends to tip on the narrow reamer shaft.

Fig. 78 Should the reamer head become stuck in the medullary canal, removal is simple. The ram is slid over the protruding end of the reamer guide. Either the reamer guide handle or a vice grip is then used to hold onto the proximal end of the reamer guide. The flared end of the reamer guide allows one to hammer out with ease the assembly from the medullary canal.

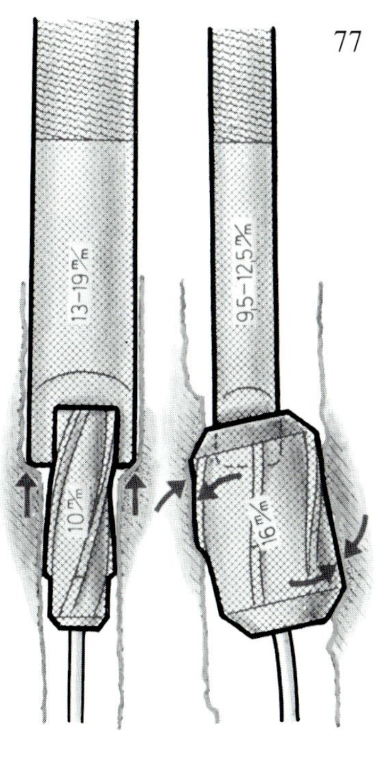

77

78

2.4.4 AO Insertion Assembly for the Tibial and Femoral Nails

In inserting the tibial nail, the nail guide must protrude through the hole in the back part of the tubular portion of the nail. When putting together the assembly for the insertion of the tibial nail one must insert the curved driving piece between the threaded conical bolt and the cannulated weight guide, and make certain that the weight guide and the nail point in the same direction.

Fig. 79 *AO insertion instrumentation for the intra-medullary nails. Left for the tibia and right for the femur.*

a The *intra-medullary nails.*

b The *4-mm nail guide.*

c The *threaded conical bolt* which fits into the threaded proximal portion of the nail.

d The *curved driving piece* for tibial nails.

e The *cannulated weight guide rod* which is essential for the insertion of the femoral nail, since the nail guide must go up the weight guide during the insertion.

f The *ram* for the insertion and removal of the nails.

g The *flexible grip* for the cannulated weight guide rod. It is indispensable for nail removal.

h The *guide handle* for the nail. It allows one to hold the nail and prevent its rotation during insertion.

i The *driving head* which can be inserted onto the curved driving piece (d) for the tibia or onto the threaded conical bolt (c) for the femur. If the driving head is used, an 800-g hammer is used for the hammering in of the nail.

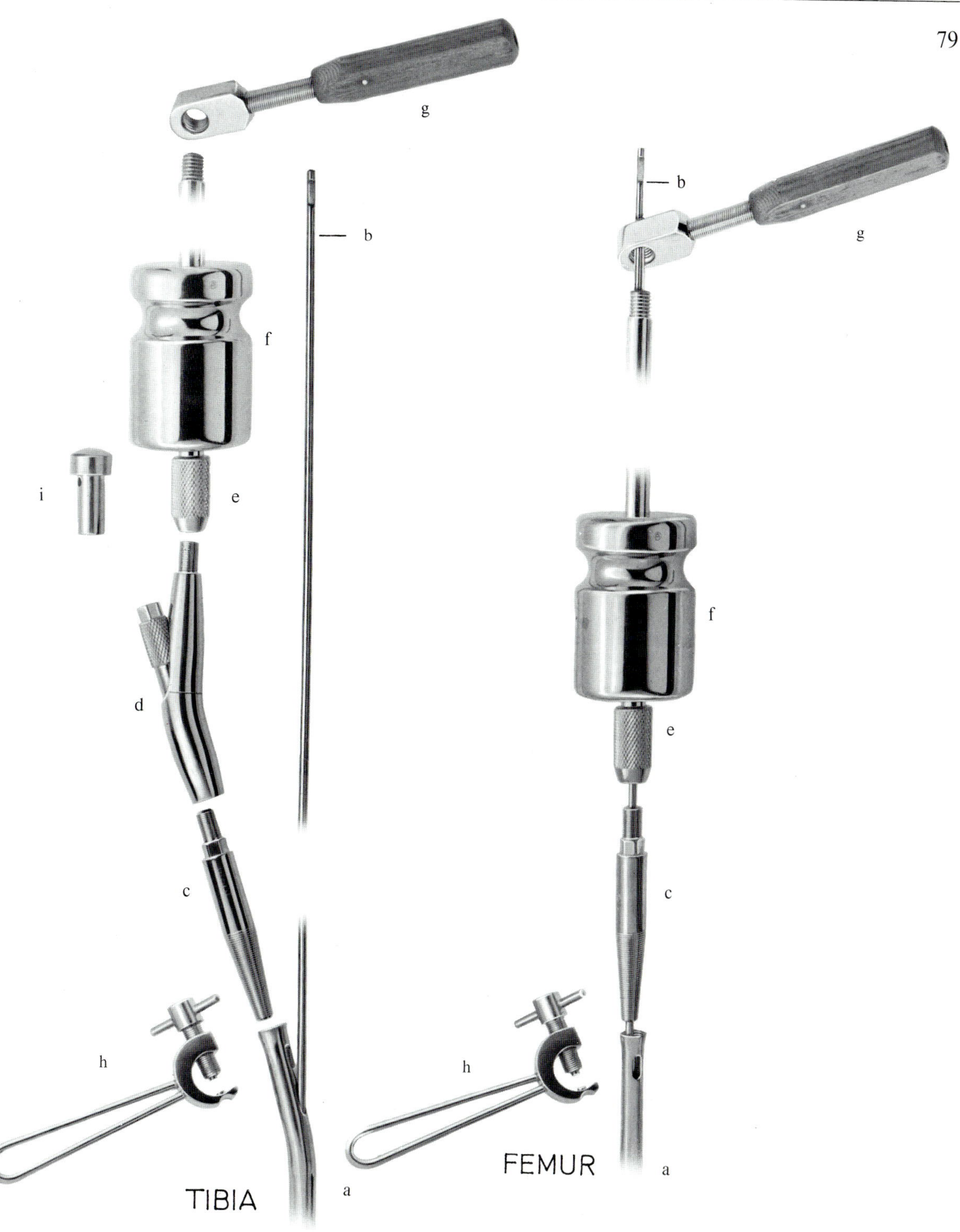
g
b
g
b
f
i
e
d
f
e
c
c
h
h
FEMUR
a
TIBIA
a

2.4.5 Complications of Nail Insertion

The nail must not be inserted over the 3-mm reamer guide. The reamer guide is too weak and too flexible. Furthermore, the flared tip could become stuck in the slit of the nail.

In nailing of the tibia, it is important to choose the conical bolt of correct dimension, otherwise it will reach too far into the nail and block the nail guide from protruding.

The curved driving piece should be used only for the insertion but not for the extraction of the tibial nails.

Remember that nail insertion should be gentle and never forced. If resistance is encountered the nail should not be forced in. The nail must be removed and the diameter of the canal reamed a further 0.5 mm–1.0 mm. Before re-insertion the size of the nail should be checked with the nail template.

Fig. 80 *Choice of the threaded conical bolt.*

a The threaded conical portion of the bolt must correspond to the diameter of the nail.

b If too small a conical bolt is chosen, it will penetrate too deeply into the nail and block the exit of the 4-mm nail guide through the back of the tibial nail.

Fig. 81

a The curved driving piece makes it possible at the time of nail insertion to hammer in the same direction as the long axis of the tibia.

b The 4-mm nail guide protects the posterior cortex of the tibia during the insertion of the nail. Also the tip of the nail has been tapered to protect the posterior cortex of the tibia. At the time of insertion the elastic deformation of the nail can reach up to 15°.

80

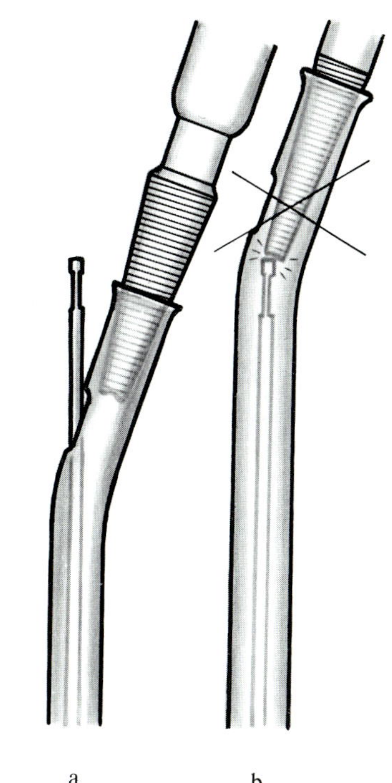

81

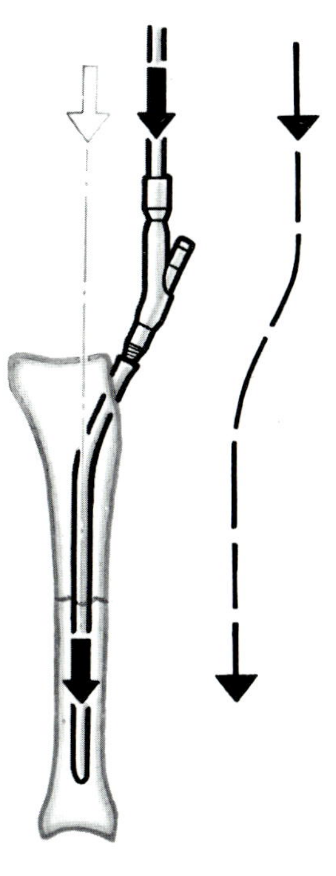

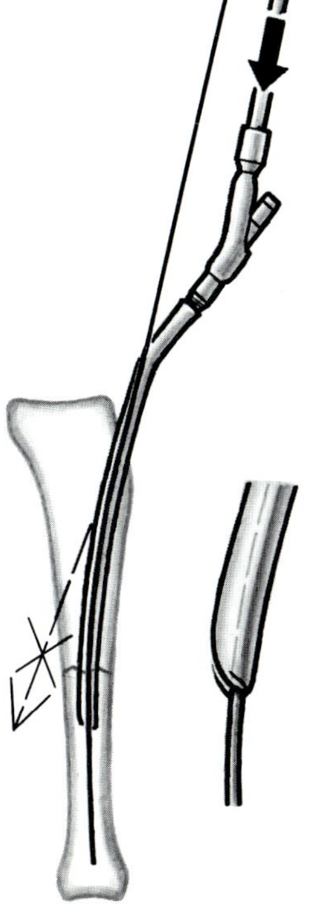

2.4.6 Technique of Open Intra-medullary Nailing of the Tibia

Open nailing of the tibia is carried out on a normal operating room table. Acute flexion of the knee facilitates the insertion of the nail and the protection of soft tissues.

Prior to surgery, measure the distance on the intact limb between the knee joint and ankle joint and subtract 3–4 cm. Once the reamer guide is inserted check the predetermined nail length against the portion of the reamer guide inside bone. The thickness of the nail can be determined only during surgery. The size of the nail should correspond to the last reamer head used. The shorter fragment should be reamed over a distance of at least 3 cm.

Fig. 82 *The steps of an open intra-medullary nailing of the tibia.*

a Make an 8 to 10-cm-long incision 1 cm lateral to the anterior crest of the tibia. Expose the fracture and reduce it. Maintain the reduction by means of a semi-tubular plate and two self-centring Verbrugge bone clamps. Make a 5 to 6-cm transverse incision at the level of the joint over the infra-patellar tendon. Bend the knee to 130°–140° and place the foot on the operating room table.

b Split the infra-patellar tendon longitudinally along the line of its fibres and insert a self-retaining rake. With the awl perforate the thin cortex above the tibial tubercle.

c Twist the awl back and forth and guide its tip first posteriorly and then in the direction of the long axis of the tibia till its cutting head disappears inside the bone.

d Insert the 3-mm reamer guide with its curved tip pointing forwards. An X-ray should now be taken to check the position of the reamer guide as well as the length to which it has been inserted.

e Thread the flexible reamer shaft with the fixed 9-mm cutting head over the reamer guide. It will cut through any bony obstruction and open the medullary canal. Exchange this reamer shaft for the 8-mm flexible reamer shaft with the 9.5-mm exchangeable reamer head. Ream the medullary canal increasing the reamer heads by 0.5 mm at a time. Protect the skin and the infra-patellar tendon with the tissue guard. If reaming to 13 mm or more, the 8.0-mm flexible reamer shaft must be exchanged for the 10-mm reamer shaft designed to take the larger reamer heads.

f At the end of reaming insert the plastic medullary tube over the reamer guide and remove the reamer guide. Irrigate the medullary canal with Ringer's solution till clear.

g Insert the 4-mm-thick nail guide into the medullary tube.

h Remove the medullary tube and then carefully and slowly insert the properly assembled 13 to 14-mm-thick tibial nail (see Fig. 79). Lift the ram 10 cm and then let it fall over the cannulated weight guide.

i Remove the Verbrugge clamps, the semi-tubular plate, the nail guide, the cannulated weight guide and the ram. Screw on the driving piece and with the 800-g hammer slowly drive in the nail till the tip is driven into the distal metaphysis and the proximal tip comes to lie flush with the cortex.
Check the accuracy of the reduction and the rotational stability. Should there be a gap at the fracture, close it by hitting firmly with a fist over the femoral condyles or by extending the knee and hitting firmly on the heel. Check the nail once again. Remove the curved driving piece and the conical bolt with the T wrench. Irrigate the nail and the wound with Ringer's solution. Drain both the site of nail insertion and the fracture.

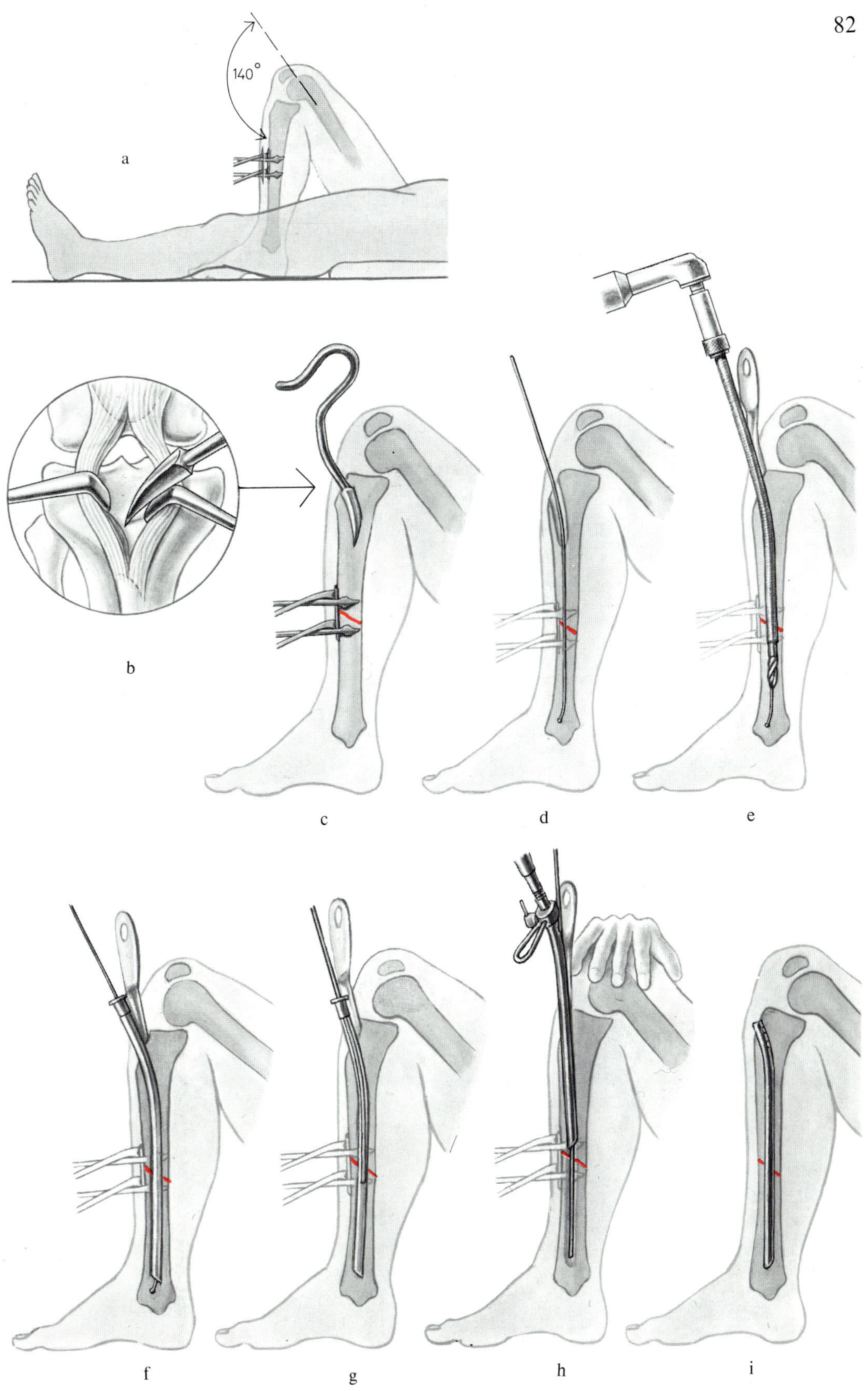
140°
a
b
c
d
e
f
g
h
i

2.4.7 Technique of Open Intra-medullary Nailing of the Femur

The point of entry of the nail should be just lateral and anterior to the tip of the trochanter. This will protect the hip joint during the reaming. We are very much opposed to the retrograde insertion of the reamer guide and to the so-called retrograde nailing. Retrograde insertion causes the reamer guide to come out medially, intra-articularly, medial to the tip of the trochanter through the superior cortex of the neck. This may damage the blood supply to the femoral head. It also opens a communication between the fracture and the hip joint which in the eventuality of sepsis could have tragic consequences.

Exposure of the Fracture and of the Greater Trochanter. The fracture is exposed through a straight lateral incision made along an imaginary line joining the posterior aspect of the greater trochanter with the lateral epicondyle. The vastus lateralis is reflected forwards. The greater trochanter is exposed by making a longitudinal cut 5–6 cm long upwards from the tip of the greater trochanter. The abductors are exposed, spread in the line of their fibres, and the tip of the trochanter is exposed. The trochanter is then opened with the awl. Occasionally it is necessary to open the medullary canal with the 6 to 9-mm hand reamers before the reamer guide can be inserted.

Reduction. Most often simple fractures can be reduced easily by increasing the deformity, hooking on the cortices and then levering the bones straight. If this fails reduction should be carried out with the aid of the femoral distractor.

Reaming is continued until one has a contact, a few centimetres long between the reamer head and bone on each side of the fracture. This will determine the size of the nail. Before insertion the size of the nail is checked with the nail template.

Length of the Nail. The length of the nail is determined pre-operatively on the normal extremity by measuring from the tip of the greater trochanter to the knee joint and then subtracting 2–3 cm. After reduction and insertion of the reamer guide the exact length of the nail is determined from the length of the reamer guide in bone, after the position of the reamer guide has been checked by means of an X-ray.

Fig. 83 *Positioning of the patient for intra-medullary nailing of the femur.* Position the patient on his side with the injured limb upwards. Flex the hip and the knee and support the back and the symphysis pubis with special rests. Make a 10 to 15-cm incision over the fracture and a 5- to 6-cm incision upwards from the tip of the greater trochanter.

Fig. 84 *Technique of open intra-medullary nailing of the femur.*

a Insert the awl just lateral to the tip of the greater trochanter and as far forwards as possible.

b Insert the reamer guide with a slightly curved tip as far as the fracture.

c Expose the fracture and carry out the reduction. Often it is necessary to increase the deformity some 40°–60°, hook on the cortices. Then, with the aid of a Hohmann retractor, whose tip has been inserted into the medullary canal of the distal fragment, the femur is levered straight. Once the fracture is reduced, the reamer guide is inserted into the distal fragment.

d Maintain the reduction with a semi-tubular plate and two self-centring Verbrugge bone clamps. Ream the medullary canal and insert the nail as described under intra-medullary nailing of the tibia. Take note of the nail assembly for the femoral nail.

83

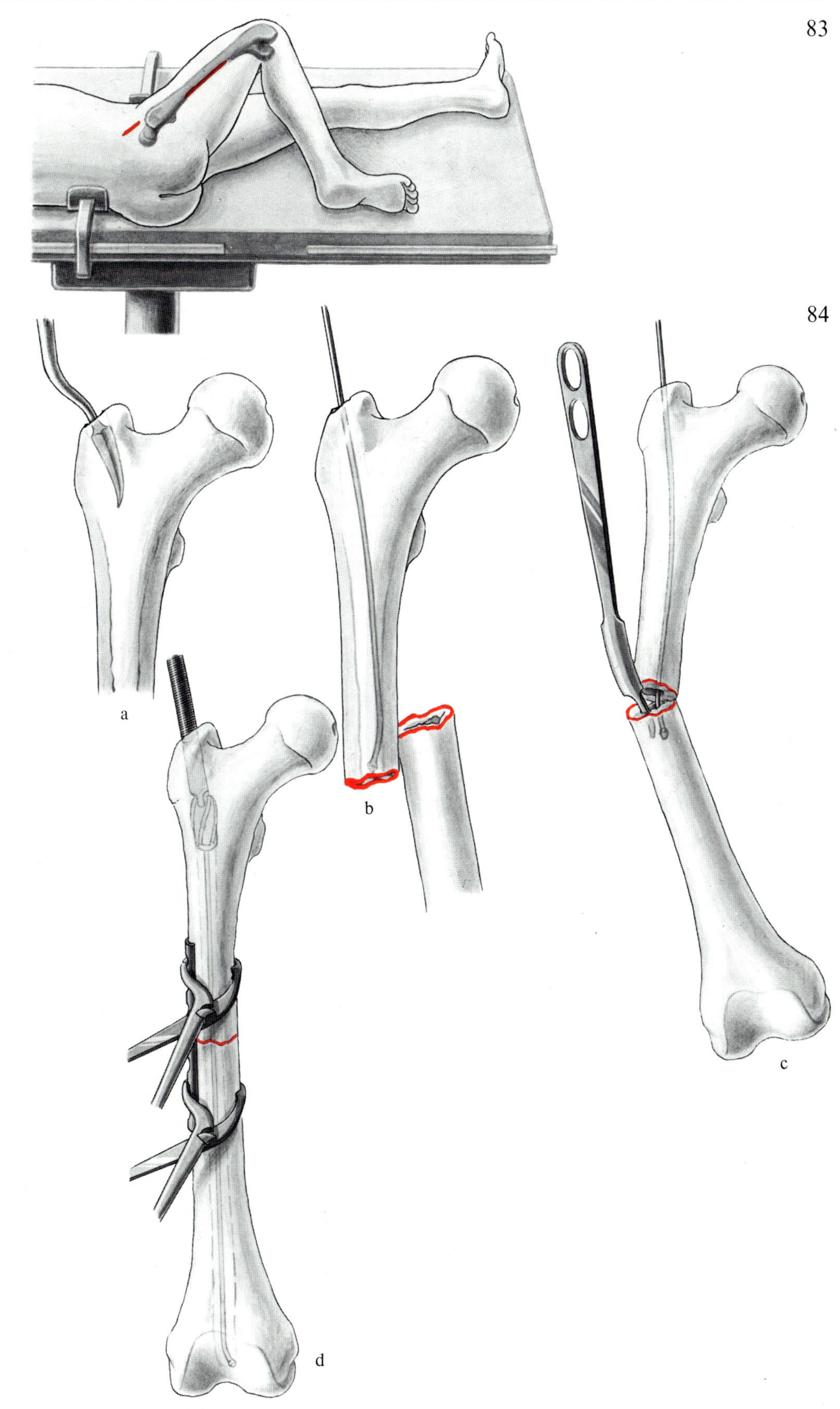

2.4.8 Femoral Distractor

We originally designed the femoral distractor for lengthening post-traumatically shortened femurs, but we have found it also most useful in carrying out the reduction of fresh fractures. It has proven to be a particularly useful aid in the reduction of fresh femoral fractures of athletic, muscular patients in whom the reduction can often be very difficult because of muscle spasm and shortening.

The femoral distractor has been patterned on the HARRINGTON distractor. About 30° of axial rotational correction can be carried out, but we strongly recommend that rotational alignment be re-established before the holes for the connecting bolts are drilled. This can be done by palpating for the linea aspera. In simple fractures the connecting bolts need to be drilled only through one cortex. In metaphyseal regions, or when one is carrying out corrective osteotomies with lengthening, the connecting bolt should be inserted through both cortices. The bolts must be screwed into the near cortex in both instances. The clamps of the distractor are then fixed over the bolts as close to the bone as possible.

Fig. 85 *The femoral distractor*. The screw spindle (*a*), its movable wing (*b*), the distraction screw (*c*), the compression screw (*d*), the hinged joint for rotational correction (*e*), the fixation screw (*f*), the lever (*g*) and connecting bolt (*h*).

Fig. 86 *The use of the femoral distractor in reducing a transverse fracture of the femur.*

1 Make a longitudinal incision over the fracture 10–20 cm long, in line with the imaginary line joining the tip of the greater trochanter with the lateral epicondyle. Open the fascia and reflect the vastus lateralis forwards and expose the fracture. Begin with the proximal fragment. About 4–5 cm from the fracture drill a hole with the 4.5-mm drill bit through both cortices either at right angles or slightly obliquely to the long axis of the femur. Insert the connecting bolt (*a*) in the universal chuck (*b*) and screw it into the bone till all the thread disappears in the cortex. Insert the reamer guide (*c*) through the hole near the tip of the greater trochanter and push it into the medullary canal as far as the fracture. If the connecting bolt blocks the reamer guide in a narrow medullary canal remove the connecting bolt, turn it around and insert just its threaded portion through the near cortex.

2 Take the distal fragment and correct the rotational alignment till the two linear aspera are approximately in line. With the 4–5-mm drill bit (*d*) about 4–5 cm from the fracture line drill both cortices of the distal fragments, drilling parallel or slightly convergent to the previously inserted connecting bolt. Insert the second connecting bolt (*e*).

3 The wing tips (*f* and *f'*) of the distractor are now passed over the parallel bolts making certain that the hinged connector (*g*) is loose. The holding screws (*h* and *h'*) are now tightened. The distraction screw (*i*) is now turned on the threaded spindle. The compression screw (*k*) has to be opened widely. As soon as the correction is obtained, the hinge joint (*g*) is tightened.

4 The fragments can now be brought out to length with great ease and the fracture reduced. Once the fracture is reduced the reamer guide is pushed into the distal fragment. Maintain the reduction with a semi-tubular plate and reduction clamps. Remove the distractor and the connecting bolts, and ream the medullary canal to the desired size.

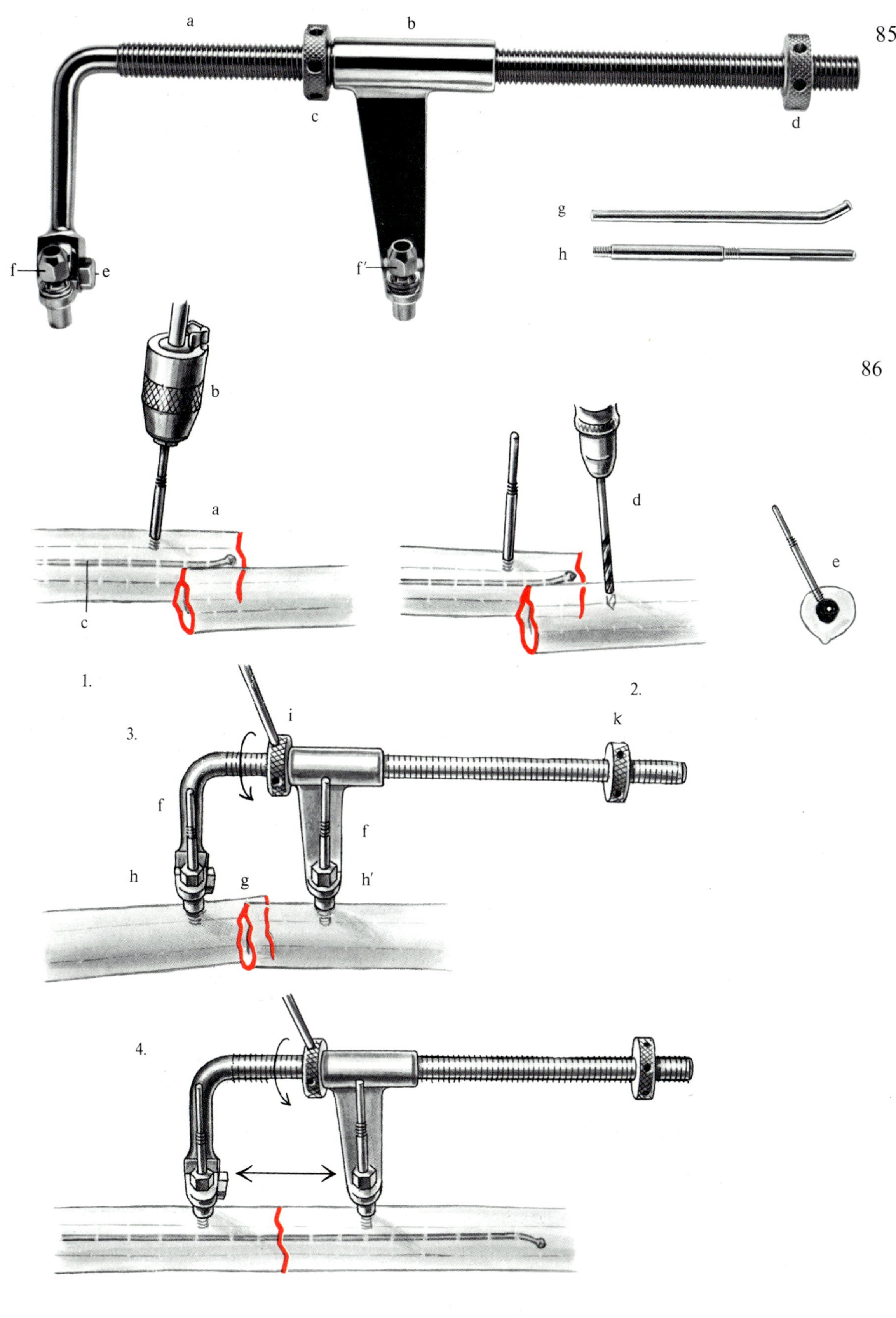
85
a
b
c
d
g
h
f
e
f′
86
b
a
c
d
e
1.
2.
3.
i
k
f
f
h
g
h′
4.

Use of the Femoral Distractor in a Comminuted Fracture of the Femoral Shaft

Comminuted fractures of the femur can often be stabilized by combining the intra-medullary nail with screws or with a plate. The details of this difficult procedure will now be demonstrated. It is most important not to devitalize the comminuted fragments and separate them from their muscle attachment. If the procedure is carried out correctly, the devitalization of these fragments is minimal.

Fig. 87

a Make a postero-lateral incision the length of the femur; do not split the muscle along the line of its fibres but reflect it forwards. Do not touch the comminuted fragments. Expose only the proximal and distal main fragments. Deliver both fragments into the wound and ream both with the intra-medullary reamers for a distance of 4–10 cm to a diameter of 13–15 mm. Do not enter the greater trochanter with the reamer.

b Expose the greater trochanter as previously described (Fig. 83) and perforate it with the awl; perforate it with the hand reamer if necessary, insert the reamer guide and ream the remaining portion of the proximal fragment with the flexible reamer shaft and corresponding reamer heads. Insert the nail but stop 3–5 cm from the fracture. Insert the nail guide into the distal fragment.
With the 4.5-mm drill bit, drill a hole in the near cortex about 2–3 cm from the fracture. Insert into this hole a 4.5-mm Steinmann pin about 15 cm in length. It will serve as a directional guide. Carry out an approximate reduction to correct the rotation. Check the reduction by palpating for the linea aspera. Drill the second fragment with the 4.5-mm drill bit drilling parallel or slightly convergently to the Steinmann pin in the proximal fragment.

c Remove the Steinmann pin and insert the nail into the distal fragment.

d Into the prepared 4.5-mm holes insert the short threaded portions of the connecting bolts.

e Insert the wings of the distractor over the connecting bolts and tighten the connecting screws.

f Slowly distract the fragments by turning the distraction screw. The comminuted fragments, because of their soft-tissue attachment, are now pulled into the gap around the nail. Occasionally one has to turn some of the fragments through 180° before they fall into place. As soon as the rotational alignment of both fragments has been corrected, the hinge joint is tightened with the open spanner. Once the reduction of the comminuted fragments has been achieved, the distraction screw is turned back. The tendency of the femur to shorten will clamp the fragments together. If necessary one can increase this axial compression by tightening the compression screw.
If necessary, to increase stability, the comminuted fragments can be fixed with lag screws, or a long narrow plate can be fixed laterally just above the linea aspera. The plate is fixed with short screws through only one cortex. There must be at least two screws in each of the main fragments gripping at least three cortices. The distractor and the connecting bolts are then removed. The nail is inserted so that its proximal end is flush with the tip of the greater trochanter.
In comminuted fractures, cancellous bone grafting of the cortex opposite the plate should be carried out at the end of the operation.

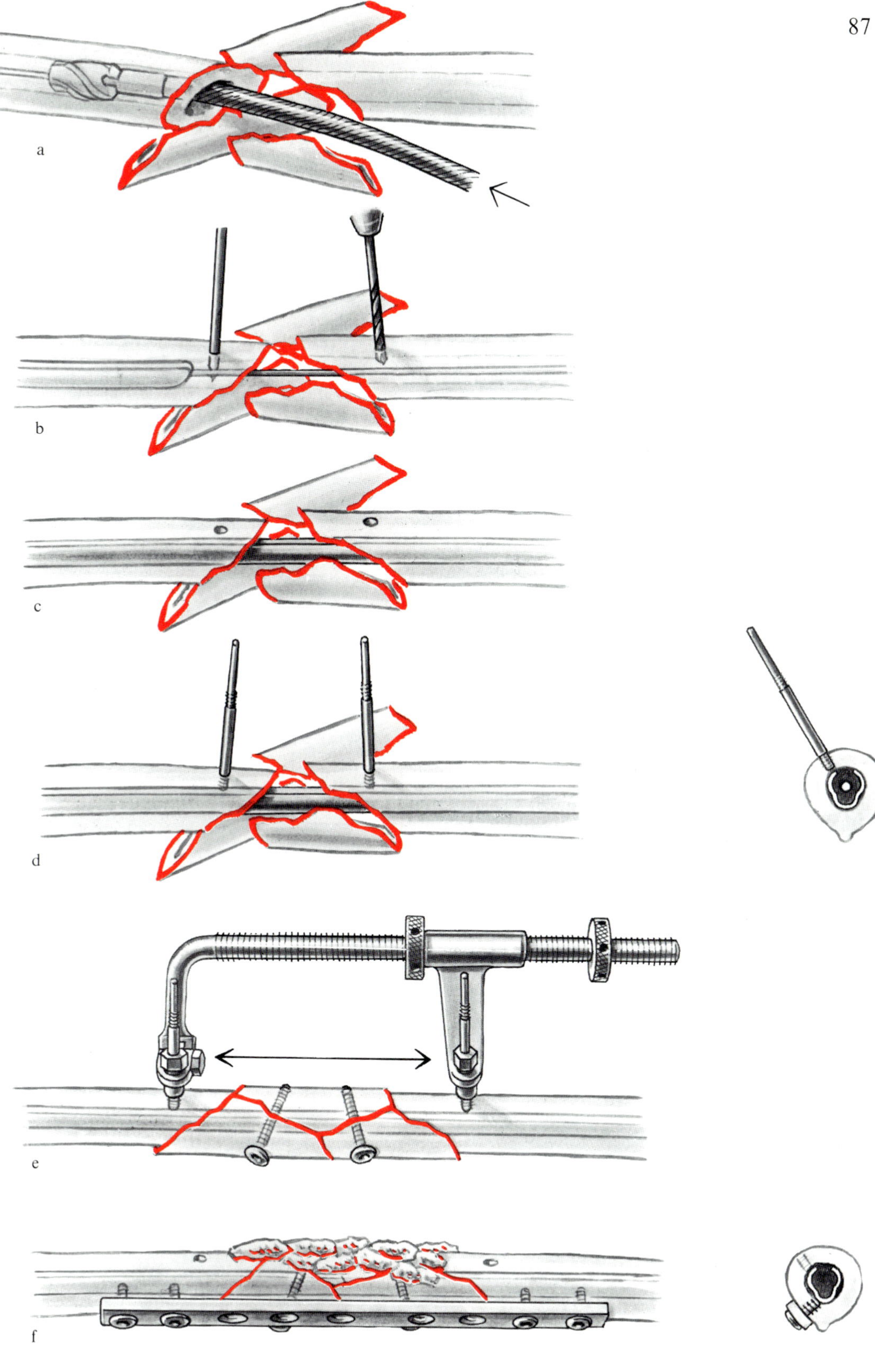
a
b
c
d
e
f

2.4.9 Closed Intra-medullary Nailing of the Tibia and Femur

Closed intra-medullary nailing must be carried out on a special traction table with the aid of an image intensifier which will allow a visual check of the important phases of the procedure. These are: the reduction of the fracture, the insertion of the reamer guide into the distal fragment, the check of the distance of the tip of the reamer guide from the adjacent joint, the insertion of the first reamer head into the distal fragment, the nailing, and finally the reduction at the end of the procedure.

Note:

1 Unsatisfactory reduction of the femur: At times if a reduction cannot be achieved, it may be necessary to expose the fracture and fix it provisionally with a narrow DCP.

2 Closed intra-medullary nailing without a special fracture table: Flex the knee acutely to 120°. Reduction is then achieved simply by longitudinal traction on the ankle. Lateral angulation is seldom necessary. The reamer guide can then be passed into the distal fragment without difficulty. When the tip can be felt to abut against the hard bone in the distal metaphysis, the reamer guide has been correctly inserted. The reduction is checked after the nail insertion by making a small incision over the fracture. This allows one to remove all the reamer debris at the same time and insert a suction drain at the fracture.

If at the end there is a gap at the fracture, it can be closed by extending the knee and delivering a few blows to the heel.

Fig. 88 *The technique of closed intra-medullary nailing of the tibia with the aid of an image intensifier (according to* WELLER*).*

(*a*) Position the patient supine on the traction table and bend the knee to an angle of at least 90°–100°. Apply traction to the leg by means of a pin passed through the os calcis (*b*) or by means of a leather shoe (*c*). Correct the rotational alignment to match the normal side. The trans-malleolar axis is usually in 20°–25° of external rotation with respect to the knee joint axis (5°–35°). The medullary nailing proceeds then as already described on page 116. We recommend that suction drainage be inserted at the fracture also in closed intra-medullary nailing. The trocar of the drains is inserted through a stab proximal to the fracture and is brought out distally.

Fig. 89 *The technique of closed intra-medullary nailing of the femur with the aid of an image intensifier.*

(*a*) The patient is positioned on his side on the traction table. Traction is achieved through a trans-condylar pin (*b*) with the knee in flexion (this is useful as it relaxes the gastrocnemii and their deforming force and permits an accurate check of rotation). In closed nailing of pseudarthroses, traction is applied through a leather shoe (*c*).

The intra-medullary nailing is carried out as already described on page 118. A slight curve of the tip of the reamer guide will facilitate the threading of the reamer guide into the distal fragment. Often manual external manipulation of the fragment becomes necessary. The curve at the tip of the reamer guide will also facilitate the centring of the reamer guide in the distal metaphysis.

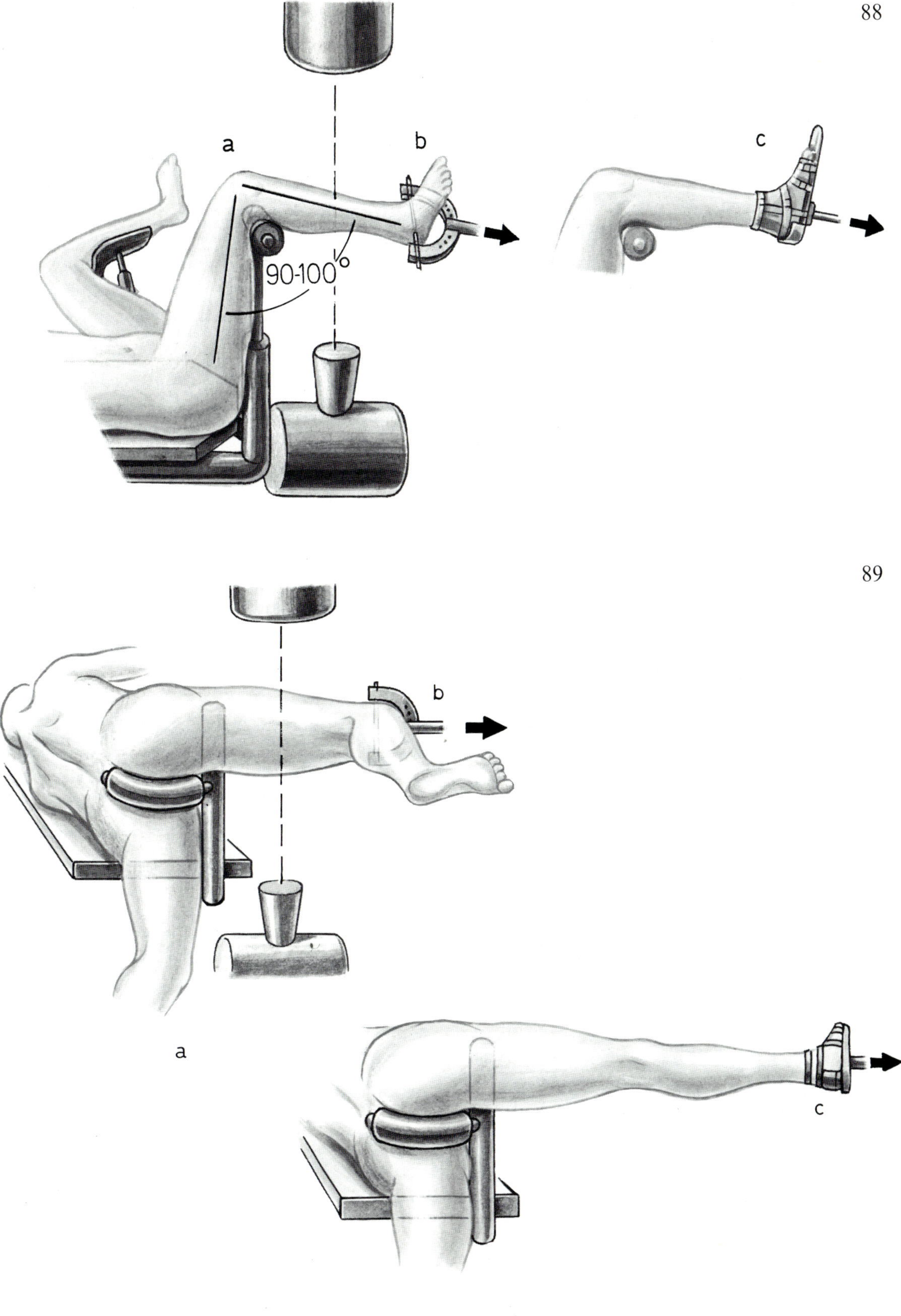
88
a
b
c
90-100°
89
b
a
c

2.5 External Fixators

2.5.1 Instrumentation

External skeletal fixation has become an integral part of modern orthopaedic surgery. External fixators have proven themselves most reliable in the fixation of lengthening osteotomies of the femur and of the tibia, correction osteotomies through the proximal and distal tibia, and in arthrodeses of the knee and ankle. They have also proven themselves most useful in the treatment of infected pseudarthroses, since they permit fixation at a distance from the area of sepsis, and in the treatment of grade II and grade III open fractures particularly since the introduction of the more stable new external fixators.

The first AO external fixator was designed by M.E. Müller in 1952. With the early device, stable fixation could be achieved only if one applied axial compression and then only over short distances. The newly developed external fixators are much more stable and permit the bridging of far greater distances. Axial malalignment can be corrected by means of the adjustable clamps and hinged connectors. The instrumentation does not permit rotational correction. This has to be done by re-inserting the proximal Steinmann pin. This is a small disadvantage of this instrumentation but with practice, even in comminuted fractures correct rotational alignment can be achieved on the first attempt.

Axial compression is achieved by means of two compression devices. These are slipped over the end of the tubes and fixed to the tubes. Turning of the compression screw exerts axial compression by pressure on the clamps which hold the Steinmann pins. In open fractures axial compression is applied only after the first 3–4 weeks have passed.

Note: The old AO external fixator with the threaded rods (Fig. 327b) can be used either as distraction or compression devices if applied to the ends of the Steinmann pins.

Instrumentation available on request: Tubes 600 mm for bridging of joints; a protection stirrup with two clamps; triple clamps for triangulation.

Instruments required in addition to the standard instrumentation for external fixation: small airdrill, a 20° triangular angle guide and occasionally suspension rings and spacers for suspension and traction ropes.

Fig. 90 *External fixator, the standard instrumentation.*

1 *Implants*

a Steinmann nails 5.0 mm, 150, 180, 200 and 250 mm long.

a′ Corresponding protection caps.

b Steinmann nails 4.5 mm of similar length.

b′ Corresponding protection clamps.

c Steinmann nails 4.5 mm with a central 5-mm thread of 150-, 180- and 200-mm length.

d Schanz screws threaded right to the tip with the thread and shaft diameter of 5 mm, 170 and 200 mm long.

e Tubes of 100, 150, 200, 250, 300, 350, 400 and 450 mm length and 11 mm diameter.

2 *Standard instruments*

a Standard clamp.

b Single adjustable clamp.

c Double adjustable clamp.

d Compression device.

90

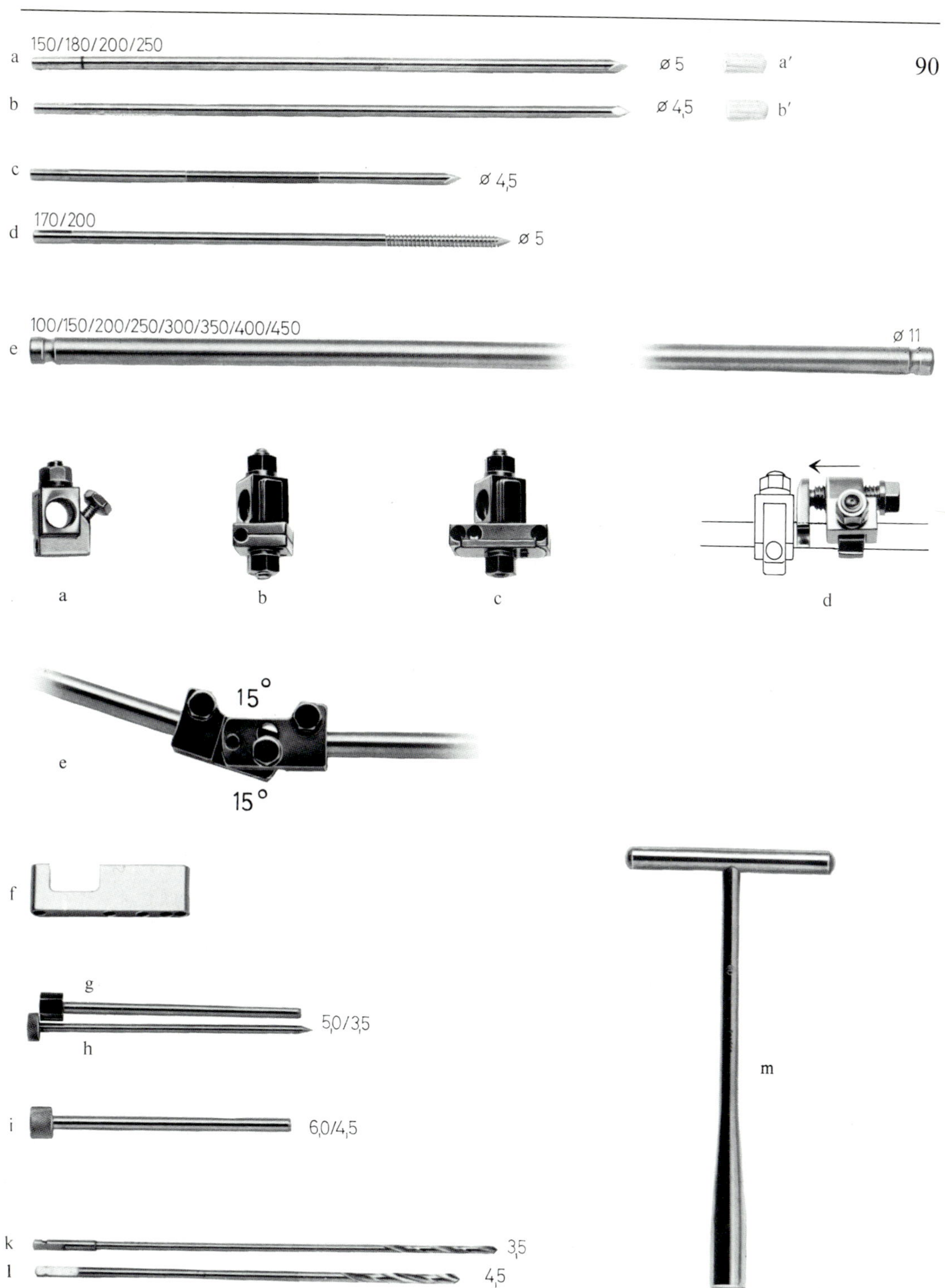

e Hinged connector which allows axial correction of 15° in each direction.

f Drill guide with holes spaced at 16, 18, 30, 32, 42, 52 and 58 mm with a total length of 67 mm.

g Drill sleeve with a 5-mm outer and 3.5-mm inner diameter, 80 and 110 mm long.

h Corresponding trochar 3.5 mm in diameter and of similar length.

i Drill sleeve with a 6-mm outer and a 4.5-mm inner diameter, 110 mm long. The 4.5-mm Steinmann nail serves as its trochar.

k, l Extra-long drill bits 3.5 and 4.5 mm with rapid coupling.

m Socket wrench SW 11 mm

2.5.2 Correction of Rotational Alignment with the External Fixator

Note: The tibial fixator provides good stability even without axial compression. Axial compression is achieved either through the compression devices or by means of the old threaded fixators which are fixed to the protruding ends of the Steinmann pins. In order to increase the stability of the fixation, two further Steinmann pins can be inserted close to the fracture. Whenever possible a lag screw should be inserted across a fracture and occasionally the external fixator can be combined with a short plate (see Figs. 262 and 264).

Fig. 91

1 *The first Steinmann pin.* Use the drill sleeve with the 5-mm outer and 3.5-mm inner diameter. Take the shorter 80-mm one with its corresponding trochar, insert it 3 cm proximal to the ankle joint in front of the lateral malleolus and at right angles to the tibia. Push it in until the tip abuts on the bone. Remove the trochar and with a 3.5-mm drill bit, drill a hole through the tibia. Insert a 5-mm, 180-mm-long Steinmann pin with the universal chuck on the T handle. In case of hard cortex, instead of the 3.5-mm drill sleeve, use the 4.5-mm drill sleeve with a 4.5-mm Steinmann pin as its trochar, and then drill the bone with the 4.5-mm drill bit.

2 *The second Steinmann pin.* This 5-mm Steinmann pin is inserted 3 cm from the knee joint, in exactly the same way as described above. Thread two standard clamps over the distal Steinmann pin with the tightening screw facing out. Use the double adjustable clamp proximally. Join the clamps with tubes of sufficient length, apply traction, reduce the fracture and tighten the clamps. Check the rotational alignment. Bend the knee to 90° and check the degree of external rotation and compare it to the uninjured limb. If present, measure the rotational malalignment. For instance, an internal rotation of 10°, instead of an external rotation of 10° corresponds to a 20° malrotation which has to be corrected.

3 If we have, for instance, a 20° rotational malalignment, we dismantle the external fixator by loosening the clamps and removing the tubes in order to correct it. We then drill a new hole 1 cm distal from the proximal Steinmann pin through the tibia. This is done with the aid of the 20° triangular angle guide. The new hole is drilled at 20° to the first Steinmann pin. This will correct the rotational malalignment.

4 *The third Steinmann pin.* Slip the drill guide over the protruding end of the distal Steinmann pin. Choose a hole and then pass the 3.5-mm drill sleeve with its trochar through the hole and through a standard clamp which has been previously threaded over tube. *Note:* Prior to the drilling of the hole for the third Steinmann pin, correct any anterior or posterior angulation. Push the drill sleeve with its trochar until it abuts against the bone, remove the trochar and with the 3.5-mm drill bit, drill a hole through the tibia. Insert a 4.5-mm Steinmann pin through the standard clamps. It is better to use the 4.5-mm rather than the 5-mm Steinmann pin, since the 4.5-mm pin is more flexible and allows for some correction of a slight malposition.

5 *The fourth Steinmann pin.* Proceed in very much the same way proximally, but do not use the drill guide. Use the double clamp as your drill guide. Loosen the double clamp and insert the drill sleeve with its trochar through the double clamp but take care not to tighten it too much. The drill sleeve has thin walls and is easily damaged. Drill the fourth hole and insert the fourth Steinmann pin. If there is a varus of valgus deformity, it can be corrected through the rotation possible through the proximal adjustable double clamp. An anterior or posterior angulation can be corrected by changing the long tubes for two shorter ones and then joining them with the hinged connector.

6 In the epiphyseal and metaphyseal region, two Steinmann pins can be inserted through simple clamps arranged in such a way that one Steinmann pin passes dorsally and one ventrally. The 30-mm holes of the drill guide serve here as a guide for their insertion.

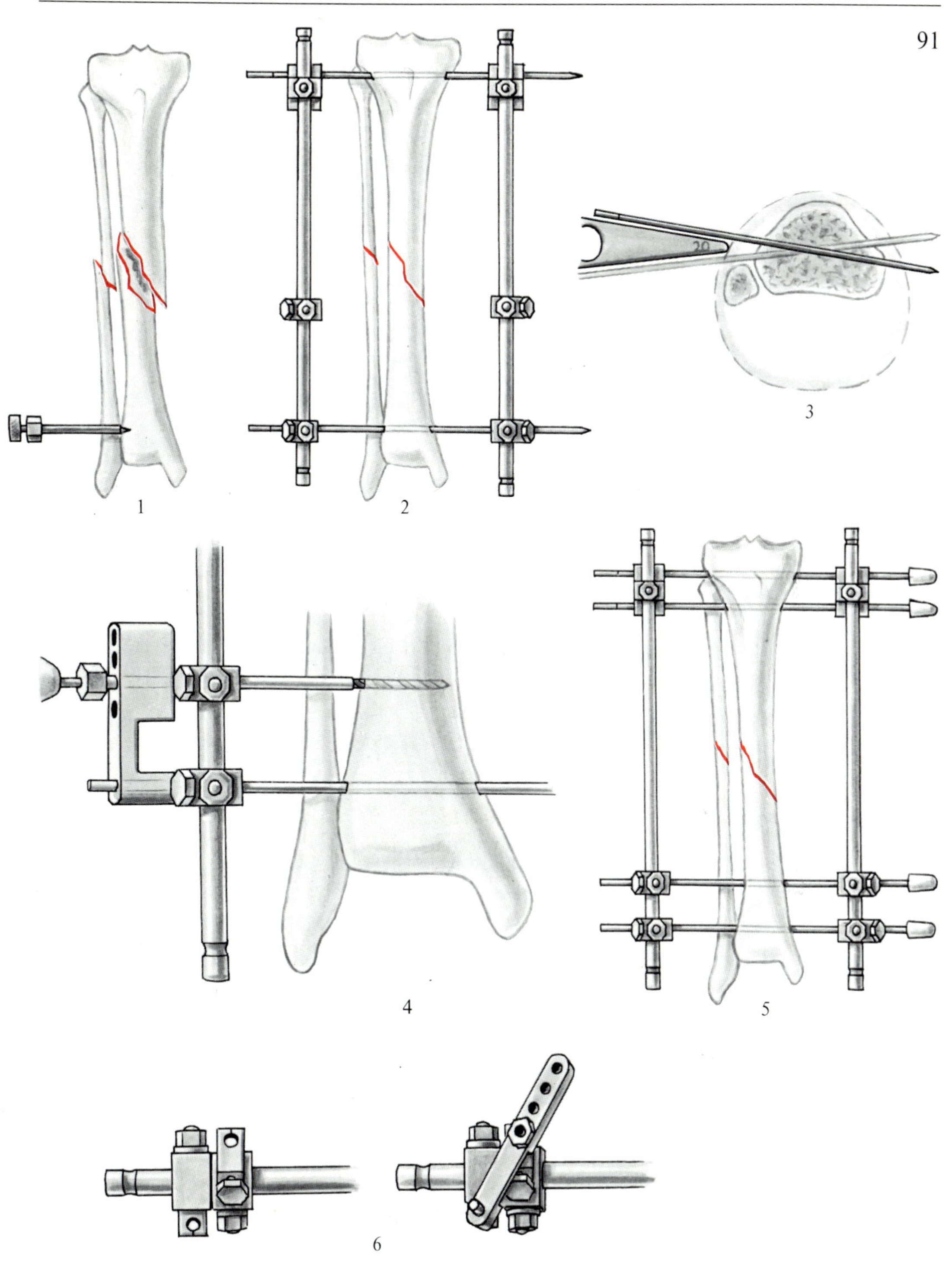
20
1
2
3
4
5
6

2.5.3 External Fixator and Compression

Absolutely stable fixation with the external fixators is possible only in cancellous bone when the distance between the Steinmann pins is no more than 3–4 cm and when there is an axial compression of over 50 kp. Cortical bone can never be stably immobilized by means of an external fixator even when supplementary configurations, such as triangulation, are employed.

Indications: These are infected pseudarthroses, arthrodeses, proximal and distal tibial osteotomies and open fractures. In treating comminuted open fractures (see p. 310) we aim at first for a reduction and for maintenance of the reduction. After the first 4 weeks, one can begin to compress the young bridging callus. At this time, if necessary, one can still carry out certain correction of varus or valgus, or anterior or posterior angulation.

The Removal of the External Fixators. When the clamps are loosened on one side the bent Steinmann pins tend to spring back. This can be not only quite painful but can also lead to a fracture of the bony bridge, if the latter is not very strong. We recommend, therefore, that, at the time of removal, the compression devices be used and the compression gradually reduced before loosening the clamps.

The old threaded external fixators proved their worth as a method for the fixation of arthrodeses and metaphyseal osteotomies. They did not provide sufficient stability when used for the fixation of open fractures.

Fig. 92 *Examples of the use of the external fixator.*

a Infected pseudarthrosis of the tibia and fibula.

b High tibial osteotomy.

c Supramalleolar osteotomy.

d Knee arthrodesis.

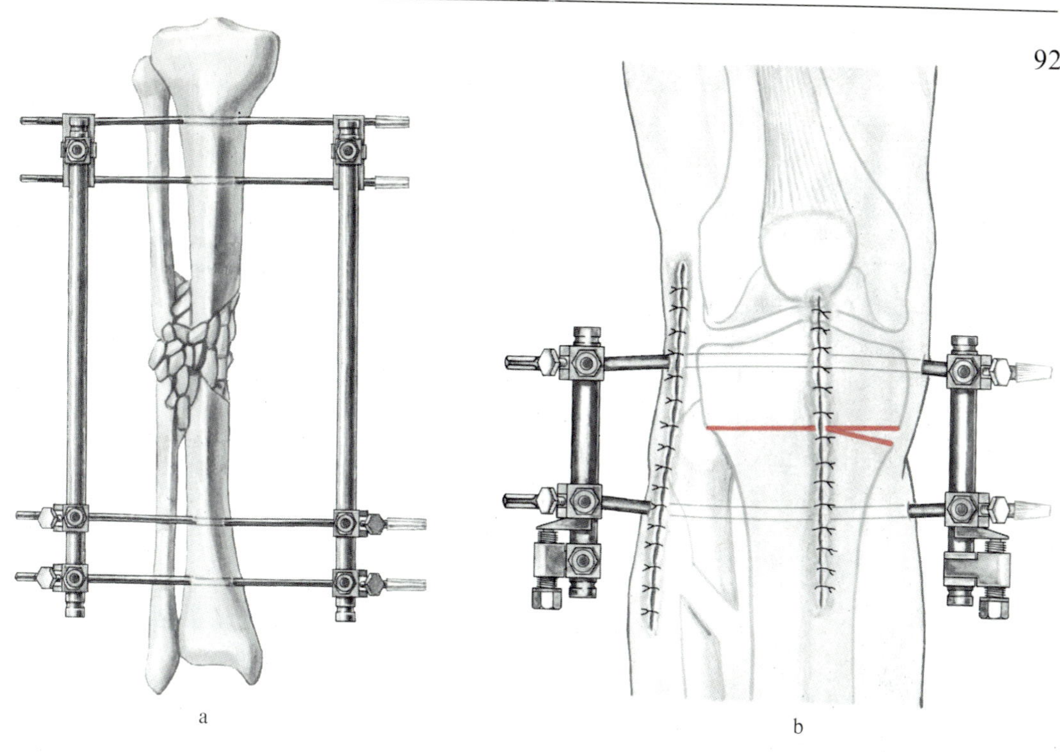

a b

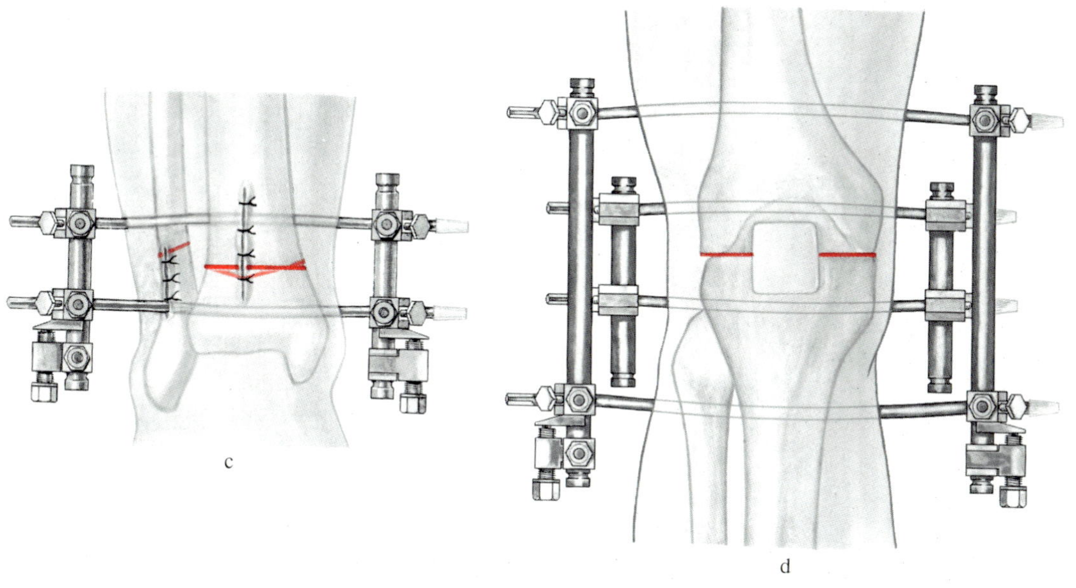

c d

2.6 Composite Internal Fixation

Composite internal fixation is the combination of plates with bone cement to secure internal fixation. It is used for the fixation of pathological fractures, such as might occur through metastatic deposits or through severely osteoporotic bone because in such bone screws and intramedullary nails do not provide adequate fixation. To carry out a composite internal fixation, the medullary canal is filled with bone cement. While the cement is still soft, the plates and screws are inserted.

We recommend composite internal fixation of osteoporotic bone only in elderly patients. In younger patients we attempt to impact the fragments and improve the stability of the fracture by changing its biomechanical configuration. An example of this is provided in Fig. 183. A valgus reduction of the proximal fragment of the femur was carried out.

This considerably reduces shear and increases the stability of the fixation.

Fig. 93 *Technique of the composite internal fixation of a pathological fracture, the result of a metastasis.*

a Use a broad and a narrow DCP.

b Reduce the fracture and with an oscillating saw, cut a 7 to 10-mm-wide, long slit in the bone. Curette out the medullary canal.

c Insert the narrow plate into the medullary canal and lay the broad plate along the cortex. Drill the holes and insert the screws through every second hole of the broad plate. In this way, the holes of the broad and narrow plate can be lined up. Remove the screws and the plates and over-drill the holes in the near cortex with a 4.5-mm drill bit.

d Insert the cement while it is still liquid into the medullary canal.

e Insert the narrow plate into the medullary canal into the soft cement mass, lay the broad plate along the lateral cortex and insert the screws.

Fig. 94 *Composite internal fixation of a comminuted inter-trochanteric fracture in the presence of severe osteoporosis.*

a Curette out the neck and head and the proximal medullary canal. Line the bones with a thin layer of cement and let it harden.

b Reduce the fracture and carry out the internal fixation, e.g. with an angled plate and screws. Remove the implants and fill the medullary canal with cement, reduce the fracture and re-insert the plate and the screws into the cement before it polymerizes.

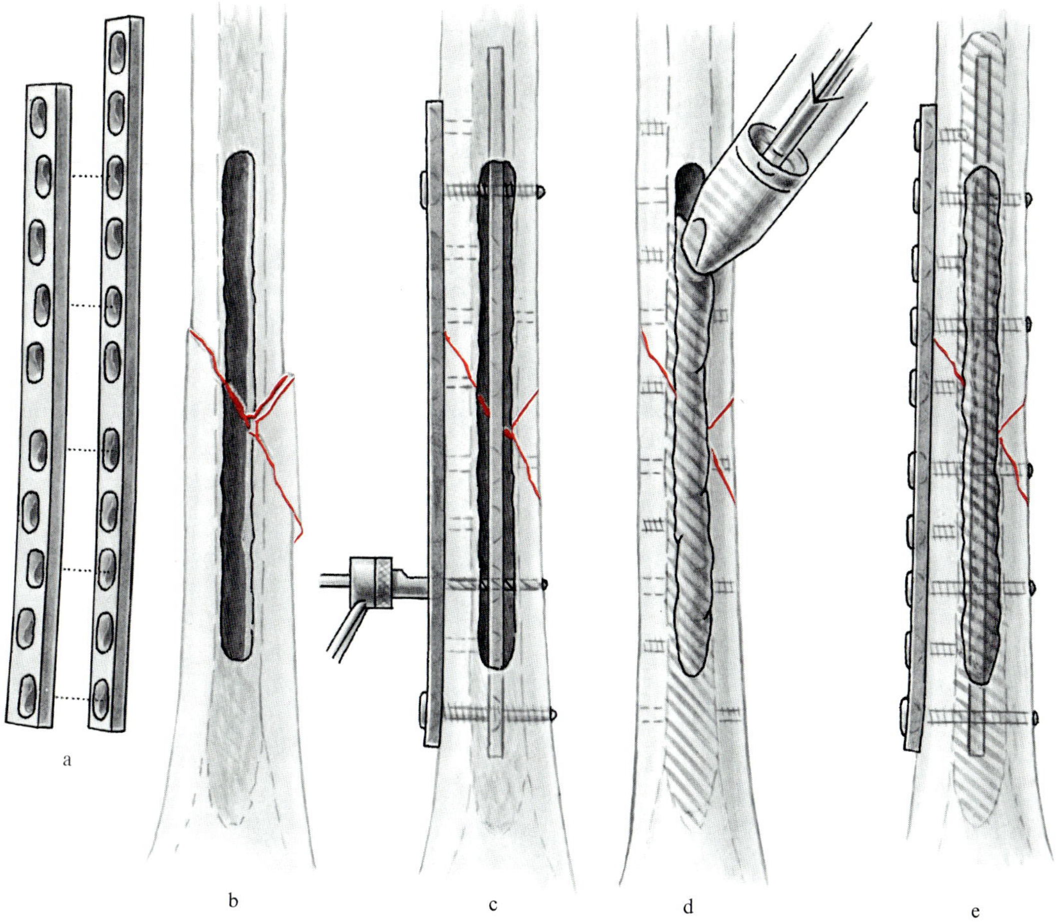

93

a b c d e

94

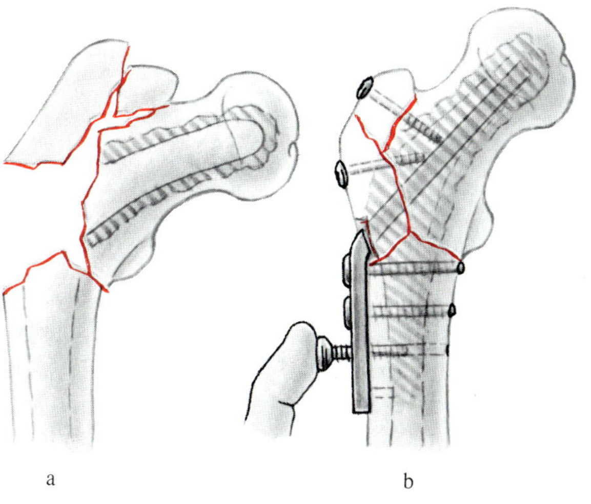

3 Preoperative, Operative and Postoperative Guidelines

3.1 Organizational Prerequisites

The asepsis of the operating rooms must be guaranteed. A high degree of professional expertise and discipline of the members of the surgical team – the doctors, the nurses and the orderlies – is far more important than technical refinements. The best index of asepsis is the infection rate, which should not exceed 2% for elective cases and closed fractures.

In order to be sure of success it is also important to have a sufficient supply of all the necessary implants and instruments – not only the specific instruments for internal fixation (see p. 20) but also all the general surgical instruments. Soft tissues must be handled with the utmost care and delicate instruments are required. The state of the instruments both for bone surgery and general surgery should always be checked. Prior to a procedure, the sharpness of the chisels, drill bits and taps, and the faultless functioning of drills, oscillating saws etc., must be ensured.

3.2 Priorities in the Assessment and Treatment of Injuries

In order of importance, priorities are: blood loss, head injuries, thoracic injuries and abdominal injuries. Of great importance in the treatment of the poly-traumatized patient is the prophylaxis of the so-called trauma lung (wet lung or shock lung). A *patient with multiple injuries* should be considered to have respiratory insufficiency until blood gases have proven the contrary. Our experience has shown that in patients with multiple injuries, immediate stabilization of long bones greatly facilitates and improves intensive care. RÜEDI and WOLFF have shown that immediate internal fixation does not increase in any way the danger of embolization, particularly of fat emboli.

The *age of the patient* must be taken into account when deciding on the indications for surgery. Internal fixation is rarely an emergency procedure in a child, but quite often it is in the adult. Bed rest for the elderly patient can often be equated with a death sentence. For them, internal fixation is quite often a life-saving measure.

We would like to direct the reader once again to the Introduction where we discuss *indications* for internal fixation. In subsequent chapters while dealing with specific fractures, we will indicate briefly the indications for internal fixation. We would like to take this opportunity, however, to point out that internal fixation is a difficult and demanding undertaking and should be carried out only by adequately trained surgeons who have the necessary equipment and the proper environment.

3.3 Timing of Surgery

Internal fixation can be carried out either prior to the development of swelling or when the swelling has subsided. If an open reduction and internal fixation is indicated, there are many reasons for carrying it out as an emergency procedure. Some of these reasons are medical, such as rendering the limb painless through stable fixation and in this way preventing the so-called fracture disease. Other reasons are psychological. The injured patient who is free from pain and able to move the limb freely soon after the trauma, is much less likely to suffer from post-traumatic neuroses. Certain fractures have their specific temporal indications. Patients with malleolar fractures rapidly develop osteoporosis of the malleoli. This is particularly true in elderly patients. This makes a delayed internal fixation considerably more difficult. We consider fractures of the femoral neck in children, teenagers and the adult as surgical emergencies, if avascular necrosis and subsequent necessity for prosthetic replacement of the femoral head is to be avoided. In very difficult fractures such as the intra-articular fractures of the proximal and distal tibia and the central fracture dislocation of the hip, we delay surgery till we have assembled all the necessary help and equipment. In hospitals with good aseptic technique, patients who have delayed internal fixation appear to have no increase in the incidence of sepsis when compared with the group operated on as an emergency. In treating comminuted fractures of the femur, there are sound arguments for preliminary traction and delayed surgery. The hyperaemia of the inflammatory process leads to a much better blood supply to the soft parts and bony fragments. With delay there is also an increased danger of thrombophlebitis. Our own preference is to carry out early surgery for fractures of the femur.

3.4 General Guidelines for Operative Procedures

3.4.1 Planning

The fracture should be carefully analyzed with the aid of an intact model bone or a preoperative drawing and a careful plan of the internal fixation should be formulated (see Fig. 60). The operating room personnel should have careful instructions about which instruments and implants to prepare for the procedure. Careful preoperative planning removes much of the danger of rushing about during an operation.

3.4.2 Preparation of the Operative Field

Extensive shaving of the patient is unnecessary and can be dangerous if it is done on the day before surgery. It is enough, just prior to surgery, to shave about a hand's breadth of the area where the incision is to be made. In elective cases, we recommend Phisohex skin preparation to begin on the day before surgery and be repeated twice, 8 hours apart.

Surgery on a limb should be carried out under pneumatic tourniquet control. The pneumatic tourniquet must have a gauge. It can be left inflated on an arm or leg for 2 hours, as long as there are no acute or chronic circulatory problems preoperatively. We use a transparent plastic sheet to cover the skin.

3.4.3 The Operation

In diaphyseal regions of bone, the skin incisions are made parallel to Langer's lines. As a rule straight incisions are made along lines which have fixed points of reference. Around joints, the pattern of blood supply and the need to gain adequate exposure, lead most often to longitudinal cuts which cross Langer's lines. We irrigate with Ringer's solution every 15 min throughout the procedure. We do this for two reasons. First, the irrigation physically removes bacterial contamination both from the air and from the operating team. Secondly it moistens and prevents the drying out of tissues. We prefer suction rather than sponging away blood and tissue fluids. A sucker with progressively greater diameter of tubing will overcome the annoyance of frequent plugging.

Once the soft-tissue dissection is completed, the exposure of the fracture itself is carried out with the aid of the different bone levers. Because they are levers, one must remember that they multiply the force and, therefore, must be handled with care and have their position changed often to prevent tissue necrosis from pressure. Extensive periosteal stripping is to be avoided. Good exposure of the fracture can usually be achieved with periosteal stripping of only 2–3 mm along the fracture edges.

First, the whole fracture is exposed and the fracture surface is carefully cleansed of clot and tissue. The configuration of the fracture is then carefully analyzed and the position of the gliding holes and thread holes of the lag screws carefully planned (Figs. 19, 20, 37) as is the reduction. The reduction must be carried out atraumatically. When the occasion arises, particularly in the femur, the distractor should be used (Figs. 86 and 87).

The accuracy of the reduction is checked visually and, in parts hidden from sight, with the small sharp hook. The reduction is maintained with the different reduction clamps (Fig. 10) and with the self-centring, bone-holding clamps. Whenever possible, circumferential stripping of the bone should be avoided. In certain comminuted fractures temporary cerclage wiring can be most useful (see Fig. 97).

Fig. 95 *Covering the skin of the calf with a plastic drape.*

Fig. 96 *The AO reduction clamp* makes it possible to grip the anterior and posterior edge of the tibia without extensive exposure. This type of temporary fixation should be used whenever possible because it keeps devitalization of the bone to a minimum.

Fig. 97 Occasionally, in dealing with certain torsional butterflies associated with spiral fractures of the tibia, cerclage wiring may be necessary, but beware of devitalization of the butterfly.

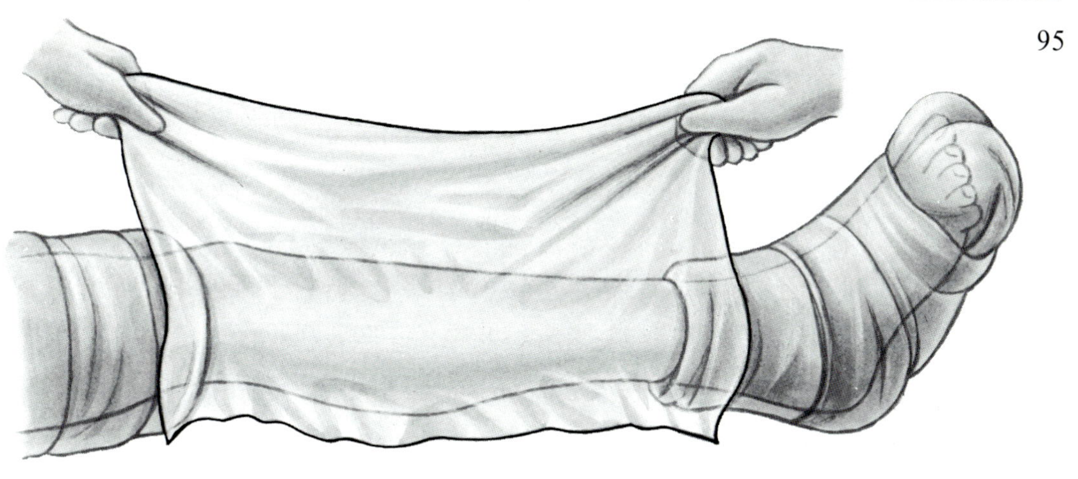
95

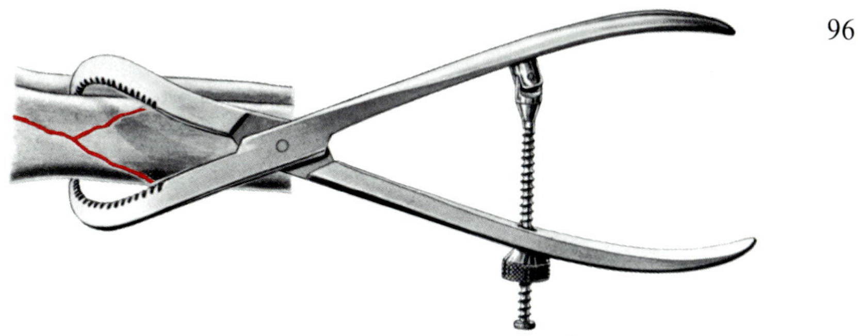
96

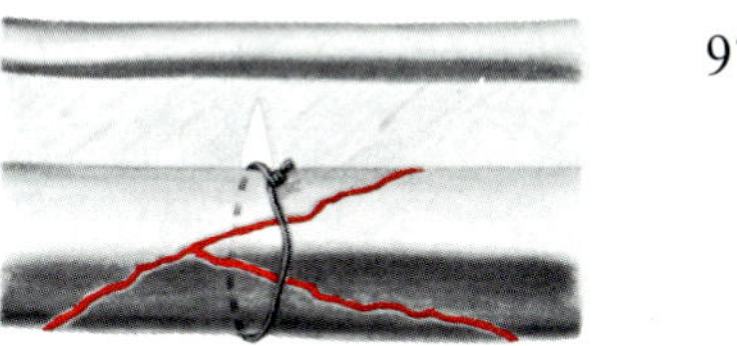
97

3.4.4 Autologous Cancellous Bone Grafts

The fate of a plate used in internal fixation is often determined by the state of the cortex opposite the plate. The smallest primary or secondary defect of the opposite cortex leads to cyclic bending of the plate and movement of the fracture. This results in resorption and finally in a fatigue fracture of the plate. Therefore, whenever there is a defect in the cortex opposite the plate, it must be bone grafted. Furthermore, we feel that whenever a femur is plated, it must be bone grafted. We also feel that bone grafting should be carried out whenever the cortex opposite the plate is extensively denuded of soft tissue.

The various donor sites are shown in Fig. 98. Cortico-cancellous grafts are best taken from the wings of the ilium. The greater trochanter is an excellent source of pure cancellous bone. For bone grafting of the opposite cortex, we can use either cancellous or cortico-cancellous bone (taken from the inner table). Whenever the bone graft is used for buttressing we prefer larger cortico-cancellous chips, which are then inserted into the defect.

Fig. 98 *Donor areas of bone for grafting.*

a The most common donor site is the inner table of the anterior wing of the ilium. The skin incision is made either 2 cm laterally or medially to the iliac crest. A curved gouge is used. The long, parallel cortico-cancellous strips are usually cut up into smaller pieces approximately 15 mm in length and 5 mm in width. Experience has shown that these are incorporated more quickly than the long cortico-cancellous strips. Gel-foam is used for haemostasis and is inserted into the defect between the bone and the overlying iliacus muscle.

b When the patient is prone, we obtain bone from the outer table of the posterior portion of the wing of the ilium. The skin incision is made laterally to the iliac crest. After incision of the fascia and retraction of the muscle, expose the bone. A large number of stout, cortico-cancellous strips of bone can be taken.

c The greater trochanter is an excellent source of pure cancellous bone.

Fig. 99 A 1-cm-wide, straight gouge is most useful for the taking of bone grafts.

98

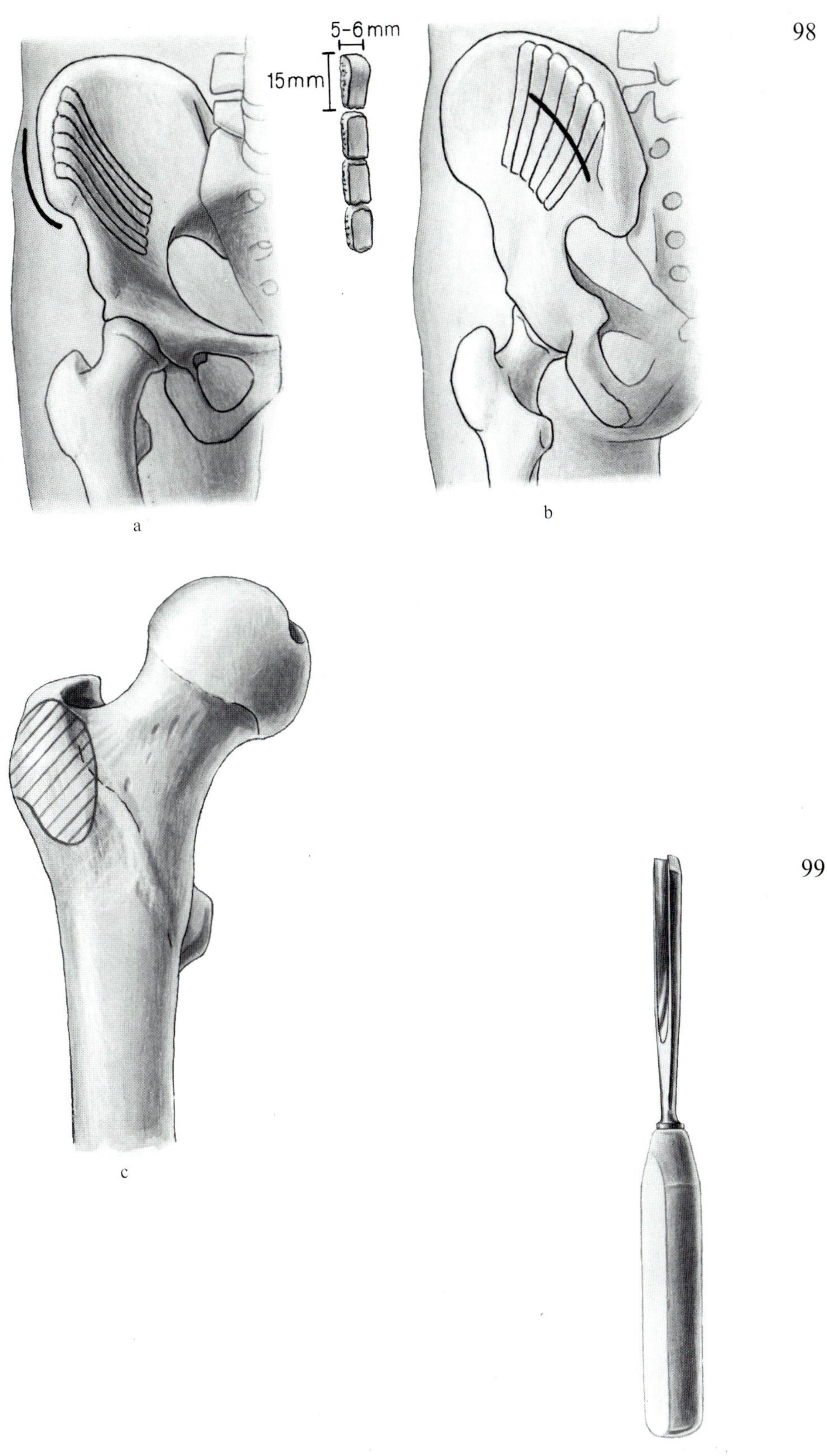

a

b

c

99

3.4.5 Wound Closure

The insertion of the internal fixation does not bring the operation to a close. The handling of the soft tissues and the wound closure will often decide the success of the internal fixation. The fascia is closed loosely with absorbable suture (Dexon, Vicryl). Closed fascial compartments such as the anterior tibial compartment should be left open. Once the deep fascia is loosely re-approximated, no further sutures are inserted, particularly not into the subcutaneous tissue planes. We use suction drainage to prevent the formation of haematomas or seromas. The Redon drains should be perforated and be 3–4 mm in diameter. Usually only three to five perforations remain open and permit the suction drainage to function. The "Ulmer drains" have holes of various sizes which distribute the suction effect over a greater distance.

Skin closure is carried out with vertical mattress sutures. This ensures a good approximation of the dermis. It gives a tight closure and a good cosmetic result. These sutures can be inserted, interrupted or continuous. For the sutures we use 4–0 nylon or rarely 3–0 nylon.

Releasing incisions should be used rarely and never when the skin and soft tissues have been contused. The classic releasing incision in the leg is a posterior mid-line incision with extensive mobilization of the medial and lateral skin flaps. Skin grafting of the posterior defect is usually unnecessary. Wound contracture and a few "steri" strips lead to rapid closure.

3.4.6 Release of the Tourniquet

The tourniquet is usually released after closure of the skin and opening of the suction drains. In cases with extensive soft-tissue dissection, where one is not certain of haemostasis, it is best to release the tourniquet prior to closure so that a meticulous haemostasis can be obtained.

Fig. 100

a The Donati suture.

b The Donati suture as modified by Allgöwer. It goes through the dermis only on one side and in this way protects the skin edge of a slightly devitalized flap.

Fig. 101

a Modified Redon suction drainage with disposable container and outer spring.

b The "Ulmer drain" of Burri which has holes of gradually diminishing diameters.

Fig. 102 *The double U splint*. This splint is particularly useful after fractures of the tibia and after malleolar fractures.

a Two plaster slabs are applied with the knee and ankle at 90° (if possible with the patient lying on his stomach).

b The double U splint protects the heel and allows for early active dorsiflexion of the ankle.

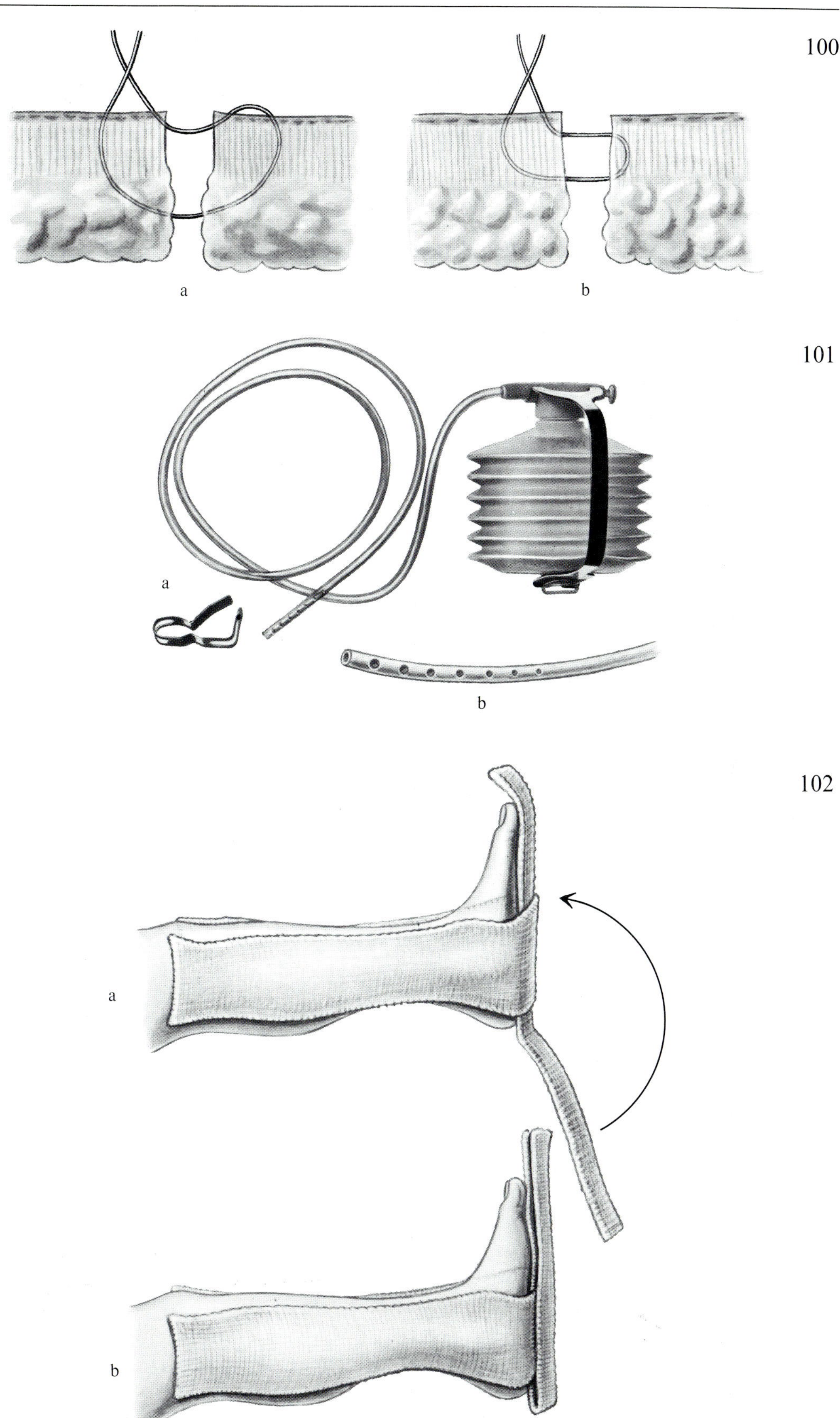

100
a
b
101
a
b
102
a
b

3.5 Postoperative Positioning and Treatment

Postoperative positioning and treatment is designed to overcome four dangers. Postoperative *haematoma* is prevented by the careful insertion and subsequent supervision of suction drainage. *Oedema* is prevented by elevation of the limb. *The moist chamber of the dressing* is overcome by removal of the dressing and leaving the wound open 24–48 hours after surgery. *Fracture disease* is prevented by early active mobilization of the limb.

These aims can be achieved only if the internal fixation is so stable that external plaster fixation becomes superfluous. The standard postoperative positioning of the limbs is shown in Fig. 103a, b and c. Note the absence of plaster. We encourage the patient to begin, within the first 24 hours, active mobilization of foot, knee and hip under supervision. The patient should also perform light resistive exercises with the injured limb to preserve as much as possible the normal pattern of activity. Physiotherapists can be of considerable help in achieving these goals.

There are Certain Exceptions to the Cast-free Postoperative Treatment. Fractures involving the ankle joint are splinted (U splint) at 90° to prevent equinus. Active dorsiflexion is encouraged after the first 24 hours. In the arm we use plaster fixation to aid with elevation but in such a way that early active mobilization is possible. Occasionally, plaster fixation must be used to protect a very difficult internal fixation or the internal fixation in uncooperative and unreliable patients. The joints, however, must be mobilized before the limb is put in plaster.

Fig. 103 *The three standard postoperative positions.*

a Fractures of the tibia are elevated on a Braun frame which is padded with sponge rubber. The knee is flexed to 45° and the ankle kept at 90°. The splint is so arranged that the foot is higher than the knee and the sole rests firmly against the foot rest. The limb is loosely tied to the splint.

b Fractures of the middle and distal femur are positioned in this way for the first 4–6 days. The limb is positioned on a right-angled splint, padded with sponge rubber. The knee and foot are kept at 90°. The limb is loosely tied to the splint.

c Positioning of the limb after fractures of the forearm or hand. The elbow is bent to 80° or 90° and the limb is suspended from an intravenous pole by means of a sling. The hand is kept in supination.

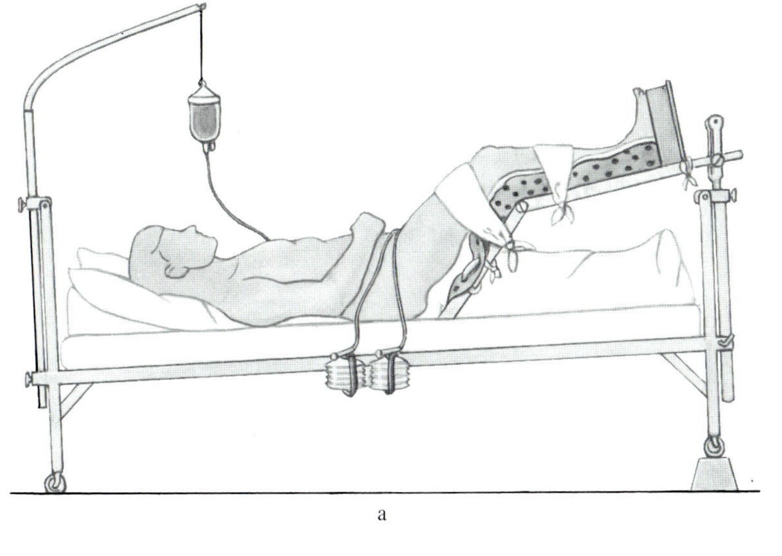
a

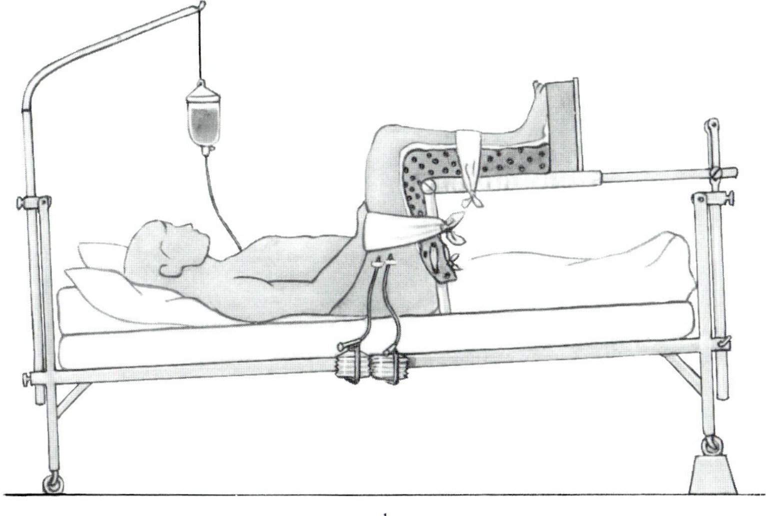
b

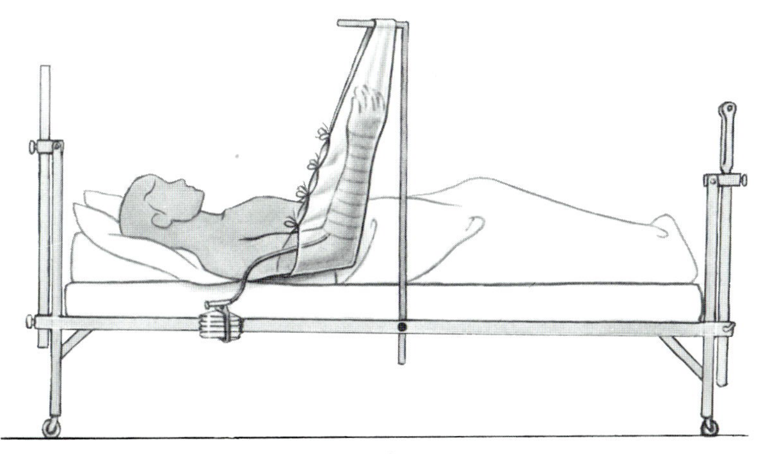
c

3.5.1 Guidelines for Weight Bearing

After the first 4–5 days, when it has been determined that wound healing is progressing satisfactorily, the patient should begin active mobilization of the arm or, if he has a fracture of the leg, he should begin to learn partial weight bearing. A patient with a fractured leg, whose fracture is stably fixed, may begin with weight bearing of 10–20 kg. The patient should practise this every day with a step-on scale or more simply with his own bathroom scale. Complete non-weight bearing with crutches with the hip and knee flexed leads to circulatory stasis and rapid disuse osteoporosis. As a rule, the patient should use two crutches for 2–3 months. If there is radiological evidence that bone healing is progressing satisfactorily (see p. 146), the patient should progress to one crutch which is used on the side of the fracture for a further month or two.

3.5.2 Secondary Plaster Fixation

A plaster applied immediately after surgery leads to permanent joint stiffness. Once the full range of movement of the adjacent joints has been achieved, plaster fixation, as a rule, will not lead to any permanent loss of movement. For fractures of the tibia in uncooperative and unreliable patients, we use a lightly padded, long leg plaster and for malleolar fractures a lightly padded, below-knee plaster.

3.5.3 Education of the Patient

The danger of stable internal fixation is that it is completely pain free and that it gives the patient the feeling after the first 8–10 days that his limb is perfectly normal and that it can be used without any restrictions. In order to achieve a full active mobilization of the limb and partial weight bearing, some patients must be encouraged to increase their level of activity. Others on the other hand must be slowed down. All patients, however, must learn the local "danger signals" which may appear during the period of bone consolidation after an internal fixation. These are *swelling, redness, local heat* and *pain*. The patient must know that swelling means more elevation. Elevation does not mean sitting in a chair with his foot propped up on a stool! The patient must be told that redness and pain signify either instability or infection and that, should these develop, he must seek immediate medical attention.

3.5.4 Prophylactic Antibiotics

For many years, the AO has been against the use of prophylactic antibiotics in fracture treatment because we feel that it leads to the development of antibiotic-resistant bacteria. Despite this, our infection rate is below 2%. Certain misconceptions have arisen about the use of prophylactic antibiotics. Earlier on, the discussion usually centred on whether these should be used during or immediately after surgery. BURKE from Boston, in his animal experiments, was able to show that in order to achieve true antibiotic prophylaxis to protect against intra-operative contamination, the bacteria at the time of contamination must come in contact with a certain concentration of antibiotic in the blood and in the tissues. Therefore, in order to speak of a true antibiotic prophylaxis, the patient must receive the antibiotic intravenously, preferably prior to surgery or at least at the start of the operation. This is the only way that one can ensure an adequate antibiotic concentration in the blood and in the tissues at the time of surgery.

In the light of these facts, our position is as follows: in general we are opposed to antibiotic prophylaxis, but there are certain exceptions. We use antibiotic prophylaxis if we must reoperate, particularly if there has been some problem with wound healing and if there is some question of possible sepsis, or if we are about to carry out a difficult internal fixation which will take a long time, such as a difficult central fracture dislocation of the hip. If we choose to carry out antibiotic prophylaxis, then our drugs of choice are the cephalosporens. We begin with the prophylaxis on the day before surgery or intravenously at the time of induction. During the operation, the patient receives 4 grams antibiotic every 2 hours intravenously. We continue with antibiotic prophylaxis for the first 48 hours and then stop.

3.5.5 Prophylaxis of Thromboembolism

There are three drugs we can choose for prophylaxis of thromboembolism. These are dicumerol, heparin and dextran.

Dicumerol has the advantage of relatively easy regulation, dose administration and length of activity, but it is not suitable for prophylaxis of early embolic phenomena. Dicumerol is used in patients who require anticoagulant therapy longer than the first few postoperative days. These are patients who require prolonged bedrest and who are over 20 years of age.

We have found low-dose heparin and dextran to be about equally effective in the prevention of thromboembolic complications in musculo-skeletal surgery. Heparin is administered subcutaneously twice daily in doses of 3000–5000 units. It is given on the day of surgery and is continued until the patient is completely mobilized. Heparin injections, however, are unpleasant for the patient and occasionally have caused bleeding problems because the absorption rate from the subcutaneous depot is difficult to control.

We have found dextran to be the simplest prophylactic agent to use. We use 500 ml of 60,000 molecular weight dextran. We begin with the dextran intravenously right after the induction of the anaesthetic. It is given again on the day after surgery as well as on the third postoperative day. The only drawback to dextran is the occasional anaphylactoid reaction (administer hydrocortisone 1 g i.v. and nor-adrenalin). We have seen anaphylactoid reactions very rarely under anaesthesia and rarely after the first dose.

3.5.6 Radiological Follow-Up of Bone Healing

A radiological examination at 2 months is useless unless there are symptoms which suggest problems. The most important X-ray is taken at 4 months because by that time most fractures should have united. Those which have not will clearly show the beginnings of complications. In the usual course of events, if healing is progressing normally, the fracture lines begin to vanish from the eighth week onwards and disappear completely 3–4 months after surgery.

The following changes are radiological warning signals: *widening of the fracture gap* and the appearance of a cloudy, poorly defined callus, the so-called *irritation callus*. Widening of the fracture gap means that resorption at the fracture is outstripping bone formation. Irritation callus means mechanical instability. Both mean that loading must be reduced. These radiological signs usually accompany the clinical signs of pain and redness. They demand elevation and a temporary and complete relief of loading. This usually results over the next 4–6 weeks in consolidation of the fracture line and in the change of the irritation callus to a clearly defined *fixation callus*. If one is able to catch the warning signals and intervene in time, the fracture will usually have united by the end of four months. Histologically, however, remodelling of the bone goes on for the next year or two. If the fracture has not united within 4 months, we speak of *delayed union*. After 8 months, we speak of a reactive or atrophic *non-union* or *pseudarthrosis* (see p. 335). If delayed union is diagnosed during the fourth month and properly treated, pseudarthroses can be completely prevented. Occasionally this requires a second operation (see p. 334).

Fig. 104 *Normal course of healing of a tibial fracture fixed with lag screws and a neutralization plate.* After 17 weeks the fracture line is barely visible. At 55 weeks the bone architecture appears normal. Note that healing is taking place without any radiologically visible callus.

Fig. 105 *The appearance of cloudy irritation callus as a warning signal.*

a At 5 weeks following internal fixation there is no callus.

b Six weeks later we see the appearance of a cloudy, poorly defined irritation callus, the result of mechanical instability at the level of the fracture.

c Six weeks later after unloading and intermittent elevation the irritation callus has changed into a fixation callus.

d Seven months later the fracture is united and the callus is undergoing remodelling.

104

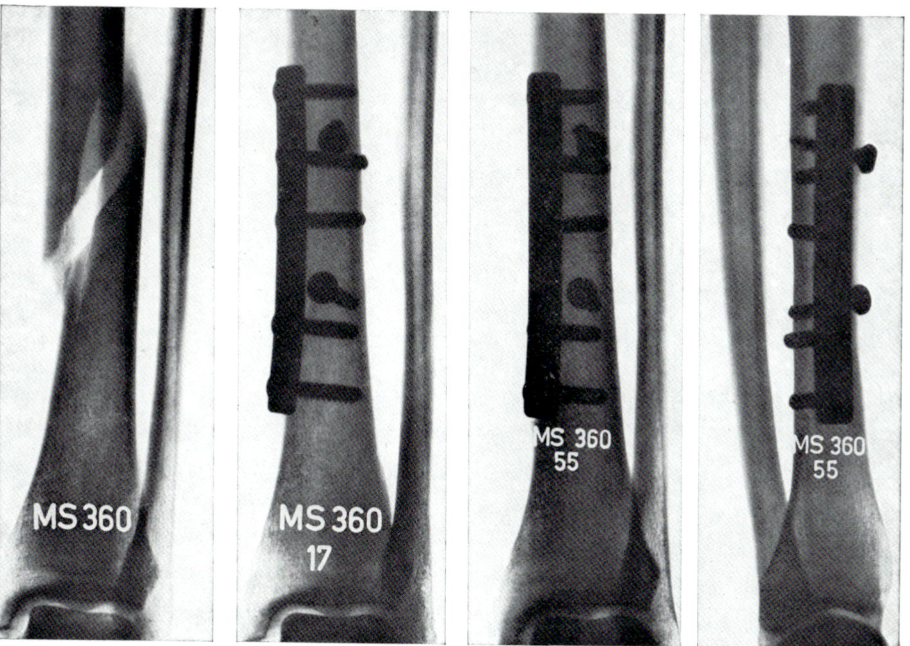

105

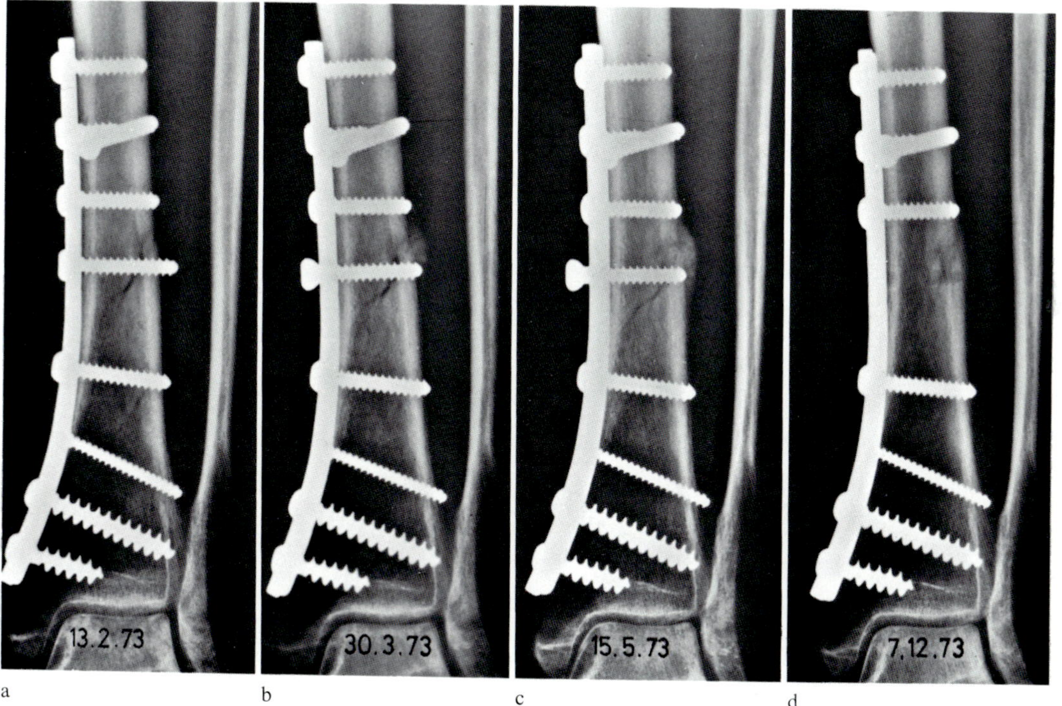

a b c d

3.6 Removal of Implants

We believe that implants should be removed because the bone under an implant never becomes biomechanically normal. This is because of the difference in the modulus of elasticity between bone and implants. There are exceptions to this rule of implant removal. Implants in non-weight-bearing bones, particularly the humerus as well as implants around the hip in elderly patients are left undisturbed. Single screws in metaphyses can also be left alone.

3.6.1 Timing of Implant Removal

Implants should never be removed before the architecture of the bone has become radiologically normal. The pre-stress of the implant slowly dissipates as bone healing and bone remodelling takes place (see p. 12). This results in some loosening. This is advantageous because it results in physiological loading of the bone and with it a return to normal architecture. This process, however, requires time.

Single screws can be removed from metaphyseal areas after 3–6 months. Minimal removal time for the combination of screws and plates from diaphyses is as follows: Tibia 1 year, femur 2 years, forearm and humerus 1.5–2 years. Intra-medullary nails are never removed under 2 years.

3.6.2 Technique of Metal Removal

The cosmetic result is better if the whole incision is reopened. The edges of bone heaped-up around the plate should be left untouched (Fig. 107b). Soft tissues must be handled with care and suction drainage should be used to prevent haematoma. The patient can usually be discharged the day after removal of the suction drains. Screws alone from metaphyses can be removed on an out-patient basis.

3.6.3 Safeguards After Metal Removal

A diaphysis loses 50% of its torsional resistance from the mere insertion or removal of a single screw. In animal experimentation, this reduction in strength lasts 1–2 months. The exact duration in man is unknown. We recommend normal use of the limb but do not allow any athletic activities for the first 3 months and no extreme levels of athletic activities for at least 4 months after metal removal.

Fig. 106

a The uniform bowing of a normal limb under eccentric loading.

b A partial stiffening of a diaphysis, the result of plating.

Fig. 107

a Severe osteoporosis of the cortex (cancellization), the result of *stress protection* in double plating.

b The reaction of cortex to plating. Porosis of bone immediately under plate (this disappears with time) and the development of ridges on each side of the plate (*arrows*). These ridges strengthen the bone considerably and should never be chiselled away at the time of implant removal.

106

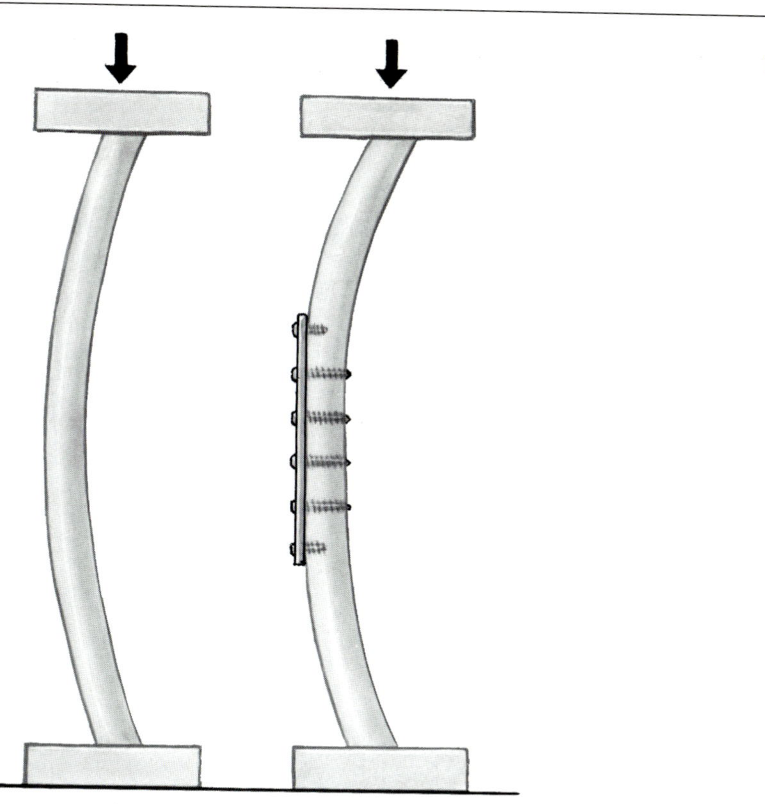

107

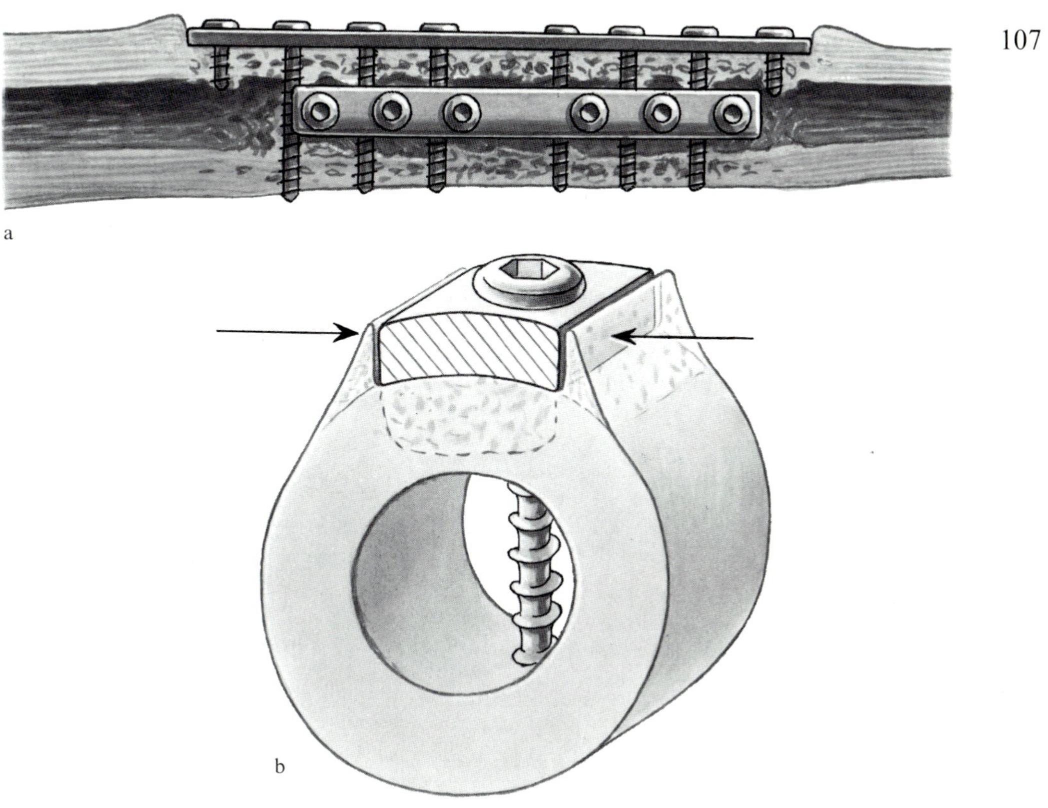

Removal of AO Intra-medullary Nails

There are certain details to be observed when removing the AO femoral and tibial intra-medullary nails (see Fig. 108).

Note: For removal of nails of other manufacturers (e.g. original Küntscher nail) we have special hooks in the standard instrumentation cassette for intra-medullary nailing.

If a nail should break inside the medullary canal, its distal end can be removed with the special extraction hook, but prior to this the intra-medullary canal has to be enlarged by reaming right down to the level of the nail fracture.

Fig. 108

a The tibial nail is removed without the interposition of the curved driving piece between the weight guide and the conical bolt.

b With the curved driving piece the force would be directed against the tibial plateau and not to the point of exit.

c The broad contact surface between the conical bolt and the threaded portion of the nail leads to a very firm grip between the two and an even, centrally directed force, at the time of removal. In contradistinction, a hook in an oval hole gives rise to an uneven force distribution and tearing of the oval hole if the nail is firmly seated in bone.

d With time the HERZOG curve of the tibial nail becomes surrounded by bone. This presents a considerable resistance to nail removal and often leads to considerable deformity of the tibial nail at the time of its removal, so that it often looks like a banana when it comes out.

e The extraction of the femoral nail.

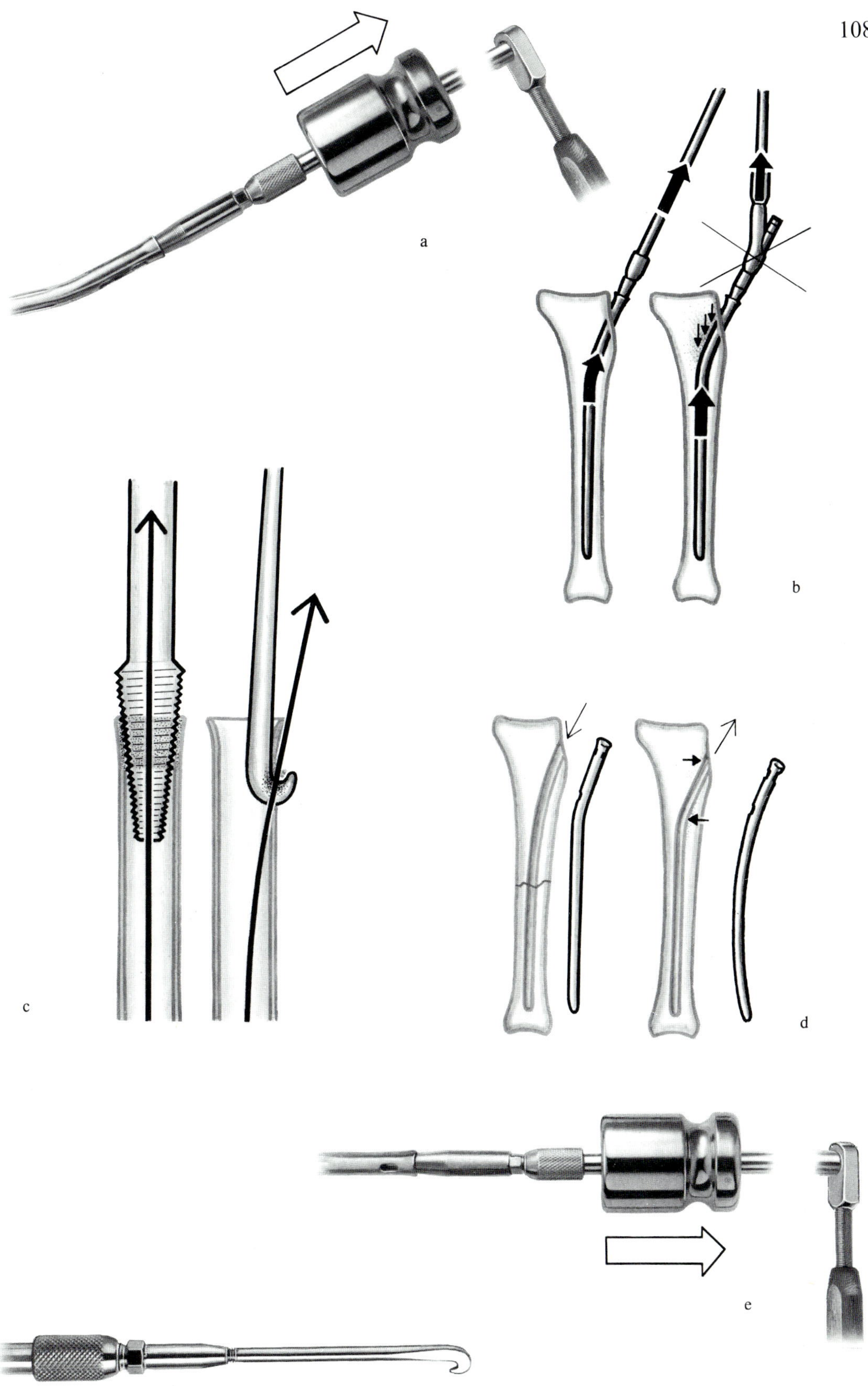
a
b
c
d
e

3.7 Postoperative Complications

3.7.1 Haematomas

After 12–18 hours a haematoma becomes an excellent culture medium for bacteria. Therefore, haematomas should be removed. If the haematoma is liquid, it can be aspirated. Usually, however, it is best to take the patient to the operating room, reopen the wound in its entirety, carefully evacuate every bit of the haematoma, copiously irrigate the wound and then carry out, once again, a primary wound closure. Cultures are taken of course during the procedure. Prior to closure, suction drainage should be inserted to prevent recurrence.

3.7.2 Postoperative Pain

Stable internal fixation results usually after the first 24 hours in a painless extremity. Postoperative pain, therefore, should be a warning signal and should suggest to the surgeon that he is dealing either with a partial muscle necrosis or an infection. One must watch the limb very closely and check frequently its motor, sensory, and circulatory status. Rise in pressure in a closed, anterior or posterior compartment of the tibia can result rapidly in ischaemic muscle necrosis if the rise in pressure is not immediately relieved by a fasciotomy as soon as the symptoms appear. The classic complication is the anterior tibial compartment syndrome which follows terribly severe ischaemic pain in the leg. Irreversible muscle damage occurs after 4–6 hours.

3.7.3 Infections

It is most important that early infection, which occurs in the first few days after internal fixation be properly recognized and treated. Swelling, redness and pain must be recognized as signs of infection and be aggressively treated. The solution is debridement. Debridement means a thorough reopening of the operative wound and painstaking removal of all potentially infected haematoma with excision of all necrotic tissue, particularly subcutaneous fat, fascia and muscle. After a thorough debridement, there are two possibilities. If the fixation is still absolutely rigid and stable, then the implants are left in (tighten all screws!). Set up an irrigation system for a few days. If the implant is loose and not providing stability, then it must be removed because it will act only as a foreign body and propagate the infection. A new stabilization of the fracture, however, must be performed. This is usually carried out by means of the external fixator. In certain cases, as for instance in the femur, the restabilization may well be carried out with an even bigger implant. We recommend also in these cases primary closure of the wound after the setting up of a suction irrigation system.

The Suction Irrigation System. This is carried out with Ringer's solution containing a suitable antibiotic. The antibiotics which we use most of the time are those which cannot be used systemically, such as neomycin, bacitracin and polymixin (at present we use either polybactrin or nebacetin as the antibiotic of choice in our irrigation systems). The first few hours of a suction irrigation system usually determine its success or failure. Therefore, one must watch closely that the system is functioning and that the outflow equals the inflow. A plugged outflow is dangerous because it endangers the viability of tissues.

We carry out irrigation with the antibiotic solution for 4–5 days. Thereafter, we irrigate with pure Ringer's solution, checking for bacterial growth in the outflow. Irrigation for more than 10 days is rarely if ever indicated.

3.7.4 Refractures

We define a refracture as a fracture which occurs in the region of previous fracture of a normal bone which appears to have undergone sound union both clinically and radiologically. The early failure of an implant or plate fracture cannot be included in this definition, since in these the fracture had probably never united in the first place. However, we include under this definition fractures which occur near implants through bone which has been weakened by remodelling.

Most refractures result from premature implant removal or improper implant removal (the chiselling away or the breaking of the ridge around plates, see Fig. 107b). Refractures occur also, either through the previous fracture gap or through remodelled bones (sclerotic, osteoporotic or deformed) a certain distance from the initial fracture. Occasionally, a refracture may occur in the presence of an implant such as a cerclage wire, a screw or an intra-medullary nail.

In the first instance, that is, after premature implant removal, trauma is usually trivial; whereas in the second the required trauma is often considerable and the refracture may occur as late as 2 years after plate removal.

In our documented series, we have encountered an incidence of 1%–1.5% of refractures after implant removal.

There is particular danger of refracture after double plating of diaphyses. We feel, therefore, that double plating should be avoided whenever possible. If it is unavoidable, because of bone loss or because of very strong deforming forces (subtrochanteric) then the metal removal should be done in two stages, removing one plate at a time 6 months apart, and bone grafting at the time of removal of the first plate.

A particular type of refracture is the transverse fracture occurring at the junction of the relatively rigid plated segment, and that of the normally elastic diaphysis. For this reason, we aim at a gradual transition from the rigid to the less rigid segment of bone by inserting one or two short screws which go through only one cortex.

3.8 Plate Fractures

Pohler and Straumann carried out an exhaustive analysis of 270 cases of plate fractures. All fractures of AO implants were *fatigue fractures* which were the result of an improper internal fixation which did not observe biomechanical rules of stability. Usually, instability was the result of a deficiency of the opposite cortex. This gave rise to alternating bending moments. Fatigue fractures of AO implants are always the result of cyclic bending stresses. Every metal, no matter how thick, has a characteristic number of cycles which are required to produce a fatigue fracture.

Fig. 109 *The different causes of cyclic bending stresses.*

a The plate is stressed only in tension, and the bone in compression.

b By over-contouring the plate, a considerable medial gap can be created, which will close under dynamic loading. This results in cyclic bending of the plate.

c Cyclic bending of a plate in the presence of a medial buttress (opposite cortex) deficiency. The centre of motion is within the plate.

d Loss of a segment results in cyclic bending of the plate, which rapidly fatigues and breaks.

109

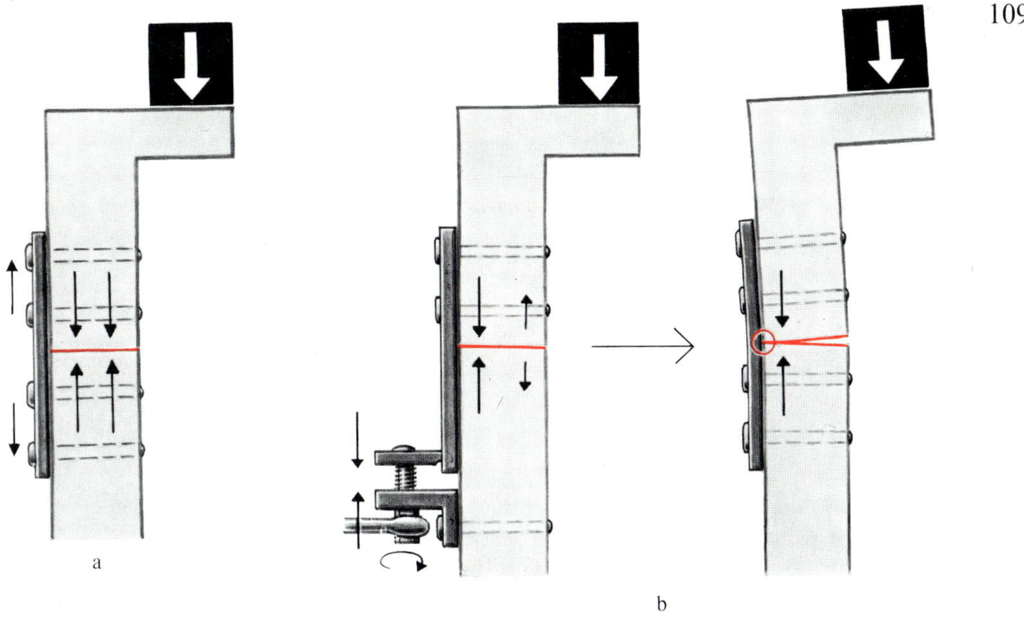

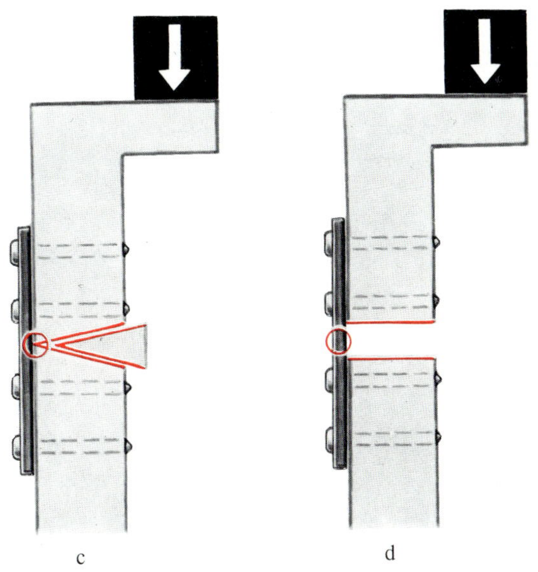

Fig. 110 *Examples of deficiency of the opposite cortex which results in cyclic bending of the plate and the resultant plate failure.*

Fig. 111 *Measures designed to overcome the cyclic bending of a plate.*

a Apply a second plate to the opposite cortex (see Fig. 206/C3) or

b Cortico-cancellous bone grafting of the opposite cortex (medial buttress). After 6 weeks, these will have consolidated enough to take on the function of a second plate.

110

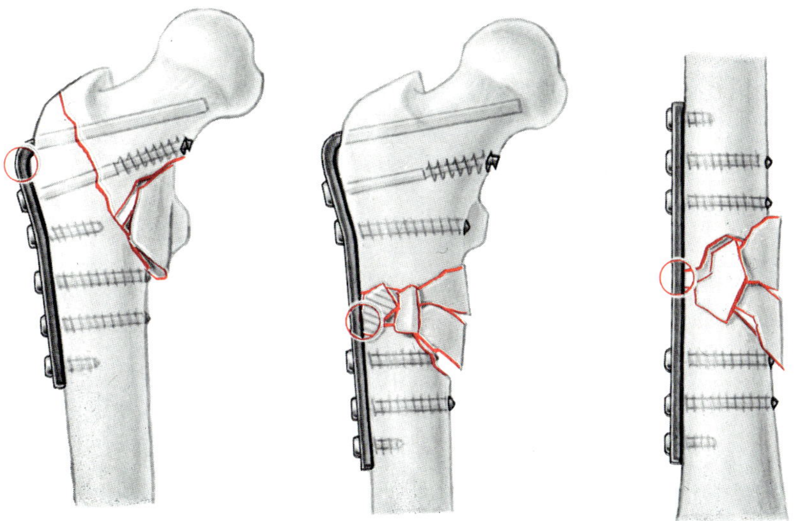

111

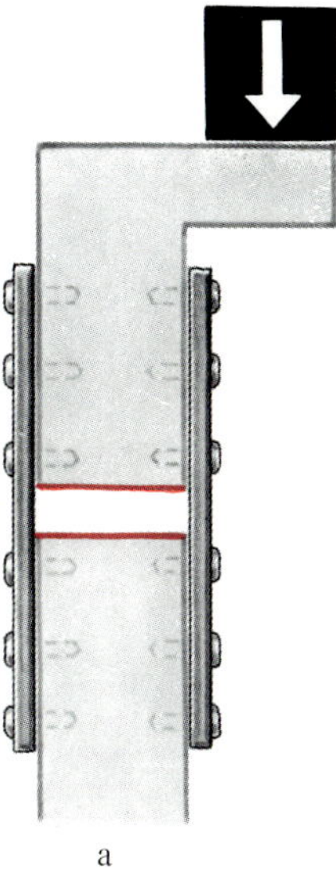

a

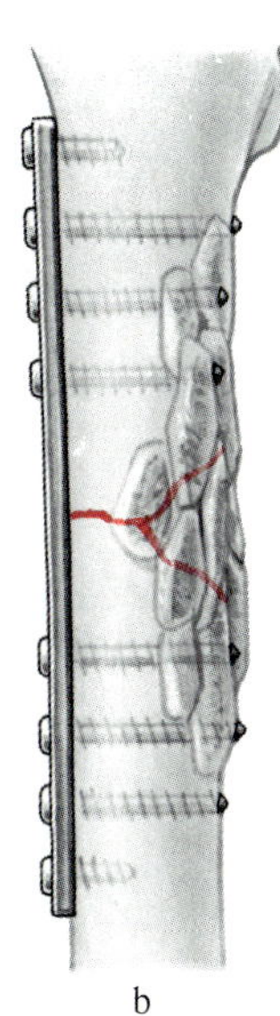

b

3.8.1 Removal of Broken Screws

We have designed the necessary instrumentation to remove every type of broken screw, whether this be due to the destruction of the hexagonal recess at the time of screw insertion or screw removal, or due to destruction of the cruciate slot, or be the result of a fracture of a screw head, or the fracture of the screw in its threaded portion. The principle of screw removal consists of gripping the broken screw with a screw of opposite threading (see Fig. 113). If the hexagonal recess or the cruciate slit are destroyed, the screw head is first removed with a special drill. The screw is then removed as shown in Fig. 113.

3.8.2 Removal of Broken Intra-medullary Nails

Nails, like plates, are subject to fatigue fractures, the result of cyclic loading. To remove the distal portion of the nail, remove the proximal portion and insert into the distal remnant the 3-mm reamer guide. Ream out the proximal medullary canal to a size 2 mm greater than the nail. Remove the reamer guide and insert the extra-long extraction hook, which should hook around the tip of the nail. The extraction hook is screwed into the weight guide and extraction proceeds in the normal fashion.

Fig. 112 *Instrumentation for removal of broken screws.*

- a Specially hardened drill bit for drilling away of damaged screw heads.
- b Hollow reamer and extraction bolt for the 3.5-mm screws.
- c Hollow reamer and extraction bolt for the 4.5-mm screws.
- d Hollow reamer and extraction bolt for the 6.5-mm screws.
- e Gouge.

Fig. 113 *Extraction of a broken screw.*

- a Open the screw tract either with the countersink or a gouge depending on where the screw is broken. This will centre the hollow reamer over the screw remnant.
- b Turning the hollow reamer with the centring pin *counter-clockwise,* ream away the bone to permit the insertion of the extraction bolt. While reaming make sure that the reamer and screw point the same way. Remove the centring pin when you reach the screw fragment. Continue to ream with the hollow reamer till you have exposed 1 cm of the screw thread.
- c Insert the extraction bolt with the left thread over the screw fragment.
- d Remove the screw fragment with the extraction bolt.

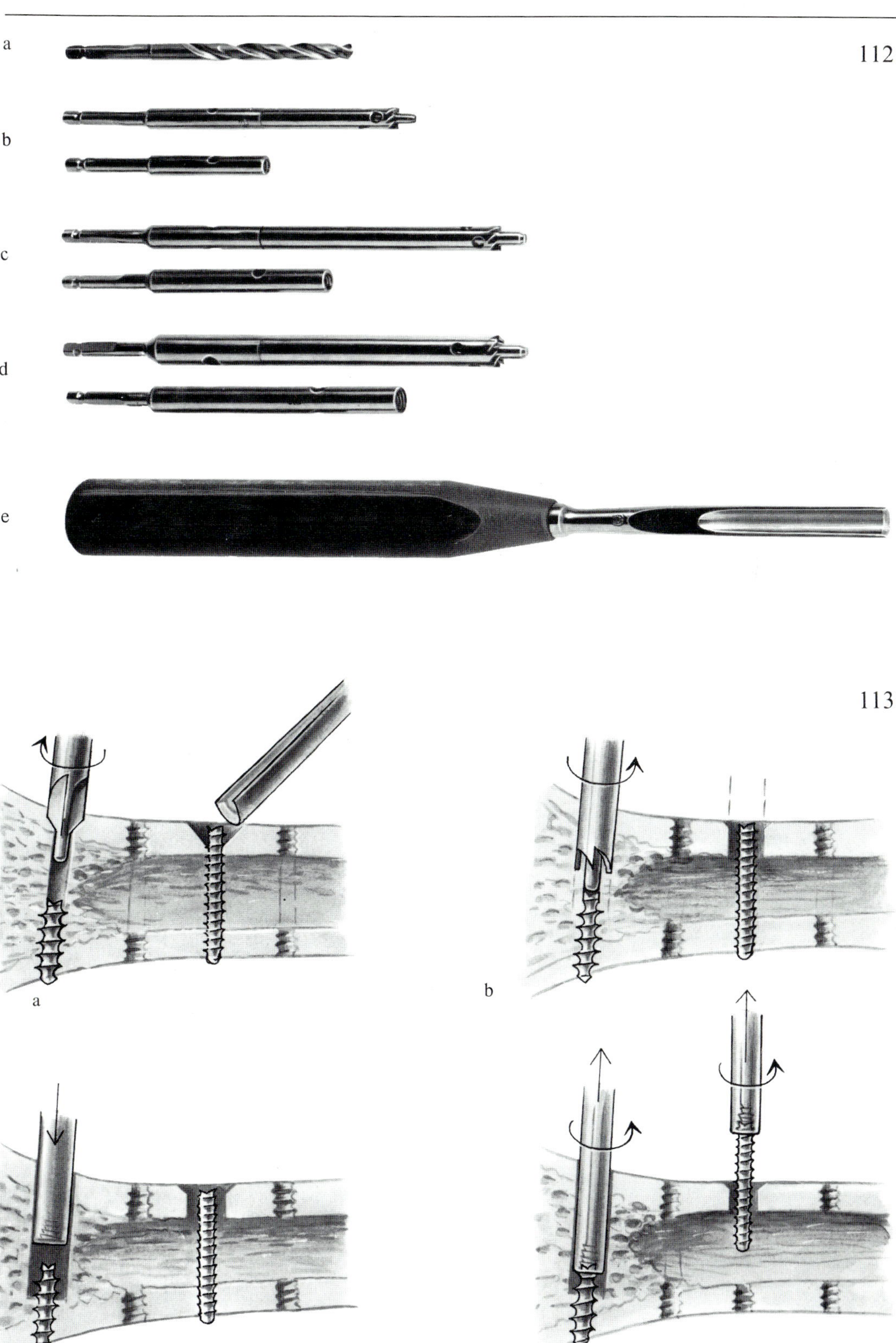
a
b
c
d
e
112
113
a
b
c
d

SPECIAL PART

Internal Fixation of Fresh Fractures

Introduction

This special part will deal with proven methods of internal fixation for most common fractures. These methods are all based on the principles of internal fixation discussed in Part I. We will also discuss special surgical techniques, the most reliable surgical approaches and post-operative care.

This section will deal first with the procedures as they apply to closed fractures in the adult. This is followed by a discussion of the specific problems of open fractures and of fractures in children. At the end there is a separate section which deals exclusively with reconstructive surgery.

Division of Long Bones into Segments

We divide long bones into three regions: proximal, shaft and distal; and into seven unequal segments: proximal joint and distal joint, proximal metaphysis and distal metaphysis, and proximal shaft, middle shaft and distal shaft.

Fig. 114 *Division of the long bones.*

a Humerus: Proximal humerus (head and tuberosities), humerus shaft, distal humerus (supracondylar and transcondylar).

b Forearm: Proximal radius (head and subcapital), radius shaft and distal radius (supra-articular and trans-articular). Proximal ulna (olecranon, coronoid process, sub-articular), shaft of ulna and distal ulna.

c Femur: Proximal femur (head, neck and pertrochanteric), shaft of femur, distal femur (supracondylar and transcondylar).

d Leg: Proximal tibia (tibial plateaus and proximal metaphysis), shaft of tibia, distal tibia (supramalleolar, pylon and transmalleolar). Proximal fibula (head and subcapital), shaft of fibula, distal fibula, supramalleolar and transmalleolar.

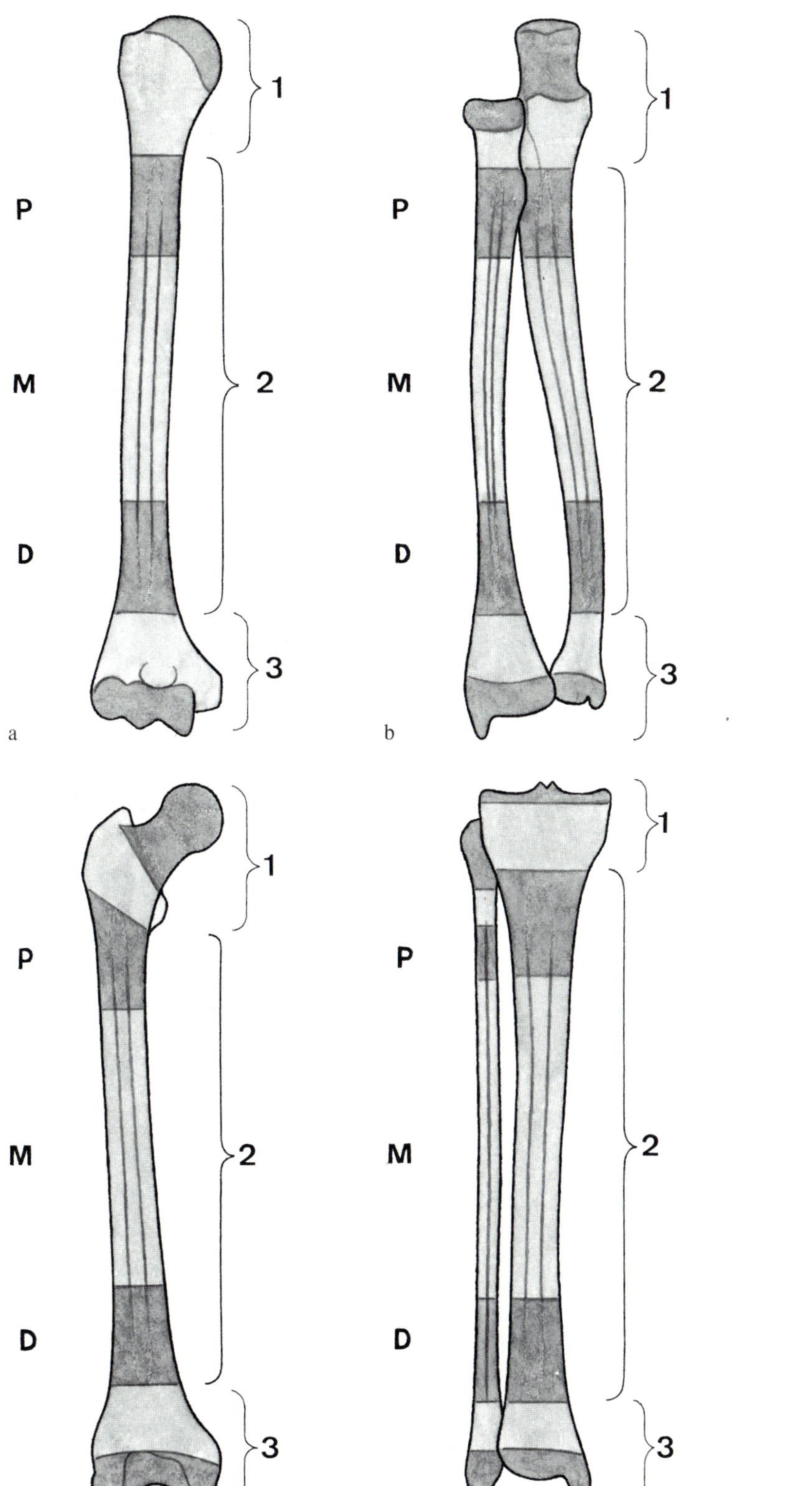
1
P
M
2
D
3
a
1
P
M
2
D
3
b
1
P
M
2
D
3
c
1
P
M
2
D
3
d

1 Closed Fractures in the Adult

1.1 Fractures of the Scapula

Most fractures of the scapula are treated non-operatively. An open reduction may become necessary for displaced fractures through the neck or for fractures of the glenoid with displacement of the articular fragments. The surgical approach depends on the mechanism of injury. Fractures of the posterior portion of the joint are more common than of the anterior portion.

Posterior Approach: At times the limited approach between the infraspinatus and teres minor (see Fig. 115) is all that is required. If full exposure of the scapula is required, we recommend the approach of JUDET (see Fig. 116). In this approach the deltoid, the infraspinatus and teres minor, together with the overlying skin, are retracted laterally and distally. The branch of the suprascapular nerve to the infraspinatus is identified in the notch and preserved. The capsule is opened posteriorly. Usually, reduction can then be carried out without any difficulty. The fixation is secured with a one-third tubular plate. The screws have an excellent purchase in the thick axillary border of the scapula (Fig. 117).

Anterior Approach: Straight incision parallel to the joint 6–7 cm long, centred between the clavicle and the axilla. The subscapular tendon is divided. Occasionally it is necessary to osteotomize the coracoid to increase the anterior exposure of the joint. Prior to osteotomy, it is best to drill a 3.2-mm hole through the coracoid to facilitate its subsequent fixation with a malleolar screw (Fig. 118).

Post-operative Care: Valpeau immobilization until healing of the incision, then active mobilization. A fracture or an osteotomy of the acromion is stabilized with tension band wire and one or two 4-mm cancellous screws.

Fig. 115 *The approach between the infraspinatus and teres minor.*

Fig. 116 *The approach of* JUDET: The deltoid and the infraspinatus are both divided 1 cm from their insertion into the spine of the scapula and vertebral border. They are then reflected together with the teres minor distally and laterally, exposing the scapula. The neurovascular bundle to the infraspinatus lies close to the insertion of the muscle at the level of the joint. It must be identified and protected.

Fig. 117 *Comminuted fracture of the posterior glenoid rim* (JUDET's approach). Accurate reduction and then internal fixation with the one-third tubular plate and five small 3.5-mm cortex screws. The screws obtain an excellent purchase in the thick axillary border of the scapula. Note the painless range of movement 2 weeks after surgery. Only 1 week of hospitalization was required.

Fig. 118 *Tangential anterior fracture of the antero-medial lip of the glenoid.* Internal fixation with two 4-mm cancellous screws. The approach is portrayed in Fig. 122_1 with osteotomy of the coracoid and division of the subscapular tendon.

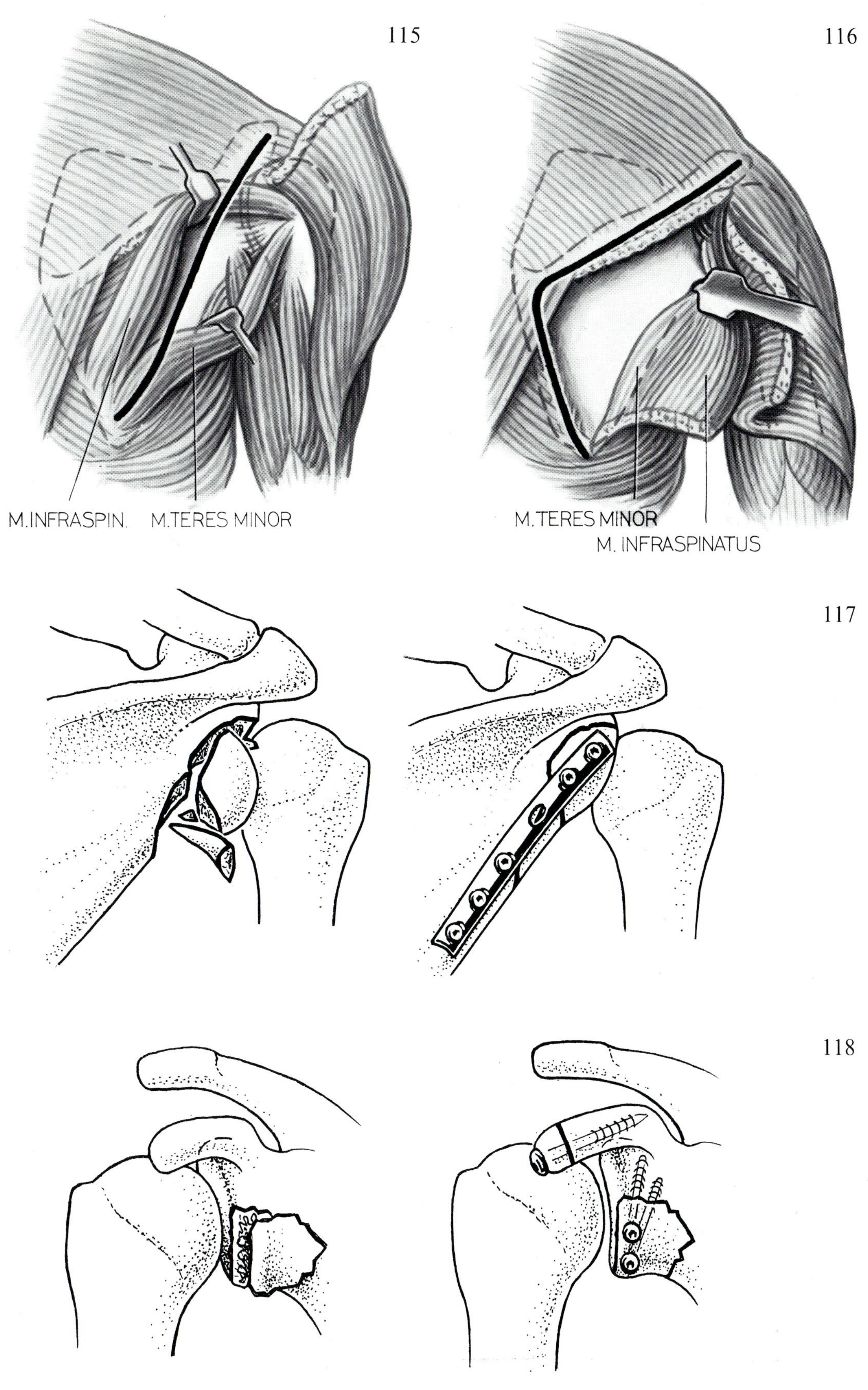
115
116
M.INFRASPIN.
M.TERES MINOR
M.TERES MINOR
M. INFRASPINATUS
117
118

1.2 Fractures of the Clavicle

Fractures of the clavicle heal with and without immobilization. The results of open reductions are less than satisfactory. The incisions spread, at times develop keloids and become ugly, and not infrequently the fractures themselves after open reduction go on to non-union. Therefore fractures of the clavicle are best treated non-operatively. Open reduction is indicated if the fragments are markedly displaced or compress the brachial plexus, or if the fracture goes on to delayed union. Internal fixation is most often carried out with either the six-hole semi-tubular plate or the small DCP. If the clavicle is very slender it is preferable to resort to the six-hole one-third tubular plate for the internal fixation. The approach to the clavicle is either through a supra-clavicular or infra-clavicular incision. The plate is applied either on the superior or inferior surface, depending on the approach.

Fractures of the outer end of the clavicle, on the other hand, often require internal fixation. If the fracture involves the acromioclavicular joint we prefer for fixation a tension band wire in combination with Kirschner wires, as illustrated in Fig. 121. Transverse fractures of the outer end of the clavicle just medial to the acromioclavicular joint, with or without joint involvement, are best fixed by a combination of a tension band wire and a malleolar screw. The malleolar screw is inserted across the acromioclavicular joint in a lateral to medial direction. The coracoclavicular ligaments are either repaired or replaced. If these are sutured, then one must suture the conoid ligament with the arm in maximum abduction and the trapezoid ligament with the arm in maximum adduction. This takes the slack off the ligaments as they are being sutured.

Surgical Approach: We prefer the more aesthetic supraclavicular incision (Fig. 119). The curved infraclavicular incision, which is made one to two fingerbreadths below the clavicle, should be used only if it is unavoidable.

Fig. 119 *Surgical approach to the clavicle.*

a Curved infraclavicular incision 2–4 cm below the clavicle.

b Supraclavicular incision one fingerbreadth above the clavicle. It gives a better scar and is therefore aesthetically preferable.

Fig. 120 *Transverse fracture of the clavicle with a small butterfly.* Fixation with a one-third tubular plate applied to the inferior surface of the clavicle.

Fig. 121 *Fracture of the outer end of the clavicle with extension into the acromioclavicular joint.* First stabilize the two small fragments with two Kirschner wires. Complete the fixation with a tension band wire.

119

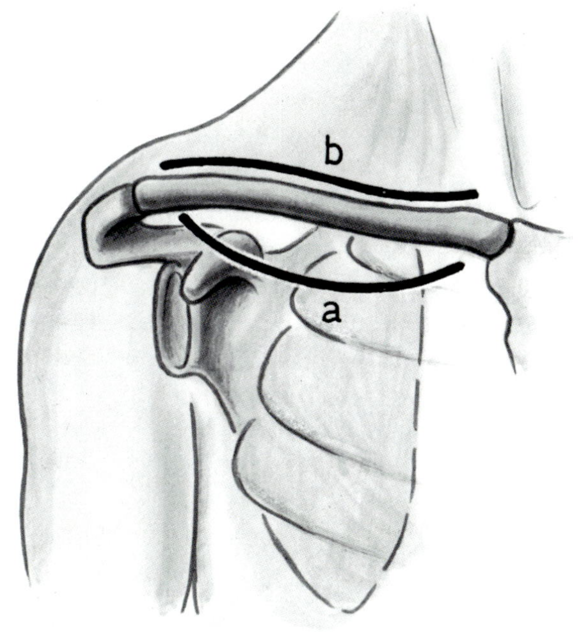

120

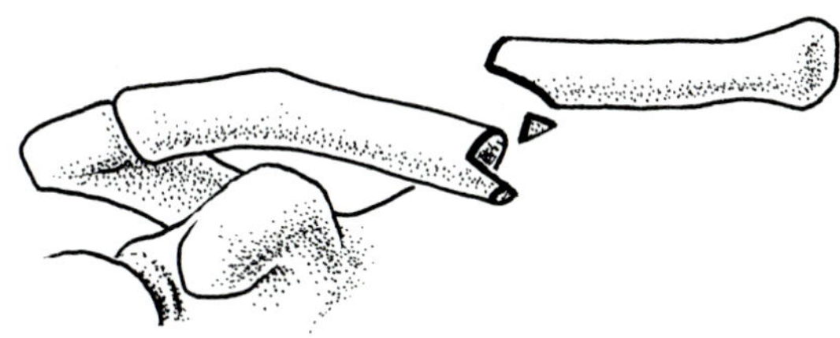

121

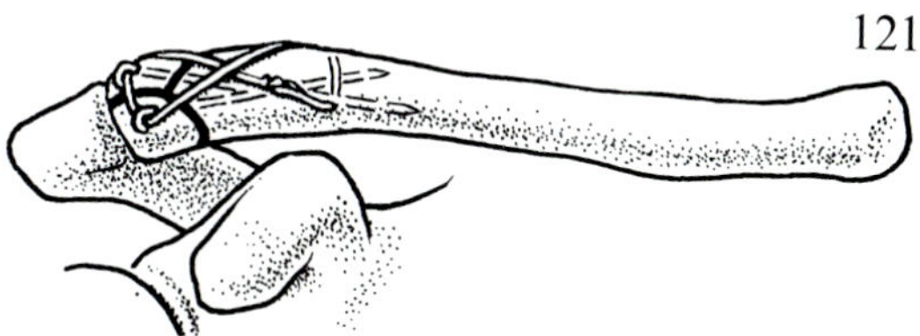

1.3 Fractures of the Humerus

Fractures of the proximal two-thirds of the humerus are most often treated non-operatively. Fractures of the distal third of the humerus, on the other hand, are most often treated operatively. For indications for operative intervention for the different fractures we refer the reader to the corresponding sections of the manual dealing with the specific fracture in question.

Surgical Approaches: The proximal half of the humerus is approached anteriorly whereas the distal half of the humerus is approached posteriorly. The epicondyles are approached through a medial or lateral incision as required.

Fig. 122 *Three surgical approaches to the different portions of the humerus.*

Incision 1: An extensile, cosmetically acceptable approach to the proximal third of the humerus. The incision begins at the acromioclavicular joint and runs along the anterior axillary border somewhat distal to the inferior edge of the pectoralis major. From here it can be extended as needed along an imaginary line to the medial epicondyle. The deep fascia is opened along the medial border of the biceps. (Take care not to injure the musculocutaneous nerve!) Expose the shaft along the medial bicipital sulcus.

Incision shown as *dotted lines*: For muscular patients we prefer this to incision 1, even though it is cosmetically less acceptable because it spreads and is often complicated by keloid formation. The shaft is exposed along the lateral bicipital sulcus and humeral head in the region of the anterior third of the deltoid.

All difficult reconstructions of the humeral head require detachment of the deltoid from its insertion. This should be done by dividing the muscle 1 cm from its attachment to the clavicle. (Beware of damage to the axillary nerve, which is most susceptible to pressure!)

Incision 2: Standard approach to the humeral shaft. Incise along the lateral bicipital sulcus with longitudinal splitting of the brachialis in line with its fibres and anteriorly where it is thinnest. (The radial nerve runs along the intermuscular septum under the brachialis. It is not necessary to expose the nerve).

Incision 3: The posterior approach to the humerus. Split the long and lateral head of the triceps. Where they join in the middle, one will come directly upon the neurovascular bundle containing the radial nerve just about in the middle of the humerus. A tape can be placed around the neurovascular bundle without the nerve having to be isolated. The distal portion of the triceps together with its tendon, is split in the middle to expose the humerus.

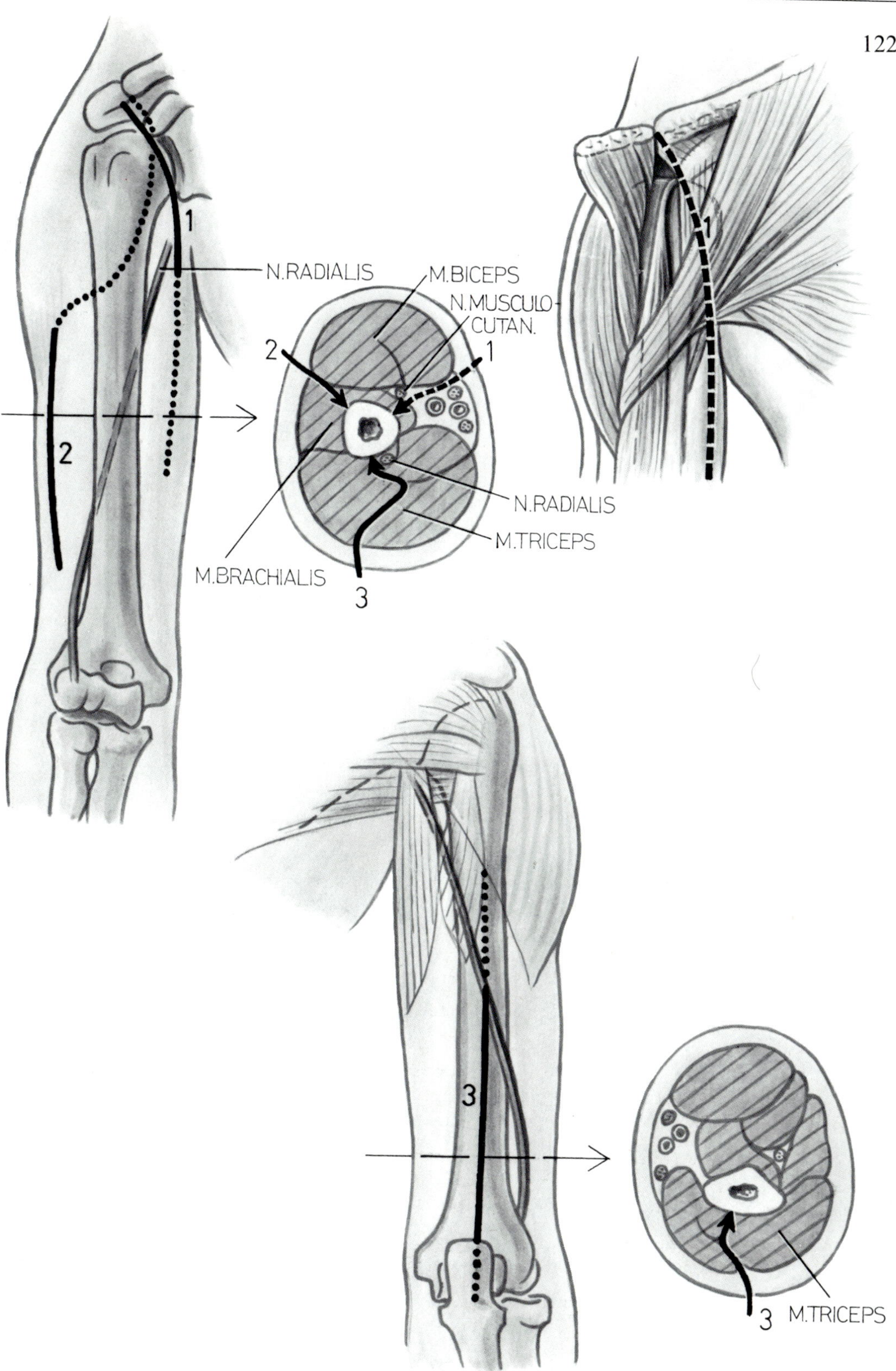
1
N.RADIALIS
M.BICEPS
N.MUSCULO-CUTAN.
2
1
2
N.RADIALIS
M.TRICEPS
M.BRACHIALIS
3
1
3
3
M.TRICEPS

Whether the patient is positioned supine or prone, one must make certain that the elbow can be flexed to 140°.

Fig. 123 *Standard approach to the distal humerus without osteotomizing the olecranon.*

a Straight posterior incision.

b Cross-section through the elbow joint taken at the base of the olecranon. The ulnar nerve must be identified at the start of the procedure, a tape must be placed around it, and it must be retracted out of the way.

c The triceps tendon aponeurosis is now incised to form a tongue-shaped flap which is reflected distally. The joint is opened widely and two small Hohmann retractors are inserted, one on each side. Flexion of the elbow to 140° or 150° will expose the trochlea and the capitellum.

d The radial nerve crosses the upper portion of the incision; care must be taken not to damage it.

Fig. 124 *Approach with osteotomy of the olecranon (rarely necessary).*

a Prior to carrying out the osteotomy, make two 1.5-mm drill holes in the olecranon as shown. Make the osteotomy at the level of the trochlea directing it upwards and forwards so that it is slightly oblique.

b Once the osteotomy is carried out, flexion of the elbow to 140° gives an excellent exposure of the whole joint.

c At the end of the procedure the olecranon is carefully reduced and stabilized with two Kirschner wires, which are inserted into the pre-drilled holes. A tension band wire places the osteotomy under axial compression.

Fig. 125 *An approach with an oblique extra-articular osteotomy of the olecranon.* Special tension band fixation is not necessary, since displacement of the small oblique fragment is unimportant.

123

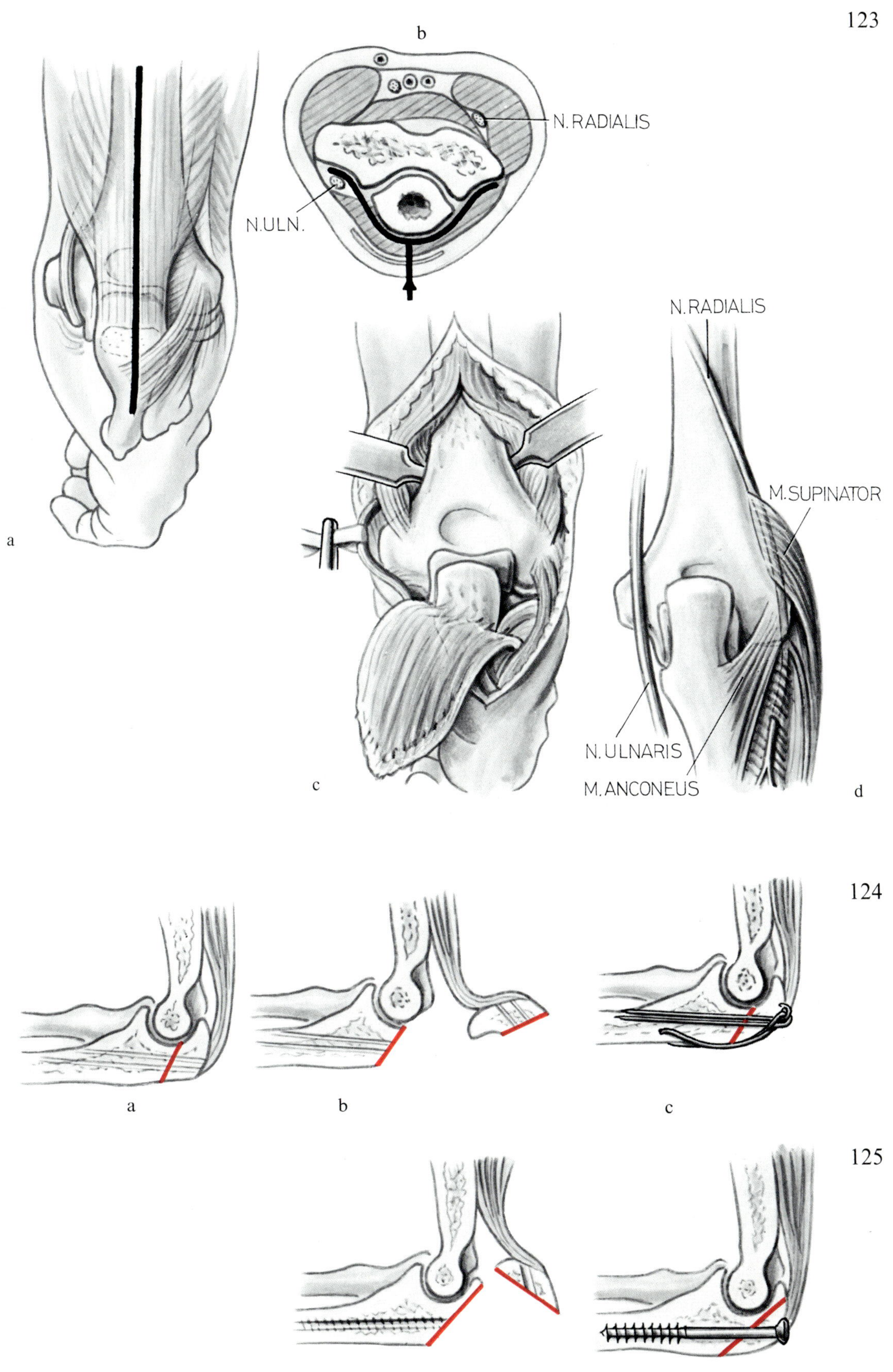

a b c d

124

a b c

125

1.3.1 Proximal Humerus

Indications for Surgical Treatment: Most fractures of the proximal humerus can be treated non-operatively. We consider the following to be indications for surgery:

- Fracture dislocations (axillary and subcoracoid)
- Comminuted fractures of the humeral head in patients under 50 years of age
- Subcapital transverse fractures with complete displacement of the shaft and no contact between the fragments
- Avulsion fractures of the greater tuberosity with displacement and subacromial impingement
- Irreducible epiphyseal plate fracture separations (interposition of the tendon of long head of biceps!)

Approach: See Fig. 122.1.

Internal Fixation: Different means of internal fixation are illustrated in Fig. 126. Avoid malreduction (improper rotation or adduction). Repair or reconstruct the rotator cuff. Elevate compression fractures of the head. Because of loss of cancellous bone due to impaction it is necessary at times, when reconstructing the head, to bone graft with cortico-cancellous bone.

Post-operative Management: The patient is up and about on the day of surgery, with the arm immobilized from 2 to 4 days. He then starts active mobilization which consists of pendulum exercises, rotation exercises and passive abduction of the arm, and active, assisted exercises. Anxious patients should be fitted with an abduction splint.

Fig. 126

a *Axillary fracture dislocation of the humeral head* which was reduced and stably fixed with a T plate. The fracture of the anterior lip of the glenoid was repaired with two malleolar screws.

b *A comminuted fracture* of the humeral head fixed with the cloverleaf plate (note that a portion of the cloverleaf plate was resected in order to make it fit).

c *A fracture through the surgical neck* with complete lateral displacement of the shaft and interposition of the long head of biceps tendon. Note the anatomical reduction and fixation with two small cancellous screws on either side of the bicipital groove. Open the joint and make certain that the tendon is in its normal anatomical position.

d, e *Avulsion fracture of the greater trochanter with subacromial impingement.* Reduction and internal fixation either with cancellous screws or a tension band.

f *Reduction of a fracture with complete anterior and medial displacement of the humeral shaft.* This fracture is often associated with interposition of the capsule and biceps tendon. The approach is simple because the displaced humeral shaft usually splits the interval between the deltoid and pectoralis and comes to lie subfascially or subcutaneously. Reduce and fix with a T plate.

a

b

c

d

e

f

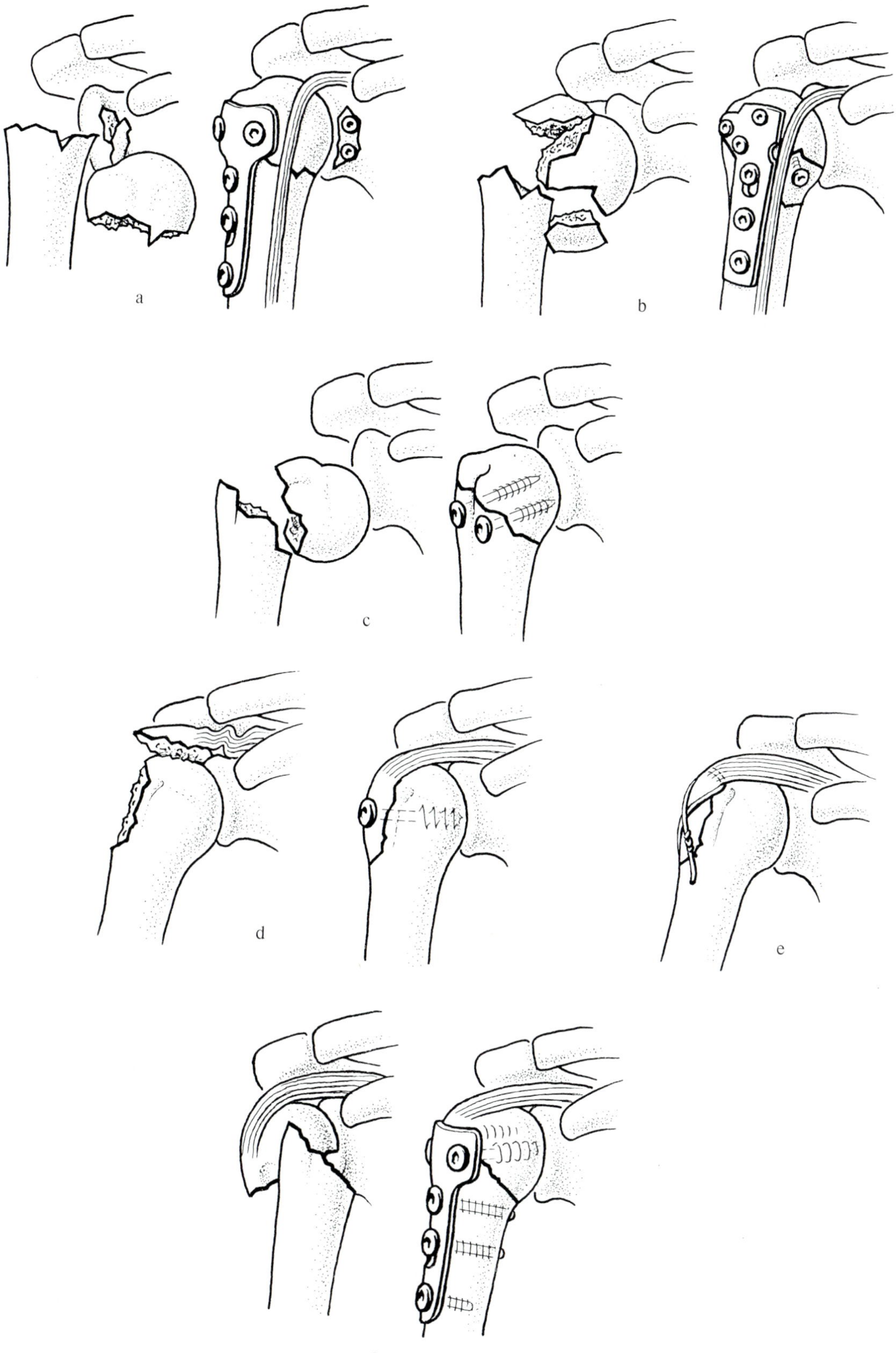

1.3.2 Fractures of the Shaft of the Humerus

Non-operative treatment consists almost exclusively of Desault or Valpeau bandaging. Clinical union usually takes place in 3–4 weeks. Thereafter, active mobilization is instituted.

Indications for Operative Treatment

- Bilateral humerus shaft fractures
- Multiple fractures in a patient requiring I.C.U. therapy
- Secondary radial nerve palsy
- Primary radial nerve palsy associated with fractures where transection of the nerve is most likely
- Painful delayed union
- Transverse fractures of the shaft
- Open fractures (see the corresponding section)

Surgical Approaches: See Fig. 122 – incision 2 and incision 3.

Internal Fixation

Spiral fractures with butterflies (very rare): These should never be treated with lag screw fixation alone. The lag screw fixation must be combined with a broad six- to eight-hole neutralization plate.

Short oblique fractures: They are an ideal indication for the broad six- to eight-hole DCP which is applied in such a way as to exert static compression. The short oblique fracture should be crossed with a lag screw (see also Fig. 129).

Transverse fractures: Compression plating with a broad DCP as for short oblique fracture. These can also be fixed with stack nails as described by HACKETHAL.

Note: Exposure of the radial nerve is not necessary in approaches 1 or 2. Its location along the intermuscular septum should always be kept in mind. In approach 3, as well as in all lateral approaches to the distal humerus, it is necessary to carry out the approach from the normal to the affected area with identification of the radial nerve.

Fig. 127 Because of the thin cortices, we do not recommend screw fixation alone for fractures of the humerus.

Fig. 128 Comminuted fractures are only very rarely operated on (see text). Here one must combine lag screws with neutralization plate (DCP).

Fig. 129 *Transverse and short oblique fractures.* Use a broad DCP combined with a lag screw across the fracture (see also Fig. 148).

Fig. 130 *Comminuted fracture of the humerus with multiple small fragments.* A six- to eight-hole DCP combined with a cancellous bone graft applied to the cortex opposite the plate.

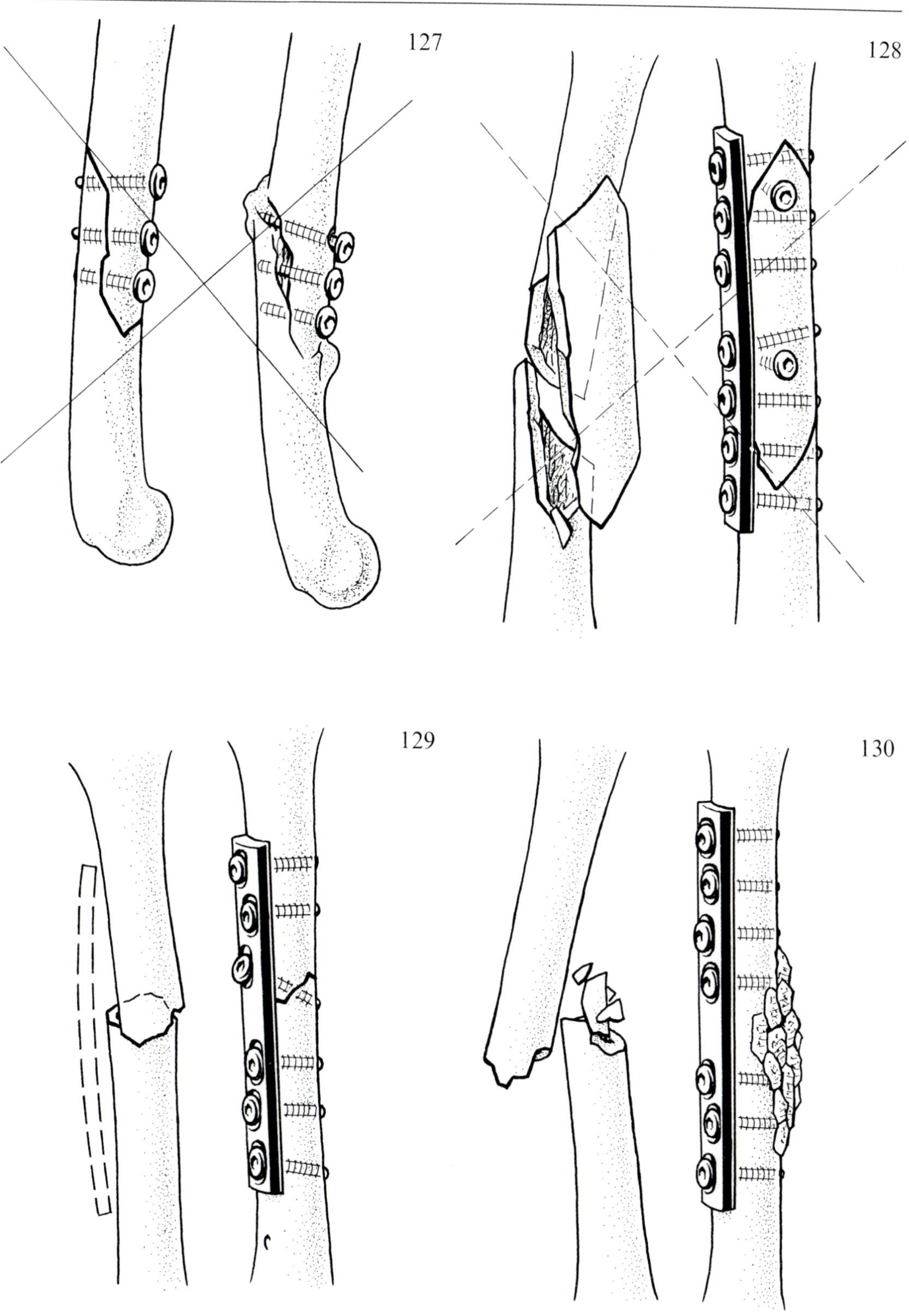
127
128
129
130

Classification: These are divided according to the classification of M.E. MÜLLER into three groups similar to the classification of fractures of the distal femur.

A Extra-articular fractures: avulsion fractures of collateral ligaments (epicondyle), simple supracondylar fractures, comminuted supracondylar fractures.

B Intra-articular fractures of one condyle: fractures of the trochlea, fractures of the capitellum, tangential fractures of the trochlea or capitellum.

C Bi-condylar fractures: Y fractures, Y fracture with comminuted supracondylar fracture, compression fractures and comminuted fractures.

Fig. 131

A_1 *Avulsion of epicondyle* (medial epicondyle trapped in the joint after dislocation of the elbow joint).

A_2 Simple supracondylar fracture.

A_3 Comminuted supracondylar fracture.

B_1 Fracture of the trochlea.

B_2 Fracture of the capitellum.

B_3 Tangential fracture of the trochlea.

C_1 Y fracture.

C_2 Y fracture with supracondylar comminution.

C_3 Comminuted fracture.

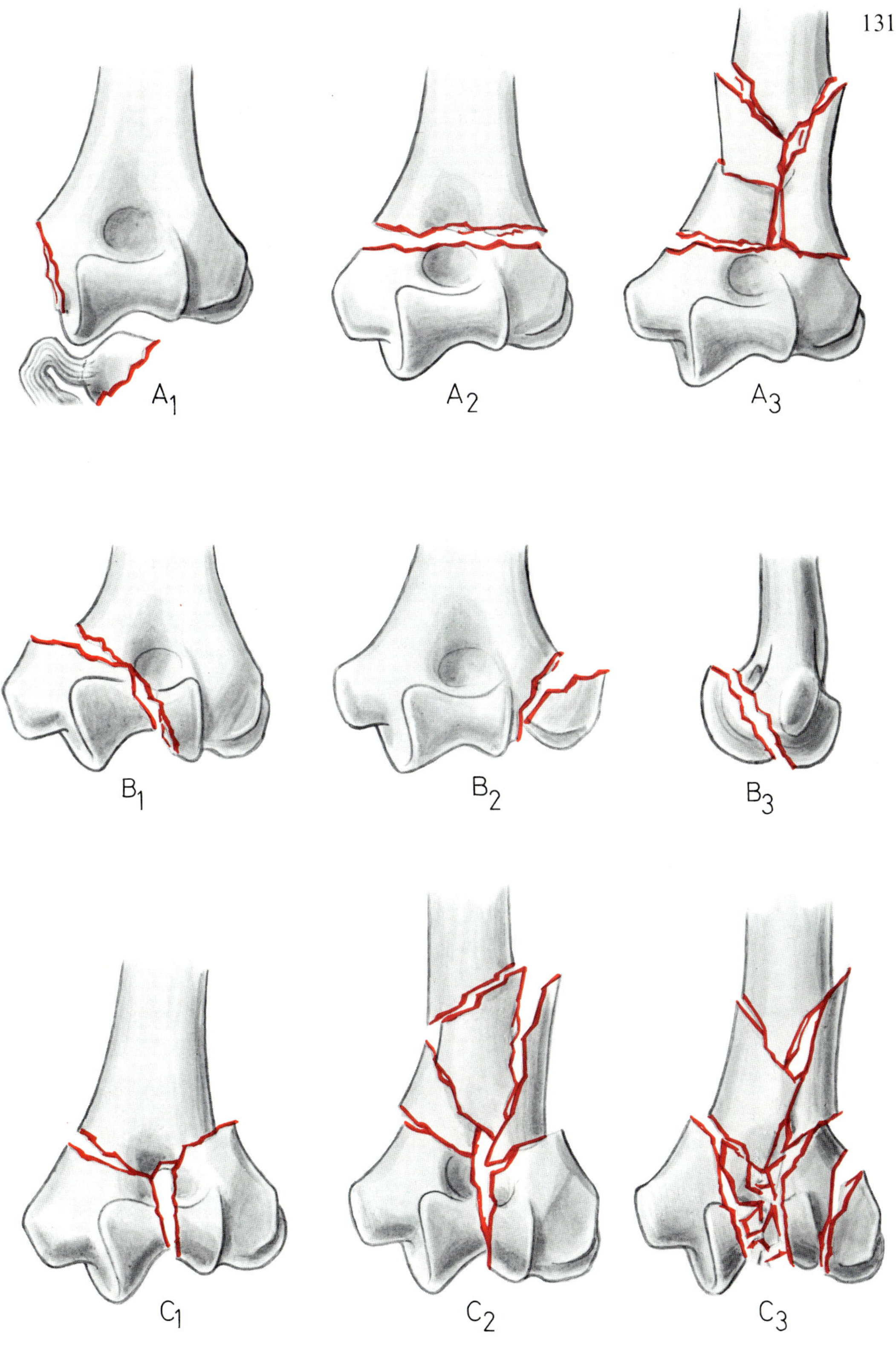
A1
A2
A3
B1
B2
B3
C1
C2
C3

1.3.3 Extra-articular Fractures of the Distal Humerus (Type A)

Because of the considerable leverage on these fractures, delayed union with resultant elbow joint stiffness is commonly the result of non-operative treatment. We feel therefore that, under most circumstances, these fractures should be treated operatively.

Surgical Approach: See Fig. 123. The standard posterior approach is used except for isolated fractures of the medial or lateral epicondyle which are best approached through short corresponding medial or lateral incisions. On the medial side, step 1 is the isolation and protection of the ulnar nerve.

Avulsion Fractures of the Epicondyle: Internal fixation is carried out by means of a lag screw (small cancellous or malleolar screw) which should take purchase in the opposite cortex. To prevent splitting of the fragments, before reduction, drill a hole through the middle of the fragment by drilling from the fracture aspect towards the intact cortex. Use either the 2-mm or 3.2-mm drill bit, depending on the screw to be used.

Internal Fixation of the Distal Extra-articular Fractures of the Humerus: Basically these are fixed very much like the fractures of the middle third. In distal fractures the semi-tubular plate should be used and as many of the screws as possible should be inserted as lag screws (see Fig. 133). If desired a DCP can also be used here most successfully.

Fig. 132 *Dislocation of the elbow with intra-articular entrapment of the medial epicondyle* (A_1). The fixation is carried out with one or two 4-mm cancellous screws or a 4.5-mm malleolar screw.

Fig. 133 *Distal short oblique fractures* (A_2). Fixation with a semi-tubular or even a one-third tubular plate. Insert as many of the screws as possible as lag screws across the fracture. Consider anterior transposition of the ulnar nerve.

Fig. 134 *Comminuted supracondylar fracture* (A_3). This short solid distal portion of the shaft is best fixed with a DCP or a T plate applied on its posterior aspect. If the bone is slender, use for fixation the narrow or even the small DCP.

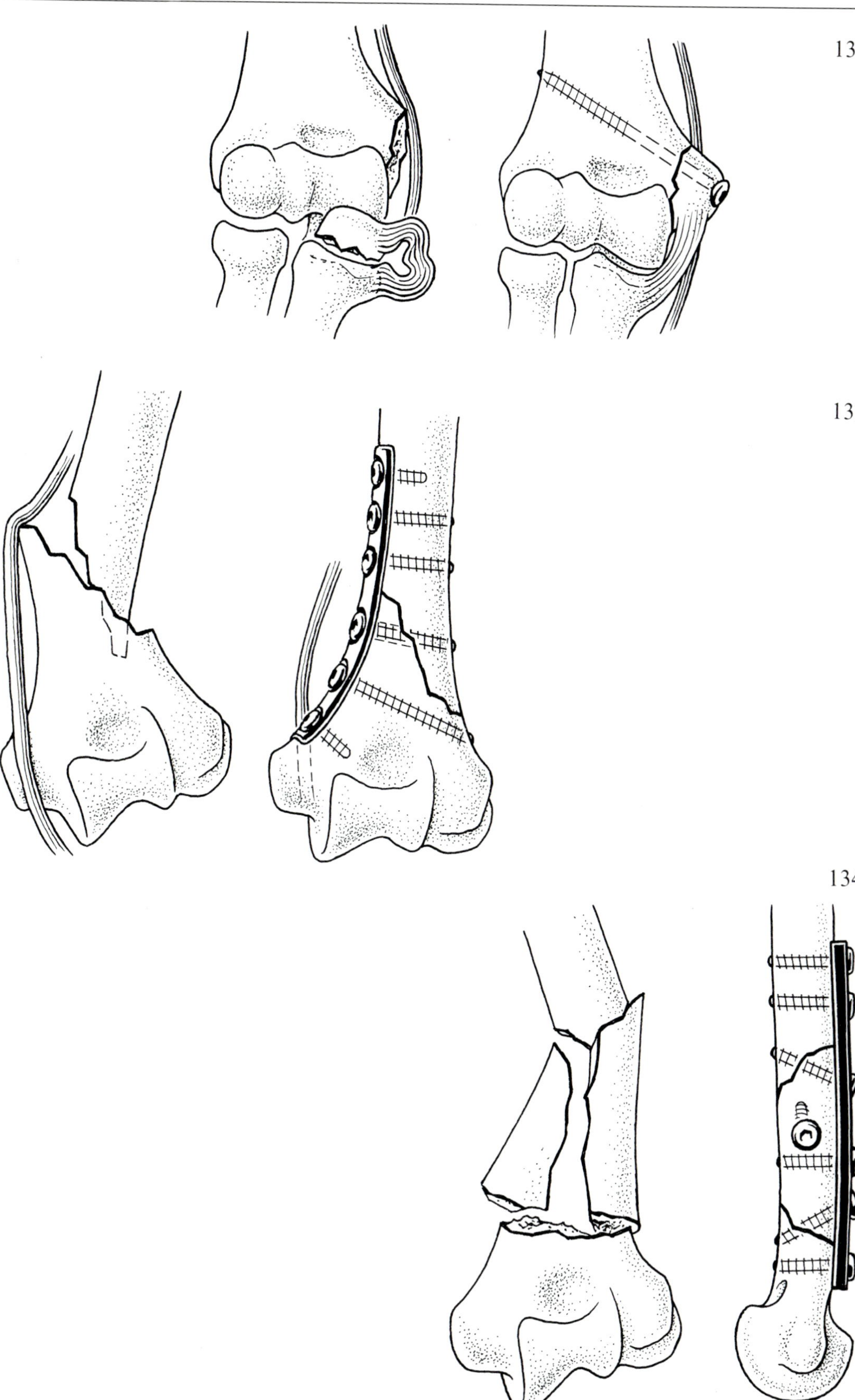
132
133
134

1.3.4 Distal Intra-articular Fractures of the Humerus

The surgical approach is posterior with or without osteotomy of the olecranon. Only in rare circumstances should one carry out an intra-articular osteotomy of the olecranon (see Fig. 124). Most important is the reduction and stable fixation of the trochlea. Tangential fractures of the trochlea can only be approached through an intra-articular osteotomy of the olecranon.

The steps of reduction and internal fixation of a Y fracture are represented in Fig. 137 a–d.

When dealing with the very badly comminuted intra-articular fracture begin by reconstructing the trochlea. Sacrifice and remove the little intra-articular fragments. It is most important to establish the normal anatomical width of the trochlea. All fragments are tapped and fixed with a cortex screw used here as a fixation screw. After fixation, the elbow is put through a range of motion and the reduction is carefully examined. If necessary the width of the trochlea is re-adjusted. The next step is the fixation of the supracondylar component of the fracture. At the end all bone defects must be grafted with cancellous bone. We feel that a primary arthroplasty with sacrifice of both epicondyles is contraindicated. Make certain that whenever you carry out internal fixation of the distal humerus you do not obstruct the olecranon fossa and coronoid fossa with implants.

Post-operative Management: On the fourth post-operative day, begin with *active* mobilization. In intra-articular fracture of the elbow, only active mobilization should be allowed. All passive mobilization and physiotherapy are dangerous and should be avoided because they frequently result in myositis ossificans and joint stiffness.

Fig. 135 *Fracture of the capitellum.* Fixation with two 4-mm cancellous screws. Method of reduction and fixation is the same as described in Fig. 137a–c.

Fig. 135 *Isolated fracture of the trochlea.* First expose and protect the ulnar nerve. Prior to reduction, pre-drill holes in the ulnar fragment by beginning from the fracture aspect and drilling through the middle of the fragment. The fixation is best carried out with two lag screws, with a malleolar and a small cancellous screw, as shown in the figure.

Fig. 137 *The reduction and internal fixation of any Y fracture.*

a Expose and protect the ulnar nerve. Next expose the fracture and pre-drill a 2-mm hole in the radial fragment. Insert an appropriate Kirschner wire into the hole.

b Reduce the trochlea and maintain the reduction with the pointed reduction forceps. Expose the radial aspect of the capitellum, remove the Kirschner wire, re-insert the 2-mm drill bit into the hole and drill the ulnar fragment of the trochlea. Tap both fragments with the 4-mm tap, which is inserted in a radio-ulnar direction.

c Fix the trochlea with a 4-mm cancellous screw (in porotic bone use a washer). Reduce the supracondylar fracture.

d Fix the supracondylar fracture with two semi-tubular plates or with two malleolar screws which should grip the opposite cortex.

Fig. 138 *Reduction and fixation of a comminuted intra-articular fracture.* Where a part of the trochlea is missing, use a cortex screw as a fixation screw in order to maintain the width of the trochlea. The supracondylar component of the fracture is fixed as described in Fig. 137d or at times with a Y plate. Finally fill the trochlear defect with cancellous bone.

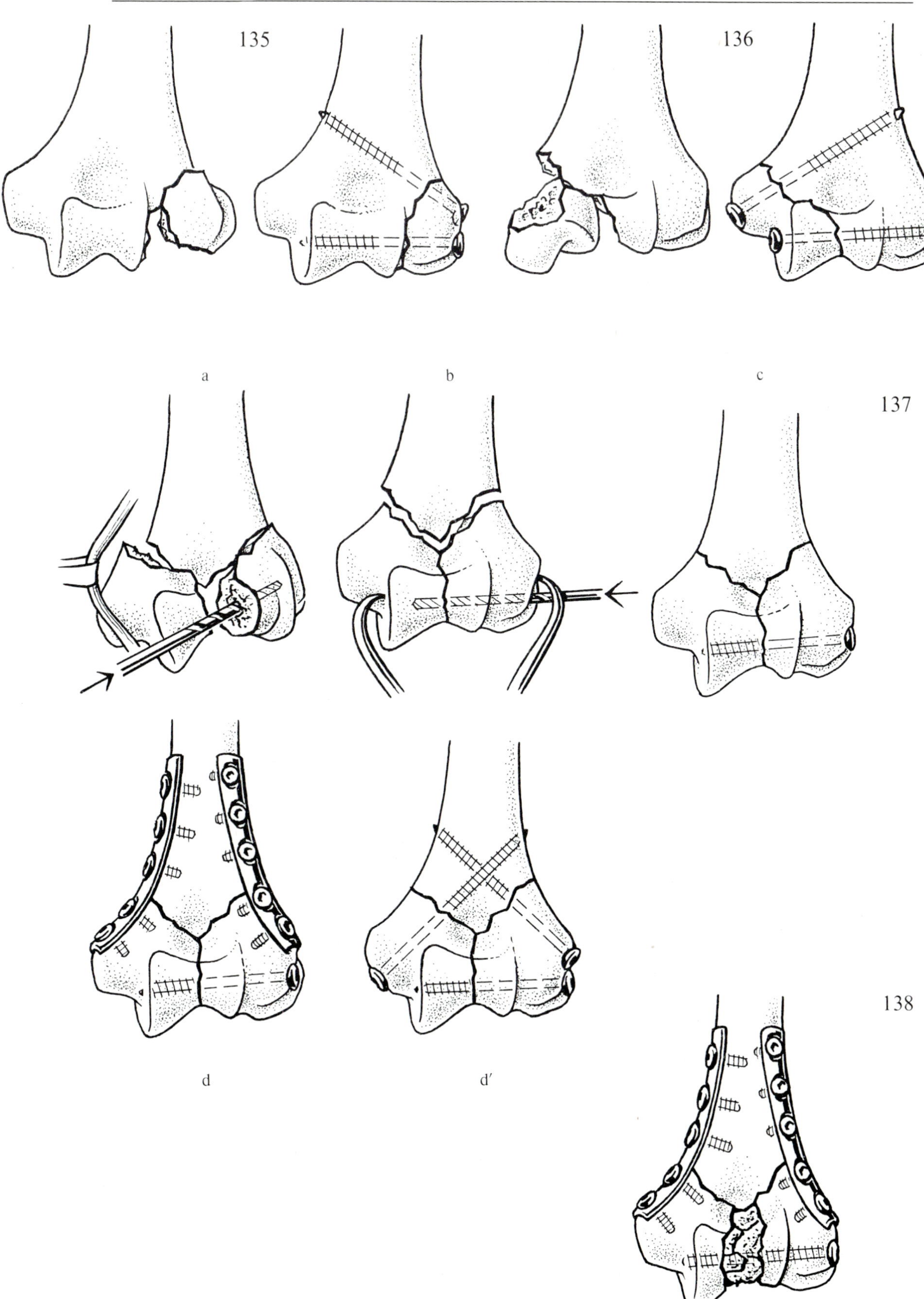

135
136
a
b
c
137
d
d′
138

1.4 Fractures of the Forearm

The radius describes a 160° arc about the ulna. This is the reason why in fractures of the radius normal pronation and supination is achieved only after anatomical reduction and restoration of the double bow of the radius. Bone healing of both the radius and ulna is slow, probably because of the small contact surfaces at the fracture. Because of this slow healing, stable fixation of the fragments is very important. If the interosseous membrane is injured there is considerable danger of cross-union, particularly if the treatment is closed and the fragments immobilized in plaster.

The medullary canal of the radius is narrow and bowed. Intramedullary wiring gives no rotational stability. Intermedullary nailing with reaming straightens the radius and destroys its physiological curvatures.

Plates without pre-stress (e.g. the LANE plate) lead very frequently to delayed callus formation and pseudarthroses. A carefully contoured pre-stressed plate, a compression plate, provides excellent stable fixation of the fragment and permits early active mobilization without the above-noted complications.

Shaft Fractures: In fractures of the radius, to prevent straightening of the bows of the radius, the plate is carefully contoured and applied on the dorsoradial side. In its middle third the radius is round. The semi-tubular plate is ideal for the internal fixation of this segment because it fits snugly the tubular segment of the bone and because it is easily contoured to fit the physiological bow of the bone. However, it can be used only for internal fixation of those fractures which have an intact opposite cortex. The semi-tubular plate and the DCP allow for axial compression of the fragments without the use of the tension device. This is of considerable advantage when dealing with the radius because longitudinal exposure is often a problem. Screws inserted into the main fragments must grip at least four cortices, preferably five, on each side of the fracture. Therefore we prefer the five- and better still the six-hole plates for internal fixation of fractures of the forearm.

Fractures of the proximal third of the radius are approached through the posterior approach of BOYD. This approach allows identification and protection of the posterior interosseous nerve (see Fig. 141). Fractures of the middle third of the ulna can be fixed just as well with the narrow round-hole plate as with the DCP. Where the ulna is small, we have found the small DCP with the 3.5-mm cortex screws most useful.

When dealing with shaft fractures of both the radius and ulna, begin with the simpler of the two, which most commonly is the ulna. After a reduction carry out a temporary fixation. Definitive fixation is carried out only after the second bone is reduced, provisionally stabilized and after pronation and supination have been tested and found to be normal. At this stage little imperfections can be easily corrected. At times, in order to achieve absolutely accurate anatomical reduction, the provisional fixation of the first bone has to be taken apart. In comminuted fractures of both bones, particularly of the proximal half, we prefer the posterior approach of THOMPSON (see Fig. 141 d). This approach gives not only an excellent exposure of both bones but also of the damaged interosseous membrane. An accurate reduction of the devitalized fragments is frequently impossible and a primary cancellous bone graft must be carried out. Care must be taken not to place the graft on the interosseous membrane, for that may lead to a cross-union. When dealing with comminuted fractures particular attention must be paid to the careful reconstruction of the physiological bow of the radius and to the correct rotational alignment of the fragments. Reduction of both bones out to length will prevent problems with the proximal and distal radioulnar joint.

The great advantages of anatomical reduction and compression plating of long bones are particularly evident in fractures of both bones of the forearm. The method, however, is not simple. Avoid prolonged surgery with increased complications by careful attention to anatomical detail and by carrying out the proper exposure. Damage to both sensory and motor nerves should also be avoided.

Periarticular Fractures: Fractures of the *proximal ulna* are often comminuted. We have found the semi-tubular plate particularly useful for their internal fixation because the plate fits closely to the contour of the proximal ulna (see Fig. 43b). Fractures of the olecranon are avulsion fractures and should therefore be stabilized by means of a tension band.

Fractures of the radial head can be fixed if an exact anatomical reduction is carried out. Screw heads which come to lie under the annular ligament should be countersunk. Comminuted fractures of the head are best treated with early resection of the radial head.

Fractures of the distal radius, particularly the Smith-Goyrand fracture, are best fixed with the small T plate.

Post-operatively apply a light dressing and elevate the limb (see Fig. 103c). If you suspect a compartment syndrome because of swelling, which is not uncommon as a result of direct trauma, remove all dressings. Note that in compartment syndromes the distal pulse frequently remains unaltered. If removal of the bandages does not result in an immediate improvement, remove all sutures, open all deep fascial compartments and leave the wounds open.

Fig. 139 *Surgical approach to the olecranon:* Make a slightly curved incision on the radial side of the olecranon aiming for the subcutaneous border of the ulna (*1*). Reflection of the anconeus is rarely necessary. Opening of the joint capsule in order to inspect the joint surface is necessary only in comminuted fractures or compression fractures of the olecranon. The broken line (*2*) outlines the approach to the proximal ulna as well as to the proximal one-third of the radius (see also Fig. 141 d). The *broken line* outlines also the approach to the radial head.

Fig. 140 *Surgical approach to the head of the radius.*

a A straight dorsal skin incision which begins at the lateral epicondyle and runs 4–5 cm distally. It is approximately 1 cm radial to the subcutaneous border of the ulna.

b Split the extensors in line with their fibres. Beware of damage to the posterior interosseous nerve, which runs in the substance of the supinator from front to back about 4–6 cm distal to the radial head.

139

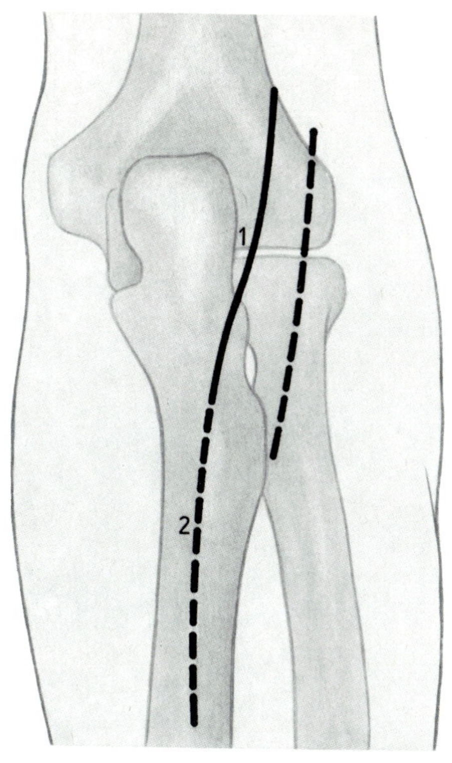

140

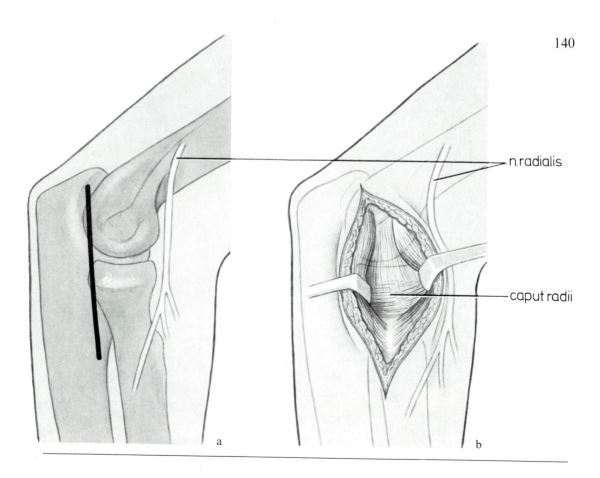

n. radialis

caput radii

a

b

By and large we prefer the straight dorsal approaches directly over the corresponding bones.

THOMPSON's extensile exposure is particularly useful in dealing with badly comminuted fractures of the proximal and middle thirds of both the radius and ulna. This approach also permits the exposure and repair of the disruptions of the interosseous membrane.

Note: See page 194 for the volar approach to the distal radius as well as for a cross-section of the distal third of the forearm.

Fig. 141 *The posterior surgical approaches to the forearm.*

a The incisions.
(1) The extensile exposure of BOYD-THOMPSON: proximal approach as for fractures of the olecranon which is then extended distally by directing the incision towards the styloid process of the radius.
(2) The approach to the shaft of the ulna is between the flexor carpi ulnaris and extensor carpi ulnaris.
(3) The approach to the middle and distal third of the radius is between the extensor digitorum communis and the extensor carpi radialis brevis.

b Cross-section through the middle of the proximal third of the forearm, taken to illustrate the approach to the radius from the ulnar aspect as well as the position of the plate on the radius and ulna.

c Cross-section through the middle third of the forearm, illustrating the surgical approaches to the radius and ulna.

d The extensile approach of BOYD-THOMPSON. It illustrates the approach to the proximal radius after partial transection of supinator and reflection of the supinator together with the posterior interosseous nerve. More distally the approach is developed between the extensor digitorum communis and extensor carpi radialis brevis.

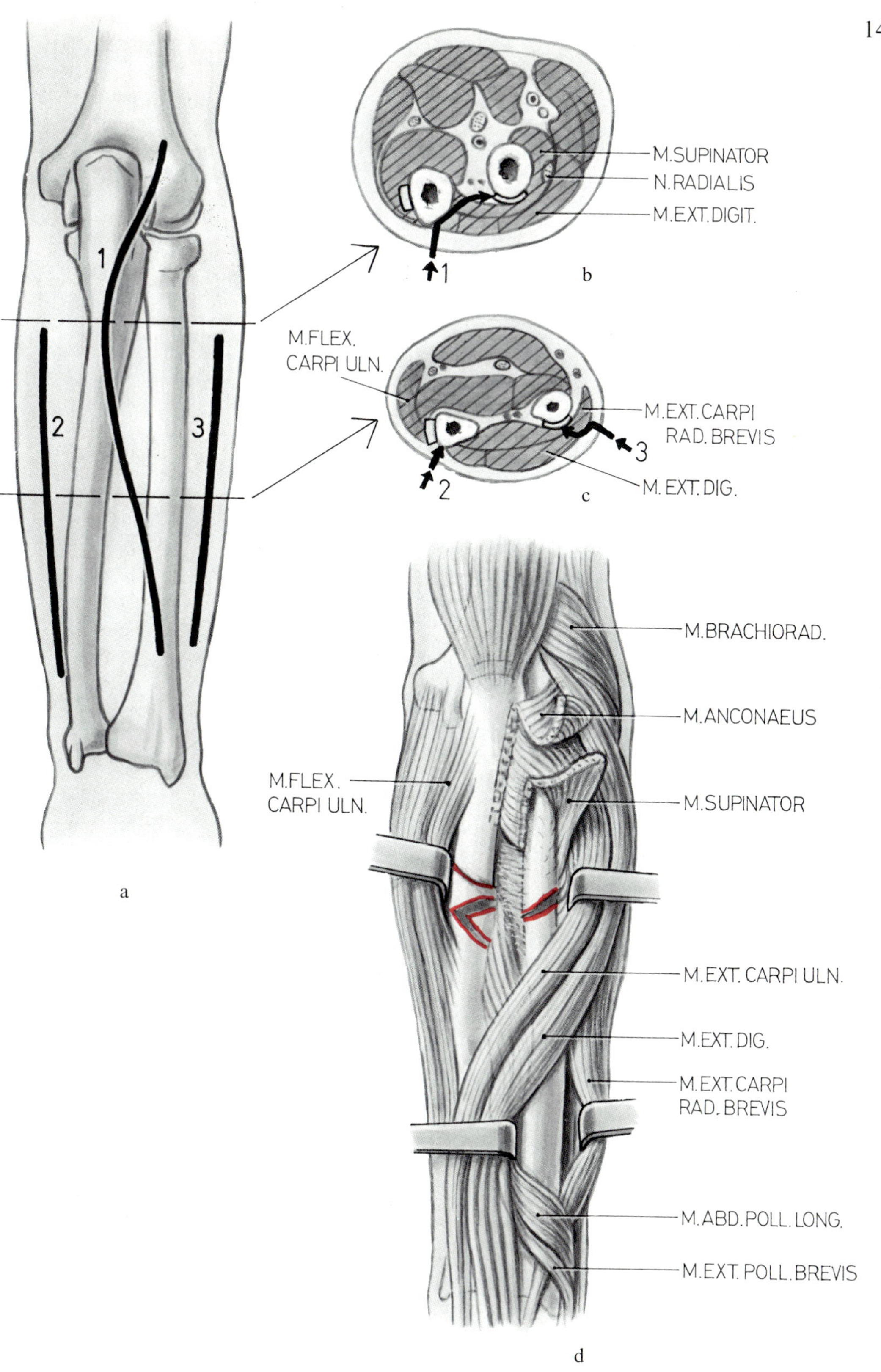
1
2
3
M.SUPINATOR
N.RADIALIS
M.EXT.DIGIT.
1
b
M.FLEX.
CARPI ULN.
M.EXT.CARPI
RAD.BREVIS
3
2
M.EXT.DIG.
c
a
M.BRACHIORAD.
M.ANCONAEUS
M.FLEX.
CARPI ULN.
M.SUPINATOR
M.EXT. CARPI ULN.
M.EXT.DIG.
M.EXT.CARPI
RAD.BREVIS
M.ABD.POLL.LONG.
M.EXT.POLL.BREVIS
d

1.4.1 Fractures of the Olecranon

Transverse and Avulsion Fractures of the Olecranon

Stable internal fixation is achieved by means of two Kirschner wires and a tension band wire (WEBER, see Fig. 26). Stable fractures without comminution may be fixed with the tension band wire alone, without the added rotational stability provided by the Kirschner wires.

Surgical Approach: see Fig. 139.

Technique: Carry out an accurate reduction and maintain it with reduction forceps with tips. Drill in two Kirschner wires through the olecranon in line with the long axis of the ulna. Next, drill a hole transversely through the distal fragment just below the subcutaneous border of the ulna. Use a 1-mm-thick wire for the tension band. Thread this wire through the transverse hole in the distal fragment, then cross it and pass it around the protruding ends of the Kirschner wires. Place the wire under tension and secure it so that the tension will be maintained. Shorten the Kirschner wires, bend over the protruding ends to form little hooks, and drive them in over the tension band wire into bone.

Oblique Fractures: Carry out a reduction and maintain it either with two Kirschner wires or with reduction forceps with tips. Carry out the internal fixation either with a 6.5-mm cancellous lag screw, which is inserted in the direction of the medullary canal of the distal fragment or with two 3.5-mm cortical lag screws and a tension band wire.

Comminuted Fracture of the Olecranon: In dealing with comminuted fractures of the olecranon which involve the joint, the most important step is the careful reconstruction of the joint surface. The reduction is then best protected either with a semi-tubular or with a one-third tubular plate. All areas of comminution as well as defects in bone are bone grafted with cancellous bone.

Fig. 142 *Transverse fracture of the olecranon.* The combination of Kirschner wires with a figure-of-eight tension band wire. This particular fracture could have been stabilized with a tension band wire alone.

Fig. 143 *A comminuted fracture.* One fragment was reduced and fixed with a 2.7-mm lag screw. The remaining fragments were then reduced, the reduction maintained with two Kirschner wires and axial compression was then achieved with the tension band wire.

Fig. 144 *A comminuted fracture* fixed with a one-third tubular plate and 2.7-mm lag screws.

Fig. 145 *Oblique fracture of the olecranon associated with a comminuted fracture of the proximal ulna.* Note the fixation of the oblique fracture with lag screws, some of which go through the semi-tubular plate, and note the cancellous graft in the region of the comminution of the shaft.

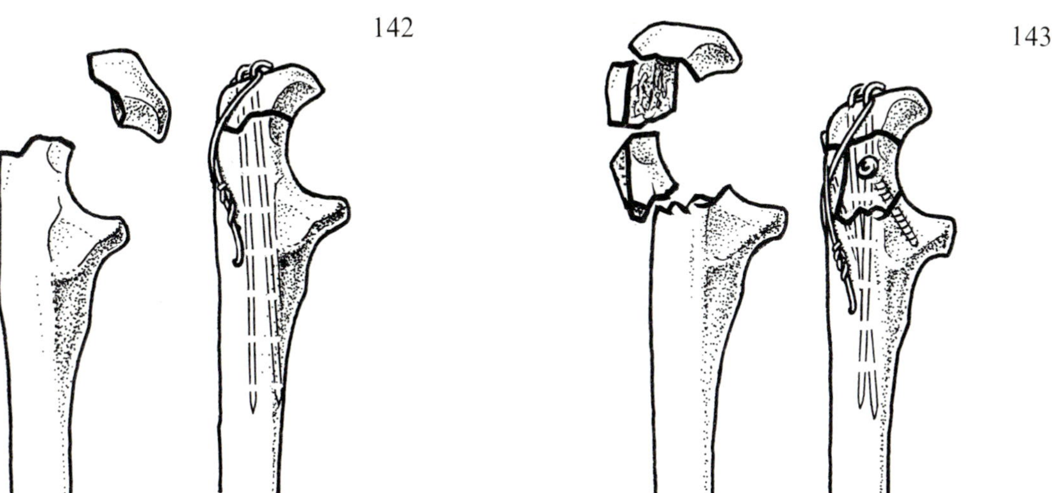
142
143

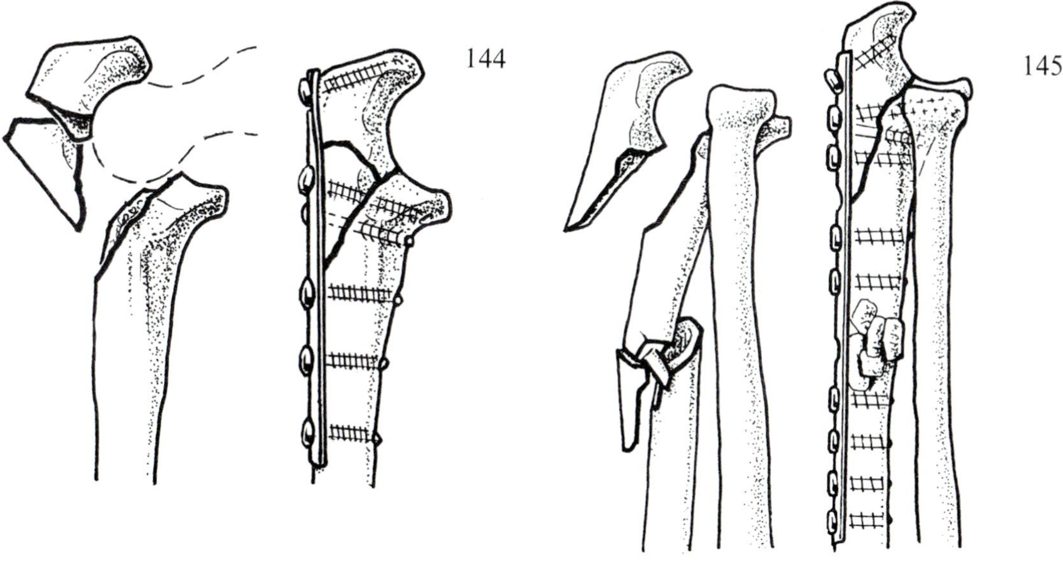
144
145

1.4.2 Fractures of the Radial Head

Only wedge fractures lend themselves to an open reduction and stable fixation either with a 4-mm cancellous screw or with 2.7-mm lag screws. Comminuted fractures and depressed fractures, if necessary, are best treated by means of resection of the radial head. If the insertion of the brachialis is avulsed together with the coronoid process, then the coronoid process must be reduced and fixed as shown in Fig. 148. Small avulsion fractures of the coronoid can be tackled only through an anterior approach. Larger fragments, particularly if associated with fractures of the radial head, may be approached from the ulnar side. A pseudarthrosis of an avulsed coronoid process often leads to habitual dislocation of the elbow joint.

Fig. 146 *A wedge fracture of the radial head* fixed with the 4.0-mm cancellous screw.

Fig. 147 *A fracture through the middle of the radial head* without comminution, fixed here with two 2.7-mm lag screws. The screws have been countersunk so that they will not irritate the overlying annular ligament.

Fig. 148 *Fracture of the coronoid process with a comminuted fracture of the radial head.* The coronoid process has been reduced and fixed with two lag screws. The radial head has been resected through its neck just proximal to the insertion of the biceps tendon.

Fig. 149 *Monteggia fracture.* Through an ulnar approach, reduction and stable fixation of the ulnar fracture as well as repair of the torn annular ligament have been carried out. In this approach it is necessary to transect only the proximal portion of the anconeus.

146

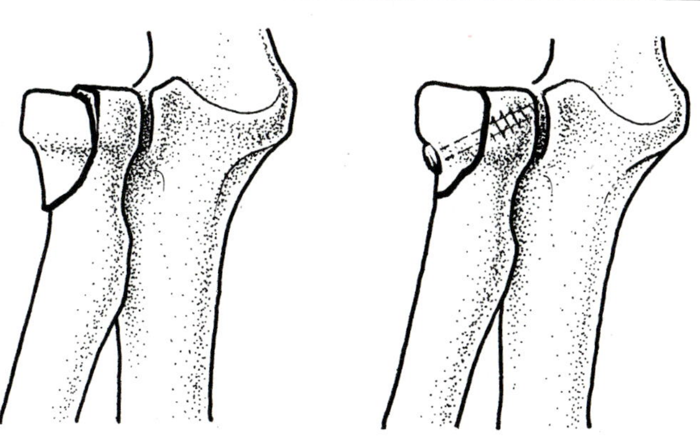

147

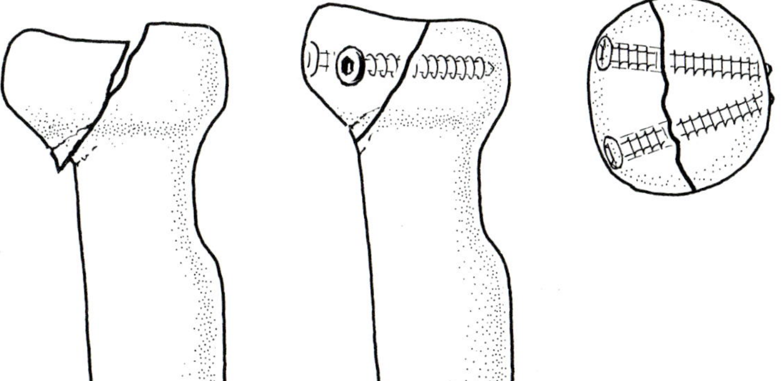

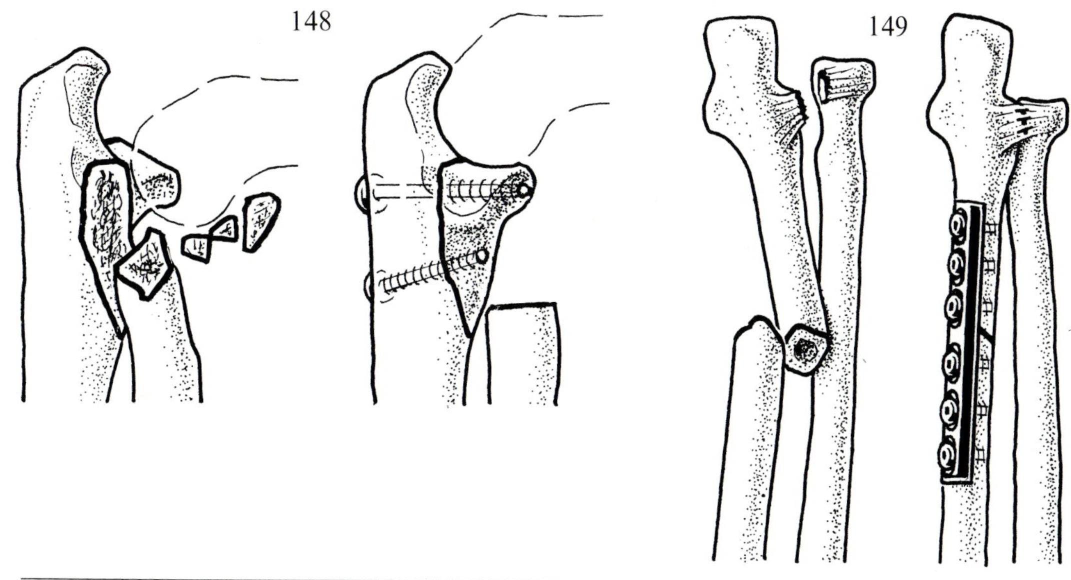

1.4.3 Fractures of the Shaft of the Radius and Ulna

The following is the technique of open reduction and internal fixation for fractures of both bones, if the fractures are through their middle third or somewhat more distally:

The surgical approach for transverse or oblique fractures is through two separate incisions [Fig. 141 (2, 3)]. In comminuted fractures we recommend the approach of THOMPSON [Fig. 141 (1)]. Expose and examine both fractures. Reduce the ulna because it is usually the simpler fracture and therefore easier to reduce. The ulna is then fixed either with a DCP or a narrow round-hole plate. The plate is fixed to the proximal fragment with two screws and to the distal fragment with a clamp. Attempt the reduction of the radius. Frequently the temporary fixation of the ulna must be released before the radius can be reduced which should be done in maximal supination. Stabilize both fractures provisionally and check the pronation and supination. If the range of movement is normal and the reduction satisfactory, complete the internal fixation of the radius and ulna. Whenever possible, insert a lag screw across the fracture. All comminuted fractures must be bone grafted with cancellous bone. Care must be taken not to place the bone graft on the interosseous membrane.

Post-operative Care: The patient gets up on the first post-operative day and begins active mobilization of the limb. In badly comminuted fractures which could not be fixed with compression plates because it would have resulted in shortening, the upper limb should be splinted in plaster for 3–4 weeks before movement is commenced.

Fig. 150 *A simple fracture of the ulnar shaft.* Transverse and oblique fractures of the shaft of the ulna are fixed with a small six-hole DCP.

Fig. 151 *A spiral fracture of the radius.* A 2.7-mm lag screw has been inserted in addition to the lag screws and semi-tubular plate.

Fig. 152 *Fractures of both bones of the forearm.* Transverse fractures of both the radius and ulna reduced and fixed with two small six-hole DCP.

Fig. 153 *Comminuted fractures of both the radius and ulna.* The fractures were approached through the THOMPSON approach. Note the use of the small DCP on the radius and ulna. The torn interosseous membrane was re-sutured and the defect in the radius was bone grafted.

Fig. 154 *Galeazzi fracture.* The fracture of the radius is fixed either with a DCP or with a semi-tubular plate. The avulsed ulnar styloid is reduced and fixed with a 4-mm lag screw.

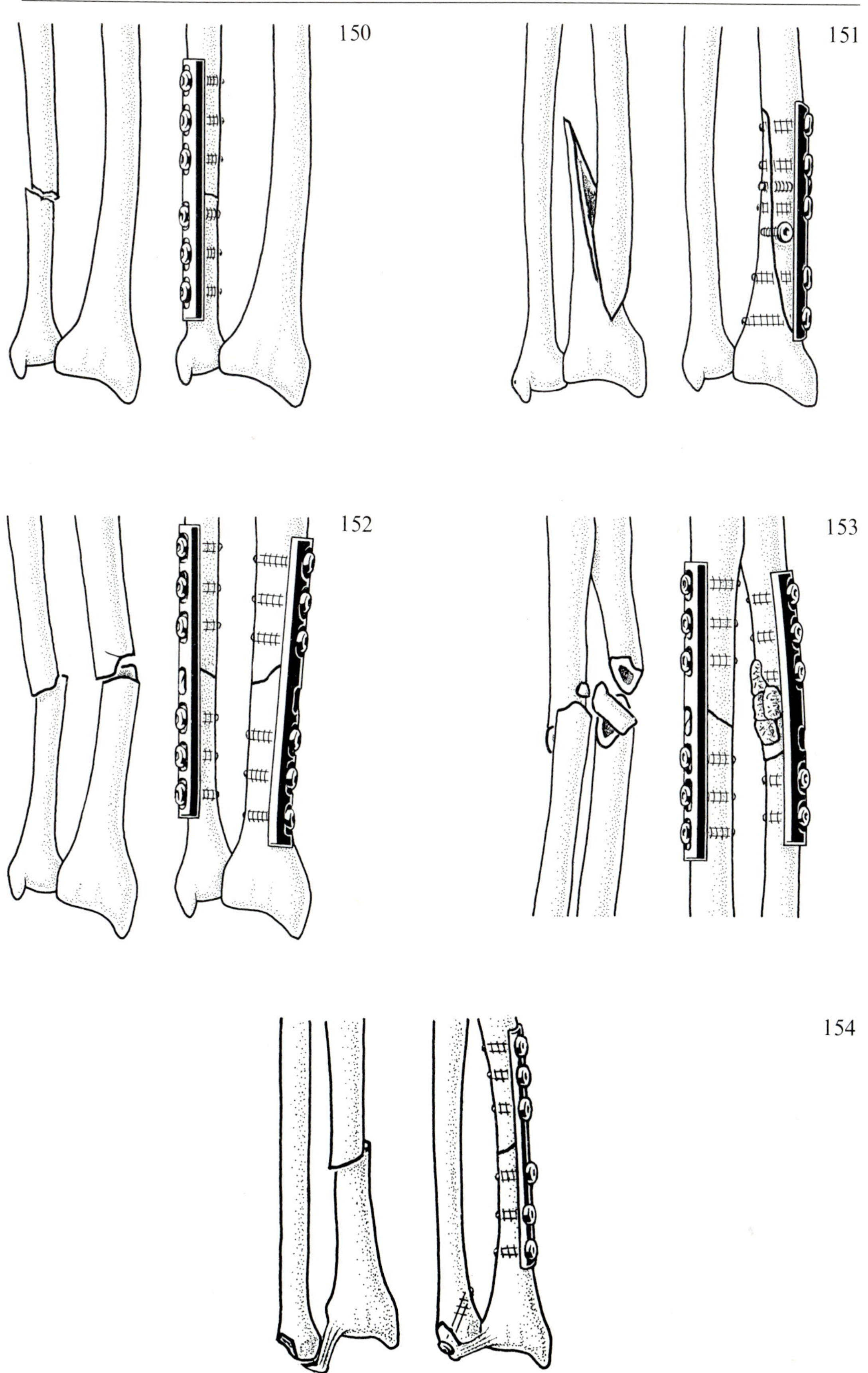
150
151
152
153
154

Anterior Approach to the Distal Radius

This approach is used for the reduction and fixation of the flexion fracture of Smith-Goyrand.

Fig. 155 *Anterior approach to the distal radius.*

a Anterior incision radial to the easily identified tendon of flexor carpi radialis. The incision begins at the most distal flexor crease and extends proximally for a distance of 6–8 cm. The deep fascia is divided in line with the skin.

b Cross-section – the *arrows* outline the approach.

c The tendons of the flexor carpi radialis and flexor pollicis longus are retracted to the ulnar side and the radial artery to the radial side. The pronator quadratus is now divided at its radial insertion and reflected. The cut ends of the muscle are marked with sutures.

d Once suitable retractors are inserted, the fracture can be easily exposed and analyzed.

Fig. 156 *The technique of internal fixation of a Smith-Goyrand fracture.* Extend the wrist over a hard roll (a). The plate is fixed first to the proximal fragment with two screws. The fracture is thus buttressed and reduces almost spontaneously (b). The volar fragment is fixed with cancellous bone screws (c).

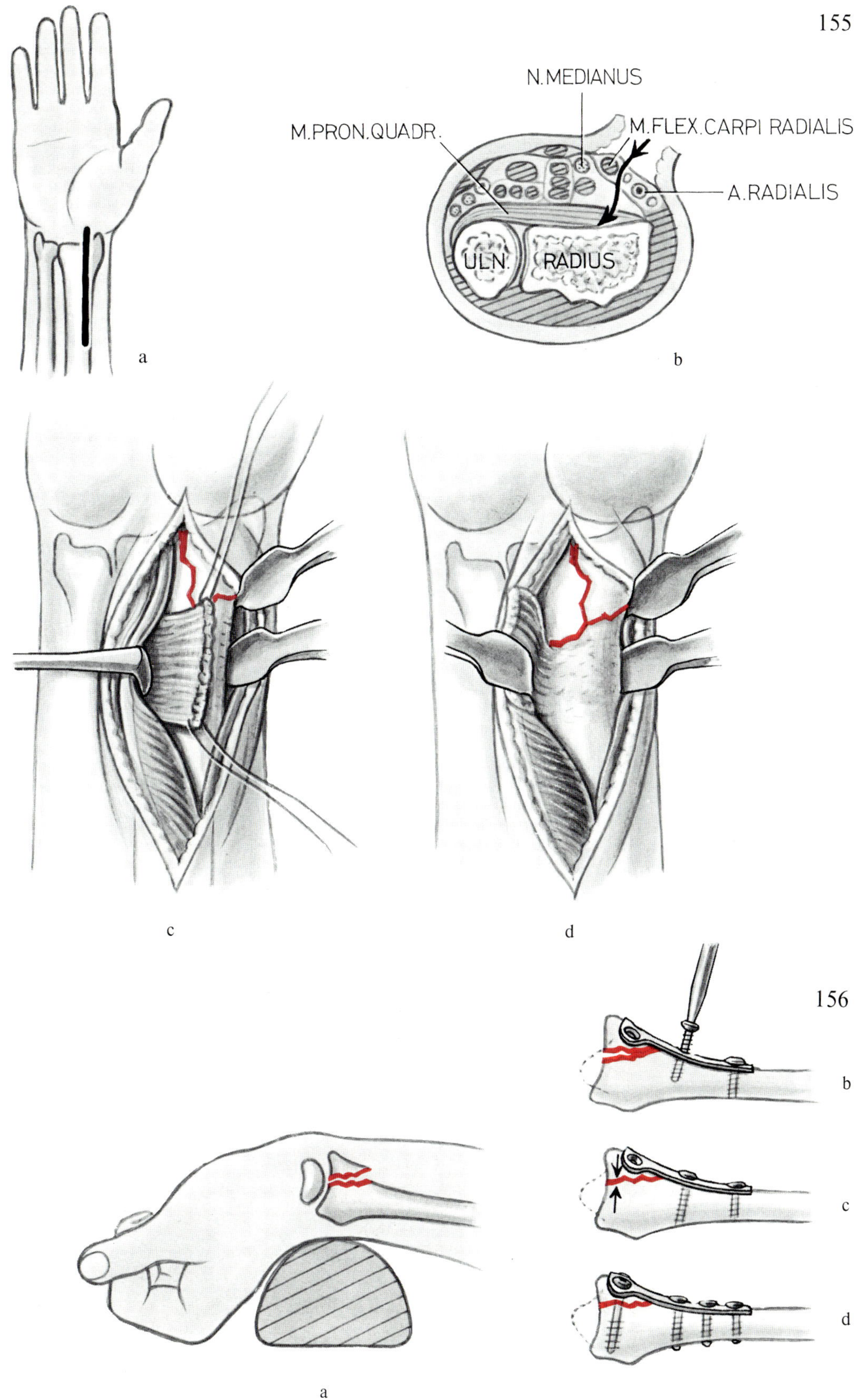
155
N. MEDIANUS
M. PRON. QUADR.
M. FLEX. CARPI RADIALIS
A. RADIALIS
ULN.
RADIUS
a
b
c
d
156
b
c
d
a

1.4.4 Intra-Articular Fractures of the Distal Radius

Colles fractures and intra-articular fractures of the distal radius.

In young patients, if closed reduction fails to restore joint congruency, length and axial alignment, an open reduction and internal fixation are indicated. This applies particularly to intra-articular fractures and to unstable open fractures of the distal radius. At times, reduction and fixation can be achieved only with a small external skeletal fixator and 3-mm Schanz screws.

The technique of open reduction and internal fixation of comminuted and compressed fractures of the distal radius.

The procedure is carried out under axillary block with tourniquet control. Make a straight 6–7-mm-long incision over the lateral border of the radius. Identify and retract dorsally the dorsal cutaneous branch of the radial nerve. The tendons of the extensor pollicis brevis and abductor pollicis longus are retracted volarly. Examine the fracture. Elevate the fragments and reduce the joint surface. Fix the sagittal split of the distal fragment with a lag screw. Reduce the transverse component of the fracture and stabilize it with two or three Kirschner wires as recommended by WILLENEGGER (Fig. 158) or with a small T plate. If the compression of the metaphysis has resulted in a defect, fill it with cancellous bone prior to the carrying out of the internal fixation. Shorten and bend over the protruding ends of the Kirschner wires. In teenagers the wires may be left protruding through the skin. Immobilize the arm in a long arm cast for 4 weeks. Remove then the Kirschner wires and begin with active mobilization.

The Flexion Fracture of Smith-Goyrand: These, if oblique, are unstable and should be openly reduced and fixed. Most commonly the dorsal cortex of the radius remains intact (for exposure and technique see page 194).

Fig. 157 *Irreducible fracture of the distal radius.* The displaced fragment may be caught among the tendons going to the thumb and a closed reduction fails. The fixation is carried out either with a Kirschner wire or a small cancellous screw. The approach is between the tendons of the extensor pollicis brevis and the extensor carpi radialis longus. Identify and retract the dorsal branch of the radial nerve.

Fig. 158 *An impacted Colles fracture with intra-articular extension and joint incongruity in a young patient.* The approach is between the tendons of the extensor carpi radialis brevis and extensor pollicis longus. Divide the extensor retinaculum, reduce the joint surface and fix the fragments with a 4.0-mm cancellous screw. Reduce the transverse metaphyseal fracture, making certain that the distal articular surface of the radius subtends an angle of 25° in the AP plane and is in 10° of volar angulation. Cancellous bone graft is added to fill the dorsal defect and the fixation is completed with Kirschner wires or at times with a small T plate.

Fig. 159 *The three types of Smith-Goyrand fracture.*

a This fracture is treated non-operatively. Operative intervention is required if reduction in extension fails.

b Small volar fragments and an oblique fracture.

c A large distal volar fragment and an oblique fracture line. b and c are unstable and require open reduction and internal fixation.

Fig. 160 *Internal fixation of the Smith-Goyrand fracture* with an additional ventral sagittal split of the fragment. Use the anterior approach and fix the fracture with a small T plate.

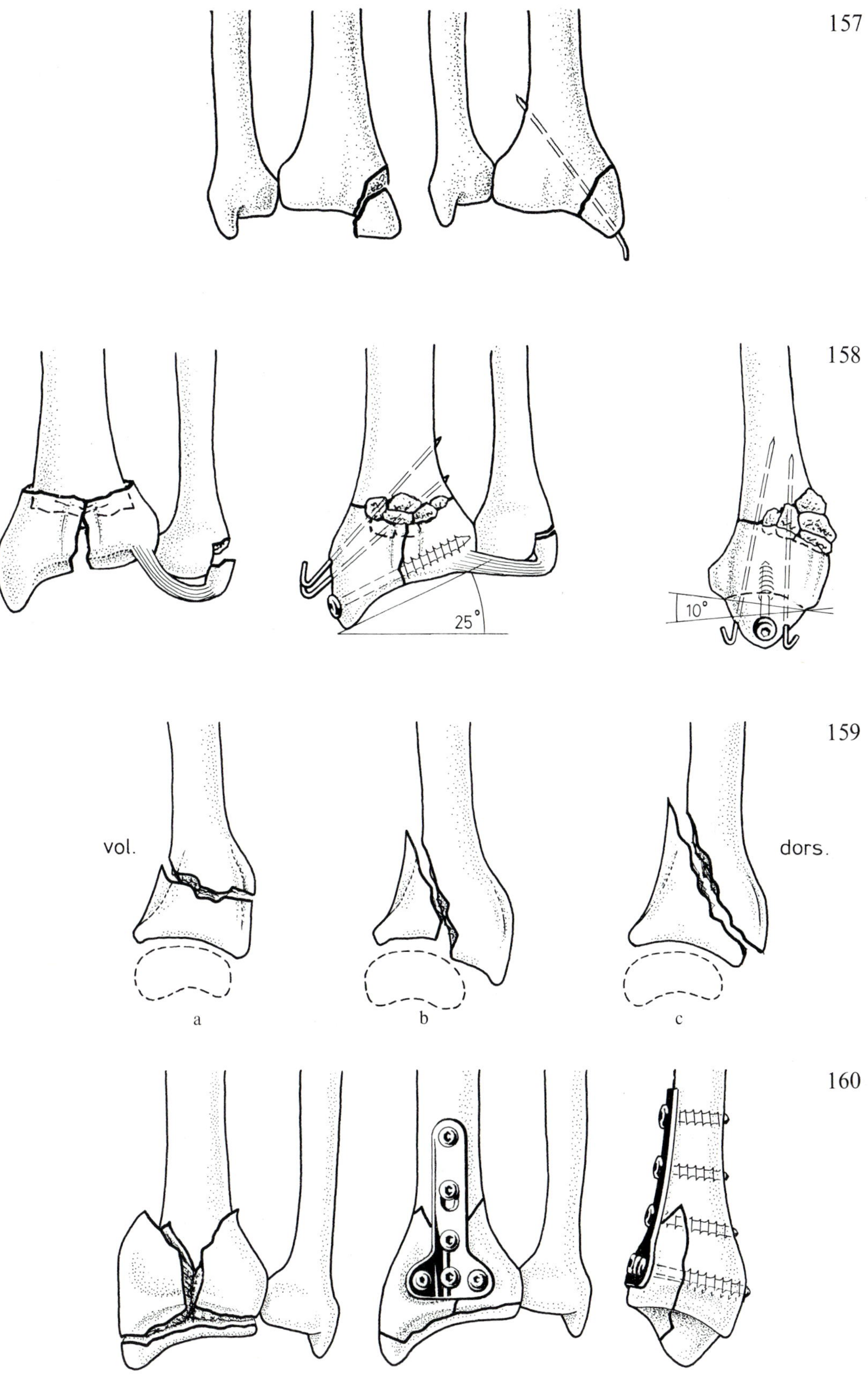
157
158
25°
10°
159
vol.
dors.
a
b
c
160

1.5 Fractures of the Hand

Most of the fractures of the hand are treated non-operatively. Indications for surgery are the irreducible fractures, fractures which cannot be maintained reduced as well as the avulsion intra-articular fractures. We have found stable internal fixation particularly worthwhile in the treatment of open fractures of the hand with concomitant injuries to tendons and nerves. Contraindications to surgery are badly comminuted fractures where internal fixation is impossible, and impaired blood supply to the part.

Surgical Approaches: Prior to making the incision draw its position on the skin. Aim for the best cosmetic result and no interference with function. On the dorsum of the hand a single straight incision can be used to expose two metacarpals. If the metacarpophalangeal joint has to be exposed at the same time, extend the incision to create a flap. Phalangeal fractures are exposed through the classic dorsolateral incision.

Fig. 161 *Surgical approaches for internal fixation of fractures of metacarpals and phalanges.*

1 The approach to fractures of the scaphoid (only for shear fractures).

2 The approach to the metacarpophalangeal joint of the thumb.

3 Dorsal, straight incisions with extensions to form flaps distally for exposure of metacarpals *2,3* or *4,5*. The exposure of such a metacarpal fracture (*3,1*).

4 The approaches for fractures of the phalanges:
The standard approach to the metacarpophalangeal joint and proximal phalanx. With this incision do not elevate skin flaps if at all possible.

5 The lateral approach to the extensor mechanism:
Flexion of the PIP joint permits exposure of the volar structures as well as exploration of the flexor sheath. Proximal transection of the lateral collateral ligaments of the PIP joint permit full exposure of the joint surfaces. At the end of the procedure, the ligament must be re-attached to the bone.

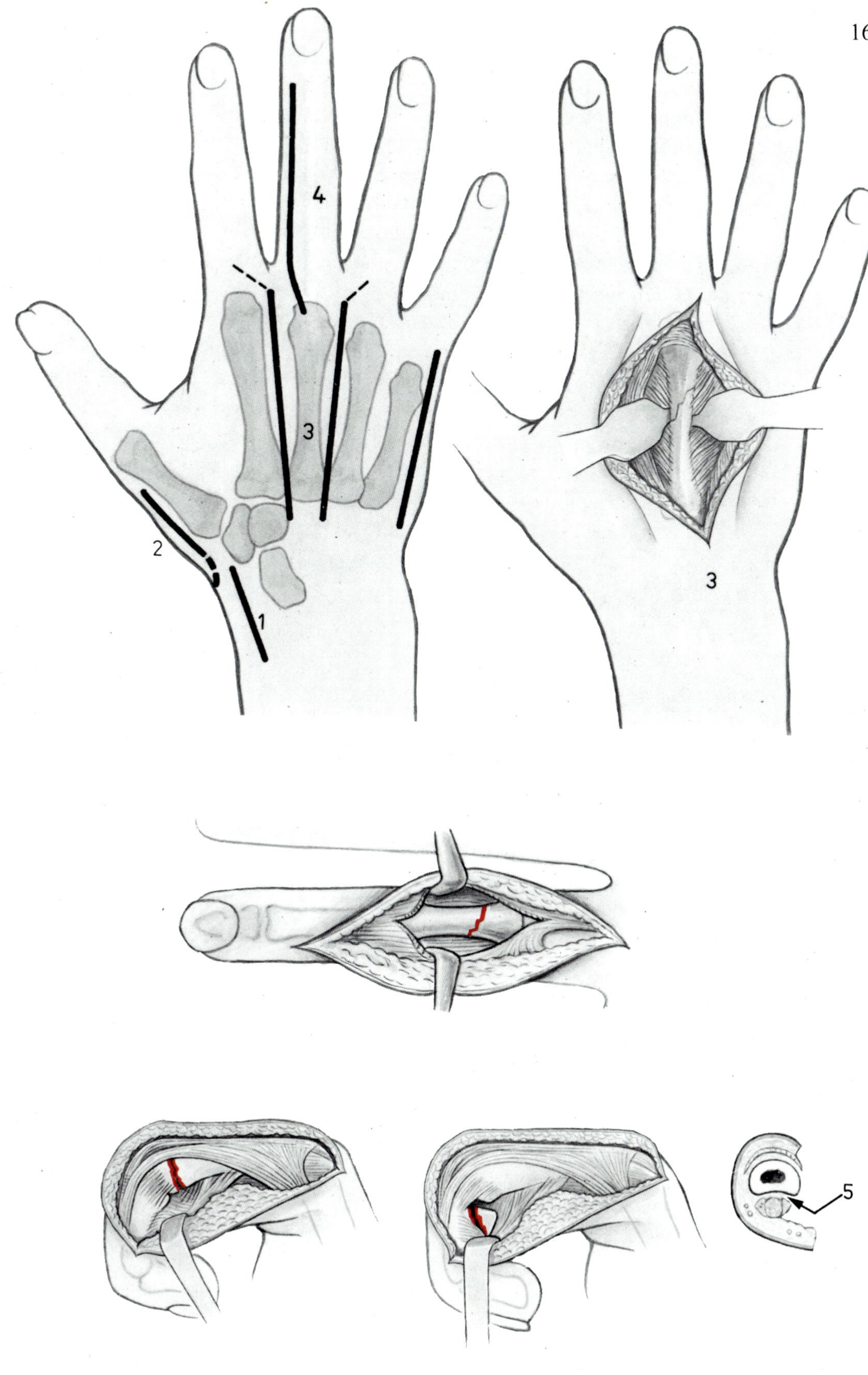
4
3
2
1
3
5

In order to achieve good functional results, the surgery must be meticulous and atraumatic. One must take particular care in the handling of the easily damaged flexor sheaths, veins and nerves.

Most of the fragments are small and short and often difficult to expose. Provisional reduction is often difficult and the internal fixation even more difficult. The fractures come under the influence of strong bending and shearing forces, particularly during the period of post-operative mobilization. This demands the most careful choice of implants, which must be adequate for the job and which should be used wherever possible as tension bands.

Post-operative Care: Of prime importance is the elevation of the limb. In addition to suction drainage, we quite often use the classic compression dressings and conventional drainage. The degree of early mobilization has to be geared to the stability of fixation achieved at surgery. Most often sufficient stability is achieved to permit early active mobilization. At times, because of danger of overload and subsequent loss of reduction, we use removable splints for a few weeks.

Metal is removed once the bone architecture has returned to normal. This requires a minimum of 4 months. Lysis of adhesions such as tenolysis can be carried out at the time of metal removal.

Fig. 162 *Examples of typical internal fixations of the hand.*

a Stabilization of a displaced vertical shear fracture of the scaphoid. The 4-mm cancellous screw is introduced through the dorsoradial aspect of the bone.

b The internal fixation of a Bennett fracture with a 4.0-mm cancellous screw.

c Tension band fixation with a small T plate of a transverse fracture through the base of the first metacarpal. This fracture is dealt with in exactly the same way as the Rolando fracture. In the latter, the articular extension would have been stabilized previously with a lag screw.

Fig. 163 *Typical examples of internal fixation of metacarpals and phalanges.*
The transverse and oblique fractures through the shaft of metacarpals 1, 2 and 5 are stabilized with one-quarter tubular plates. Where the fractures approach the joint, they are stabilized with the little T plates or with the 2.7-mm screws.
For screw fixation of spiral fractures of the shafts or intra-articular avulsion fractures of the phalanges, we use 2.7- or 2.0-mm cortex screws. Tiny avulsion fractures are stabilized with the 1.5-mm cortex screws.

162

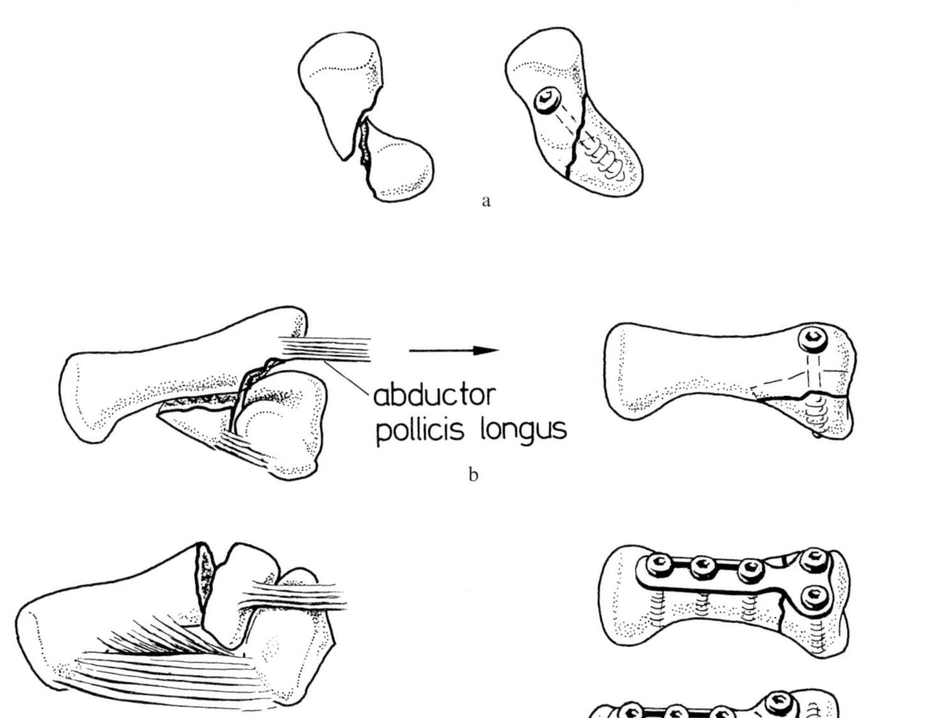

163

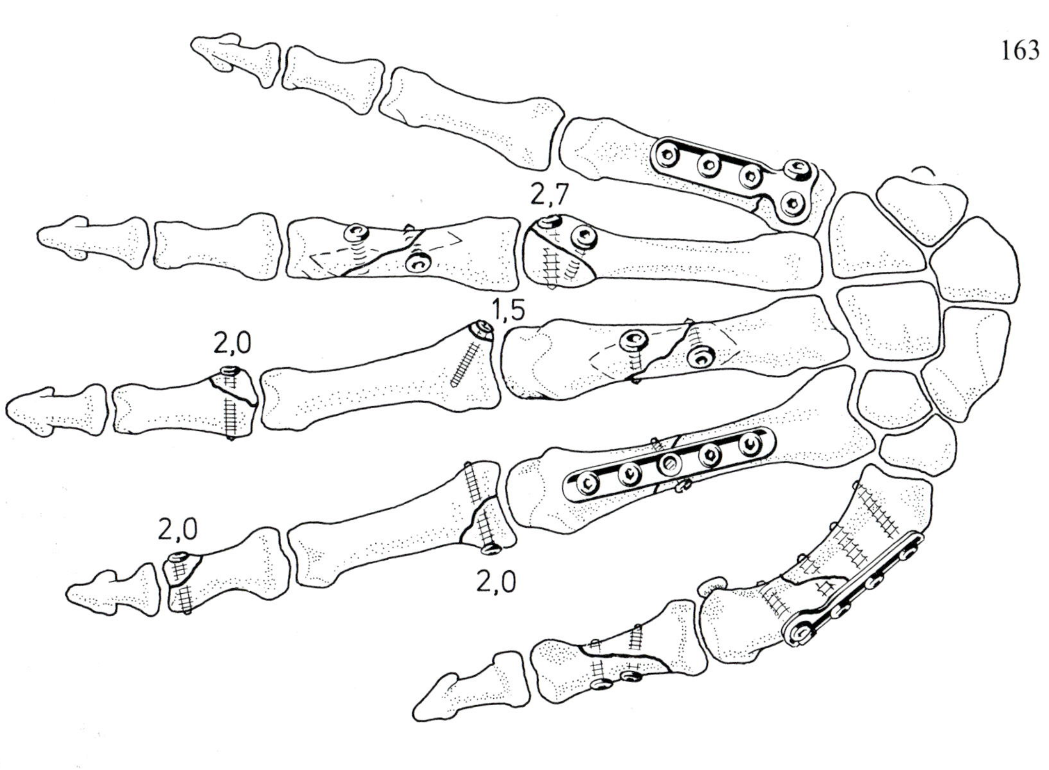

1.6 Fractures of the Acetabulum

Because of incongruity displaced acetabular fractures result in early post-traumatic osteoarthritis. Anatomical reduction and stable internal fixation of the important weight-bearing portions of the acetabulum improves the prognosis of the displaced fractures (Judet and Letournel). The internal fixation of the acetabular pillars (see below) still represents one of the most difficult tasks in hip surgery.

The operation should be performed within the first days following trauma after appropriate preparation and resuscitation of the patient. The surgeon must have the necessary skill and experience to tackle such a fracture. Concomitant anterior or posterior dislocations of the hip should be reduced as an emergency and maintained in skeletal traction through a supracondylar Steinmann pin until reduction and fixation of the fracture are carried out.

We prefer to position the patient on his side. This permits mobility of the leg which is essential at times to allow or to ease the reduction of the fracture. The lateral position also allows a simultaneous anterior and posterior approach without having to close one incision and then re-position the patient. We have found supracondylar traction to be extremely useful in these procedures. The knee should be kept flexed throughout the whole procedure. This takes tension off the sciatic nerve and allows its retraction. Simple posterior lip fractures can be operated on with the patient in the prone position.

1.6.1 Surgical Approaches

In order to decide whether to use the anterior or posterior approach, one must carefully evaluate the four standard X-rays of the acetabulum (see Fig. 164). Occasionally the acetabular roof with the adjacent wing of the ilium is best exposed through an osteotomy of the greater trochanter and upward reflection of the abductor musculature. We have found, however, that the best approach to both acetabular pillars is through a simultaneous anterior and posterior approach.

Fig. 164

a *The posterolateral approach of* Langenbeck-Kocher. The incision is carried up to the tip of the greater trochanter and is then directed to the posterior superior spine of the ilium. Gluteus maximus is split in line with its fibres and retracted apart. The small external rotators are transected 1–2 cm from their insertion into the greater trochanter. This approach allows exposure of fractures of the acetabular roof of the posterior rim or of the posterior pillar all the way down to the ischium.

If the fracture extends through the wing of the ilium, it is best approached by developing the plane between gluteus maximus and medius (Müller).

b *The anterior ilioinguinal approach of* Judet-Letournel. A tape is passed around the femoral nerve and the iliac vessels and these are retracted out of the way. Care is taken not to damage the lateral femoral cutaneous nerve. This approach gives an excellent exposure of the anterior pillar both proximally and distally, of the floor of the acetabulum, of the inner aspect of the pelvis and of the superior pubic ramus. This approach permits internal fixation of the whole anterior ilio-pubic pillar as well as of the whole superior pubic ramus. Occasionally it is enough to carry out the ilio-crural approach of Smith-Petersen.

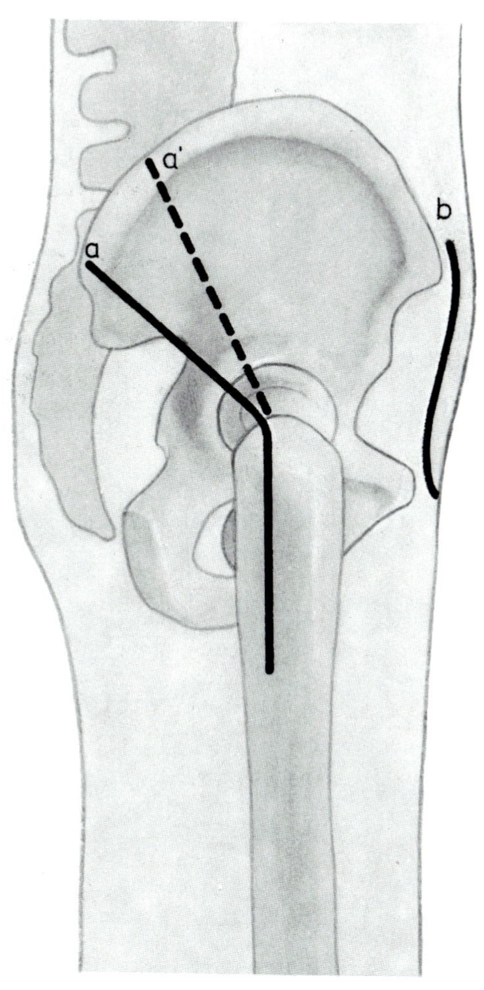
a'
b
a

1.6.2 Diagnosis of Acetabular Fractures

In the centre of the acetabulum is the Y-shaped intersection of three bones (see Fig. 165):

- Cranially or superiorly is the ilium, the acetabular roof.
- Dorsally or posteriorly is the ischium, the posterior acetabular lip.
- Anteriorly is the os pubis, the anterior acetabular lip.

If we draw an imaginary line through the middle of the os ilium (*broken line*, Fig. 165) we can appreciate better the more slender anterior pillar and the massive powerful posterior pillar of the acetabulum. To obtain a clear pre-operative evaluation of an acetabular fracture we recommend the four radiographic projections of JUDET-LETOURNEL (Fig. 166):

1. An AP projection of the pelvis allows one to evaluate both sides and rule out bilateral injuries.
2. An AP projection of the involved hip which allows an analysis of six important features (see Fig. 166a).
3. The obturator or internal oblique projection is taken with the patient rotated through 45° towards the uninvolved hip (see Fig. 166b). It permits the evaluation of the ventral or anterior pillar.
4. The ala or external oblique projection is taken with the patient rotated through 45° towards the involved hip. It permits chiefly the evaluation of the dorsal or posterior pillar.

Fig. 165 *The three bones of the acetabulum.*

A Os ilium = cranial or roof of the acetabulum.

B Os ischium = dorsal or posterior acetabular lip.

C Os pubis = ventral or anterior acetabular lip.

The *broken line* divides the anterior and posterior acetabular pillars.

Fig. 166 *The radiological diagnosis of acetabular fractures.*

Take first an AP X-ray of the pelvis. Then take:

a AP X-ray of the involved hip; the six features to be analysed are: (*1*) the ilio-pubic line, (*2*) the ilioischial line, (*3*) the tear drop of Koehler, (*4*) the anterior acetabular lip, (*5*) the posterior acetabular lip, (*6*) the acetabular roof.

b *Obturator or internal oblique exposure*, tilt the pelvis 45° to the sound side. Note (*7*) the anterior pillar; (*5*) the posterior acetabular lip; (*8*) the obturator foramen; (*6*) the region of the acetabular roof.

c *The ala or external oblique exposure;* tilt the pelvis 45° to the involved side. Note (*9*) the posterior edge of the os ischium and (*4*) the anterior acetabular lip.

165

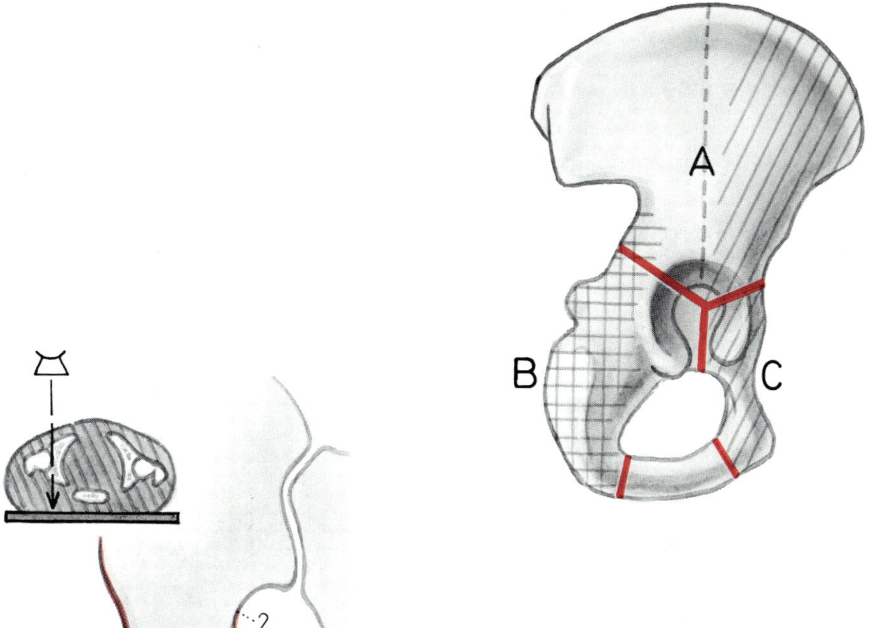

166

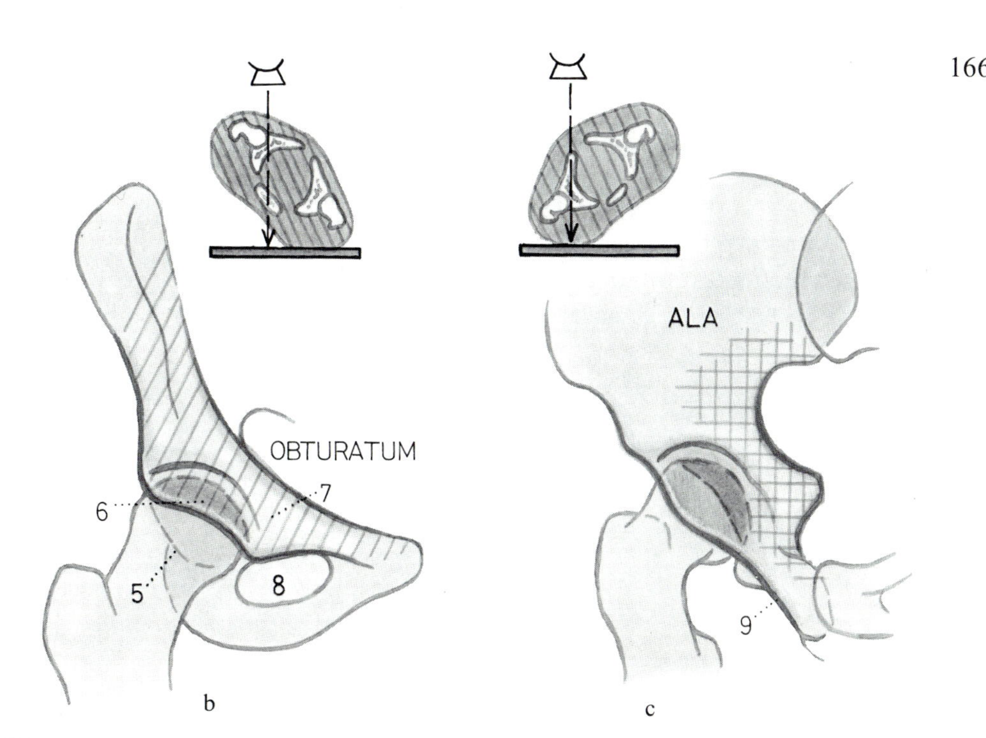

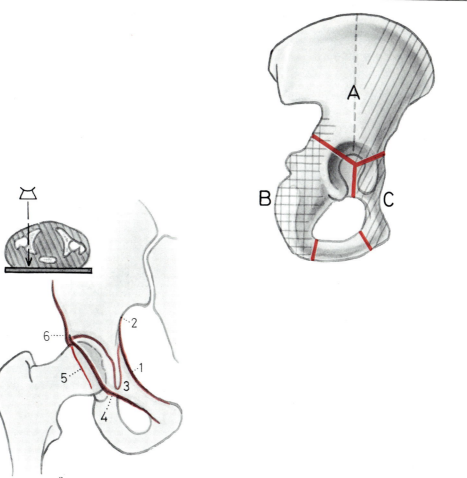

1.6.3 Classification of Acetabular Fractures

We recognize three basic patterns of acetabular fractures which may be present alone or in combination. The magnitude and the direction of the force determines the fracture pattern.

The three basic types (Fig. 167) as such are present in about 50% of the cases. These are the isolated lip fractures, the isolated pillar fractures and the transverse fracture of the acetabulum which involves both pillars.

A The isolated lip fractures

A_1 The posterior lip fracture with posterior subluxation or dislocation of the femoral head is the commonest acetabular fracture. The mechanism of the injury is a blow on the knee with the hip flexed to 90° without any abduction, i.e. the so-called dashboard fracture. With ranges of hip flexion up to 90° there is always a portion of the acetabular roof which is involved in the fracture. With flexion of the hip joint above 90° the cranial portion of the os ischium becomes involved in the fracture. The size of the fragments is best seen on the obturator or internal oblique projection.

A_2 *Fracture of the anterior acetabular lip* – this is very rare.

B Isolated pillar fractures

B_1 Dorsal or posterior pillar fracture with dorsocranial subluxation or dislocation of the femoral head.

B_2 Ventral or anterior pillar fracture with anteromedial subluxation or dislocation of the femoral head.

C Transverse fracture of the acetabulum involving both pillars

Transverse fracture of the acetabulum with central medial subluxation or dislocation of the femoral head. The mechanism of injury is usually a blow directly onto the greater trochanter.

Combined Fractures (About 50%)

Another 'type' fracture is a fracture which is composed of the basic types. The commonest of the combination fractures are the combination of A_1 and C and the oblique fracture of both pillars B_1 and B_2. Much rarer is the combination of A_2 and B_2 the so-called T fracture where the acetabular roof remains intact.

Note: In a series of 647 acetabular fractures, Judet and Letournel found the following incidence: A_1 30%, A_2 2%, B_1 and B_2 5% each, C 8%, combination A–C, combination B_1–B_2 each 20%, and combination A_2–B_2 as well as C fractures accounted for about 5% of the cases.

Fig. 167 *The basic types of acetabular fractures.*

A_1 Fracture of the dorsal or posterior acetabular lip

A_2 Fracture of the ventral or anterior acetabular lip

B_1 Dorsal or posterior acetabular pillar fracture

B_2 Ventral or anterior acetabular pillar fracture

C Transverse fracture of the acetabulum which involves both pillars

167

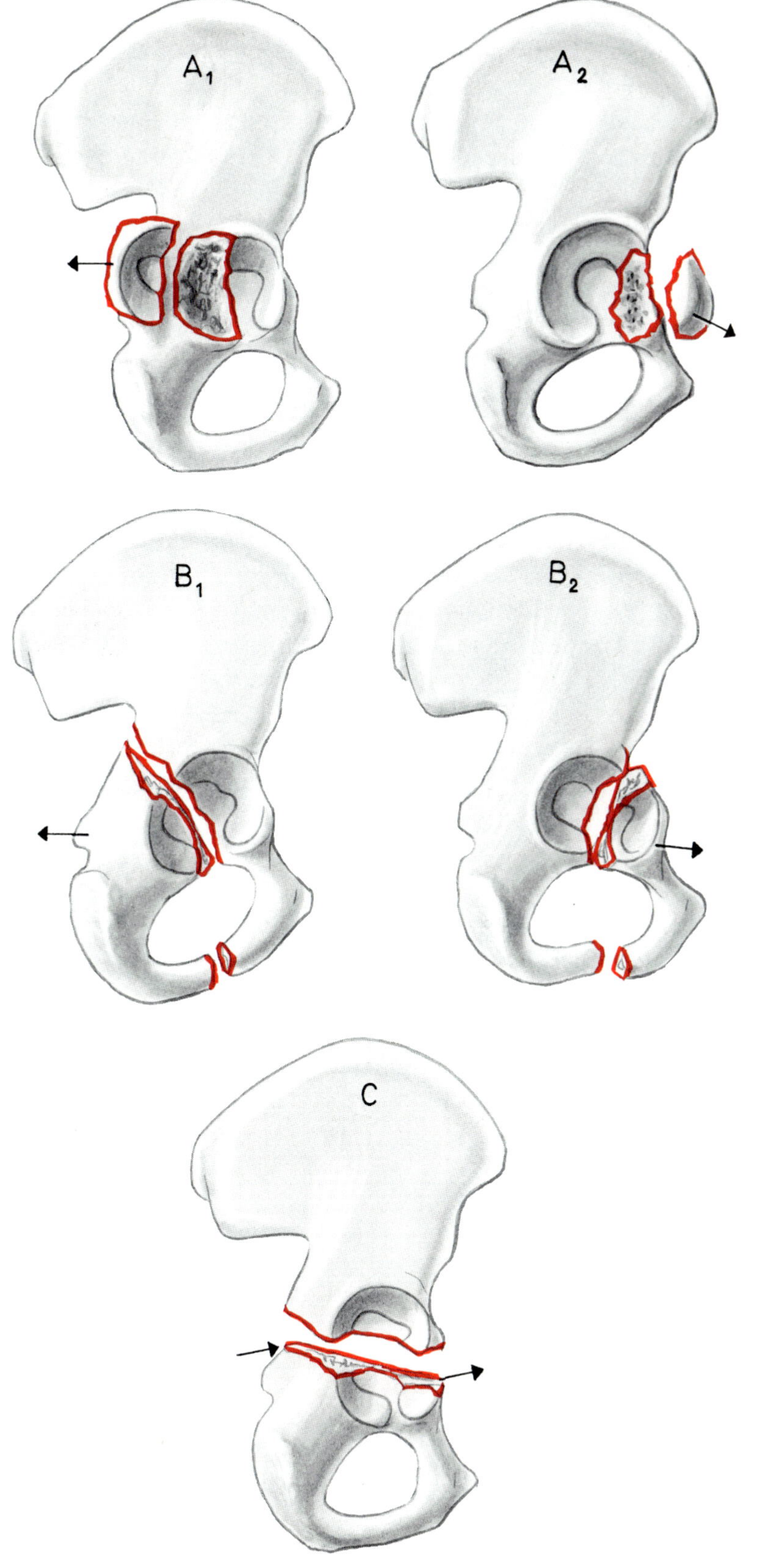

1.6.4 Technique of Open Reduction and Internal Fixation of Acetabular Fractures

Most important biomechanically is the reconstruction of the acetabular roof and of the posterior acetabular pillar. Therefore, an acetabular reconstruction must begin with the reconstruction of these two most important elements.

In fractures of the acetabular lips (A) where the fragments are larger than simple avulsion flakes of the capsule, the fragments are fixed with lag screws. Great care must, however, be taken to make certain that the screws do not enter the acetabulum. We recommend therefore that once the screws are inserted, they should be removed and the acetabular fragments re-displaced so that the position of the screw tracts can be carefully checked.

Fractures of the dorsal or posterior pillar (B_1) are carefully reduced, particularly where they involve the acetabulum, and are then fixed with a contoured six- or seven-hole DCP. Ventral or anterior pillar fractures are reduced and fixed with a contoured DCP. The DCP is applied to the pelvis along the pelvic inlet. A pure transverse fracture of the acetabulum is fixed in a manner similar to B_1.

Combination fractures are carefully analyzed and then fixed as is deemed necessary. Each component must be carefully reduced and stably fixed. In fractures of both pillars as well as in T fractures, we recommend a simultaneous anterior and posterior approach. In carrying out the reduction of these fractures we have found the reduction clamp of KNIGHT-JUNGBLUTH most useful and we have made it part of our instrumentation. Prior to reduction, two screws are inserted into the main fragments. The jaws of the clamp fasten to the screw heads in such a way that a distraction and tilting is possible and in this way the reduction of the fragments.

1.6.5 Post-operative Care

These large wounds should be thoroughly drained and closed in layers to prevent the formation of dead space. After 48 hours the drains are removed.

Three to four days after internal fixation, the patient begins active mobilization of the joint with the help of a physiotherapist. One to two weeks after surgery, the patient begins with non-weight bearing as well as with hydrotherapy in a pool.

Usually in 2–3 months the fractures have united and full weight bearing can be started.

Examples of Internal Fixation

Fig. 168 *Dorsal fracture dislocation or fracture of the dorsal pillar.* Single fragments are fixed with lag screws. Comminuted fractures require a carefully contoured plate for their stabilization.

Fig. 169 *Transverse fracture of the acetabulum.* This is stabilized with a dorsal or posterior plate very much like fractures of the dorsal or posterior pillar.

Fig. 170 *Fracture of the anterior or ventral pillar.* These are stabilized with a contoured plate applied to the ventral or anterior aspect.

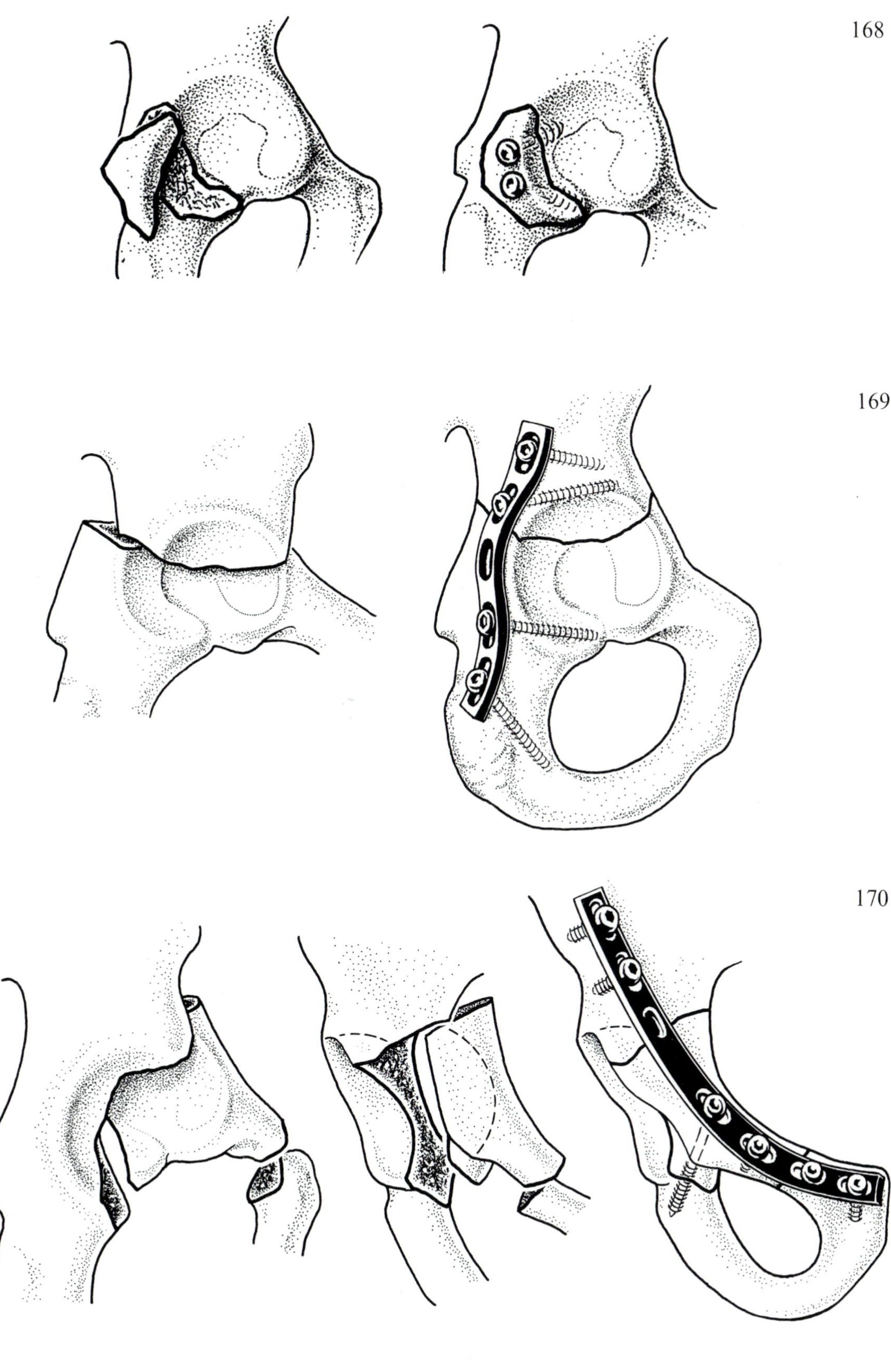
168
169
170

1.7 Fractures of the Femur

We have found the following classifications useful:
Fractures of the proximal femur with the following subgroups: Head, neck, trochanteric fractures. *Fractures of the shaft,* subtrochanteric fracture (shaft P) fractures of the middle third (shaft M) fractures of the distal femur (shaft D). *Fractures of the distal end of the femur* – either extra-articular, as for instance the supracondylar fracture, or intra-articular.

1.7.1 Fractures of the Proximal End of the Femur

1. Fractures of the Femoral Head. These are always part of a more complex injury. By and large they occur only as a result of posterior dislocations or fracture dislocations of the hip. We recognize compression fractures in young adults, avulsion fractures where the fragment remains attached to the ligamentum teres, and shear fractures. In the avulsion fractures the bony fragment may retain its blood supply through the intact ligament teres. The size of the fracture determines whether one undertakes an open reduction and fixation with lag screws. The prognosis always remains uncertain. Smaller fragments which block reduction are simply removed.

2. Fractures of the Femoral Neck. These are always intra-articular. The blood supply to the head is easily damaged and the problem with the neck fracture is avascular necrosis, which can occur in varying degrees. We recognize subcapital fractures, fractures through the middle of the neck, and fractures through the base of the neck. The mechanism of injury allows for further subclassification into abduction, adduction, and shear fractures.

Subcapital fractures and fractures through the middle of the neck. We recognize the *abduction* fracture by the valgus position and impaction of the head fragment. They are stable and do not require internal fixation. We allow our patients immediate partial weight bearing, but we forbid them to do straight leg raising. Occasionally the head is in 30° of retroversion. This malposition alone or a loss of position are indications for internal fixation (less than 5% of cases).

Adduction fractures are in varus and are often impacted posteriorly. Because of the varus, they are unstable and require internal fixation. We feel that the internal fixation should be carried out as an emergency procedure if at all possible within the first 6 hours following fracture. After the first 6 hours there is an increased danger of thrombosis of the kinked but still intact dorsocranial retinacular vessels.

We recognize the *shear fractures* by the vertical course of the fracture line. These are never impacted and are very unstable and always require surgery.

In fractures of the femoral neck, an implant alone can never neutralize the shearing forces. We feel therefore that prior to any internal fixation the fragments must be realigned and impacted in such a way that the fracture line will be under compression. To do this the head should be impacted on the cranial portion of the neck where it must sit like a 'hat on a hook' (Weber). Thus impaction of the fragments with the leg in abduction is the most important step in the open reduction and internal fixation of a neck fracture. We have found the 130° angled plate most useful for the internal fixation of the neck fractures (see pp. 66, 67 and 86). The seating chisel is inserted into the neck as far as the fracture prior to any attempt at reduction. It then serves as a lever which greatly facilitates the manipulation of the distal fragment. After reduction and impaction of the fragments, the tip of the seating chisel is driven into the lower half of the head (see Fig. 61). Further insult to the blood supply is prevented by the surgical approach and by the proper positioning of the nail within the femoral head.

We feel that a prosthetic replacement should be carried out if the neck fracture is associated with osteoarthritis of the hip or if the patient's general condition is poor and his life expectancy shortened.

Fractures through the base of the neck. These occur as a rule in young patients whose femoral neck is still filled with strong cancellous bone. These should be anatomically reduced and fixed with lag screws (for technique and examples see p. 326). Because the neck is filled with such dense cancellous bone, when one attempts to nail these there is great danger of driving the fragments apart, which may lead to tearing of the retinacular vessels and subsequent avascular necrosis. In elderly patients, these fractures are considered as part of pertrochanteric fractures.

3. Pertrochanteric Fractures: Unlike the subcapital fractures these are almost always extra-articular and are not associated with a danger of avascular necrosis. In dealing with the intertrochanteric fractures the important thing is to decide whether the fracture is stable or not, and what is the best method of achieving a stable internal fixation for the particular fracture in question.

Intertrochanteric fractures are fractures of the elderly. The average age of patients with intertrochanteric fractures is 6–12 years higher than that of patients with subcapital fractures. Surgical stabilization of these fractures allows mobilization of the patients. This prevents prolonged bed rest, which in the elderly is associated with a high mortality rate. The fixation of these fractures must be particularly stable, since the elderly patient is not able to carry out partial or non-weight bearing. The simple intertrochanteric fractures after reduction should be stabilized with either the 130° angled plate or the condylar plate. Varus is synonymous with insufficient and unstable fixation.

The unstable fractures with a single large medial fragment can be reduced and fixed with the condylar plate. The unstable four-part fractures, associated with varying degrees of osteoporosis, are best fixed in valgus with medial displacement of the shaft by means of the so-called repositioning osteotomy. We use bone cement for the internal fixation of pathological pertrochanteric fractures. The cement gives much better anchorage for the implant and allows one to carry out the so-called composite internal fixation. Prosthetic replacement of the hip joint for pertrochanteric fractures is indicated only under exceptional circumstances or when the fracture is associated with osteoarthritis of the hip.

(*Note:* Flexible intramedullary nails as well as the different types of compression screws have found their place in the treatment of intertrochanteric fractures.)

4. Fractures of the Proximal End of the Femur Associated with Ipsilateral Fractures of the Shaft: First of all one must fix the proximal fracture. If the fracture involves the proximal shaft, then both fractures can be fixed with a single plate. For a subcapital fracture we use a long 130° angled plate and for the pertrochanteric fracture a long condylar plate (see Fig. 189). If the fracture of the shaft involves the middle third then it is fixed with a broad straight plate combined with a simultaneous cancellous bone graft. Fractures of the distal third of the femur are fixed in the usual manner with the condylar blade plate (Fig. 205 A_3).

An AP and a lateral X-ray of the hip is obtained. In order to obtain the AP X-ray, the hip is internally rotated. In order to obtain the lateral one, no change is made in the position of the X-ray tube. The hip is flexed to 90° and then abducted 45° (Fig. 314). A light compression dressing is then applied to the whole limb and the leg is positioned on a sponge rubber splint with the hip in neutral. Mobilization begins on the first post-operative day with the patient sitting on the edge of the bed. On the second post-operative day the patient stands up and begins ambulation with a walker. The patient is transferred to two crutches as soon as possible. In fractures of the neck we allow only partial weight bearing of 5–10 kg during the first 3 months. The patient begins, however, with isometric muscle exercises and with active mobilization from the first day onward. We do not permit straight leg raising either forwards or to the side. After 3 months, the hip is X-rayed again and a decision is made whether weight bearing with one crutch can begin. Patients with intertrochanteric fractures are treated in very much the same way. If the fracture has a distal extension and requires considerable reflection of the vastus lateralis, then the post-operative positioning with the knee in extension could result in difficulties with knee flexion. These patients are best treated like patients with femoral shaft fractures, i.e. with the hip and knee in 90° of flexion (see p. 142). As a rule, most patients with intertrochanteric fractures begin from the third day on with partial weight bearing of 15–20 kg. Old uncooperative patients begin with almost full weight bearing whatever the instructions.

The subcapital fracture with the fracture line running through the junction of cartilage and bone runs a considerable risk of avascular necrosis because of tearing of the retinacular vessels which run posterosuperiorly along the neck to enter the femoral head (Fig. 173b). Fractures through the middle of the neck have a better prognosis as far as avascular necrosis is concerned. In the unstable adduction fractures, the head is in varus and tilted backwards. This is invariably associated with comminution of the posterior cortex and compression of the underlying cancellous bone.

The shear fracture (Fig. 172c, 178c) is the result of an axial blow to the femur. It runs vertically downwards and involves the calcar to a greater or lesser extent.

Fractures through the base of the neck are most commonly shear fractures. If badly displaced, however, they may be associated with avascular necrosis because of damage to the medial femoral circumflex artery and its branches as they course through the intertrochanteric fossa.

The Blood Supply of the Femoral Head: Most of the blood supply of the femoral head is derived from the medial femoral circumflex artery, which in the trochanteric fossa gives rise to three or four branches, the retinacular vessels. The retinacular vessels run posterosuperiorly along the neck in a synovial reflection until they reach the cartilaginous border of the head where they enter bone to supply the head.

The obturator artery gives rise to the vessels within the ligamentum teres. These as a rule supply only a small portion of bone about the insertion of the ligamentum teres. Additional blood supply to the head comes from intra-osseous vessels which run upwards from the metaphysis. In a neck fracture these, of course, are always interrupted.

The greater trochanter is supplied by an ascending branch of the lateral femoral circumflex artery. This branch anastomoses superiorly about the femoral neck with branches of the medial femoral circumflex artery.

Fig. 171 *Avulsion fracture of the femoral head attached to the intact ligamentum teres in a posterior dislocation of the hip.*

Fig. 172 *Fractures of the femoral neck.*

a Supcapital abduction fractures (impacted spontaneously).

b The medial adduction fracture.

c The medial shear fracture with involvement of the calcar.

d The shear fracture through the base of the neck.

Fig. 173 *The blood supply to the femoral head.*

a As seen from in front.

b As seen from behind (according to LANZ-WACHSMUTH: Leg, page 164).

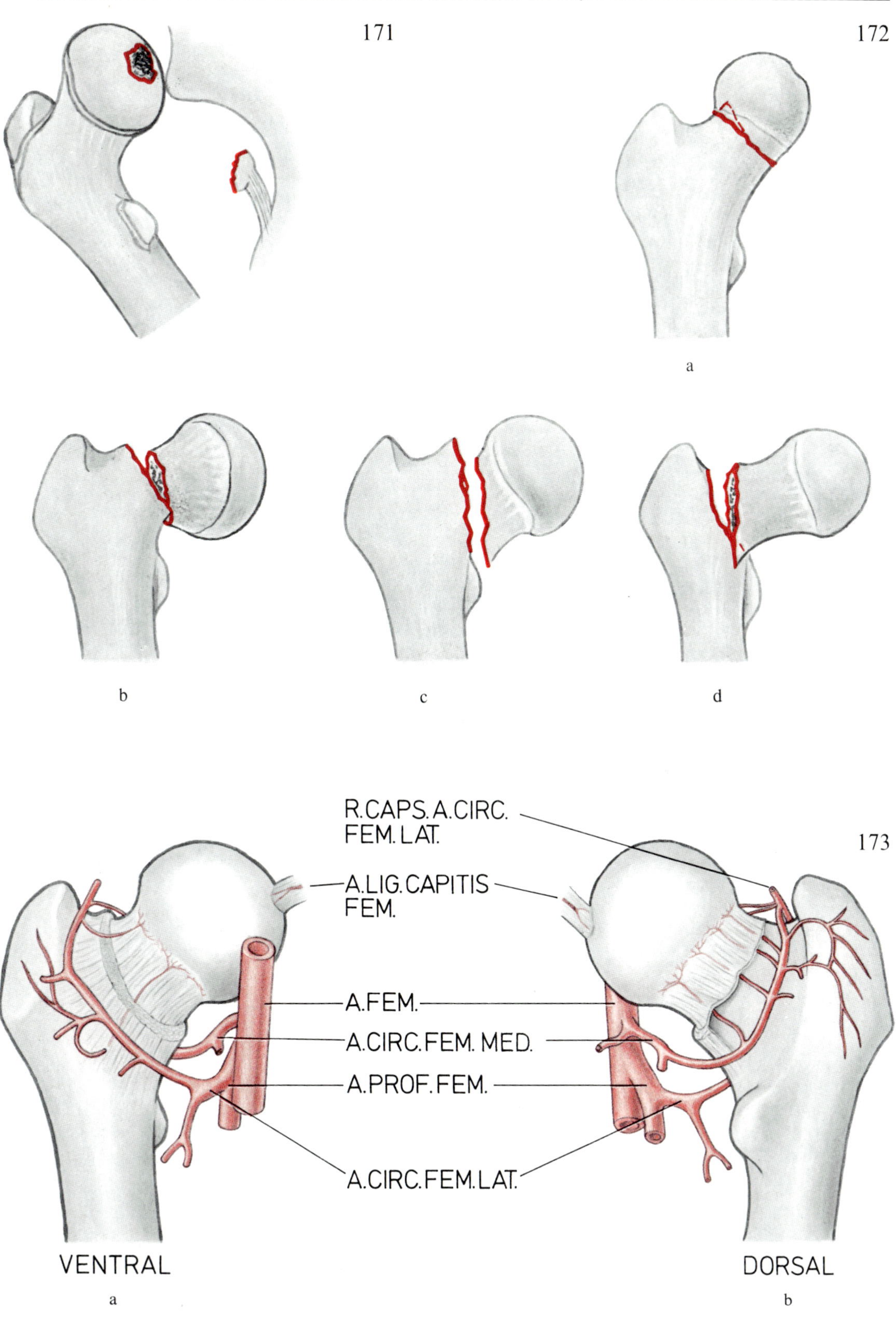
171
172
a
b
c
d
R.CAPS.A.CIRC.
FEM.LAT.
173
A.LIG.CAPITIS
FEM.
A.FEM.
A.CIRC.FEM.MED.
A.PROF.FEM.
A.CIRC.FEM.LAT.
VENTRAL
DORSAL
a
b

Fig. 174 In the adduction fracture the head is in varus and in retroversion, therefore posteriorly we find varying degrees of comminution of the cortex and compression of the underlying cancellous bone.

Fig. 175 *Principal steps in the reduction.*

a If the adduction fracture is simply anatomically reduced and fixed, one has to count on a certain degree of instability. The attitude of the fracture is such that it is subjected principally to shearing forces. Because of the posterior defect, the result of comminution and impaction, the contact between fragments is considerably reduced. Under these circumstances, no implant can overcome the adverse forces which tend to displace the fracture.

b Ideally the head should be reduced into valgus and slight anteversion. Once impacted in this position, the adduction fracture is converted into a stable abduction fracture, and rather than being subjected to shearing forces it is subjected to pure compression forces. In the axial projection we see how the tilt of the head into slight anteversion results in a much broader contact between the fracture fragments. The blade of the plate is not used to stabilize the fracture. The stability results from the reduction and impaction. The angled plate is used merely to maintain the position of the fragments.

c Ideally the tip of the blade should come to lie in the lower portion of the femoral head. In this area, the tension and compression trabecular system cross. It is the densest portion of the head where the blade will have the best purchase.

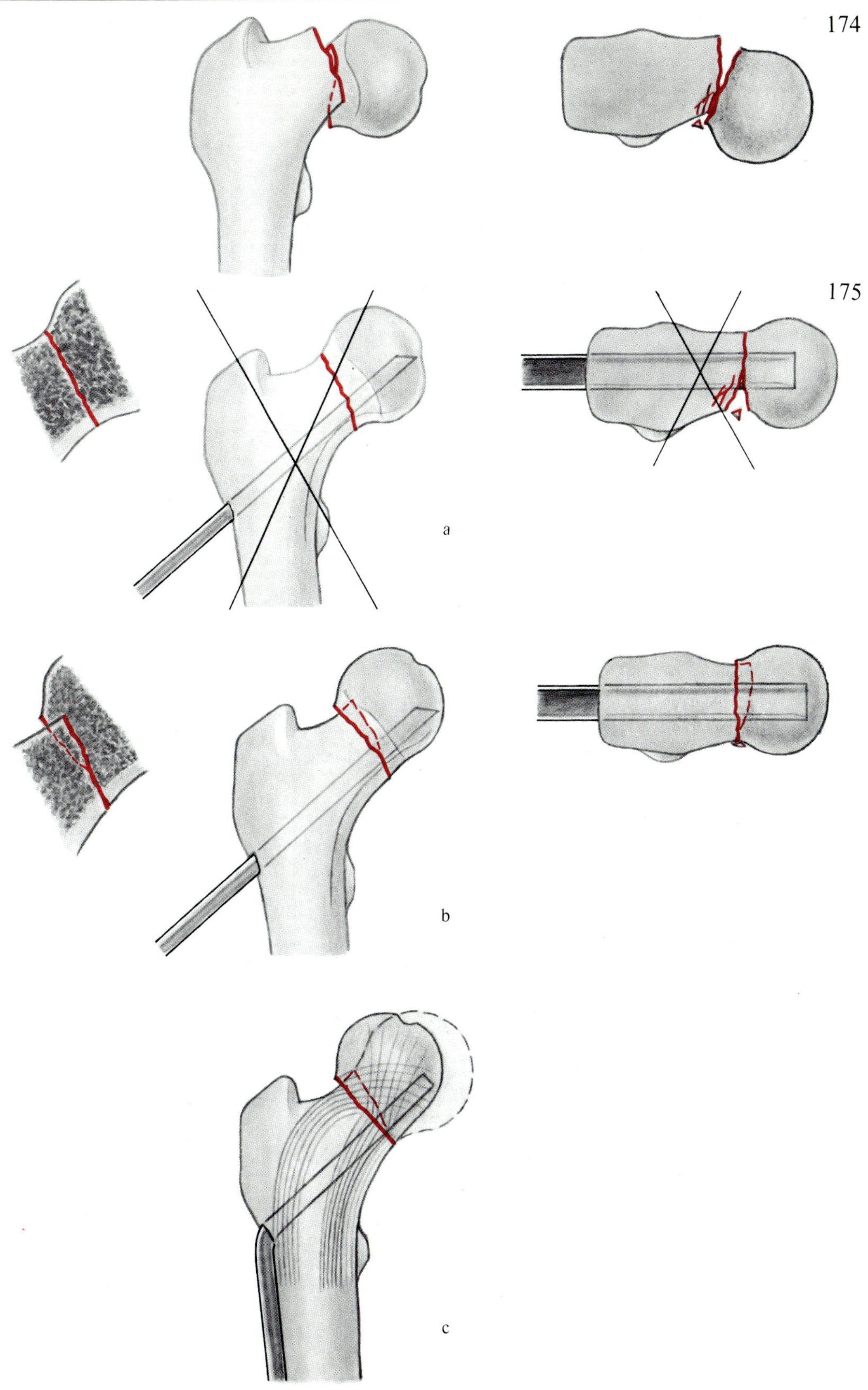
174
175
a
b
c

Fig. 176

a Use a modified WATSON-JONES anterolateral approach. Split the fascia lata in line with the skin incision. Develop the plane between the gluteus medius and minimus on the one side and the tensor fascia lata on the other. Preserve the nerve to the tensor fascia lata. Expose next the intertrochanteric region. This is done by making an L-shaped incision in the insertion of vastus lateralis and by reflecting the vastus downwards and forwards.

b The capsule is then opened anteriorly in line with the neck axis and with the aid of three Hohmann retractors, the fracture is exposed. The tip of the first 16-mm-wide long-tipped Hohmann retractor is passed over the anterior lip of the acetabulum. The second Hohmann retractor must have a short tip. It is hammered superiorly into approximately the middle of the neck. Here it cannot damage the blood supply to the head. The third Hohmann retractor is passed below the neck. The control of the head is often difficult. One can stabilize the head by hammering into its lower half the tip of an 8-mm Hohmann retractor.

c Prepare the point of entry for the seating chisel and insert the seating chisel and two 2.5-mm Kirschner wires as far as the fracture (for technique see p. 96). Remember that the blade of the seating chisel should be as close to the calcar as possible.

d In order to reduce the fracture, first of all disimpact the fracture by gentle external rotation and adduction. Then with careful traction, internal rotation and abduction, carry out an exact anatomical reduction. The reduction is greatly facilitated by having gained control of the head with the tip of the 8-mm Hohmann retractor which was inserted into the lower half of the femoral head. Stabilize the fracture by driving into the head the two 2.5-mm-thick Kirschner wires, which served also as guide wires for the insertion of the seating chisel.

e Flex the hip to 90° and check the accuracy of the reduction inferiorly and medially.

f The next important step is the impaction of the fracture. The leg is abducted and internally rotated. The internal rotation is important to prevent retroversion of the head as impaction occurs. Take the slit hammer, apply it to the base of the greater trochanter and strike it firmly with a mallet to impact the fracture. This will convert the adduction fracture into a stable abduction fracture with the head in slight valgus and anteversion.

g Flex the hip again to 90° and check the reduction at the level of the calcar. If all is well, hammer in the seating chisel a further 20 mm. This will drive it into the head. Check the hip for freedom of motion.

h Remove the seating chisel and replace it with a four-hole 130° angled plate. The blade of the plate should as a rule be 20–25 mm longer than the length of the seating chisel, measured from its point of entry to its point of exit at the fracture. Check the reduction once again and the freedom and range of movement of the hip joint.

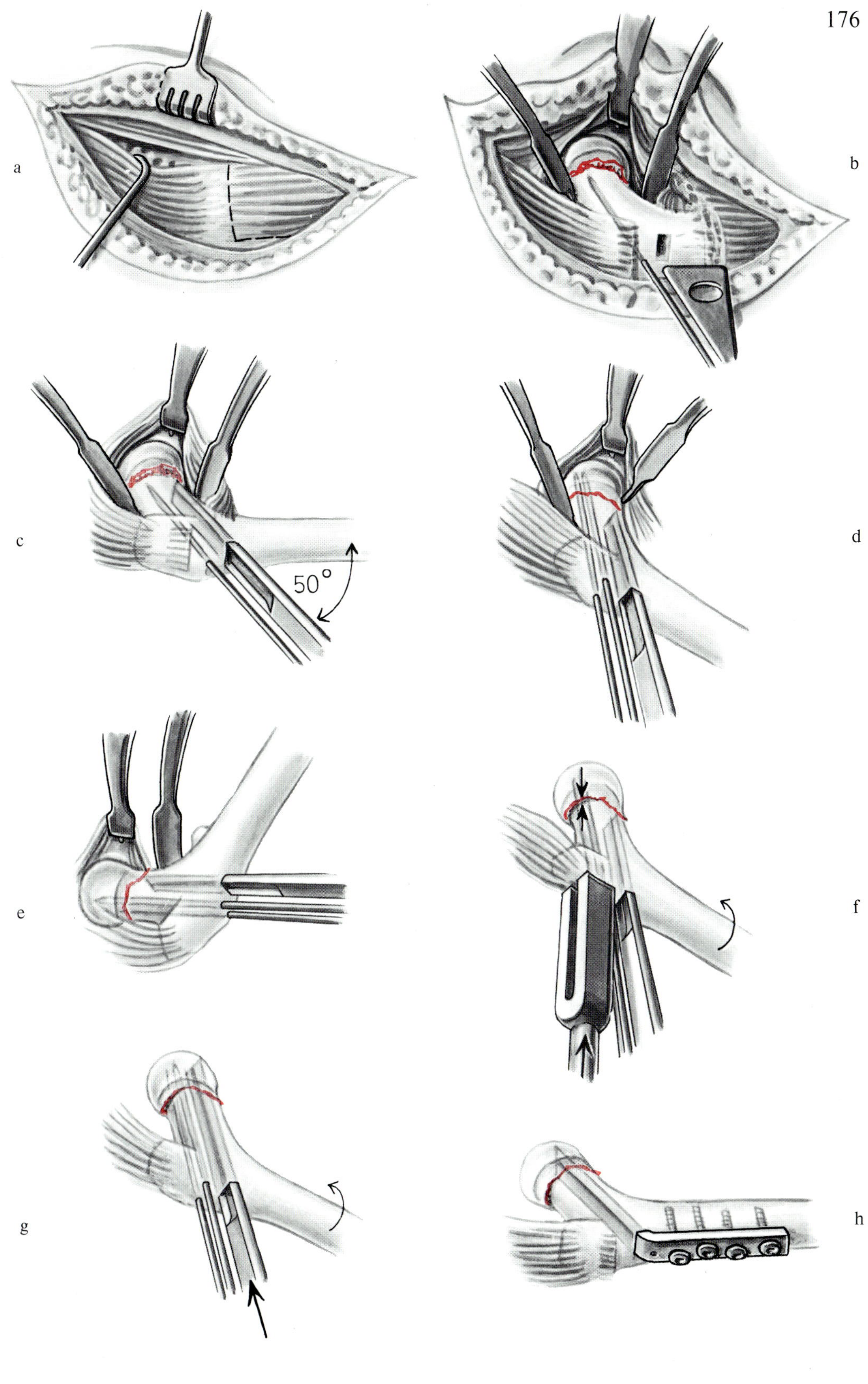
a
b
c
50°
d
e
f
g
h

Fig. 177 *An adduction subcapital fracture.*

a The fragments have been impacted with slight over-correction into valgus and anteversion. Fixation has been carried out with the four-hole 130° angled plate. The blade of the plate has been inserted into the lower half of the head beneath the intersection of the tension and pressure trabecular systems.

b In patients with a large head, further fixation is achieved with an additional cancellous screw and washer. The function of this screw is only to block rotation. Frequently one can observe some backing out of this screw as sintering occurs and further invagination of the fragments takes place. This screw should, therefore, be inserted parallel to the blade of the angled plate.

Fig. 178 *Shear fracture.*

a A small shear fracture through the middle of the neck even when the fracture plane is quite vertical is reduced in very much the same way by reduction into valgus and anteversion. After such a reduction, the tip of the calcar can be seen to protrude craniomedially. Fixation is carried out with a four-hole 130° angled plate.

b If the fracture plane is very steep, then the manoeuvre of valgus and anteversion will not be enough to neutralize the shearing forces and have the fracture under pure compression. We prefer to fix such a fracture with the doubly angled 120° angled plate and then carry out an intertrochanteric repositioning osteotomy of some 30°–40°.

c In dealing with a shear fracture through the base of the femoral neck in younger patients, prior to reduction insert three 2.5-mm Kirschner wires through the distal fragments so that they appear at the fracture. Reduce the fracture and drive the Kirschner wires into the head. Check your reduction. If all is well, replace the Kirschner wires with cancellous screws with washers.

177

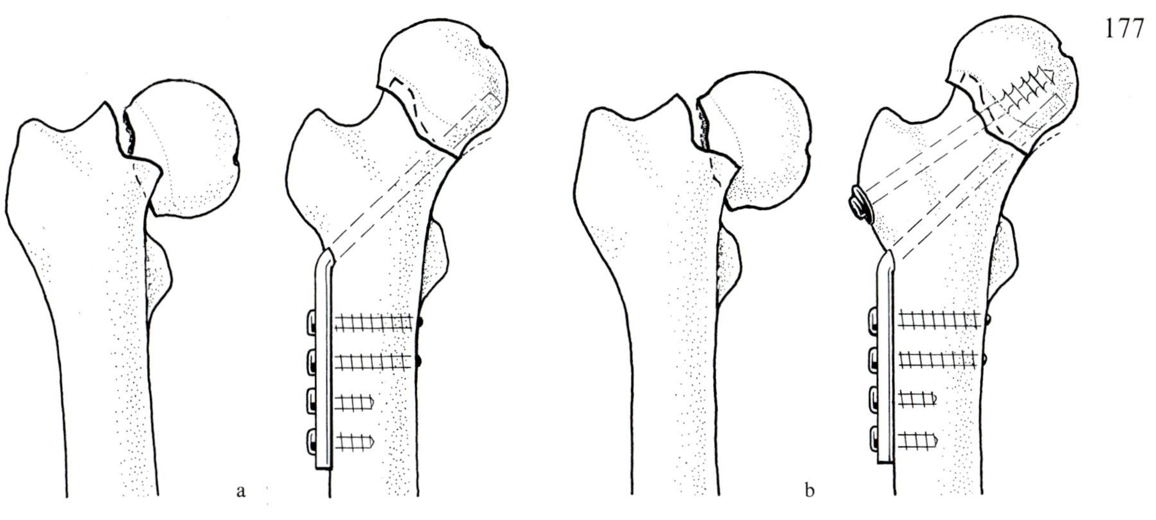

178

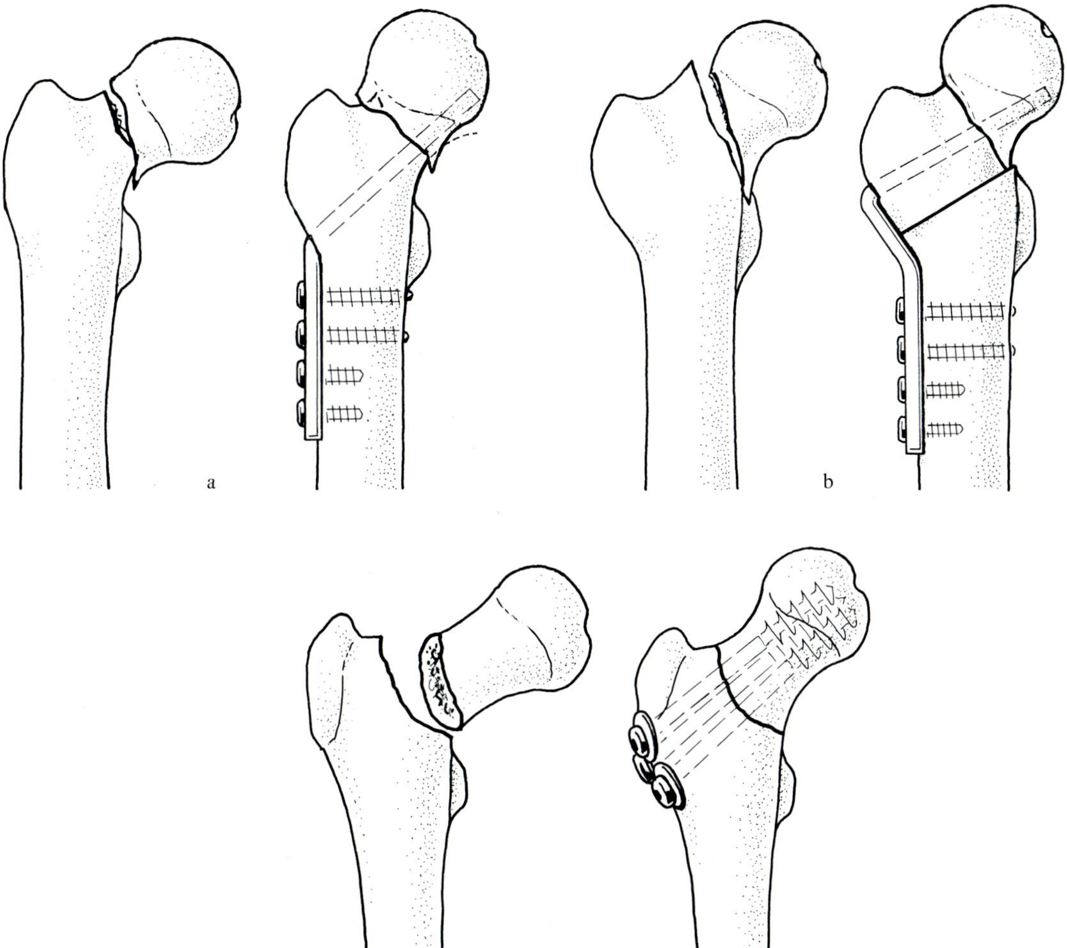

a

b

c

Classification of Intertrochanteric Fractures

Intertrochanteric fractures can be divided into the easily stabilized fractures and into the unstable problem fractures (Figs. 179 and 180). The stable intertrochanteric fractures (about 70% of all intertrochanteric fractures) have an intact medial buttress.

In the unstable intertrochanteric fractures, there is in addition to the medial fragment a large posterior fragment. These are the so-called four-part fractures, which at times may be even more comminuted than described.

Fig. 179 *The stable intertrochanteric fracture.*

a The fracture runs from the greater trochanter obliquely downwards and medially to exit just above the lesser trochanter. A good portion of the calcar is attached to the proximal fragment anteromedially. Quite commonly there is an avulsion fracture of the lesser trochanter. As a rule the distal fragment is in external rotation. Rarely the inferomedial spike of the proximal fragment is impacted into the metaphysis of the distal fragment.

b An avulsion fracture of the lesser trochanter – this type of avulsion does not result in instability because it does not weaken the medial buttress.

Fig. 180 *The unstable intertrochanteric fracture.*

a The medial fragment varies in size and reaches distally to a varying degree. As a rule it contains the lesser trochanter. If the lateral wall remains intact then the distal fragment migrates proximally because of muscle pull. Commonly there is in addition quite a large posterior fragment. Occasionally the proximal fragment contains a long medial spike made up of the calcar and lesser trochanter. This makes it into a long oblique or spiral fracture.

b If the greater trochanter is fractured, then the distal fragment is not pulled upwards.

c A badly comminuted intertrochanteric fracture has in addition to the fractures of the lesser and greater trochanter further comminution posteriorly and medially.

d The intertrochanteric fracture line is almost horizontal. Often one finds this fracture associated laterally with a further anterior or posterior fragment, and occasionally both.

e Occasionally the fracture has a reverse course beginning laterally and distally and running upwards and medially. Medially it exits above the lesser trochanter. Commonly it is associated with a fracture of the greater trochanter.
The problem with fractures (d) and (e) – about 5% of intertrochanteric fractures – is their reduction and stabilization which, because of muscle pull and fracture pattern, are particularly difficult.

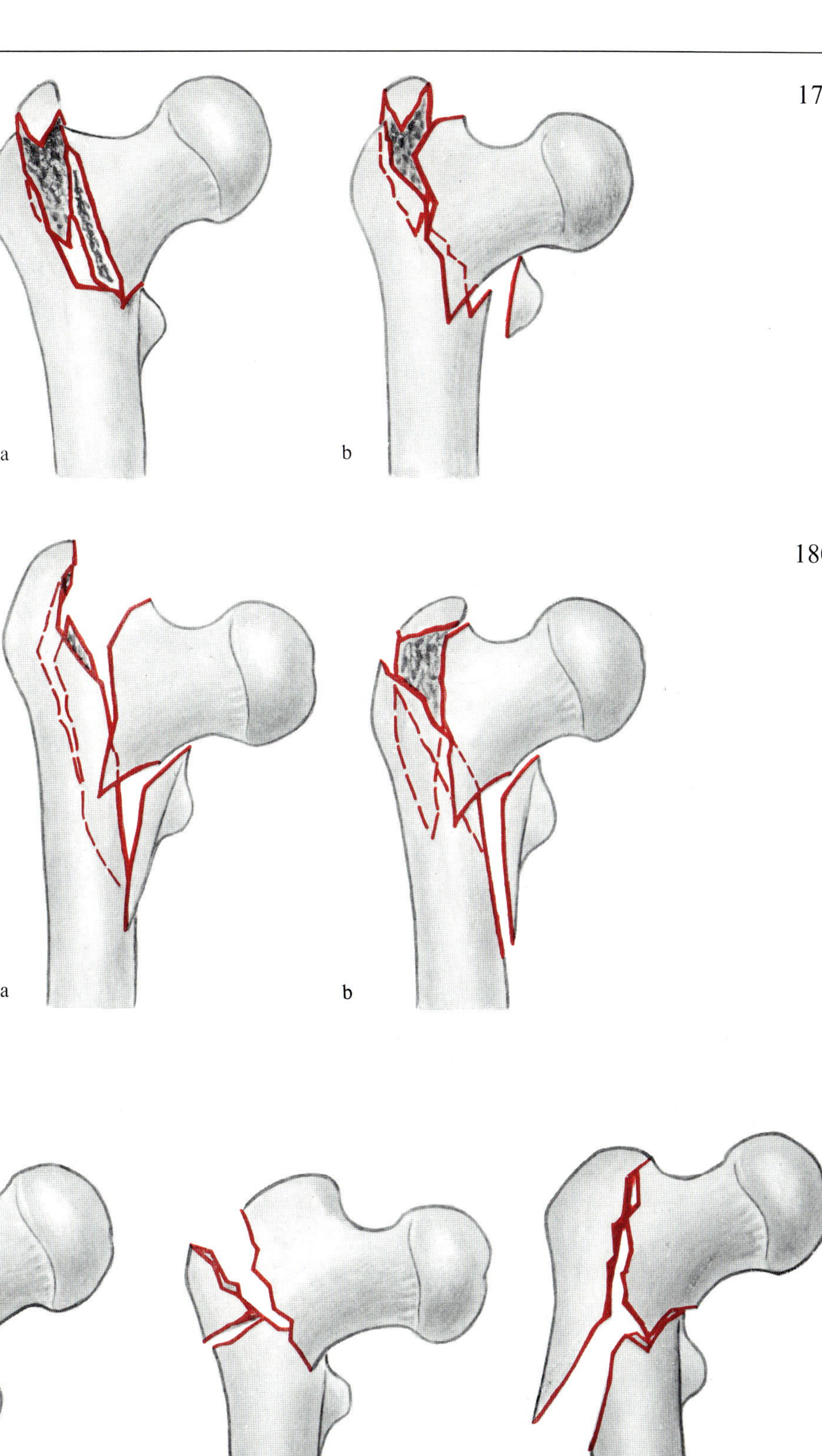
179
a
b
180
a
b
c
d
e

Technique of Open Reduction and Internal Fixation of Intertrochanteric Fractures

1. Internal Fixation with 130° Angled Plate. This method is useful for the stable intertrochanteric fracture. After reduction the fracture is fixed with Kirschner wires, the leg is abducted and the fragments are impacted. This results in valgus of the proximal fragment. The fracture is then fixed with a four- or six-hole 130° angled plate (for technique see p. 96). This method demands an intact medial buttress.

2. Internal Fixation with the Condylar Plate. Many prefer this plate to the 130° angled plate in (1). The great advantage of this plate is that one can vastly improve the purchase of the implant in the proximal fragment by the insertion through the plate of one or two lag screws into the calcar. This triangulation also brings the intertrochanteric fracture under compression. As a rule we use the five- or six-hole plate (for technique, see p. 94). An avulsed lesser trochanter is always reduced and fixed with a separate lag screw. In order to achieve a lasting stable fixation of this fracture, it must be impacted in valgus and the medial buttress must be reconstituted.

3. Valgus Medial Displacement Osteotomy with the 130° Angled Plate. This method is useful for fractures where anatomical reduction is impossible because of extensive comminution such as in the four-part fracture, or for the stable fracture in very osteoporotic bone. The key to this method is the marked valgus reduction of the proximal fragment. Commonly a trapezoid segment of bone has to be resected. This segment of bone is best determined on a pre-operative drawing. At the end, leg length should be equal. Begin with the insertion of the seating chisel into the fracture. Aim to enter the neck with the seating chisel as far cranially as possible and aim for the inferior half of the femoral head. Once the seating chisel is inserted, decide whether to proceed with the condylar plate or to perform a valgus medial displacement osteotomy. If the valgus medial displacement osteotomy is to be performed, the proximal fragment is then trimmed as planned. The leg is then rotated into neutral and the distal osteotomy is performed. This osteotomy line begins laterally and runs obliquely downwards and medially. If at all possible the first screw should be inserted as a lag screw across the osteotomy. The greater trochanter is reduced and held with a tension band wire. In this method the lesser trochanter is completely ignored.

Fig. 181

a A simple intertrochanteric fracture.

b The method of choice is internal fixation with a condylar plate after impaction in valgus. The tip of the blade has been inserted into the lower portion of the femoral head below the intersection of the tension and pressure trabecular systems. The first two cortical screws have been inserted as lag screws into the calcar.

c The internal fixation of a simple intertrochanteric fracture with the 130° angled plate after reduction and impaction of the fracture in valgus.

Fig. 182 The condylar plate should be used whenever the medial buttress has to be reconstructed as after avulsion of the lesser trochanter with a larger piece of bone. If the medial fragment is large, then as in step 1, it is reduced and fixed with lag screws to the distal fragment.

Fig. 183 *Valgus medial displacement osteotomy with 130° angled plate as a method of reduction and fixation of a comminuted intertrochanteric fracture.* Resect the tip of the calcar from the proximal fragment and a trapezoid segment of bone from the distal fragment as predetermined on the pre-operative drawing. The proximal fragment is then over-reduced by 30°–50° into valgus. As a rule we use only a 50-mm blade. The insertion of the first screw as a lag screw across the osteotomy vastly improves the rotational stability of the fixation. The *dotted line* shows the corresponding outline of the uninvolved hip joint.

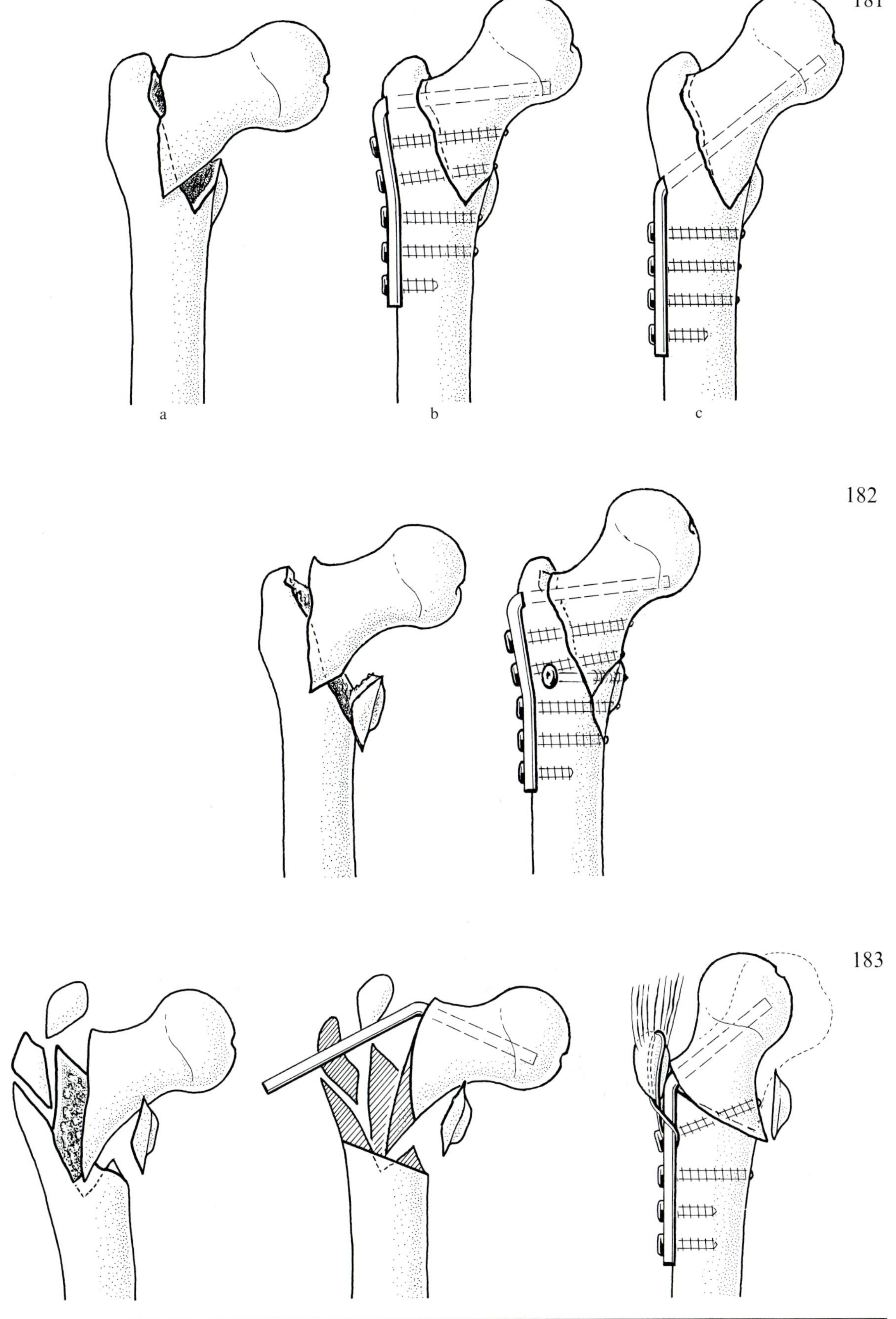
181
a
b
c
182
183

Fig. 184 *Stable intertrochanteric fracture.* Note the reduction in slight valgus with impaction. Internal fixation with the condylar plate.

Fig. 185 *Stable intertrochanteric fracture with avulsion of the tip of the greater trochanter.* Note once again the anatomical reduction and slight overcorrection into valgus with impaction and the internal fixation with 130° angled plate. The fixation was supplemented with an additional cancellous lag screw above the blade. The avulsed tip of the greater trochanter was fixed with a tension band wire.

Fig. 186 *Unstable intertrochanteric fracture with the large medial fragment.* As step 1, the medial fragment was reduced and fixed with a lag screw to the distal fragment. Then the intertrochanteric component was reduced, slightly overcorrected into valgus, impacted and then fixed with the condylar plate. The first two cortical screws have been inserted as lag screws into the calcar. Screws 3 and 4 serve also as lag screws through the plate to help in the fixation of the broken out medial fragment.

Fig. 187 *A badly comminuted intertrochanteric fracture.* Valgus reduction with medial displacement of the shaft and fixation with the 130° angled plate. The first screw through the plate has been inserted as a lag screw across the osteotomy into the proximal fragment, which increases the rotational stability of the internal fixation. The greater trochanter has been fixed with a tension band wire, which was passed around the first screw.

Fig. 188 *An intertrochanteric fracture with an additional lateral and anterior fragment.* Such an intertrochanteric fracture is difficult to reduce. The reduction can be made easier by first inserting the angled plate into the proximal fragment. The plate portion of the angled plate helps then in the reduction. An intact medial buttress is once again most important. If the lateral fragments which come to lie under the plate are large, then axial compression is used to aid in their fixation. This is done by placing the plate under tension with the tension device. The fixation of the anterolateral fragment is supplemented with a separate lag screw.

Fig. 189 *The intertrochanteric fracture with the reversed fracture line* presents problems in its reduction not unlike the subtrochanteric fracture. To facilitate matters we insert the blade of the angled plate into the proximal fragment and fix it with screws to the calcar. The reduction is then carried out about the plate. It is just as important here to reconstruct the medial buttress and if necessary use a cancellous graft. Note that the first two screws into the calcar serve as lag screws.

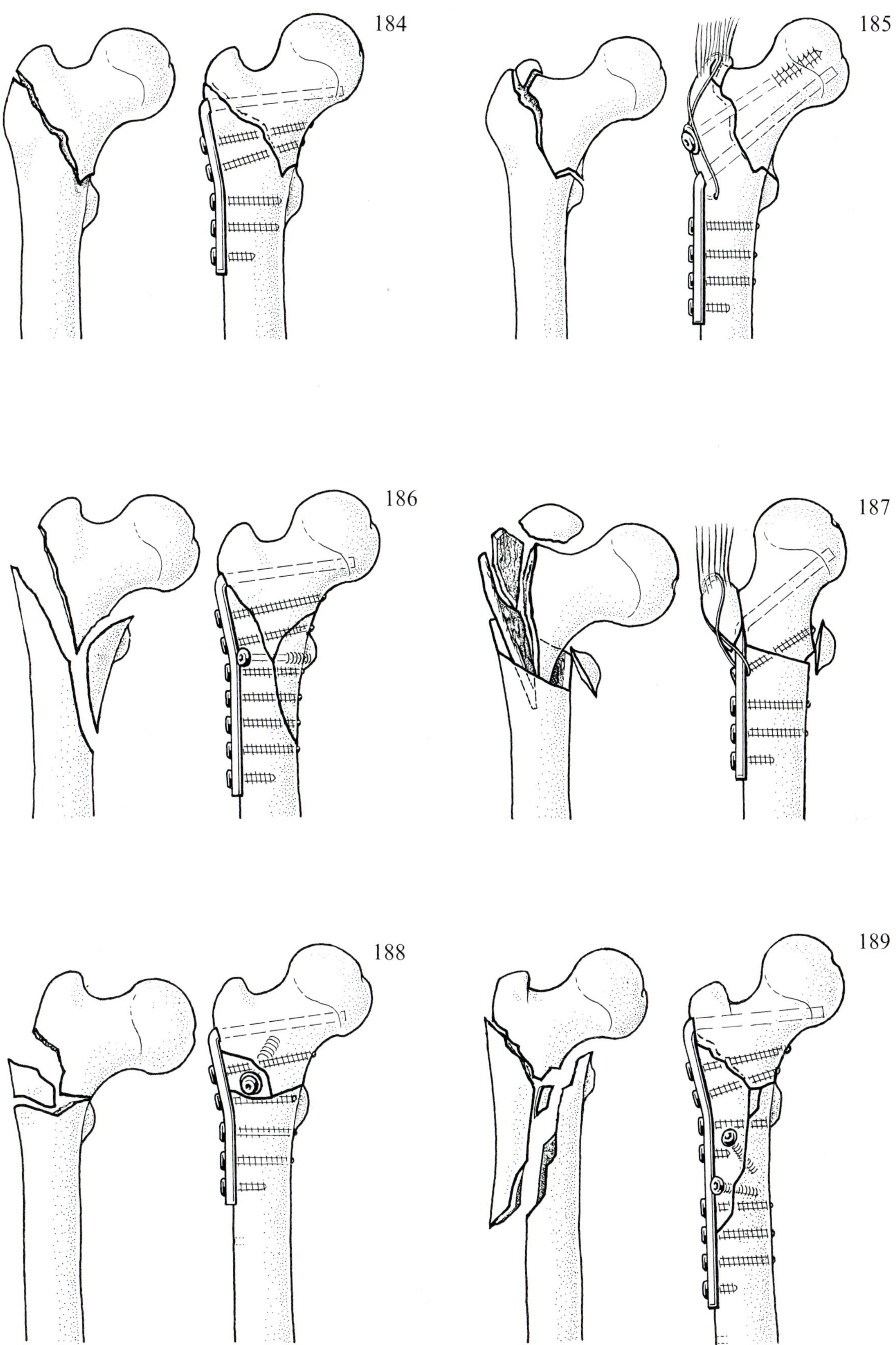
184
185
186
187
188
189

1.7.2 Fractures of the Femoral Shaft

The advantages of stable internal fixation have been particularly demonstrated in the treatment of fractures of the femur. Non-operative treatment requires traction for 2–4 months. At the end of this period of time, malunion or delayed union is not a rare occurrence and mobilization of the knee joint may take months.

As a rule, we consider every fracture of the femur as an indication for operative treatment. For fractures of the middle third we prefer intramedullary nailing with a thick nail with reaming of the medullary canal. For fractures of the proximal and distal portion of the shaft we use the condylar plate or a broad DCP. The straight plates are placed on the dorso-lateral aspect of the bone and are pre-stressed. Thus, as compression plates, they act as tension bands. Medial bone grafting is carried out if there is even the smallest defect in the cortex. To achieve the required strength the screws should traverse at least seven cortices in each main fragment.

We have found the AO distractor particularly useful in carrying out reductions of femoral shaft fractures. The combination of intramedullary nailing and plating of comminuted fractures should be carried out only if the reduction can be carried out with the distractor in such a way that all fragments retain their muscular attachments (see Fig. 87).

Positioning: For intramedullary nailing, we position the patient on his side (Fig. 83). For plating the patient is positioned either supine or on his side.

Note: In comminuted fractures it is sometimes worthwhile to keep the patient in traction for 1–2 weeks, and carry out delayed surgery. The traction should be skeletal through the tibial tubercle and the weight used should be approximately one-seventh of the body weight. The delay leads to atrophy of the musculature, which makes reduction easier. Furthermore the local hyperaemia improves the blood supply to the fragments, which aids in bone healing.

Fig. 190 *The surgical approach to the femur.*

1 The straight incision for intramedullary nailing begins at the tip of the greater trochanter and is extended upwards.

2 The proximal straight lateral incision is the standard approach to the proximal femur with reflection of the vastus lateralis. It is used for fractures or intertrochanteric osteotomies.

3 The straight lateral incision for the shaft follows an imaginary line from the tip of the greater trochanter to the lateral epicondyle. It is sometimes referred to as the mailbox incision because the vastus lateralis is lifted upwards and forwards like a lid. All perforators are carefully ligated to keep the blood loss to a minimum. An anterior transmuscular approach leads to scarring and binding down of the quadriceps and may damage the blood supply and nerve supply to the vastus lateralis.

4 The approach to the distal femur. If in addition there is a tangential fracture of the medial condyle, then a second medial parapatellar incision (5) is necessary.

190

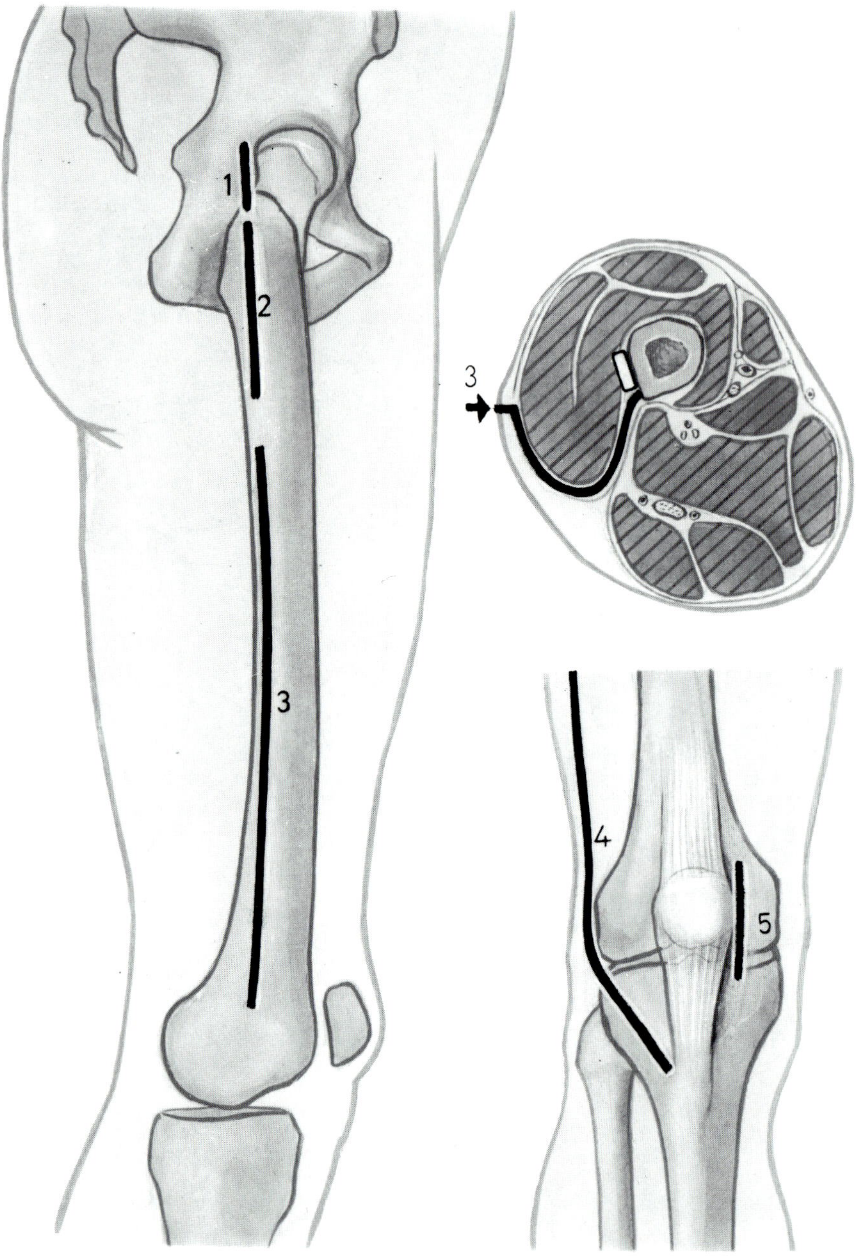

We suggest the following classification because it gives one an idea of the prognosis as well as of the difficulty in the reduction of the fracture. The simple fracture is reduced and fixed with a lag screw. A condylar plate is then used for fixation.

In fractures with butterfly fragments the butterfly fragments are first reduced and fixed with one or two lag screws as necessary. This converts them into simple fractures. The main fragments are then reduced, held reduced with bone-holding clamps and fixed with a condylar plate.

In the badly comminuted fractures the condylar plate is inserted prior to reduction into the proximal fragment. The reduction is then undertaken with the aid of the distractor. The reduction should be planned on an outline of the femur made from an X-ray of the normal side. One should draw in the different fragments and indicate the order in which the reduction and internal fixation will be carried out.

Fig. 191 *Classification of subtrochanteric fractures.*

a Simple, transverse, oblique or spiral subtrochanteric fractures.

b Lateral butterfly.

c Medial butterfly.

d Comminuted fracture.

e Shattered fracture.

191

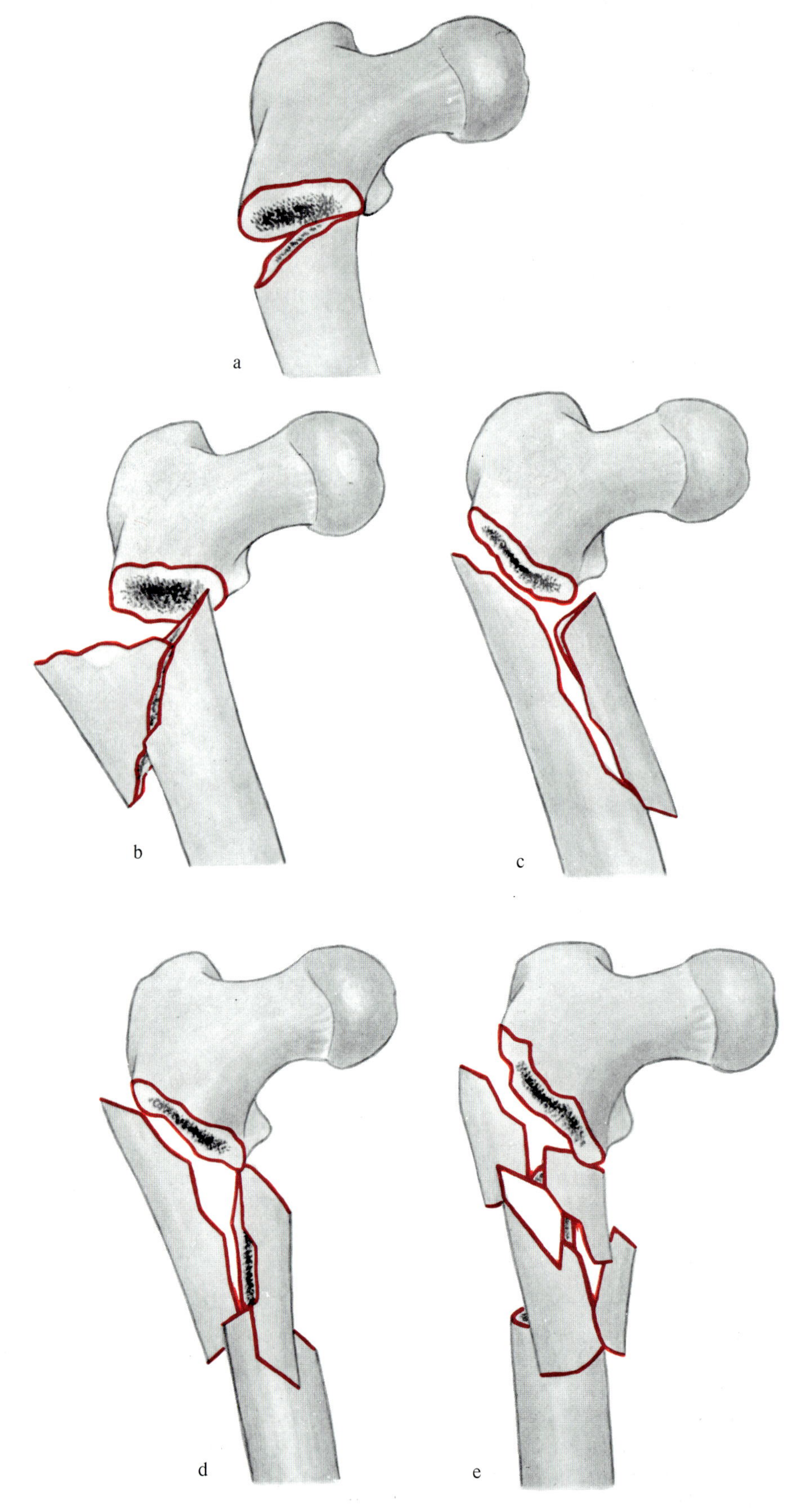

The technique of plating of a simple subtrochanteric fracture.

Make a 20-cm-long lateral incision (Fig. 160). Enter posteriorly to the vastus lateralis and reflect the muscle forwards. Insert three Hohmann retractors as shown in Fig. 176b. Expose the fracture and the femoral neck. Reduce the fracture and insert the screws from front to back. Once this is completed insert one Hohmann retractor around the calcar, the second posterior to the greater trochanter and pass the tip of a small Hohmann retractor halfway around the top of the femoral neck. Insert the seating chisel and then the condylar plate as shown in Fig. 192. That is, insert the Kirschner wires which serve as a guide to the neck shaft angle and to the anteversion. Then insert the seating chisel with its chisel guide, and then insert the condylar plate. Insert two cortex screws through the plate into the calcar. This triangulates and vastly improves the fixation of the condylar plate in the proximal fragment. Apply axial compression by pre-stressing the plate either with the tension device or by making use of the self-compressing principle of the DCP. Check the reduction, and if perfect insert all the screws. If a large butterfly is present it is first fixed to the distal fragment. The reduction of the fracture is then carried out, temporarily maintained with a bone clamp, and then fixed as described above.

Fig. 192 *The steps in the procedure according to the different fracture types are:*

A Reduction, insertion of lag screw (*1*), with the aid of the condylar plate guide, insertion of the Kirschner guide wires, opening of the cortex for the condylar plate, insertion of the condylar plate (*2*), insertion of cortex screws into calcar (*3*), and axial compression (*4*).

B_1, B_2 Reduce the butterfly and fix it either to the proximal or distal fragment with a lag screw (*1*). Reduce the main fragment and maintain temporary reduction with bone clamps. Insert the seating chisel and then the blade of the condylar plate (*2*) into the neck. Triangulate the fixation in the calcar with cortex screws (*3*) and apply compression either with the tension device or with the DCP (*4*). Insert the proximal lag screw (*5*) and then complete the fixation by inserting the remaining screws.

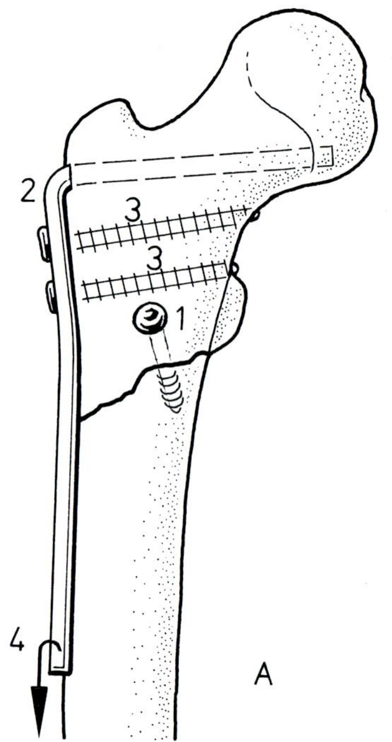
2
3
3
1
4
A

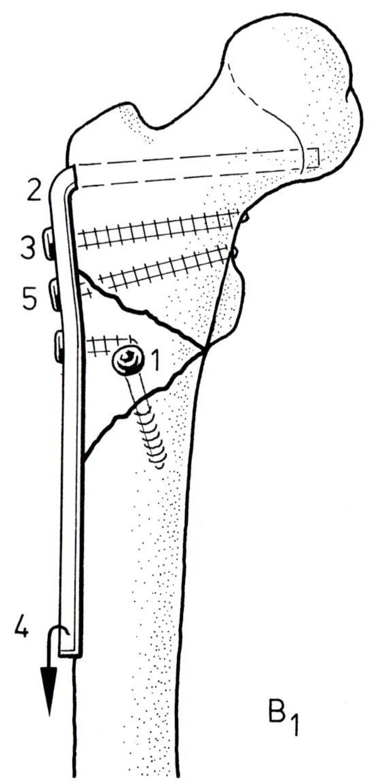
2
3
5
1
4
B1

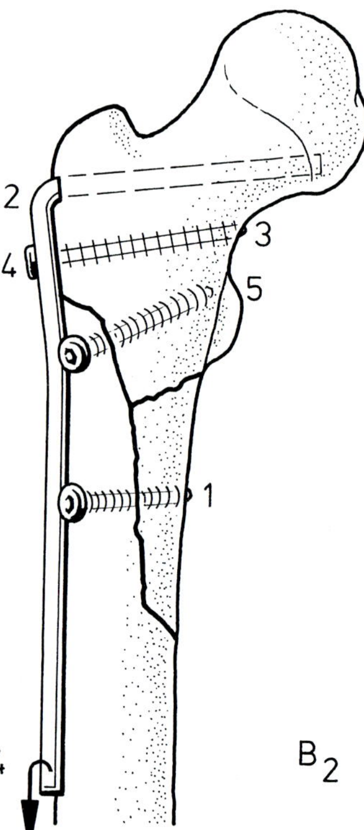
2
3
4
5
1
4
B2

Prior to reduction, insert into the proximal femur the blade of the condylar plate. The reduction is then carried out with the aid of the distractor. Pre-determine the exact position of the blade of the condylar plate in the proximal fragment from a pre-operative drawing which is made from an X-ray of the uninvolved side.

Technique: First insert the seating chisel and then the blade of the condylar plate into the proximal fragment. Fix the plate to the calcar with one or two cortex screws. Insert the proximal connecting bolt of the distractor into a hole which is pre-drilled through the plate. Distraction follows and the fracture is then more or less reduced, depending on the number of fragments and the degree of comminution. The plate is then fixed distally with a short screw and the rotational alignment is checked. Next the large fragments are screwed to the plate. At the end a copious cancellous graft must be carried out.

Fig. 193 *Steps of the operation.*

C_1, C_2 If at all possible begin by reducing and fixing the proximal lateral butterfly with a lag screw (*1*). Insert the seating chisel and the condylar plate (*2*), insert the cortex screws into the calcar (*3*) and drill the 4.5-mm hole through the second hole of the condylar plate and insert the connecting bolt (*4*). Insert the distractor (*5*) and begin the distraction (*6*). Reduce the fracture. As soon as you have achieved an axial continuity either medially or laterally, place the plate under tension. Fix the plate to the shaft with a short screw (*7*) and check the axial alignment. Insert the lag screw (*8*) and then the short and long screws (*9*) to fix the plate to the shaft and complete the fixation.

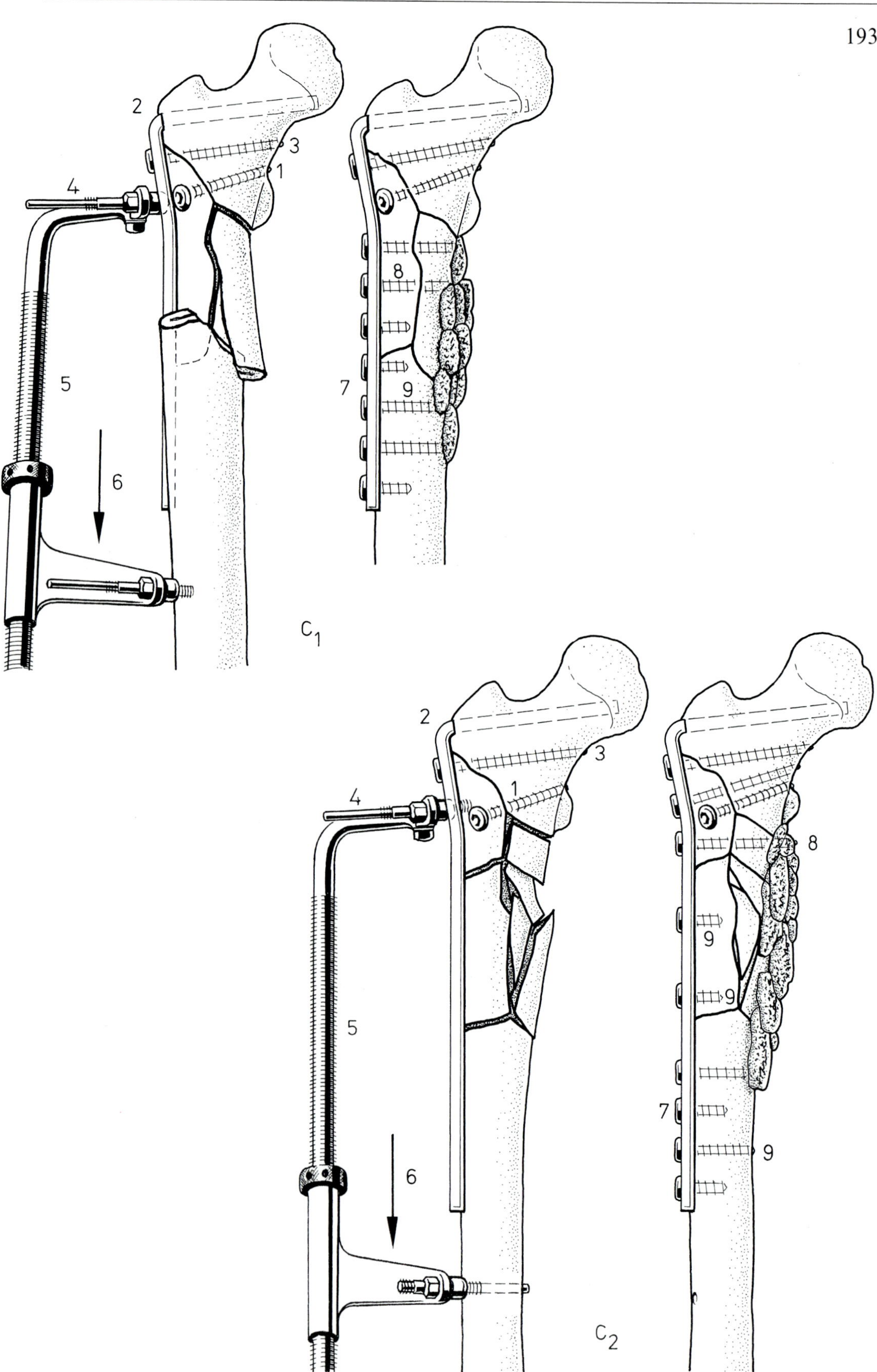
2
3
1
4
5
6
8
7
9
C_1
2
3
1
4
8
9
9
5
7
9
6
C_2

Fig. 194 *Simple subtrochanteric fracture* fixed with a condylar plate. The fracture was first reduced and fixed with a lag screw. The blade of the plate has been inserted into the neck at a point pre-determined from the pre-operative drawing. A cortical screw was inserted through the plate into the calcar to improve the fixation of the plate in the proximal fragment. The second proximal cortical screw has been used here as a lag screw. This second screw is inserted only after the plate has been placed under tension and the fracture under axial compression.

Fig. 195 *Subtrochanteric fracture with a butterfly.* The butterfly was first reduced and fixed with lag screws to the main fragment. The condylar plate was inserted at the end. Note that some of the screws are short.

Fig. 196 *A badly comminuted subtrochanteric fracture with a deficient medial buttress.* The condylar plate was first inserted into the proximal fragment and fixed with cortex screws to the calcar. The distractor was used to reduce the different fragments. These were then lagged to the plate. At the end an extensive cancellous graft was carried out medially. Weight bearing should be withheld for 6–8 weeks. At the end of this time the bone graft has usually formed a bridge medially and weight bearing can commence.

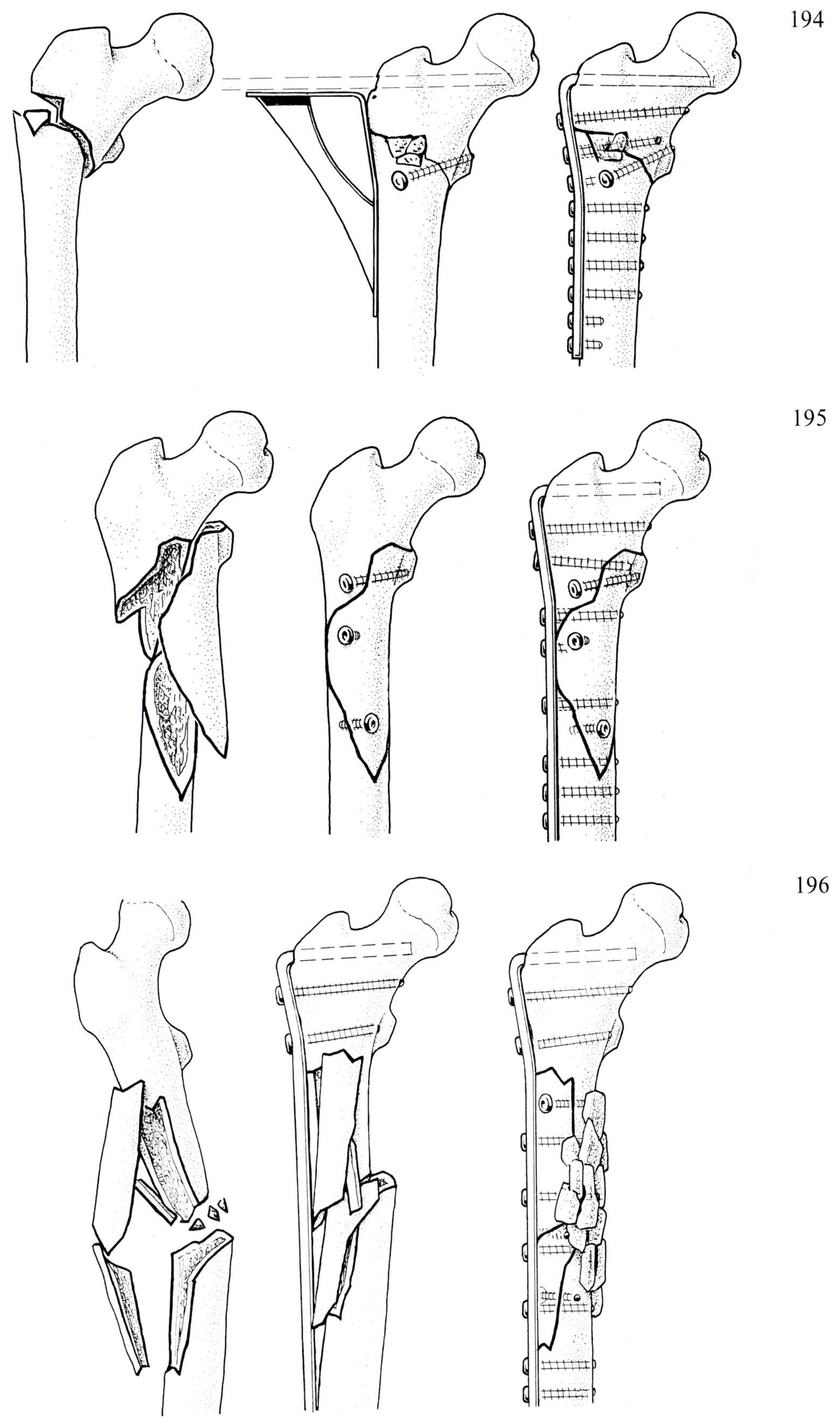
194
195
196

Fractures of the Middle Third of the Femoral Shaft

The middle third of the femur is the domain of the intra-medullary nail. The nailing is combined with reaming of the medullary canal (Fig. 84). When the nail is used to stabilize a comminuted fracture it is combined with screws with a cerclage wire or with a narrow plate (Fig. 87f). If fractures of the middle third are fixed with plates alone, then they must always be bone grafted on the side opposite the plate.

Positioning of the Patient and Surgical Approaches (see Fig. 83)

Intra-medullary Nailing: For the technique of open intra-medullary nailing see Fig. 84 and for closed intra-medullary nailing see Fig. 89.

The distractor is almost always used when one is carrying out a reduction in open nailing (Fig. 87). Approach the fracture along the inter-muscular septum and reflect the vastus lateralis forwards. In order to expose the greater trochanter, either extend the first skin incision or make a second longitudinal incision which is centered over the tip of the greater trochanter.

Post-operative Positioning: Maintain the hip and knee flexed to 90° for the first 5–6 days (Fig. 103b).

Fig. 197 *A simple fracture of the femoral shaft.* As the patient is a man, ream the medullary canal to 15-mm and insert the 15-mm new nail. For a woman it is usually enough to ream to 13-mm and insert the 13-mm nail (see p. 106).

Fig. 198 *A segmental fracture.* Both fracture lines go through the middle third of the femur. Carry out an open reduction and hold the middle fragment reduced with Verbrugge clamps. Ream only to 12–13 mm. Insert a corresponding 12- or 13-mm nail. The reaming is restricted to 12 or 13 mm. If one attempts to ream to a larger diameter one may cause the reamer to spin the middle segment, which would tear all the soft-tissue connections and deprive the bone of its blood supply.

Fig. 199 *A comminuted fracture.* For the technique see Fig. 87. For the use of the distractor in combination with an intra-medullary nail see p. 120. The medullary canal is reamed here to 15 mm and a corresponding 15-mm nail is inserted. The butterfly fragments are fixed with lag screws. The distractor is then removed and the fracture is further fixed with a narrow plate as shown in Fig. 87f.

Fig. 200 *A broad DCP used for the fixation of an oblique fracture.* Note the lag screw across the fracture and the bone grafting on the medial side opposite the plate.

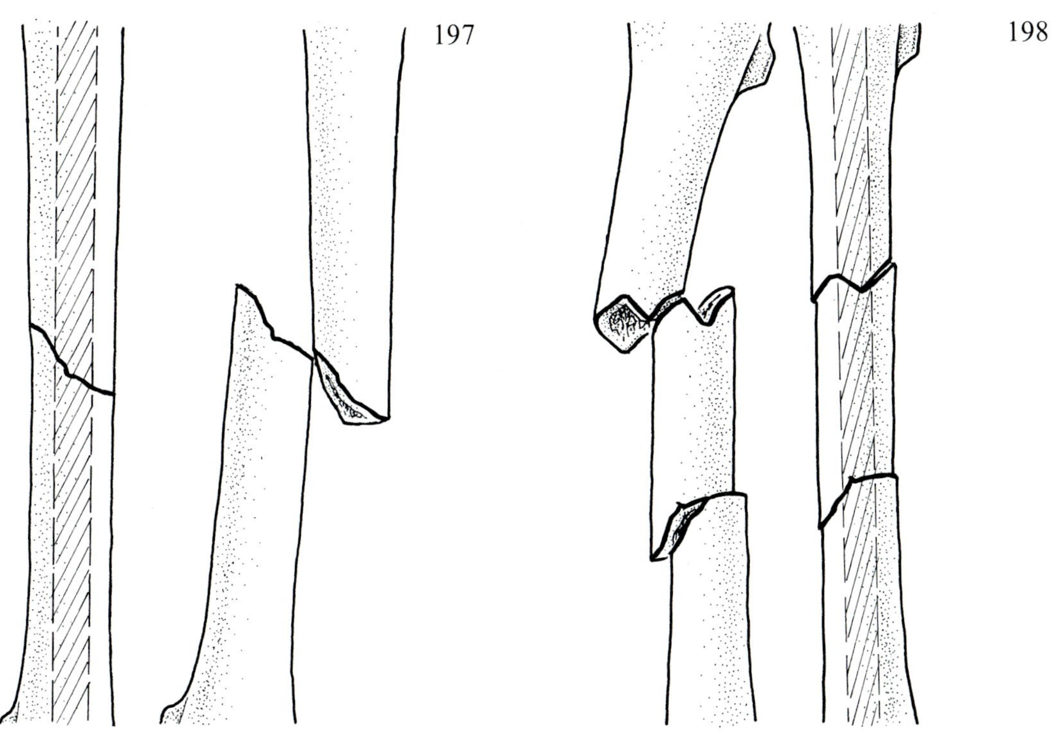

197
198

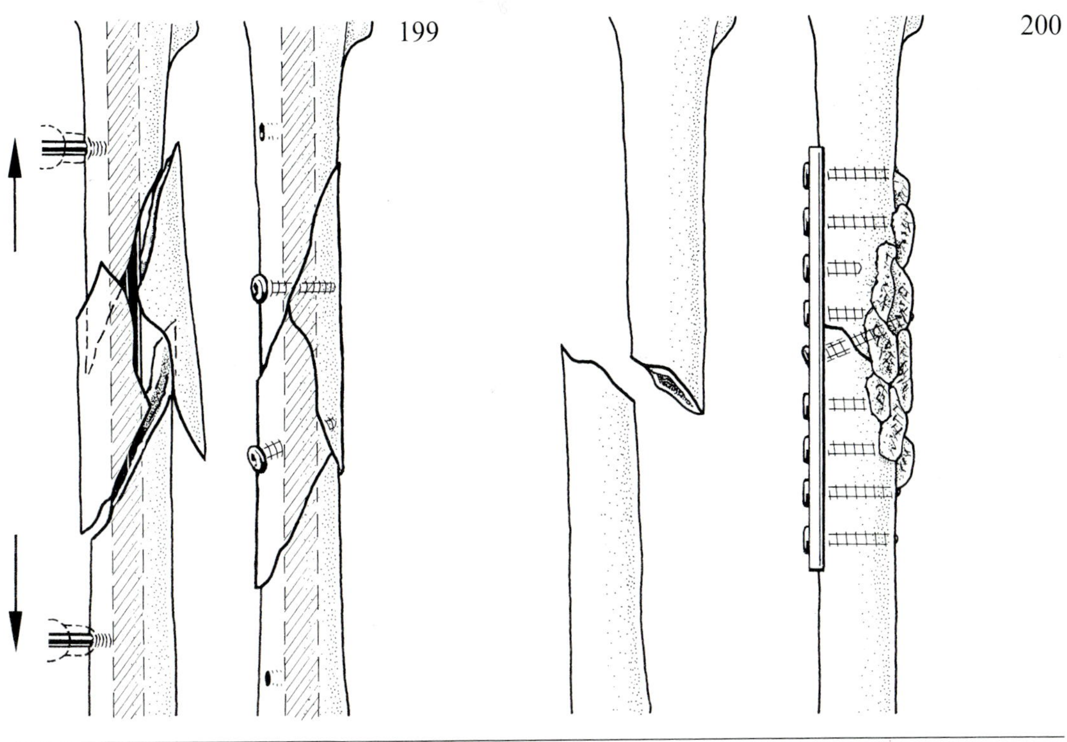

199
200

Fractures of the Distal Shaft

In the distal femur the medullary canal flares out like a trumpet. Therefore whether one reams or not the medullary nail fails to get sufficient hold in the distal fragment. In order to stabilize a fracture which has rotational stability by virtue of the interdigitation of its irregular surface, it is necessary to drive the tip of the nail within 2–3 cm of the articular surface of the knee. If this is impossible, then one must use an additional plate to gain rotational stability (Fig. 202). Under these circumstances, reaming achieves very little and should not be used. The nail used is the one that fits the isthmus. If the isthmus is very narrow, it can be reamed out somewhat to permit the use of a slightly larger nail.

If a plate is combined with an intra-medullary nail, then one must always bone graft the cortex opposite the plate. Whenever possible the fracture should also be crossed with a lag screw.

For the fixation of comminuted fractures, see "Fractures of the Distal Femur" (Fig. 205 A_3).

Fig. 201 *Transverse fracture of the distal femur.* During reduction the fragments locked together. They were stabilized with a 14-mm nail inserted after reaming of the medullary canal. Note also that the nail was inserted just short of the subchondral bone of the distal femur.

Fig. 202 *A comminuted fracture of the femur.* This fracture was stabilized with an intra-medullary nail in combination with a narrow neutralization plate. Note the cancellous bone grafting of the medial cortex opposite the plate.

Fig. 203 *An oblique fracture at the junction of the middle and distal third in a young adult.* The fracture has been fixed with a DCP combined with a lag screw across the fracture and bone grafting of the medial cortex.

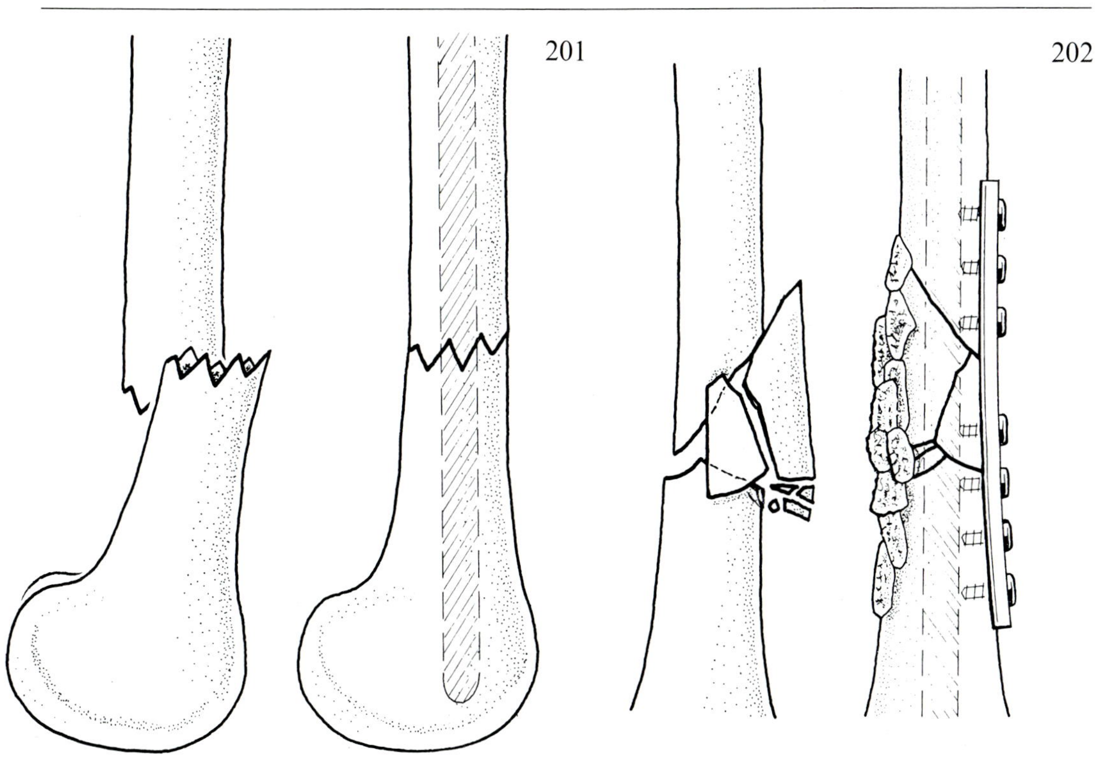

201

202

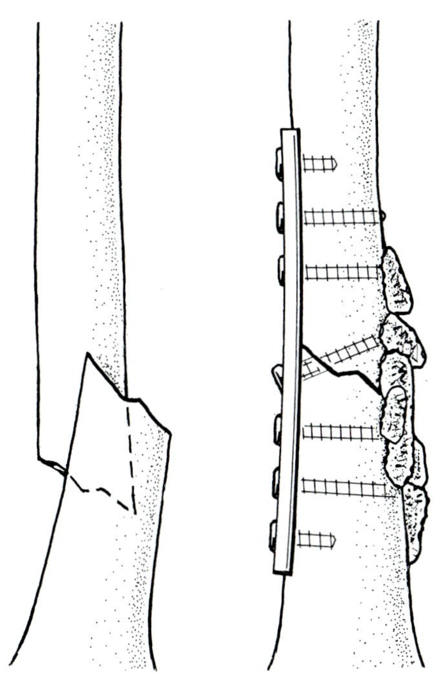

203

1.7.3 Fractures of the Distal Femur (Metaphyseal and Transcondylar Fractures)

A classification similar to the one used for the distal humerus defines the fracture, indicates its prognosis and the type of internal fixation necessary.

Fig. 204 *Classification* (according to M.E. MÜLLER).

A_1 Avulsion of the proximal insertion of the medial collateral ligament.

A_2 A simple supracondylar fracture.

A_3 A supracondylar fracture combined with a comminuted fracture of the distal shaft.

B_1 A fracture of one condyle (young adult or osteoporotic bone).

B_2 A fracture of the lateral condyle extending up into the shaft. Note that it is still attached to the tibial spine by the intact interior cruciate ligament.

B_3 A posterior tangential fracture of one or both condyles – Hoffa fracture.

C_1 Inter- and supracondylar fractures – T or Y fracture.

C_2 The bicondylar fracture combined with a comminuted fracture of the distal femur.

C A bicondylar fracture combined with a distal comminuted fracture of the shaft and an anterior tangential fracture of one or both condyles.

204

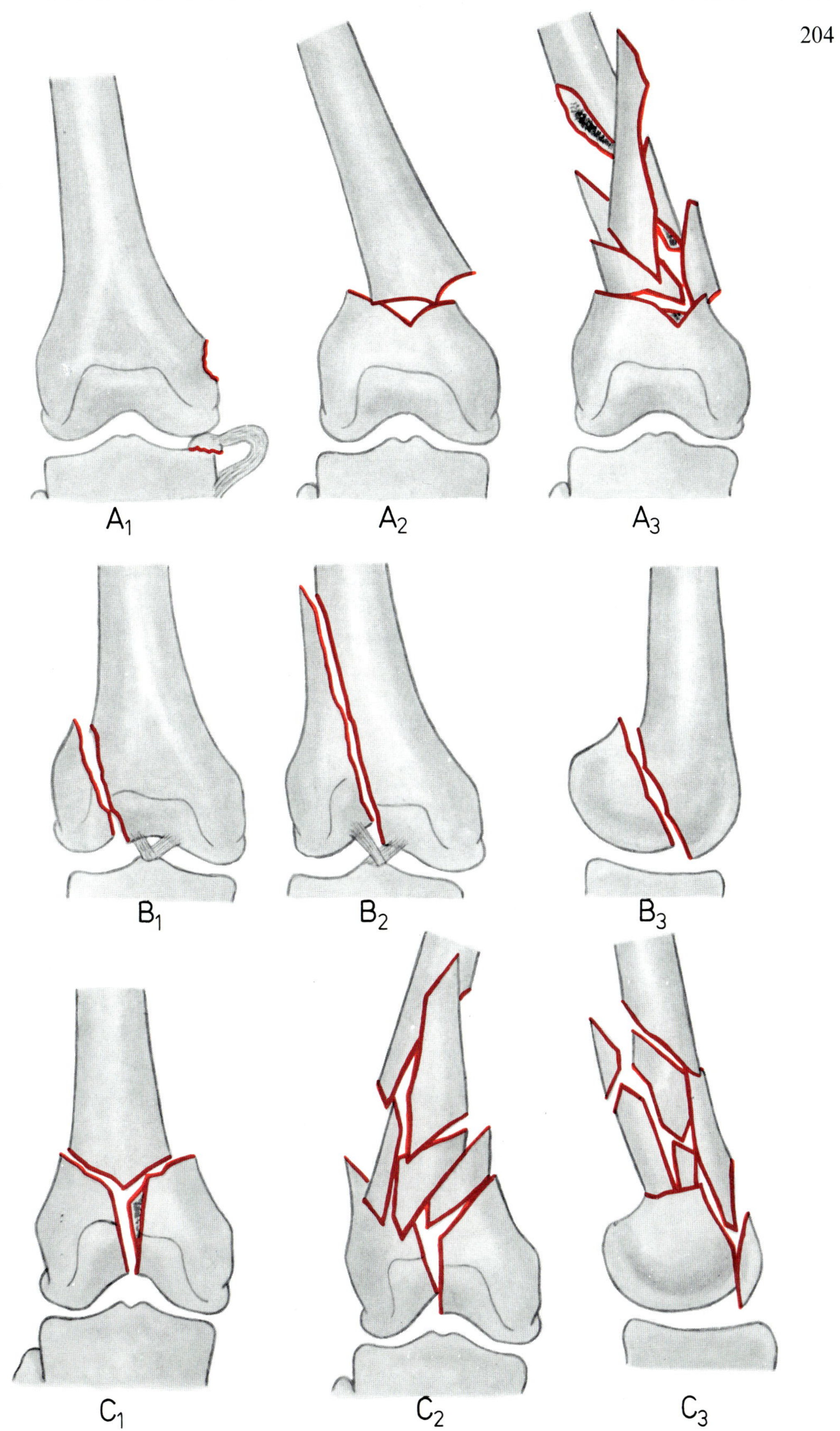

Treatment of Fractures of the Distal Femur

The use of the condylar plate together with the necessary instruments and the correct insertion of the seating chisel parallel to the knee joint axis and in line with the long axis of the femur are described on pages 98 and 100. The approaches are described on page 228.

The Positioning of the Patient: For a simple supracondylar fracture the patient is positioned supine, with a pillow under the knee. For patients with more difficult fractures, we strongly recommend the use of a knee support which can be lowered if desired. This greatly facilitates the insertion of the Kirschner guide wires above and below the patella (Fig. 69). Furthermore the reduction of the shaft portion is made very much easier by the traction one achieves as the knee joint is bent over the support.

The type of the internal fixation depends on the type of the fracture as depicted in Fig. 205.

Fig. 205 *Internal fixation of fractures of the distal femur.*

A_1 The avulsion fracture of the proximal insertion of the medial collateral ligament has been fixed with a 4.0-mm cancellous screw with a washer.

A_2 *A simple supracondylar fracture:* It is reduced and fixed with Kirschner wires. Open the cortex for the insertion of the seating chisel. With the aid of the condylar guide, insert the Kirschner guide wires and then insert the seating chisel. Use the five-hole condylar with a 50- or 60-mm blade. Be sure to supplement the fixation of the blade in the distal fragment by the insertion of one or two cancellous screws. Apply axial compression either with a tension device or by making use of the self-compressing principle of the DCP.

A_3 *Supracondylar fracture combined with a comminuted fracture of the distal shaft:* Do not attempt a reduction; position the patient supine with the knee at 90°. Insert the guide wires. Open the lateral cortex for the insertion of the seating chisel and then insert the seating chisel parallel to the knee joint with the seating chisel guide parallel to what is taken as the long axis of the femur. Insert a 9- or a 12-hole condylar plate. Forcibly bend the knee over the knee support. This applies traction to the shaft. While this traction is maintained, fix the plate with one screw to the proximal fragment. Check the rotational alignment. Do not attempt an anatomical reduction of the comminuted fragments but leave them alone with their soft-tissue attachments intact. Instead of attempting to reduce the fragments carry out very extensive cortico-cancellous bone grafting. Occasionally the comminution is so severe that one has to supplement the fixation medially with a second buttress plate such as a long T plate or DCP.

B_1 *Fracture of the lateral condyle:* In the young adult carry out an anatomical reduction and temporary fixation with a Kirschner wire. Then fix the fracture with two cancellous screws with long threads. The washers are used to prevent the sinking of the screw heads through the cortex. If the bone is osteoporotic use a T plate as a buttress plate. Reduce the fracture, fix it temporarily with Kirschner wires, carefully contour the T plate and then fix the fracture with the two cancellous screws.

B_2 *The fracture of the lateral condyle with a long proximal extension:* If the bone is very osteoporotic, reduce the fracture, keep it reduced with a reduction clamp and then carefully contour a long T plate and use it as a buttress plate. Insert whichever screws you can as lag screws.

B_3 *The tangential posterior fracture of one or both condyles – the Hoffa fracture:* Fix the fracture with the 6.5-mm cancellous screws with 16-mm threads. Insert the screws in AP direction at a right angle to the femoral shaft axis. Insert the screws as far laterally as possible to avoid the articular cartilage. If this is impossible, countersink the heads below the articular cartilage.

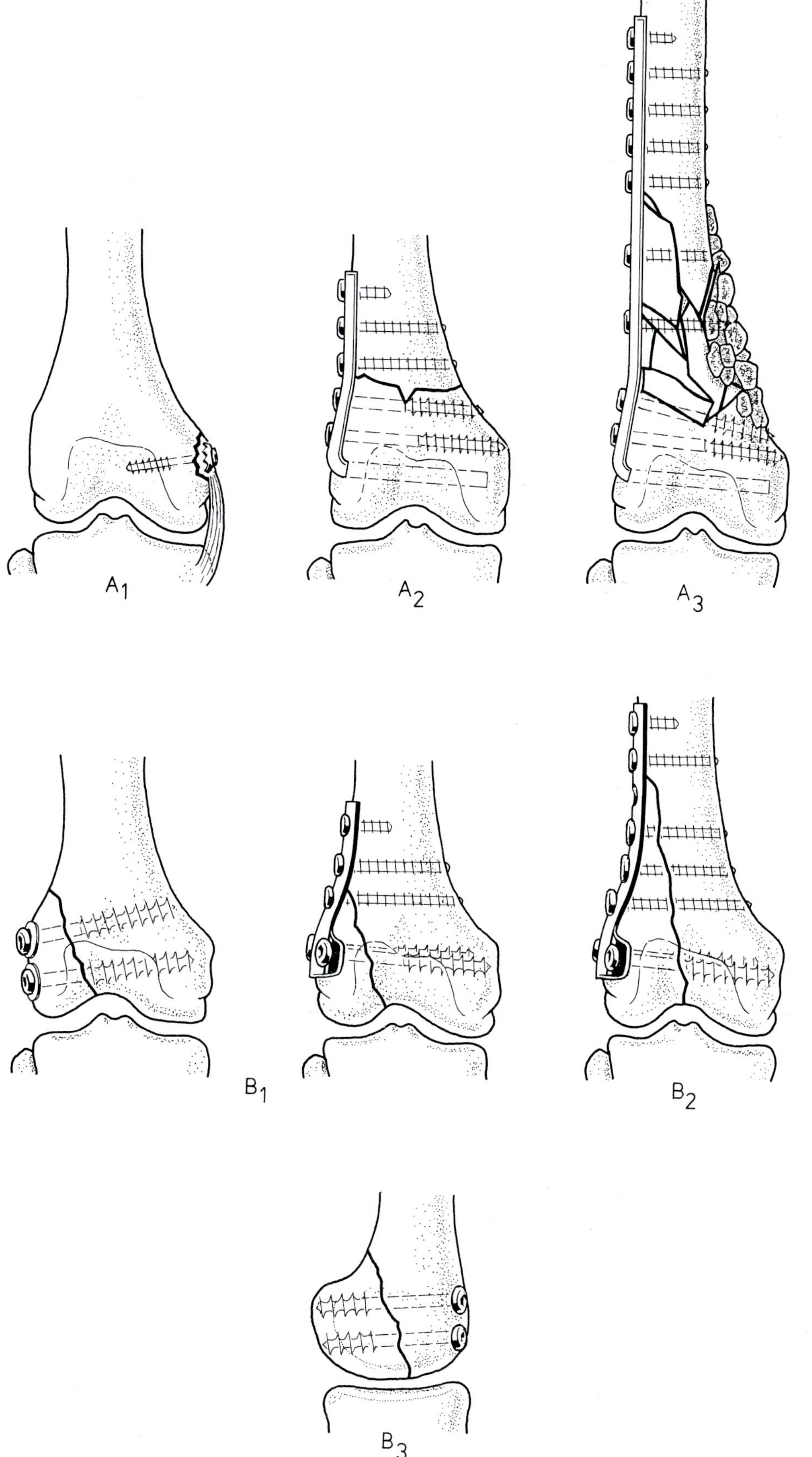
A1
A2
A3
B1
B2
B3

Post-operative Care

All fractures of the middle and distal femur are immobilized for the first 5 days with the hip and knee at 90° (Fig. 103b). On the fifth post-operative day, the patient is allowed to sit and dangle his legs and then usually within a day or two he is able to get up with crutches and begin partial weight bearing. All haematomas, if they occur, must be removed surgically.

Fig. 206

C_1 *Supracondylar fracture or the so-called T or Y fracture.* Determine the point of entry for the seating chisel. Reduce the condyles with the knee flexed to 90°. Occasionally a few comminuted cortical fragments have to be disimpacted and elevated out of the metaphysis. Temporarily fix with Kirschner wires and then carry out the fixation of the condylar fracture with two long threaded cancellous screws with washers. Take care not to obstruct the site of entry for the seating chisel and make sure that the threads do not cross the fracture. As the seating chisel is driven in parallel to the knee joint axis the assistant must provide counter-pressure medially. This fracture can also be treated in a manner similar to the simple supracondylar fracture, i.e., the supracondylar component of the fracture can be reduced prior to the insertion of the seating chisel. This makes it possible to use the condylar guide.

C_2 *Bicondylar fracture with comminution of the distal shaft.* Operate with the knee bent to 90°. Carefully reduce the joint fragments and elevate all depressed metaphyseal fragments. The condyles are reduced and fixed with lag screws as in C_1. The blade of a 9- or 12-hole condylar plate is then inserted with the blade parallel to the knee joint. With the knee forcibly flexed to apply traction the plate is fixed proximally with one screw and the leg is checked for rotational alignment and for range of movement of the knee joint. If the medial buttress is very badly comminuted or if there is an actual bone loss medially, the medial cortex must be bridged with a second long T buttress plate. At the end an extensive corticocancellous graft must be carried out.

C_3 *Bicondylar fracture combined with a comminuted fracture of the distal femur and an anterior tangential fracture of one or both condyles.* An additional medial parapatellar incision becomes unavoidable whenever there is a tangential fracture of the medial condyle. Whenever there are tangential fractures of both condyles, one can run into problems with the insertion of the seating chisel because of the AP lag screws necessary to fix the anterior tangential fragments. In these cases we use the long condylar buttress plate. Take great care to avoid a frontal or sagittal malalignment. If the medial buttress is deficient, carry out an extensive cortico-cancellous bone graft and buttress the fracture medially with a T plate. Prior to the fixation of the fracture with the lateral condylar buttress plate, reduce the tangential fractures and fix them with lag screws. Then reduce the condyles and fix them with three or four Kirschner wires. Apply the plate and insert all the lag screws through the plate. Then reduce the shaft and fix it to the plate. Finally, carry out an extensive cancellous graft as already indicated.

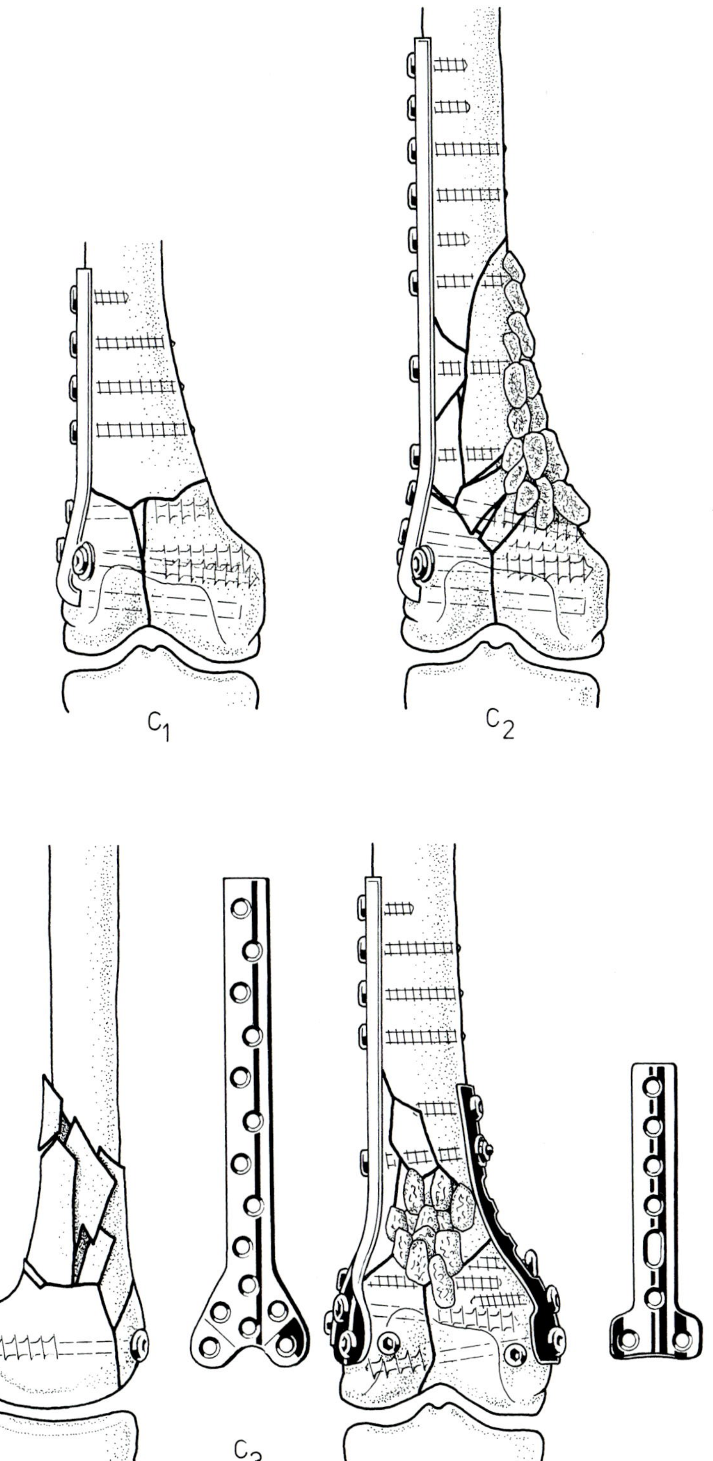
C_1
C_2
C_3

1.8 Fractures of the Patella

Pauwels used an anterior wire suture for a fractured patella in order to demonstrate the principle of tension band fixation. Tension band wiring has since become the preferred method of internal fixation for fractures of the patella. If meticulously executed, tension band wiring allows immediate mobilization of the knee, which is very important, because the patella does not begin to articulate with the intercondylar notch until the knee reaches 45° of flexion. Only at 90° of flexion and above does the patella gain full congruent contact with the condylar surfaces. In extension only the inferior pole of the patella makes contact with the articular surface of the femoral condyles.

An anteriorly placed tension band wire neutralizes all the distracting forces and converts them into forces of compression. One should aim for a slight overcorrection (Fig. 24). The parallel Kirschner wires which are inserted once reduction is accomplished greatly improve the anchorage of the tension band wire (Figs. 27a, 207).

In comminuted fractures of the patella where the middle segment is beyond salvage but the proximal third and distal poles are still intact, excise the intervening middle third, and lag the distal pole to the proximal fragment. A tension band wire completes the fixation (Fig. 209).

Pauwels demonstrated years ago that even very badly comminuted fractures of the patella, if without rupture of the joint capsule and of the quadriceps retinaculae and if protected with a loosely applied tension band wire, could be treated with immediate mobilization. The results of such treatment were surprisingly good even though some of the patellae ended up longer than normal. Thus a patellectomy can almost always be avoided.

Avulsion fractures of the inferior pole of the patella are best first fixed with a small cancellous lag screw which is then supplemented with a tension band wire (Fig. 210). Repairs of ruptures of the infrapatellar tendon are protected by a tension band wire. If the tibia is osteoporotic, the tension band wire is passed around a screw which is inserted transversely through the tibial tubercle (Fig. 211). Vertical fractures of the patella without displacement are treated non-operatively. If displaced, they should be reduced and fixed with two lag screws.

1.8.1 Tension Band Wiring of the Patella

The technique of internal fixation of a simple transverse fracture: Approach the fracture through a longitudinal lateral parapatellar incision. Examine the fracture and the articular surfaces of the patella and those of the femoral condyles. Reflect back the quadriceps tendon 2–3 mm along the fracture. Carry out an anatomical reduction and maintain it temporarily with reduction clamps. Insert the first tension band wire close to the bone, through the patellar and quadriceps tendons. The passage of the wire through the tendon is made very much easier by taking a slightly curved large-bore needle and by passing it through the tendon close to the bone at exactly the desired location. The wire is then pushed into the lumen of the needle and the needle and the wire are then withdrawn together. The second wire is inserted more anteriorly and is passed through the Sharpey fibres. The fracture is then slightly overreduced and the tension bands are tightened. We prefer the special AO wire with loop and the AO wire tightener to the twisting of the wires. Note that when the knee joint is bent the overcorrection vanishes and the fracture surfaces come under full contact and compression and remain so not only during flexion but they remain in contact during extension. Close the capsule and the retinacular expansions and insert two suction drains.

1.8.2 Internal Fixation of the Patella with Two Kirschner Wires and Tension Band

A simple tension band wiring has been shown to be exceptionally effective in treating fractures of the patella. It remains the method of choice. The one problem with it is the insertion of the wires close to the bone. If the wire is passed through the tendon some distance from its insertion, then one runs the risk of shortening the infrapatellar tendon and producing a patella baja. We have begun therefore to use more and more the combination of Kirschner wires with tension band wiring where the Kirschner wires serve as anchorage for the tension band wire. Although the method is reliable and relatively simple, it must be carried out with precision. Post-operatively the knee can be kept flexed to 90° and one can begin with immediate active mobilization of the joint. This method can be used not only for transverse fractures but also for comminuted fractures of the patella (if necessary combined with lag screws). In the latter two examples, in the comminuted and stellate fractures of the patella, it is best to flex the knee to 90° during the reduction. The quadriceps retinaculae are intact in most of these stellate or shattered fractures of the patella. If the middle third of the patella is beyond salvage, this segment can be excised, the edges of the remaining fragments smoothed with a saw and then screwed and wired as shown in Fig. 207.

1.8.3 Post-operative Care

Use a compression dressing for the first 48 hours and position the limb on a sponge rubber splint with the knee flexed to 60°. Quadriceps exercises and active extension is attempted on the first post-operative day. On the third or fourth day, the patient sits up and dangles his legs over the edge of the bed. If a haemarthrosis develops this is aspirated as necessary.

Initially we tended to be overcautious. Today, we recommend early active mobilization. If the wires are securely anchored one need not fear loss of reduction. If the adequacy of the internal fixation is in doubt, then after the initial mobilization of 8–10 days, we immobilize the knee in a plaster cylinder for a further 4–6 weeks.

Fig. 207 *Apply a Tourniquet.* Position the patient supine with the knee flexed to 90° over a roll. Only with the knee flexed to 90° or more (110°–120°) does the articular surface of the patella come into contact with the intercondylar notch. Make either a transverse or a straight longitudinal lateral incision directly over the middle of the patella. Expose the fracture and clean its edges. Elevate any articular depressions and if necessary fill the defects with cancellous bone.

a About 5–6 mm from the anterior surface of the patella, drill with the 2-mm drill bit two parallel drill holes in the proximal fragment. The distance between the two drill holes should be 20–25 mm. To ease the drilling hold the patellar fragment with the reduction clamps and tip the fragment so that the fracture surface faces you. Once the first hole is drilled, the drill bit is removed and is replaced with a Kirschner wire. This Kirschner wire then serves as a guide for the second drill hole. This ensures that the two drill holes will be parallel to one another.

b Insert two 1.6-mm-thick and 15-cm-long Kirschner wires into the two drill holes.

c Replace the Kirschner wires with 2-mm drill bits inserted in a proximo-distal direction.

d Reduce the fracture and keep it reduced with the reduction clamps with tips. Check the accuracy of reduction. If it is perfect, drill with the 2-mm drill bits the holes in the lower fragment.

e Carefully remove the drill bits and replace them with 1.6-mm Kirschner wires. With the special bending pliers, bend over the proximal end of the wire to form a hook and cut off any excess.

f Take the 1.2-mm-thick AO wire with a loop and pass it around the Kirschner wires in such a way that the loop comes to lie next to the proximal end of the lateral Kirschner wire. This will facilitate future metal removal.

g The wire is then tightened with the AO wire tightener. Its free end is then bent over to prevent slippage and the excess is cut off. The short free end is then buried deep in the soft tissues. Check the reduction once again. If all is well, adjust the Kirschner wires so that the curved ends face backwards, pull them through the patella and hammer the curved portions into bone. The distal portions of the Kirschner wires are only slightly bent so that they will not interfere with future metal removal, and are cut off about 10 mm from where they exit from the bone. The joint capsule and the quadriceps retinaculae are now carefully resutured. Post-operatively position the knee in 60° of flexion and begin with active mobilization under supervision on day one. The patient usually gets up on the fifth post-operative day.

a b c d

e f g

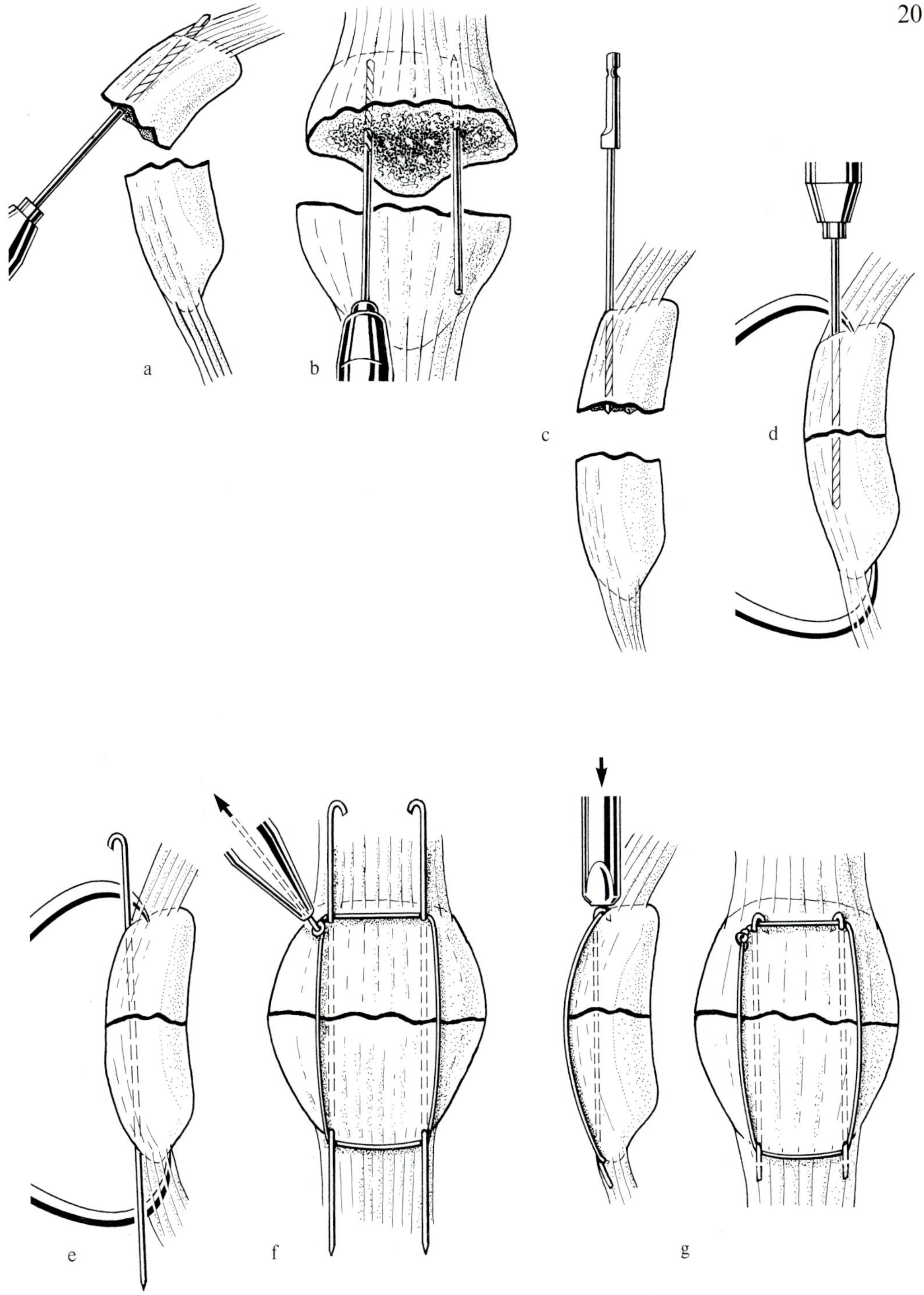

Fig. 208 *A simple transverse fracture:* The tension band wires lie on the anterior surface of the patella (for an alternative method of fixation see Fig. 207).

Fig. 209 *A comminuted fracture of the patella:* The proximal and distal fragments are cut flush with a saw to make them fit. They are then fixed with two lag screws. A tension band wire is passed through the tendinous insertions. It complements and protects the screw fixation.

Fig. 210 *An avulsion of the distal pole:* Tension band wire alone tends to tip the inferior pole into the joint. The fragments should therefore be first fixed with a lag screw. The fixation is then completed with a tension band wire.

Fig. 211 *Rupture of the infrapatellar tendon:* The ruptured infrapatellar tendon is repaired with resorbable sutures. The suture line is then protected with a figure-of-eight tension band wire which is passed through the quadriceps insertion and the tibial tubercle. If the tibial tubercle is osteoporotic, then a screw is inserted transversely through bone and the tension band wire is wound around the screw. The figure-of-eight tension band protects the repair of the tendon and makes early mobilization possible. The tension band wire should be removed in about 6 months.

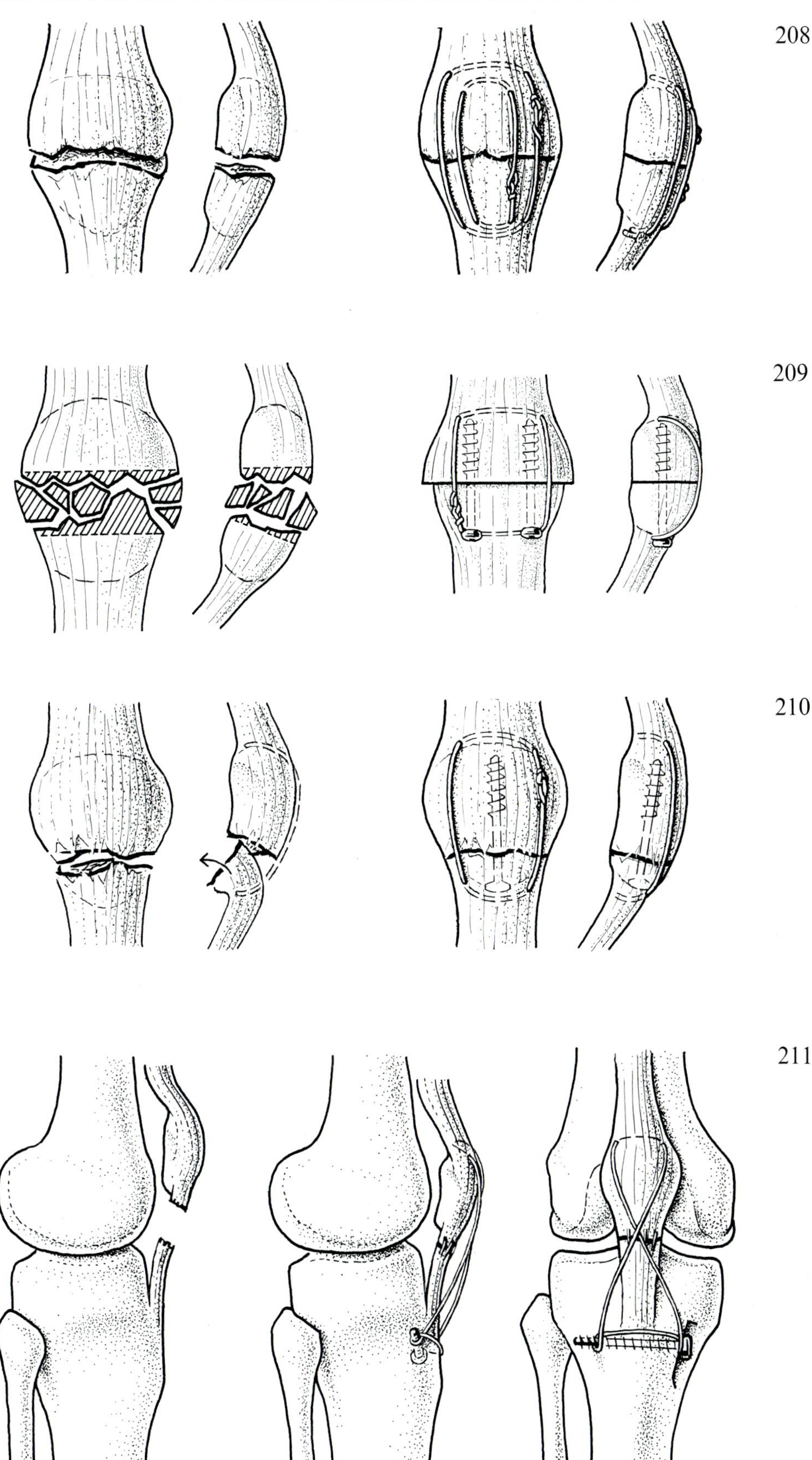

208
209
210
211

1.9 Fractures of the Tibia

The problems which a fracture of the tibial shaft will present, such as shortening, are clearly evident right from the start. If there is little shortening and the reduction is easily accomplished, then the fracture should be treated non-operatively in accordance with the principles initially laid down by BOEHLER and further developed and modified by DEHNE and SARMIENTO. DEHNE recommends a long leg plaster for the first 6–8 weeks and then transfer to the patellar tendon-bearing plaster (PTB plaster) as recommended by SARMIENTO. Unstable fractures of the tibial shaft, if treated non-operatively, require a combination of traction and plaster fixation. This often results in permanent loss of function due to malunion and joint stiffness.

Indications for Open Reduction and Internal Fixation

Open reduction and internal fixation should be considered only if the staff and environment available meet all the prerequisites outlined on page 134. We consider irreducible joint fractures an absolute indication. The following are our indications for open reduction and internal fixation of the adult tibial shaft fractures including the so-called 'boot-top fractures':

- Irreducible joint fractures are an absolute indication.
- Unstable fractures with displacement of the main fragments the full width of the shaft.
- Shortening of more than 1 cm.
- Fractures initially treated by closed reduction and plaster fixation which slip within the first few days. Most commonly these are the fractures between the middle and distal thirds. They are unstable because of muscle and tendon interposition, particularly interposition of the tendon of tibialis posterior.
- Isolated tibial shaft with a varus deformity of more than 5°.
- All shaft fractures in polytrauma patients who require treatment in an intensive care unit.
- All grade two and grade three open fractures (see p. 306).
- All segmental fractures.

Badly comminuted fractures may present tremendous technical problems if an internal fixation is attempted. Such fractures, if they do not involve joints, are best treated non-operatively or with an external fixator. If problems arise during the treatment such as a delayed union or a non-union, it usually involves only one or two fracture lines in one segment of the bone, rather than the whole extensive area of comminution, and technically they become a very much simpler problem to cope with.

There are good reasons for teaching centres to extend the scope of their indications for surgical treatment beyond the ones enumerated above. It is extremely difficult for a surgeon to teach or indeed to learn internal fixation if the only cases he operates are the difficult and complex fractures. Furthermore, such practice usually carries a much higher risk and danger for the patients.

Choice of the Type of Internal Fixation

By and large, the closed short oblique or transverse fracture of the middle third of the tibia are treated by intra-medullary nailing. All other shaft fractures are treated by a combination of lag screws with neutralization plates. All short oblique or transverse fractures which are plated should have a lag screw inserted across the fracture to supplement the fixation (see p. 74). This lag screw should preferably be inserted through the plate. The proximal and distal intra-articular fractures will be discussed in their respective sections (see pp. 258 and 280).

1.9.1 Fractures of the Tibial Plateaux

Surgical Approaches

Tibial plateau fractures are approached either through a parapatellar incision or through a curved incision whose upper limb runs halfway between the patella and the tibial plateau and which then curves downwards to run straight just lateral to the anterior crest of the tibia. The two limbs should subtend an angle of 120°. If both plateaux are to be approached simultaneously, then a third limb is added making this a tri-radiate incision (so-called Mercedes incision). The three flaps thus all subtend an angle of 120°. The three incisions intersect at a point directly over the middle of the infrapatellar tendon halfway between the inferior pole of the patella and the tibial tubercle. They must never intersect either over the patella or the tibial tubercle because of danger of wound-edge necrosis. Beware of flaps which cross the middle of the leg because few vessels cross over the midline and flap necrosis can occur.

In dealing with type II and type III tibial plateau fractures it is necessary to expose the lateral tibial plateau in such a way that its articular cartilage can be seen as well as the lateral cortex where it joins the shaft. To achieve such an exposure, the proximal insertion of the extensor muscles must be reflected and the articular surface exposed. In order to expose the articular cartilage of the lateral tibial plateau, it is necessary to divide through the capsule below the meniscus. A cuff of tissue must remain on both to permit subsequent re-suture of the meniscus. Once the capsule is incised the meniscus is retracted upwards which provides an excellent exposure of the articular surface.

Concomitant Injuries of the Capsule, Collateral Ligaments and Cruciate Ligaments

Type II and type III are compression fractures of the lateral tibial plateau. In order to be produced the medial capsule and medial collateral ligament must remain intact at least throughout the initial phase of the deformity. Once these are torn, a valgus force cannot exert any compressive force on the lateral tibial plateau. It is important to recognize that despite this about 20% of compression fractures of the lateral plateau are associated with injuries of the medial collateral ligament, which is either avulsed out of bone or torn through its substance together with a tear of the posterior capsule. Therefore once the lateral plateau is reconstructed, it is necessary to test the stability of the medial collateral ligament and capsule by applying a valgus force with the knee first in full extension and then in 30° of flexion. One must also test for meniscal injuries and for injuries to the cruciate ligaments.

Fig. 212 *Surgical approaches to the tibial plateau.*

a Parapatellar incision.

b Curved incision.

c "Mercedes" incision – this is used for the reconstruction of the whole proximal tibia. The skin, together with the subcutaneous fat, must be lifted as a single flap. The flap must be handled very gently and must not be squeezed or folded over!

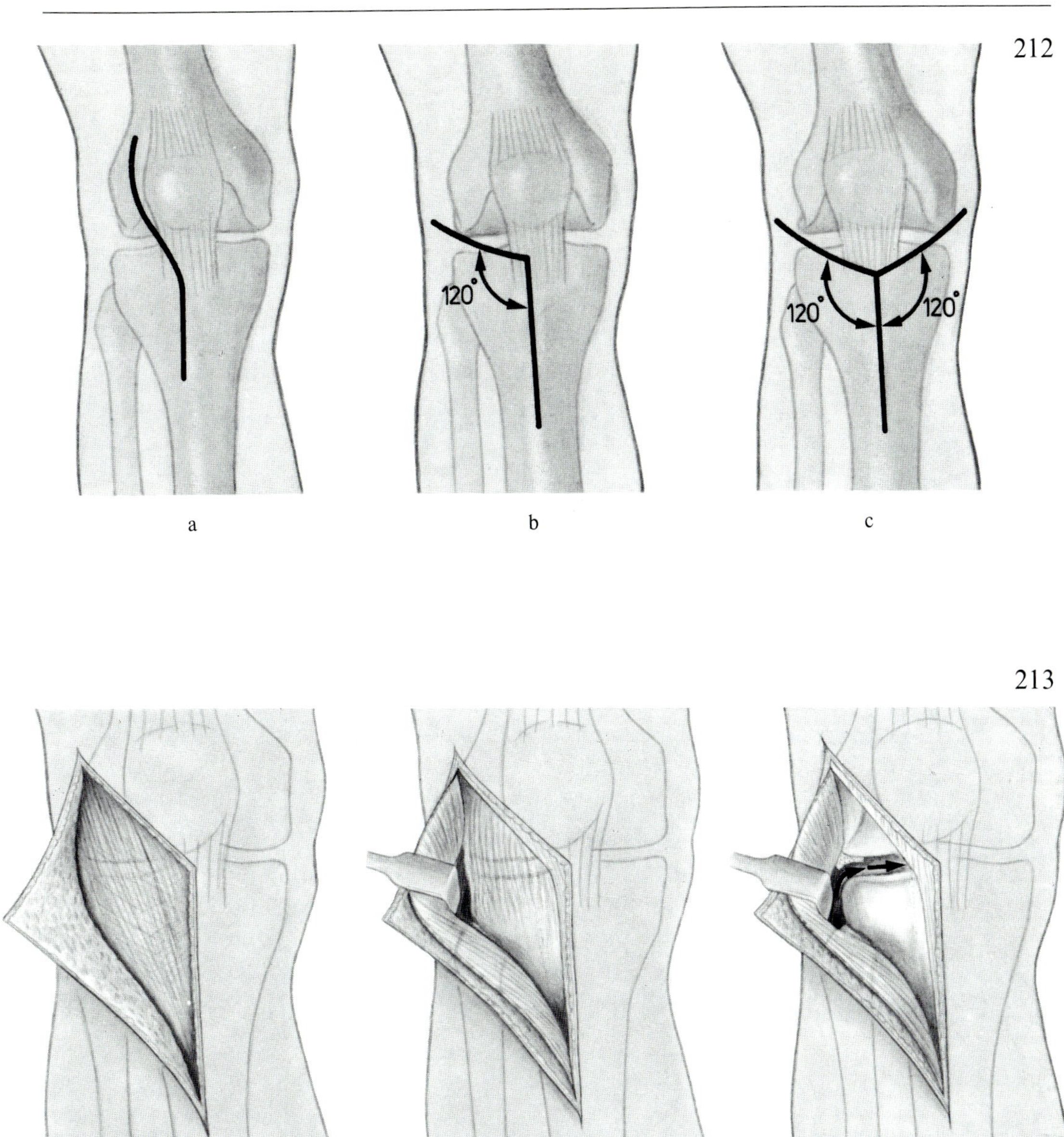

Fig. 213 *The surgical approach to a tibial plateau (lateral).*

a The raising of the flap and exposure of the iliotibial tract.

b Incision into the iliotibial tract as parallel to the direction of the fibres as possible.

c The capsule is incised between its attachment to the meniscus and the tibial plateau. Upward retraction of the capsule together with the meniscus exposes the articular surface of the plateau.

Classification of Tibial Plateau Fractures

Type I. Pure wedge fracture; these are relatively rare and occur most commonly laterally or posteriorly. If they occur medially then they give rise to a corresponding varus deformity. The fracture plane is either in the frontal or sagittal plane.

Type II. Pure depression fractures, the result of valgus overload. The lateral tibial plateau is pushed in by the lateral femoral condyle. The plateau itself is not widened.

Type III. Combination of types I and II with joint depression and fracture of the lateral cortex. On the AP projection the plateau always appears widened. In the lateral projection the anterior or posterior part of the articular surface is either intact or depressed.

Type IV. The Y and T fractures or comminuted fractures of both condyles, at times associated with fractures of the intercondylar eminence. The lateral plateau is usually the more severely damaged.

Whenever doubts exist as to the exact fracture pattern, one should resort to both AP and lateral tomography.

Internal Fixation of the Different Types of Tibial Plateau Fractures

Type I, wedge fracture (Fig. 214a). Expose the tibial condyle and after a careful inspection of the fracture lines, carry out an anatomical reduction. The fixation is carried out with two cancellous screws with washers. If one is worried about axial overload and resultant varus or valgus deformity, the internal fixation should be combined with a buttress plate. Depending on the fracture, a narrow DCP, a round hole plate, a T plate or an L plate can be used for buttressing. Such a plate buttresses the cortex and fixes the two fragments together.

Fig. 214 *The different types of tibial plateau fractures.*

a The wedge fracture – may also occur medially (type I).

b Depression fracture – almost always laterally (type II).

c Depression and wedge fracture – (type III).

d Y and T fractures – comminuted fractures (type IV).

Fig. 215 *The types of internal fixation of the different wedge fractures.*

a A lateral wedge fracture fixed under compression with the proximal cancellous lag screw and buttressed at its tip with the lower cortex screw.

b Internal fixation with a buttress plate.

c The internal fixation of a posterior wedge fracture (rare).

214

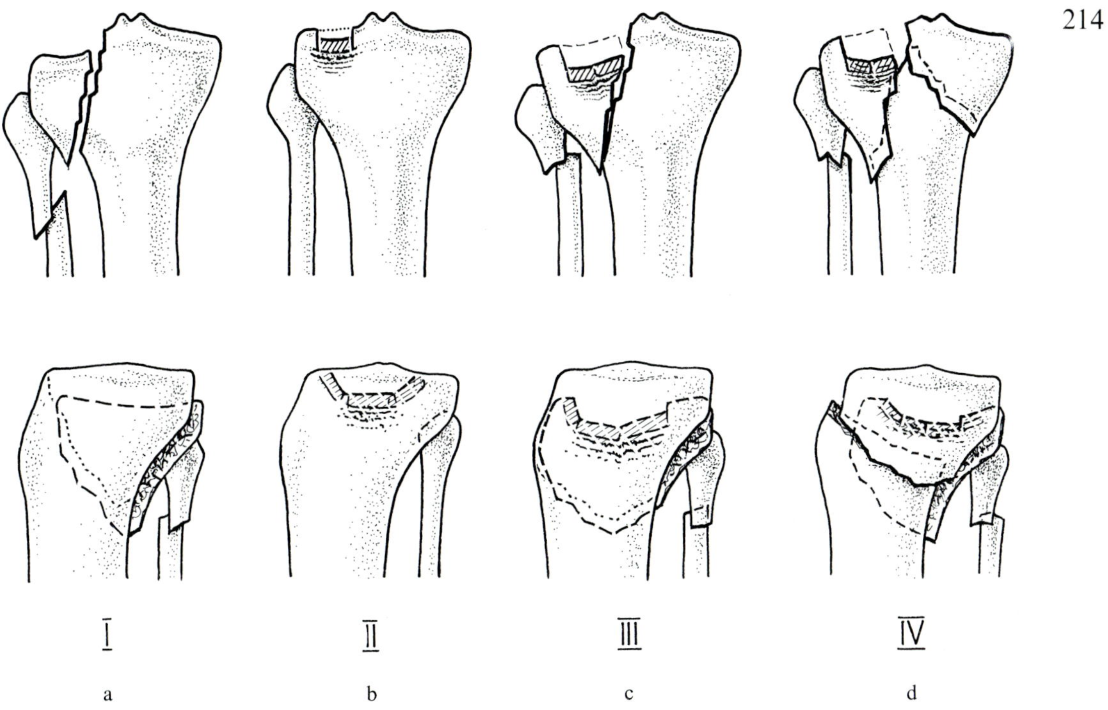

215

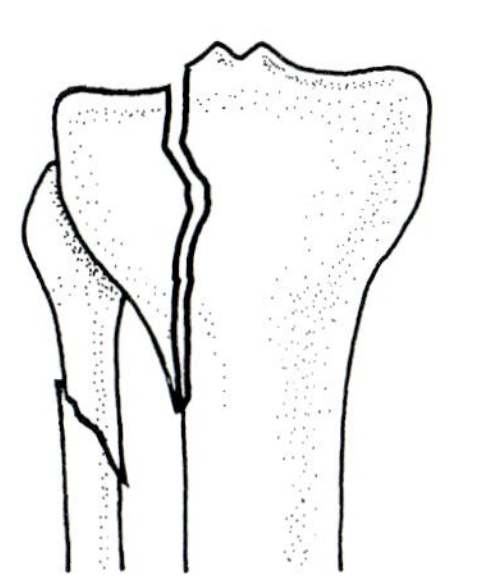

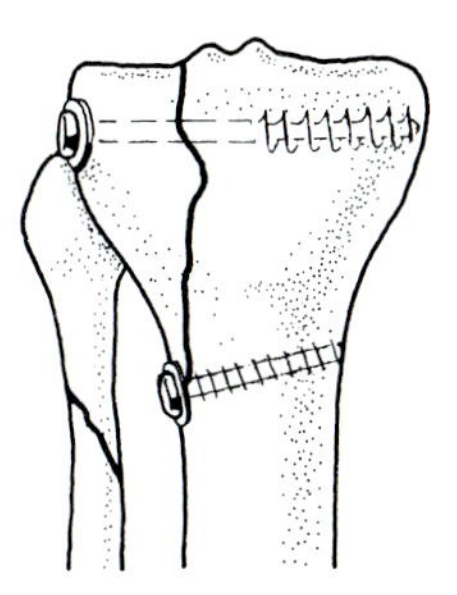

a

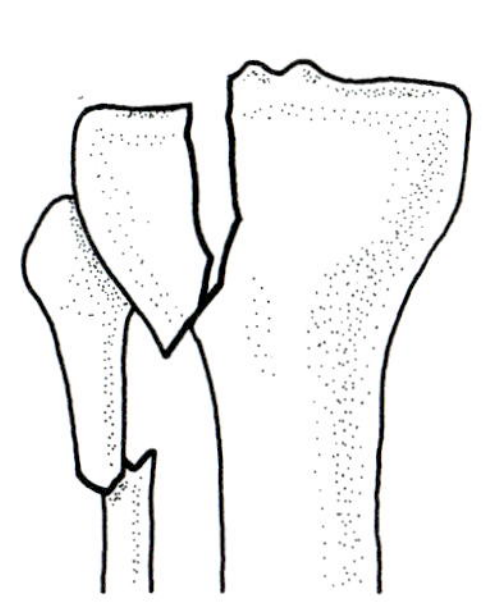

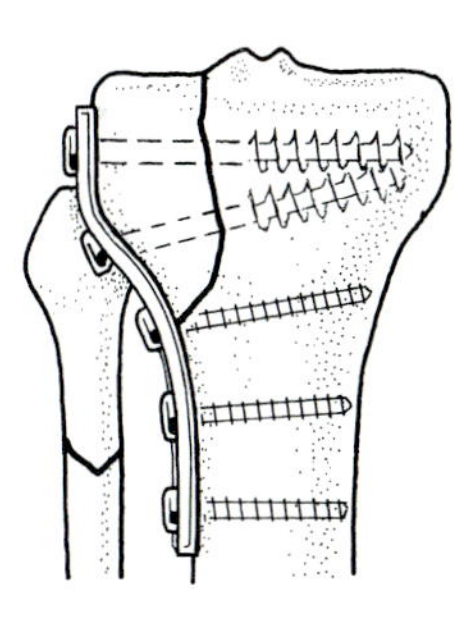

b

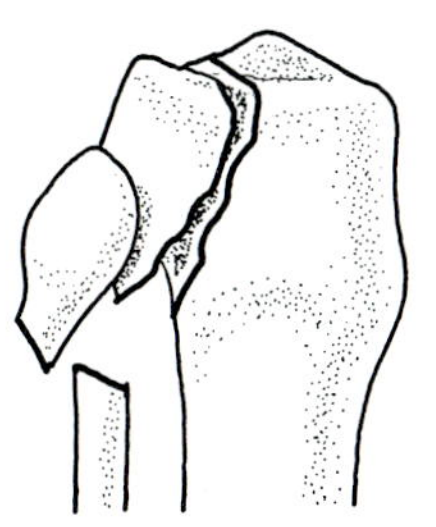

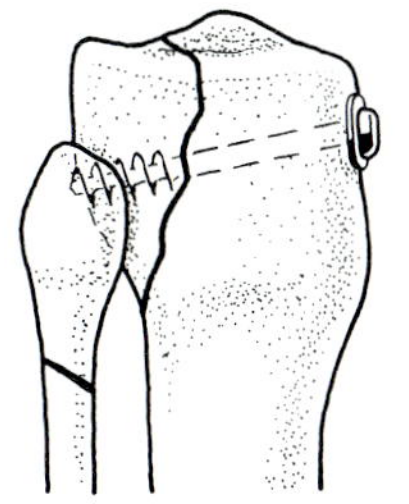

c

Type II

Fenestrate the anterior cortex of the tibia just in front of the fibula along the underside of the tibial condyle. Through this window introduce a bone punch and by pushing upwards, reduce the depressed articular surface.

The elevation of the depressed fracture results in a defect in the metaphysis which must be carefully packed with cancellous bone either from the ipsilateral greater trochanter or anterior iliac crest. It is usually enough to stabilize and protect the reduction with two transversely inserted cancellous screws with washers.

Type III – Combination of Type I and Type II (Fig. 217)

Proceed as in type II. Here it is very easy to make a fenestration. All that is necessary is to remove some of the small fragments of the thin and cracked anterior cortex. After reduction the temporary fixation of the lateral cortex is carried out either with two cancellous screws as in type II or preferably with one or two Kirschner wires. The resultant hole in the metaphysis must now be carefully filled with a cancellous graft which aids in the support of the elevated joint surface. At the end, the internal fixation is accomplished with a buttress plate. Usually all that is necessary is a carefully contoured narrow plate. The T or L buttress plates fit the lateral contour of the tibia very well (see Fig. 54) and are well suited for this purpose.

Fig. 216

a Type II with depression of the lateral tibial plateau.

b Fenestration of the lateral tibial condyle and reduction of the depressed fragments.

c The end result after grafting of the metaphyseal defect and stabilization of the fracture with one or two cancellous lag screws.

Fig. 217

a Type III – combination of wedge and depression fracture.

b Reduction and temporary fixation with Kirschner wires.

c The bone grafting with cancellous bone.

d The insertion of the buttress plate (the regular DCP, T plate or L plate can be used).

216

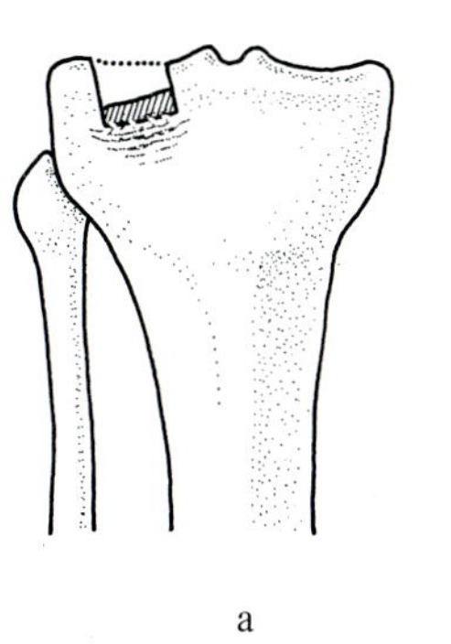

a

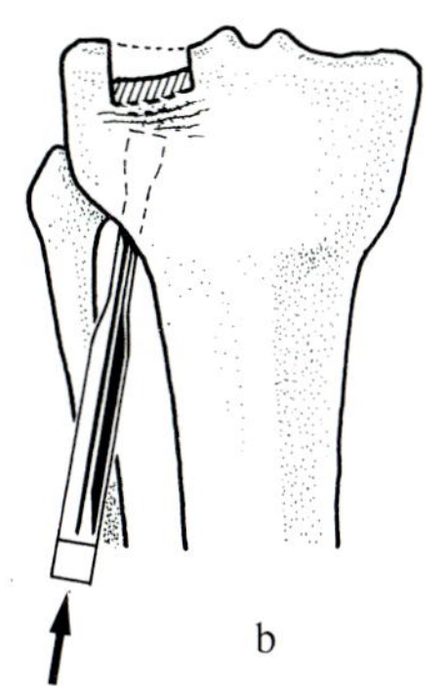

b

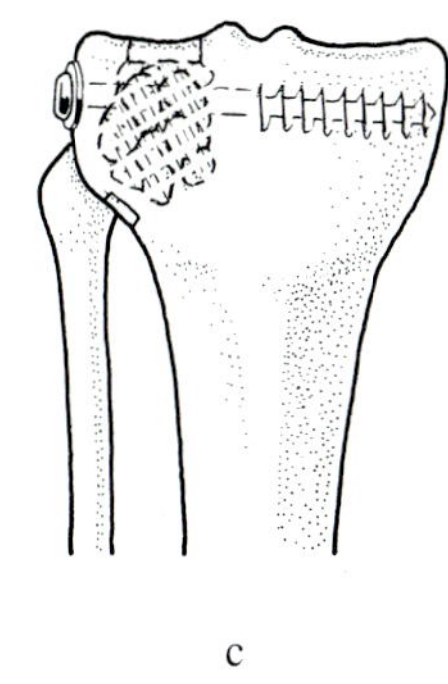

c

217

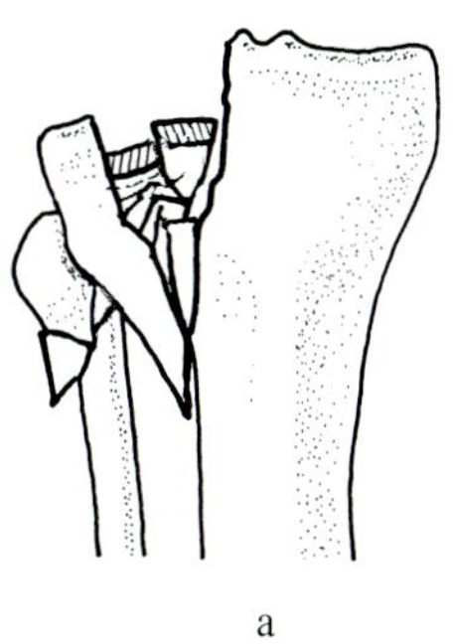

a

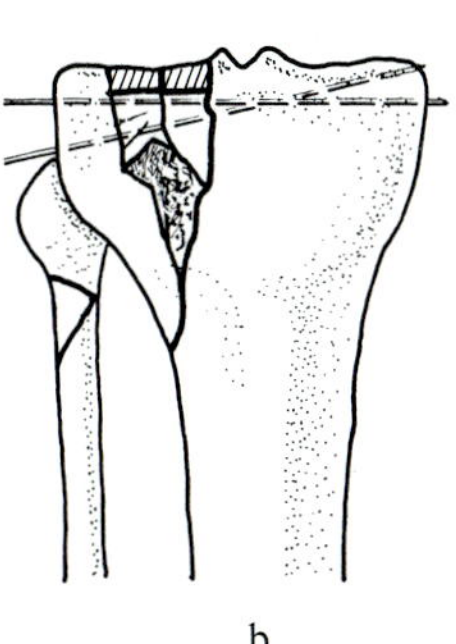

b

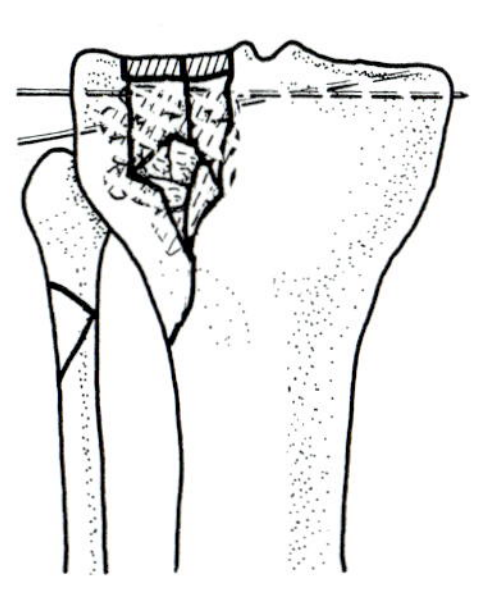

c

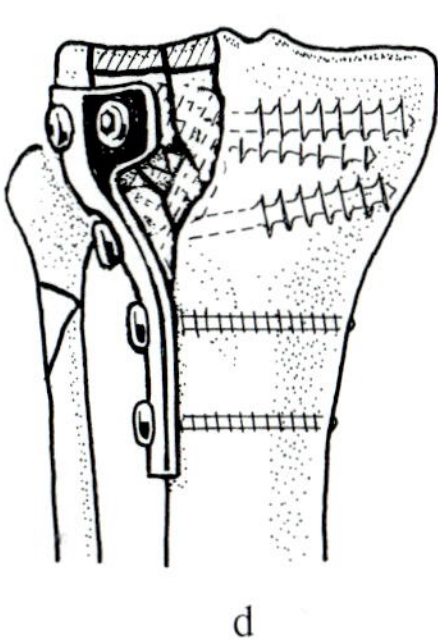

d

Type IV (Y and T Fractures and Comminuted Fractures) (Fig. 218)

Most often the medial tibial plateau escapes any joint depression. It retains the attachment of the pes anserinus and breaks out together with the intercondylar eminence. This makes it possible to fix it back to the tibial tuberosity. Laterally, on the other hand, we usually encounter an extensive comminution of the plateau. The reduction of this portion is carried out as described under Type III. For the approach we use the tri-radiate Y-shaped incision. If the intercondylar eminence is avulsed, then a much wider exposure is necessary. We Z-plasty the infrapatellar ligament and elevate it upwards together with the ligament, the infrapatellar fat pad, and the menisci (Fig. 213c).

Another way to gain the same approach is to osteotomize the tibial tubercle and elevate it together with the infrapatellar tendon. We do everything possible to preserve both menisci. Peripheral detachments are carefully re-sutured with absorbable sutures (Dexon or Vicryl). The aim of the procedure is to reconstruct the main weight-bearing portions of the joint and to interpose between them in the normal anatomical position the shock-absorbing menisci.

The steps in the reconstruction and fixation here are the same as in any intra-articular fracture. There are two important steps. The first step is to reconstruct the plateau and the second is to join it with the shaft. We find that after the reduction, it is best to fix it temporarily with Kirschner wires. For the final stabilization it is necessary to use buttress plates and occasionally screws with nuts and even bar bolts.

There are two further cardinal rules in intra-articular joint reconstructions. The defect in the metaphysis must be carefully filled with bone graft, and the damaged menisci and collateral ligaments must be repaired. We wish to draw the reader's attention once again to the fact that the medial plateau usually escapes any significant depression. Thus if the patient should develop post-traumatic osteoarthritis of the lateral plateau, a varus high tibial osteotomy can be carried out and the load transferred to the still intact medial plateau. In this way knee function can be preserved for a further number of years longer. We feel, therefore, that it is very important to make certain that the implants do not damage the medial plateau. The soft tissues must be handled carefully and atraumatically, and the flaps of this extensive exposure must be handled with great gentleness and care to prevent any wound-edge necrosis.

Post-operative Care of Patients with Tibial Plateau Fractures

The knee is nursed in 45° of flexion and the patient begins immediate assisted active exercises through an arc of 60° from full extension to flexion. Rotational exercises are begun 2–4 weeks following the internal fixation depending on the stability of the fixation achieved. After 6 weeks or so the patient begins weight bearing and can usually increase the amount quite rapidly because fractures through metaphyseal bone which have been carefully bone grafted with the defect carefully filled, usually heal very quickly. Full weight bearing commences somewhere between the second and fourth post-operative months.

218

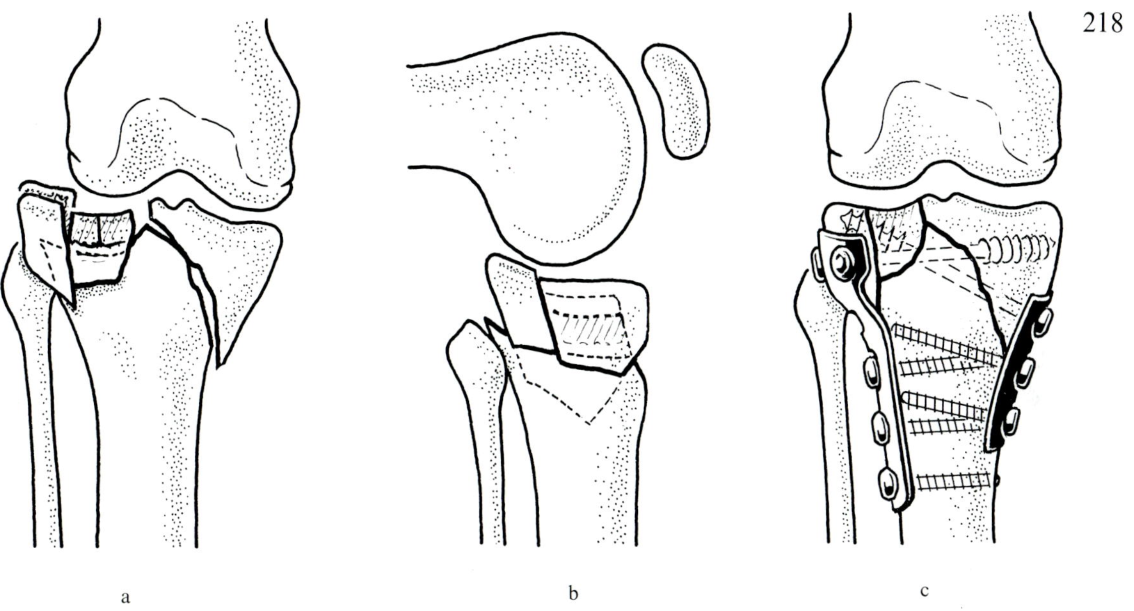

Fig. 218

a, b Type IV

c The final result of an open reduction and internal fixation with a small medial and a larger lateral buttress plate. Beware of doing any damage to the usually intact medial tibial plateau.

1.9.2 Fractures of the Shaft of the Tibia

Surgical Approaches to the Shaft of the Tibia (Fig. 219)

The commonest skin incision is made 1 cm lateral to the anterior crest of the tibia. In order to correspond to Langer's lines it must be made perfectly straight. In the supramalleolar region, if a distal exposure is necessary, the incision is curved medially and posteriorly so as to describe a gentle arc around the front of the medial malleolus. Do not hesitate to make long incisions to give you adequate exposure. A straight incision in the leg, if carefully closed, heals with a barely visible scar. The bone is exposed only on the medial side. This is done by reflecting the periosteum through the full extent of the fracture lines. The fracture surfaces are cleared with small periosteal elevators and a sucker. Once the fracture lines have been exposed, they are carefully examined, compared with the X-ray and with the pre-operative drawing of the internal fixation. Only now can the decision be made on how the definitive internal fixation should be constructed.

An incision down the middle of the medial surface of the tibia is made very rarely. It is used when one has to deal with an extensively comminuted fracture with bone loss which is going to be stabilized with double plating, i.e. by the application of a semi-tubular plate to the anterior and medial edge of the tibia.

The posteromedial incision is very rarely used.

The operative approach to open fractures is dictated by the location of the injury. The approach must be made in such a way that no additional devascularization of either the soft tissue or bone results (see chapter on Fractures).

For intra-medullary nailing we prefer the transverse incision made halfway between the inferior pole of the patella and the tibial tuberosity. The infrapatellar tendon is then split longitudinally (see Fig. 219b and the chapter on Intra-medullary nailing).

Fig. 219

a The standard approach to the shaft of the tibia for screw fixation and plating.

b The approach for intra-medullary nailing:
  (*1*) The intra-medullary nail is inserted through a transverse incision made halfway between the inferior pole of the patella and the tibial tubercle. The infrapatellar tendon is split longitudinally (for details see Fig. 82b).
  (*2*) The approach for the open reduction and removal of reaming debris.

c Cross-section of the leg illustrating the surgical approaches:
  (*1*) The standard anterolateral approach.
  (*2*) The medial approach made in the middle of the medial surface of the tibia is rarely used. It is used for double plating of shaft fractures with bone loss and for the exposure of certain metaphyseal fractures.
  (*3*) The posteromedial approach is used when one wishes to plate the posterior surface of the tibia as when one is dealing with certain open fractures.

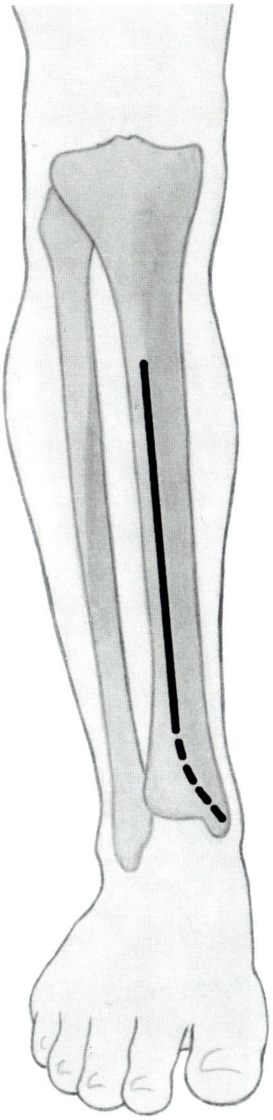

219

1
2
a 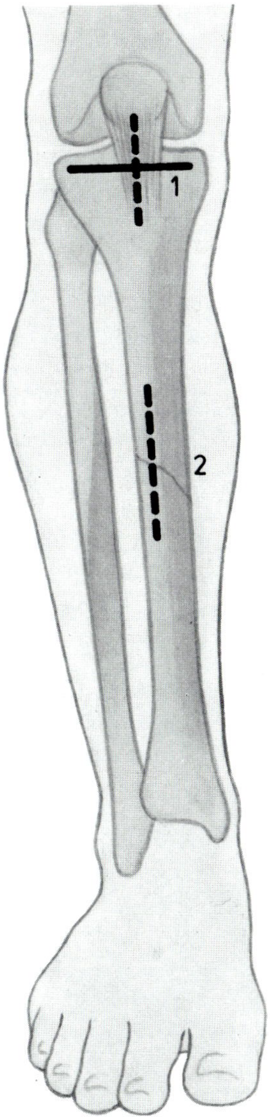
b

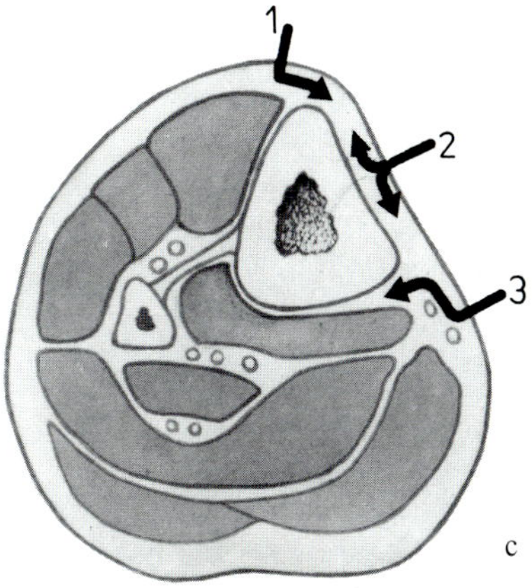

1
2
3
c

Transverse and Short Oblique Fractures of the Middle Third of the Tibia

Short Oblique Fractures: These may be defined as fractures less than twice the diameter of the bone in length at the level of the fracture. In the middle third of the tibia they are an ideal indication for intramedullary nailing (see p. 116).

Rotational stability can be improved by inserting proximally through the nail a screw in an AP direction (Fig. 220c). Distally rotational stability can be enhanced by inserting ledge wires (Fig. 220d).

Note: The further the fracture is removed from the middle third of the tibia, the less ideal an indication it is for intramedullary nailing. An intramedullary nail should not be used to stabilize spiral fractures. Under load, these tend to shorten and develop rotational deformities.

Short Oblique and Transverse Fractures of the Proximal and Distal Third of the Tibia

In the proximal third these fractures represent one of the few indications for double plating (very strong lever arms!) These are best fixed with two semi-tubular plates, one applied to the anterior and one to the posteromedial edge of the tibia.

Short oblique fractures at the level of middle and distal third of the tibial are best stabilized by the combination of lag screws which exert interfragmental compression and a plate which exerts axial compression.

Fig. 220 *Intramedullary nailing of the tibia.*

a, b Short oblique and transverse fractures: ideal indications.

c Improvement of rotational stability by the insertion proximally of a screw in the sagittal plane (the nail has a slot proximally through which the screw can be passed).

d Improvement of the rotational stability distally by the insertion of ledge wires.

Fig. 221 *High short oblique fractures.* Two semi-tubular plates are fixed with short screws to the anterior and posteromedial edge of the tibia, respectively. The anterior plate acts as a tension band (note that its proximal limit is the tibial tubercle). The surgical approach is an incision down the middle of the medial surface of the tibia (see Fig. 219c2).

Fig. 222 *Internal fixation of short oblique fractures outside the middle third of the tibia with DCPs.*
First step: Provisional reduction, drilling of the gliding hole between the main fragments, insertion of the 3.2-mm drill sleeve through the plate into the gliding hole.
Second step: Insertion of screw No. *2* in neutral position.
Third step: Insertion of the third screw as a load screw and tightening completely screws Nos. *2* and *3*.
Fourth step: Drilling of the thread hole and insertion of the lag screw between the two main fragments.

Fig. 223 *Short spiral fracture of the distal third of the tibia.* This is fixed with a lag screw which does not go through the plate, and with a neutralization plate (*Note:* any screws which cross the fracture line, even though they are used to fix the plate, should be inserted as lag screws).

a

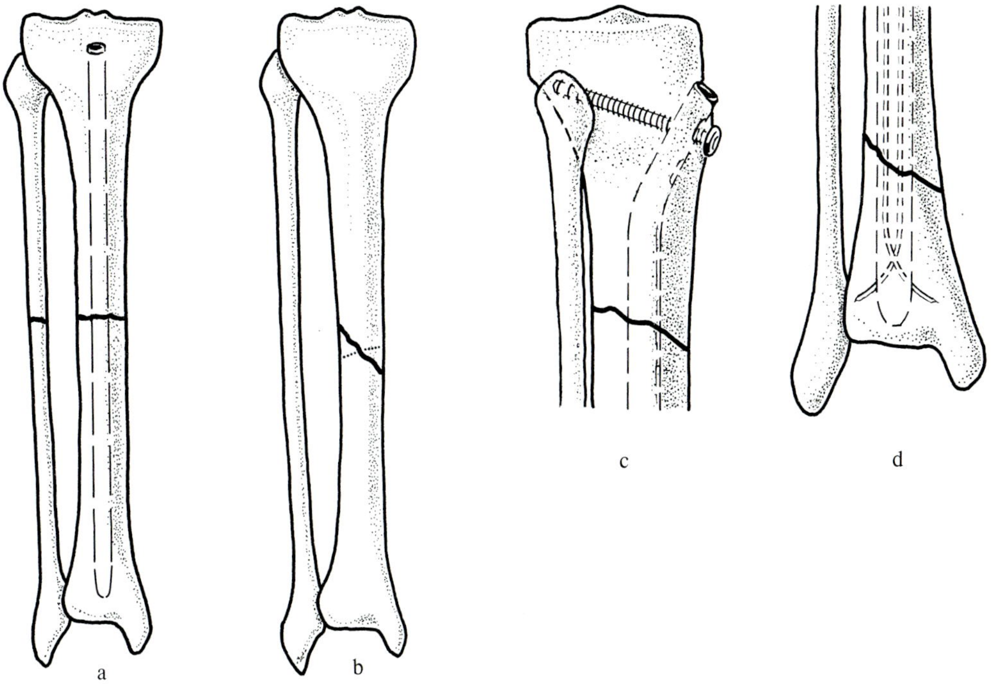

b

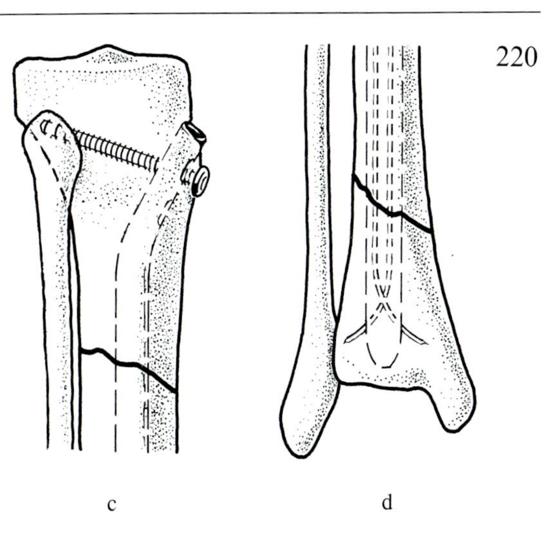

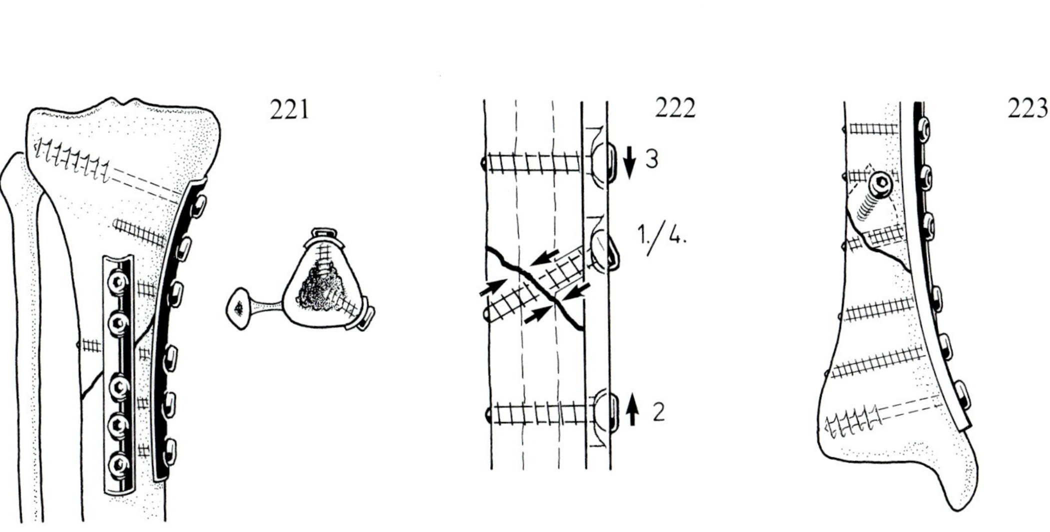

Torsion Fractures

Fracture Types

Bone which breaks due to *torsional forces* breaks into two pieces. The incipient crack runs longitudinally. Its ends are joined through the torsion spiral (Fig. 224).

Torsion combined with bending: This gives rise to a third fragment: the torsion butterfly. We distinguish three torsion butterflies: an anterior, a dorsomedial and a dorsolateral.

Steps in the open reduction and internal fixation of pure torsional fractures and fractures with torsional butterfly fragments

Cleanse and carefully inspect the fracture lines.

First question: Screw fixation or screw fixation in combination with a neutralization plate? Contact between the main fragments which is in length at least twice the diameter of the bone at the level of the fracture allows for stabilization with lag screws. For the position of the screws see Fig. 21. Fractures with torsional butterflies require a neutralization plate in addition to the lag screws. An exception to this rule is the fracture with an anterior torsional butterfly and a long contact between the main fragments.

Second question: Should the gliding and thread holes be drilled before or after the reduction? (see pp. 36 and 38).

The Reduction: Pure spiral fractures and fractures with an anterior torsional butterfly are very easy to reduce. Once reduced, they are held with a reduction clamp. Fractures with a posterior torsional butterfly are more difficult and at times require temporary cerclage wiring for provisional stabilization.

Internal Fixation: Whenever possible the internal fixation is carried out first with lag screws. The neutralization plate is added later on. If the plate comes to overlie a lag screw, the lag screw is temporarily removed and then re-inserted through the plate.

Fig. 224

a *Simple torsion fracture:* Look for incomplete fracture line and hidden torsional butterflies (*dotted lines* indicate the typical location of the hidden fracture lines). Reduction is easy with the help of the reduction clamp.

b The screws are inserted at right angles to the shaft and are staggered, since they follow the spiral of the fracture (see also Fig. 225).

Fig. 225 *Fracture with anterior torsional butterfly.*

a, b Such a fracture type seen in the AP and lateral projections.

c The torsional butterfly is fixed to each of the main fragments with a lag screw. A six-hole neutralization plate completes the fixation. Note that the third screw from the top of the plate has been inserted as a lag screw and adds to the interfragmental compression of the fracture.

d Side view.

e Cross-section: The drawing shows the main fracture line crossed by a lag screw inserted through the plate as well as by a lag screw inserted from in front and used to fix the torsional butterfly.

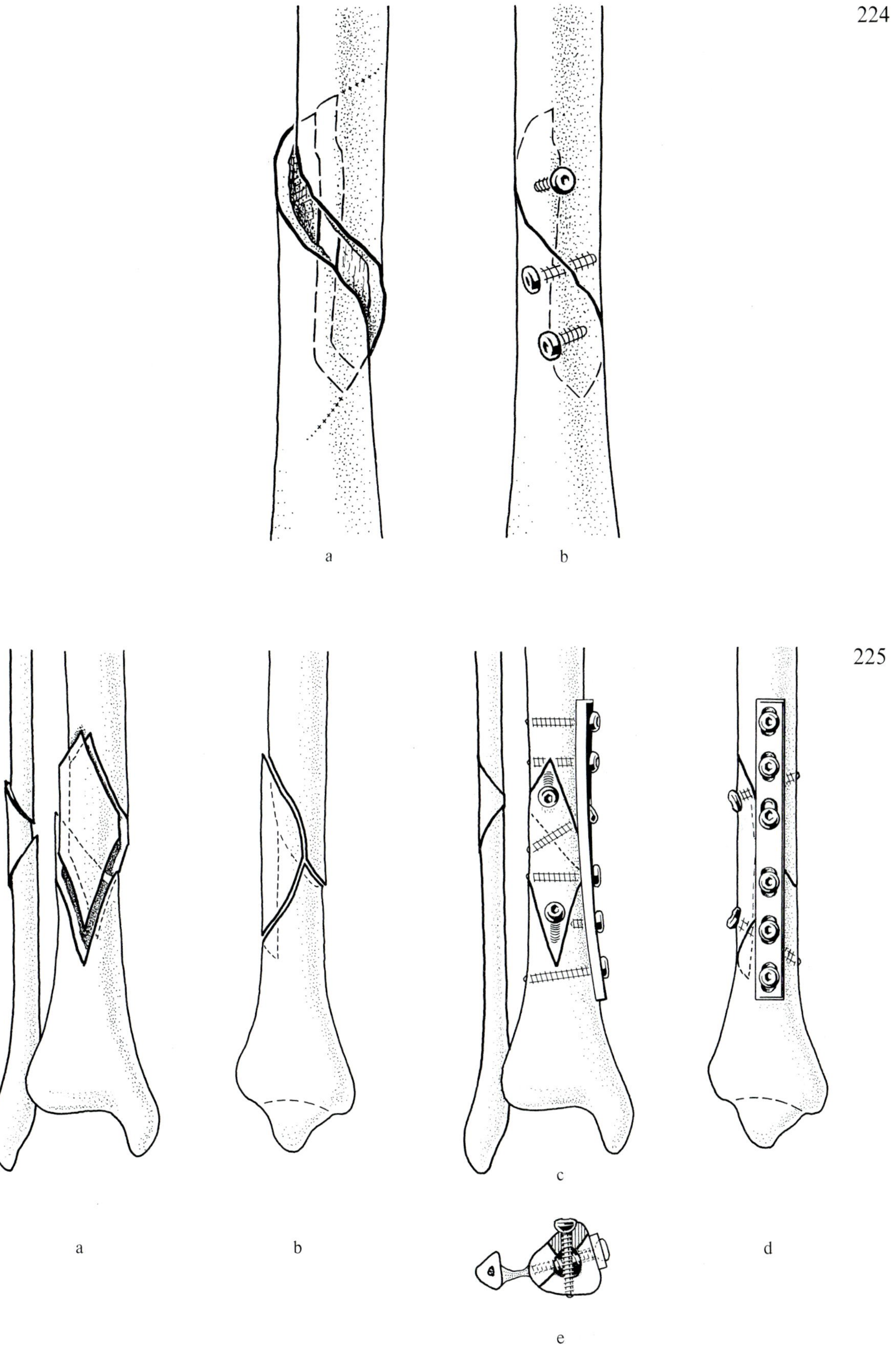
224
a
b
225
a
b
c
d
e

Position of the Neutralization Plate in Fractures of the Tibia

A tibia does not have a tension side. Loading depends on whether the leg is in the stance or swing phase. Neutralization plates can therefore be applied to the most accessible medial surface of the tibia. The following are exceptions to the rule:

- Damage to the soft-tissue envelope on the medial side
- Pseudarthroses of the tibia with varus deformity
- At times a posteromedial torsional butterfly

Every screw which is inserted through the plate and which crosses a fracture must be inserted as a lag screw. If the plate bridges a long butterfly fragment, then a number of screws may have to be left out of the plate (Fig. 229).

Note: Use only narrow plates for the internal fixation of the tibia. The prognosis of fracture healing does not depend on the cortex beneath the plate, but on the cortex opposite the plate! Bone defects in the opposite cortex and denuded avascular opposite cortex must therefore be bone grafted in order to secure bone healing.

Fig. 226 *Posterolateral torsional butterfly.*

a, b Drawing of the fracture.

c Temporary fixation with a cerclage wire permits the reduction of the 'invisible butterfly' as traction is applied to the limb.

d, e The completed internal fixation.

Fig. 227 *Posteromedial torsional butterfly.*

a, b Representation of the fracture type.

c The torsional butterfly is fixed to each of the main fragments with a lag screw which does not cross the plate. The medially applied neutralization plate joins the two main fragments.

d The torsional butterfly is fixed with lag screws inserted in an AP direction. The main fragments are joined with a laterally applied neutralization plate. This allows one to use a shorter plate and one does not have to drill such large holes in the butterfly fragment, since all the gliding holes are drilled through the main fragments.

e Lateral view of the internal fixation.

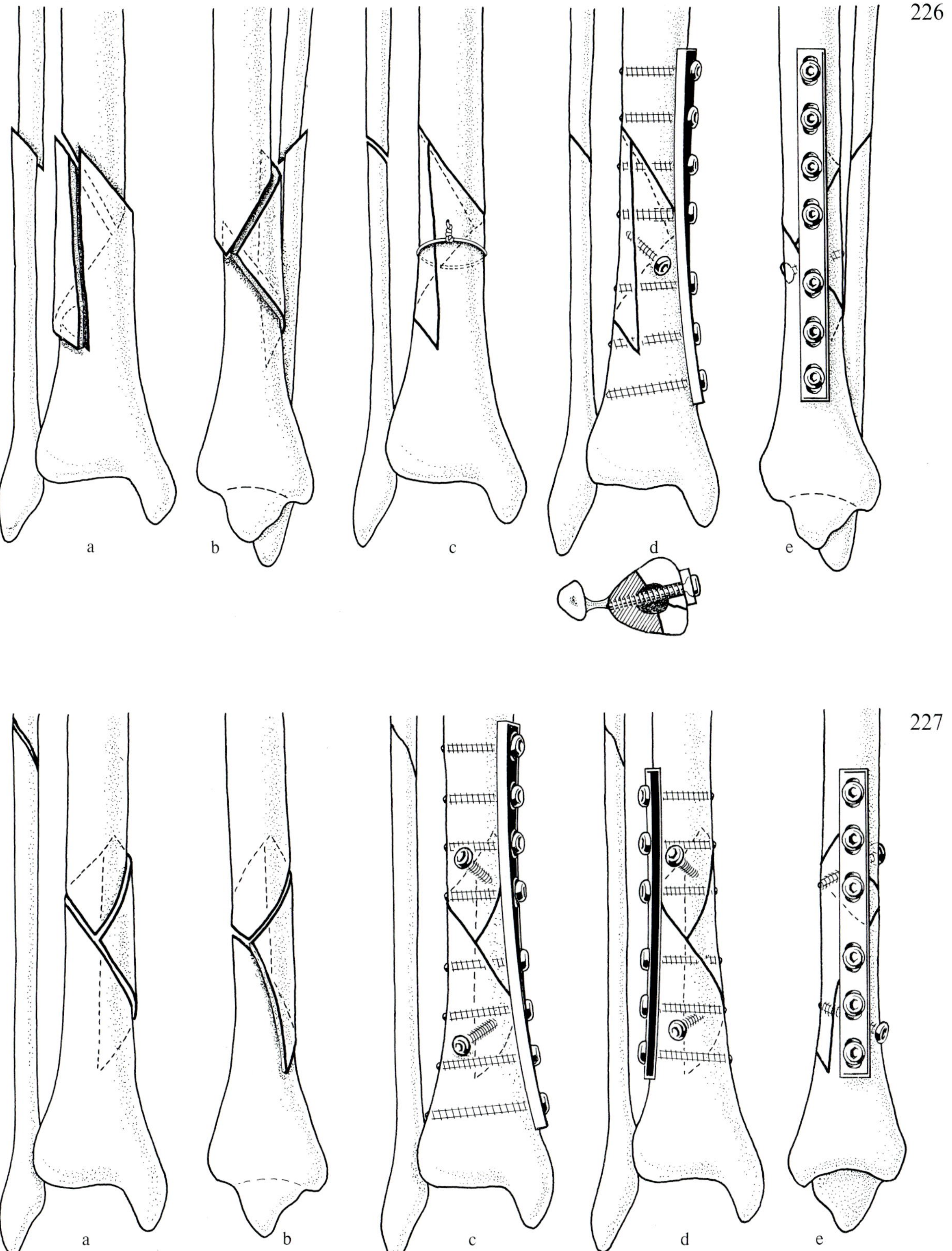
226
a
b
c
d
e
227
a
b
c
d
e

Bending Fracture with a Burst Butterfly

Butterflies which are the result of bending forces are treacherous because most of the time they lose their blood supply and break out easily during reaming for intramedullary nailing. During intramedullary nailing, if a small burst butterfly fragment breaks out it is best removed and replaced with a primary cancellous bone graft!

Internal Fixation with a Plate: Standard procedure for the main fragments in keeping with Figs. 228 and 229. Fit in the butterfly fragment and fix it with two 3.5-mm cortex screws used as lag screws.

Comminuted Fractures

The greater the comminution and the less the displacement, the more we are inclined to recommend non-operative treatment. If after 12 weeks union is not complete, then we are usually left with only *one* fracture area which is not united. This can usually be dealt with by a relatively simple type of internal fixation.

Internal Fixation of a Badly Comminuted Fracture: Begin distally and build progressively the tibia by fixing the comminuted fragments with lag screws or reduce them and fix them with cerclage wires. Bridge the whole area with a long neutralization plate. All screws inserted through the plate which bridge fracture lines must be inserted as lag screws! Bone graft all areas of bone loss and all areas of cortex suspected of having become devitalized.

Fig. 228 *Bending fracture.*

a, b Representation of the fracture.

c Reduction and internal fixation of the burst butterfly by means of two small cortical screws which do not go through the plate. The neutralization plate joins the two main fragments. One lag screw has been inserted through the plate in such a way that it lags the two main fragments together.

d Side view.

e Alternative: replacement of the devitalized burst butterfly by a primary cancellous bone graft. This technique is also recommended for intramedullary nailing of the tibia when the burst butterfly breaks out.

Fig. 229 *Comminuted fracture of the tibia.*

a, b Representation of the fracture.

c The situation after reduction and temporary fixation with a cerclage wire and two lag screws.

d Side view.

e The internal fixation has been completed with a long plate and three separate lag screws. All screws traversing the fracture have been inserted as lag screws!

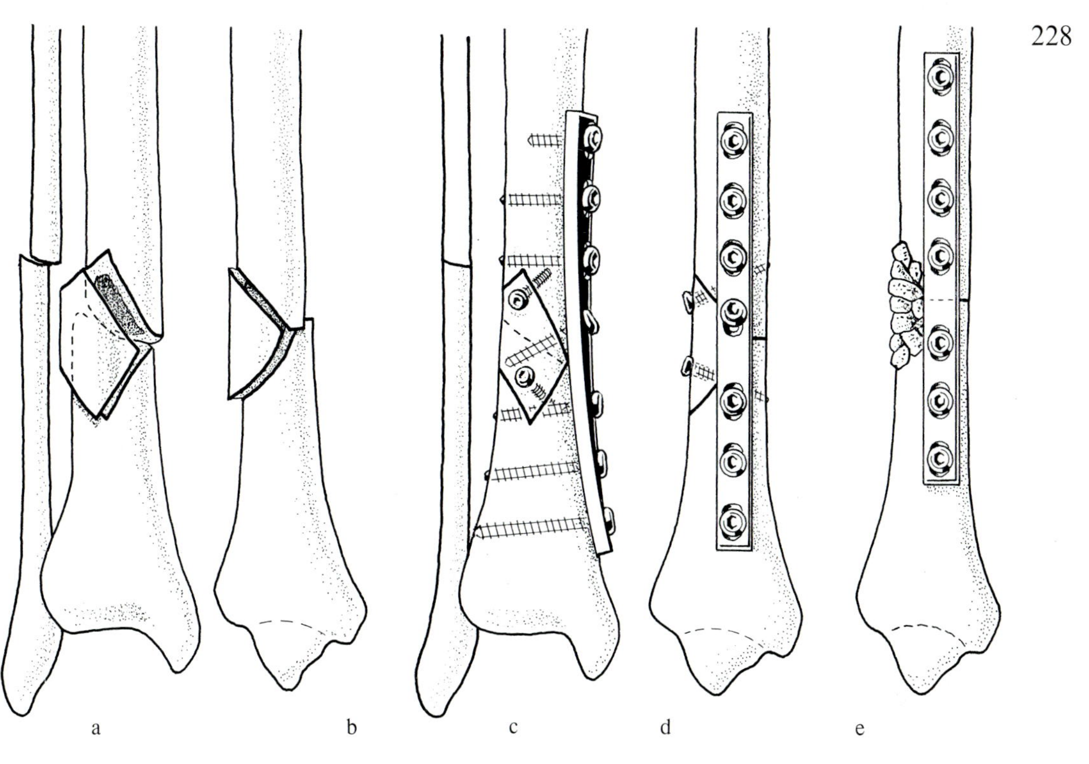
228
a
b
c
d
e

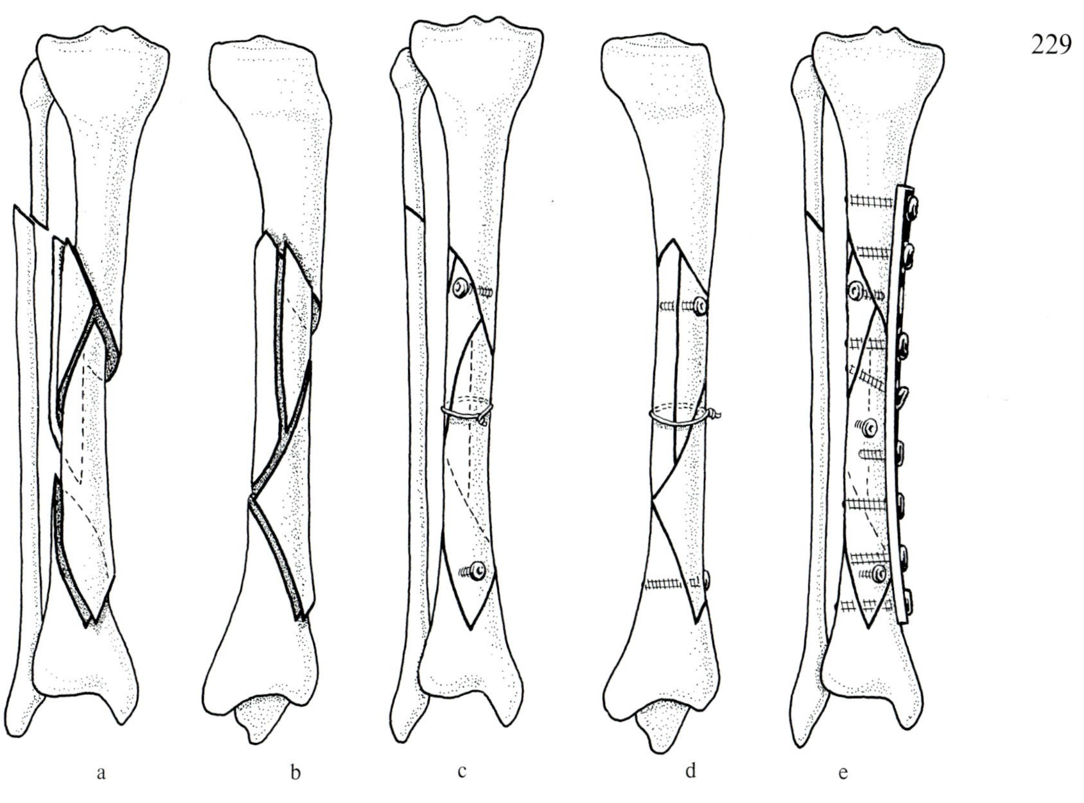
229
a
b
c
d
e

Segmental Fractures

A segmental fracture is one where one or more pieces consist of the full circumference of the bone.

Approach: Long straight incision, rarely two separate incisions.

Internal Fixation: A DCP is ideally suited for the internal fixation of such fractures, since it allows each fracture to be placed under axial compression independently from the others. Reduce the fragments accurately but restrict exposure to a minimum so as not to devitalize. Insert few screws outside the fracture area. An intramedullary nail is also a good method of internal fixation for segmental fractures. It may have to be combined with a semi-tubular plate applied along the edge of the tibia to provide rotational stability. Be careful when reaming because the intermediate fragment can lose all its soft-tissue connections and become completely necrotic.

Short, Badly Shattered Fragments of the Shaft

Localized areas of extensive comminution of the shaft can hinder considerably the reconstruction of the cortex. If such fractures are associated with damage to the soft-tissue envelope, the leg is best placed in an external fixator and the fracture either primarily or secondarily bone grafted. If the overlying skin is not damaged then the reduced shaft is stabilized with two semi-tubular plates applied to the anterior and posteromedial edges of the tibia respectively. The area of comminution and bone loss is filled with a bone graft. The approach is through a straight longitudinal incision down the medial face of the tibia (see Fig. 219c*2*).

Fig. 230 *Segmental fractures of the tibia.*

a, b Representation of the fracture type and its fixation with a medially applied plate and separate lag screws inserted across the main fracture lines. This type of fixation can be carried out only if the soft-tissue envelope is intact.

c, d The internal fixation of an open fracture or of a fracture with a damaged soft-tissue envelope. The plate is applied laterally under muscle cover.

e, f The use of an intramedullary nail without reaming to regain axial alignment of the fragments. Additional rotational stability was obtained by means of a narrow plate (possibly a semi-tubular plate).

Fig. 231 *Short, shattered segment of the tibial shaft.*

a, b Never sacrifice fragments which can be fitted into place! Small shattered cortical fragments are best removed and replaced by cancellous bone graft. In such cases, it is best to bridge the zone of comminution with two plates, either two semi-tubular plates or a DCP in combination with a semi-tubular plate. These are fixed one to the anterior and one to the posteromedial edge of the tibia.

c Semi-tubular plate with short screws fixing them to the respective edges of the tibia.

230

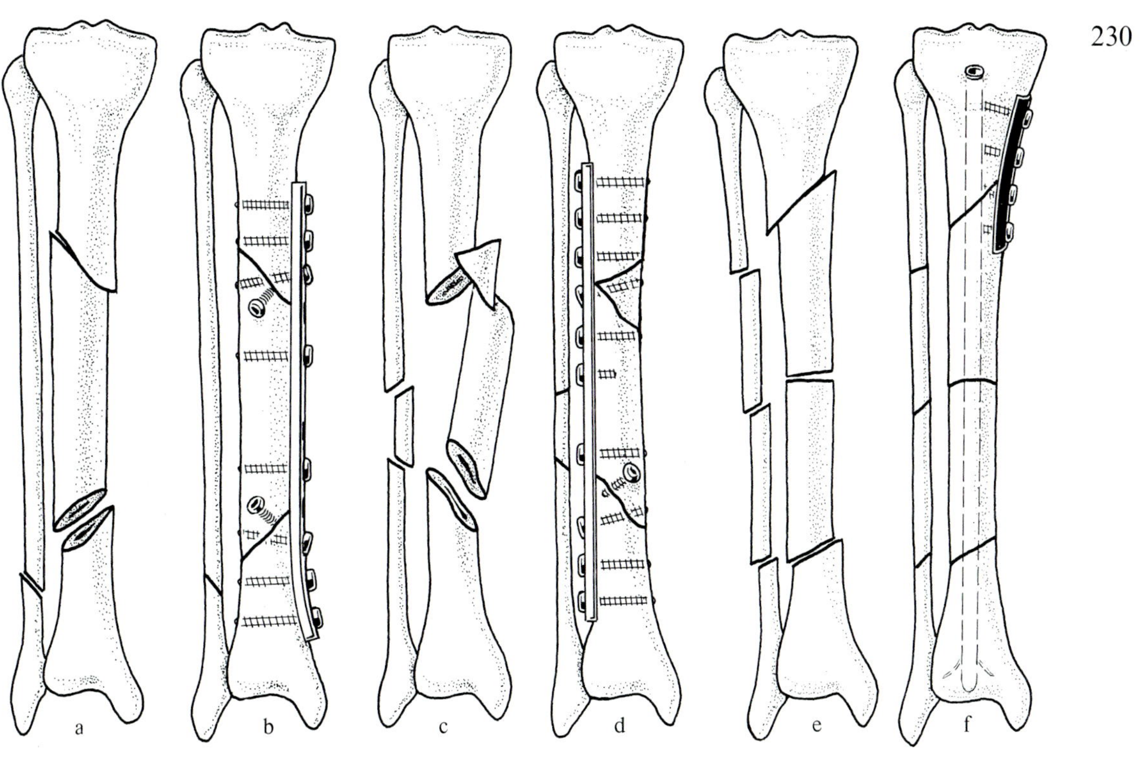

231

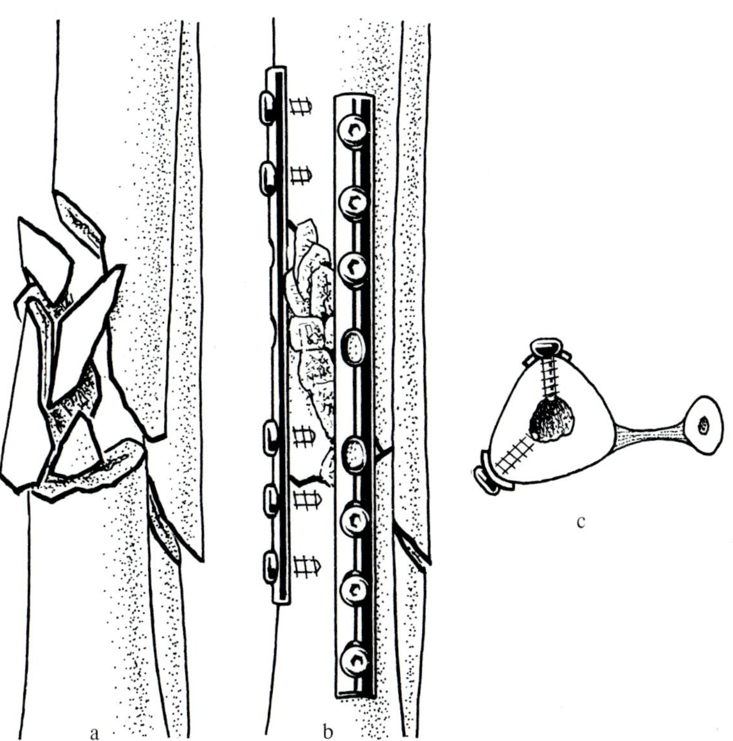

Torsional Fractures of the Distal Third of the Tibia Without Extension into the Joint

Long torsional fractures can be stabilized with lag screws alone, but the patient must be co-operative and reliable, and not obese. Fractures with torsional butterflies require the combination of lag screw fixation and neutralization plating. The neutralization plate must be most carefully contoured to fit the medial surface of the tibia. Thus the plate must be not only bent but also twisted.

A bending press is used to achieve the required bend and the twisting irons are used to achieve the required amount of twist. The bending pliers allow one to achieve the bending and twisting simultaneously. The plate is inserted into the bending pliers, at a 45° angle, so that the anterior edge is proximal and the posterior edge of the plate is distal (see Fig. 233).

The use of the DCP allows us to achieve axial compression through a plate which has been most carefully contoured to fit the medial surface of the tibia. This is achieved by inserting a screw in its load position (Fig. 232). If a round-hole plate is used, the plate must be contoured in such a way that its middle stands 1–2 mm away from the bone. As the plate is screwed to the bone, beginning at both ends and progressively working towards the middle, the bone will be placed under axial compression (see Fig. 40c).

Fig. 232 *Fracture of the distal tibia without extension into the ankle joint.* Lag screw fixation combined with a neutralization plate which was contoured to fit exactly the shape of the bone. Axial compression is achieved by the insertion of the load screw.

Fig. 233 The bending pliers allow simultaneous bending and twisting of the plate which is necessary in contouring the plate to fit the distal third of the tibia.

232

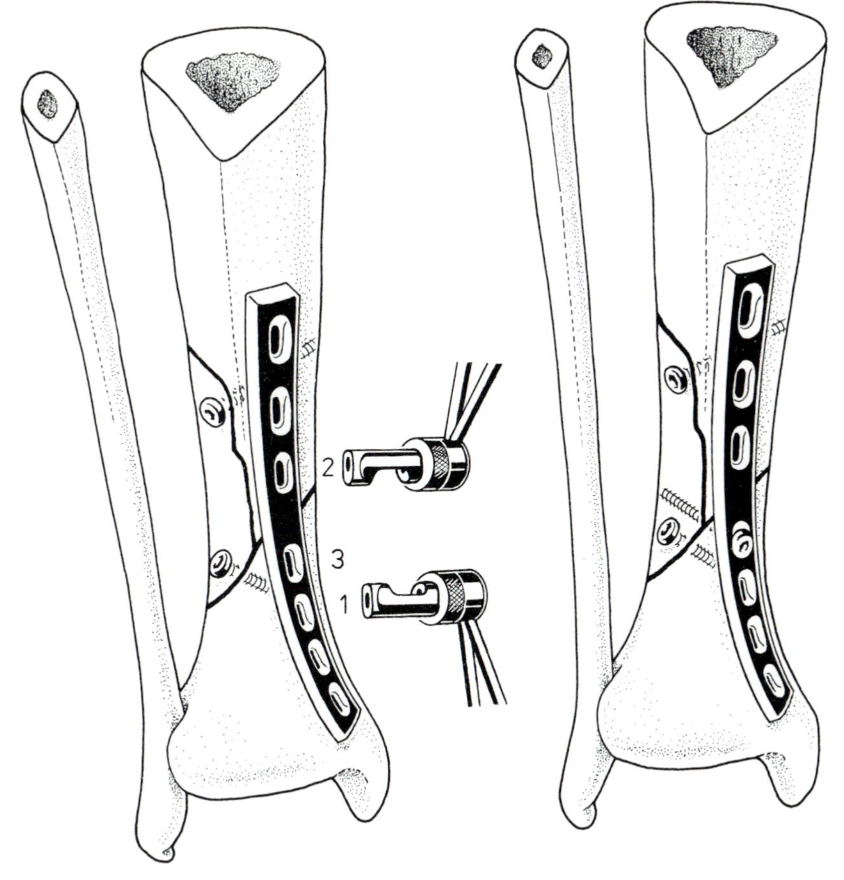

233

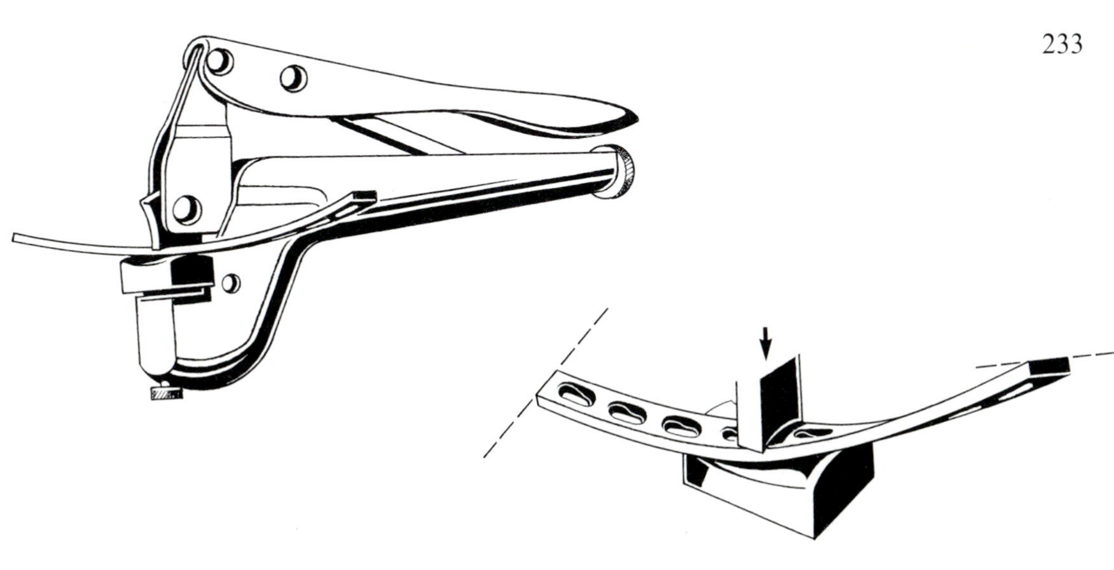

Fractures of the Distal Third of the Tibia with Extension into the Ankle Joint: "Pilon Fracture"

Approach: Make two incisions with as wide a bridge of skin anteriorly as possible (the narrowest portion should be no less than 7 cm!). The medial incision is made just lateral to the anterior crest of the tibia. It then curves medially and posteriorly around the medial malleolus. The lateral incision is made just posterior to the posterior edge of the fibula.

Classification of "Pilon" Fractures (see Fig. 235).

Post-operative Care: Early movement, delayed weight bearing! The comminuted intra-articular component of the fracture must be allowed to heal with firm union of all cartilaginous fragments. Weight bearing must therefore be withheld and is only very gradually introduced during the first three post-operative months. A U-splint is used for the first few days, to keep the ankle joint at 90°. This is to prevent an equinus deformity. While in this splint, the patient begins with active dorsiflexion of the ankle. For the long period of non-weight bearing, we recommend a patellar tendon-bearing weight-relieving caliper. (*Note:* There is danger of venous stasis, therefore begin gradually with the weight-relieving caliper!).

Fig. 234 *Surgical approach to fractures of the distal tibia (pilon fractures).* Make an incision just medial to the tendon of tibialis anterior and a lateral one just behind the most prominent portion of the lateral malleolus and fibula. Make certain that the bridge of skin between the two incisions is kept as wide as possible.

Fig. 235 *Classification of fractures of the distal tibia with intra-articular extension (pilon).*

I Fissure fracture without significant displacement.

II Fissure fracture with significant articular incongruity.

III Compression fracture with displacement of the weight-bearing cancellous segments of the metaphysis.

234

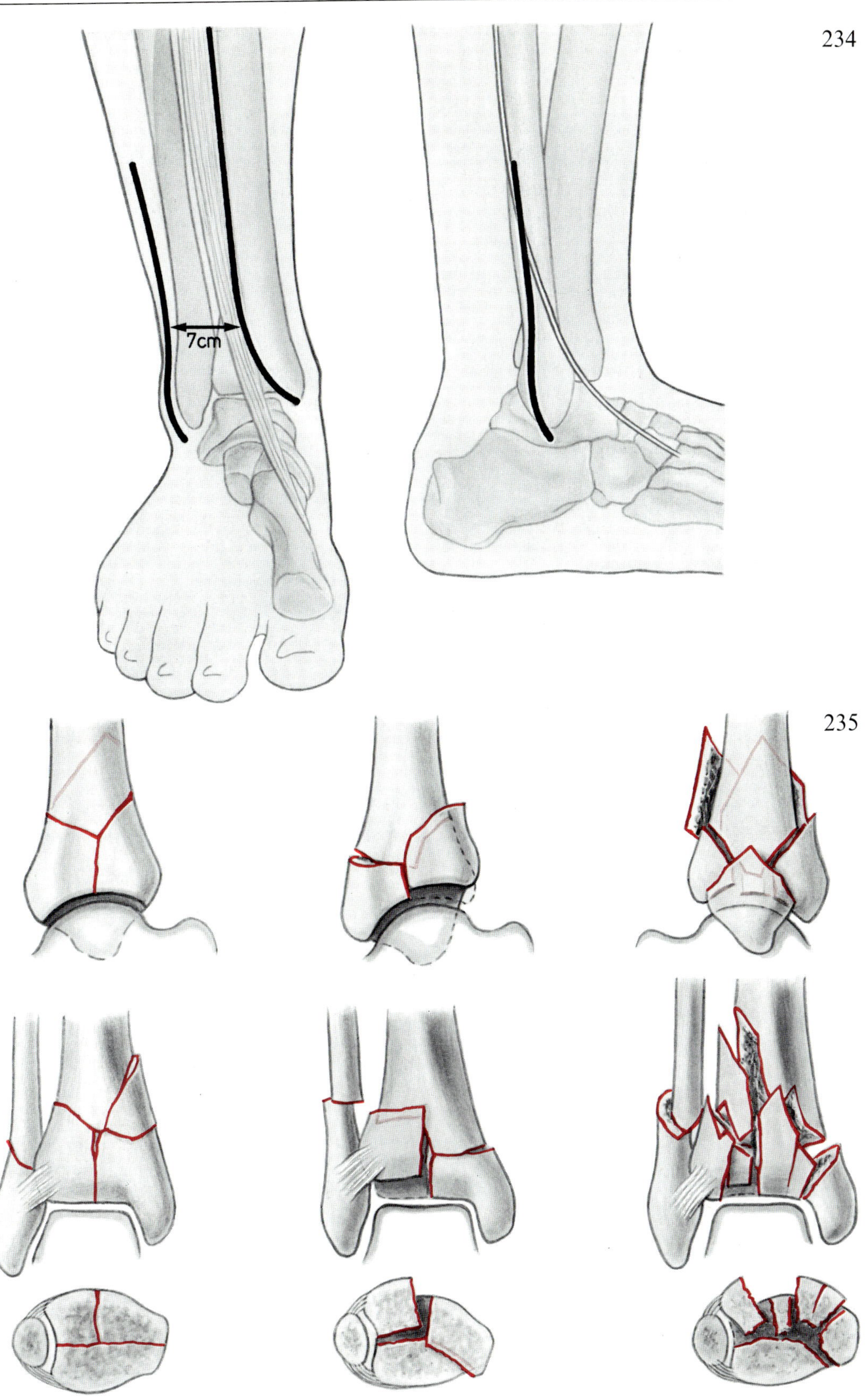

235

Steps in the Open Reduction and Internal Fixation of a "Pilon Fracture"

Preparatory Operation: Obtain cancellous bone (see p. 138).

First Step: Fibular reconstruction – in 80% of cases a stable internal fixation can be achieved with the one-third tubular plate. This manoeuvre alone in most cases brings about a considerable reduction in the displacement of the distal tibia. In about 20% of cases, a primary fibular reconstruction is not possible because of extensive comminution or it does not help in the reduction of the tibia, because the fracture is high and the syndesmotic ligaments and interosseous membranes are ruptured. In those cases the operation begins with the reconstruction of the tibia. One must search for the key main fragment because it will indicate the correct length of the tibia and allow one to begin the reconstruction.

Second Step: Reconstruction of the joint surface. The articular surface of the talus is used as a template for the articular reconstruction of the distal tibia. Temporary fixation is obtained first with Kirschner wires or if necessary with two or three screws, first from the medial side and subsequently from the lateral side, through the tubercle of CHAPUT.

Third Step: Cancellous bone graft (in order to shorten the tourniquet time, the bone graft is obtained prior to the actual reduction and internal fixation!)

Fourth Step: Medial or anterior buttress plating (T plate, cloverleaf plate or spoon plate)

Whether to bone graft prior to or after application of the plate is determined by the accessibility of the bone defect. In rare cases, where the articular surface of the distal tibia and talus are completely destroyed, one should consider a primary arthrodesis (see Fig. 272).

Fig. 236 *The four steps in the reconstruction of the distal tibia.*

a Fracture type III.

b First step: fibular reconstruction.

c Second step: reconstruction of the articular surface and its provisional fixation with Kirschner wires.

d Third and fourth steps: bone grafting of the defect with cancellous bone and then buttressing medially or anteriorly to prevent subsequent varus deformity. Whether one bone grafts before or after application of the buttress plate depends on the accessibility of the bone defect.

Fig. 237 *The spoon plate as anterior buttress plate.* At times, the anterior buttress plate can give far better fixation than the medially applied T or cloverleaf plate. The cancellous graft is sandwiched between the usually present large posterior fragment and the splintered, comminuted anterior cortex. This results in a solid reconstruction of the distal tibia.

Fig. 238 *The weight-relieving caliper* is available in four sizes. It allows early motion but delayed weight bearing. With a little practice, the patient can resume walking without the help of crutches. Weight is transmitted to the tibial condyles and to the patellar tendon. The foot is kept out of equinus by means of a spring. This relieves most of the weight from the distal tibia. The patient must begin gradually with the use of the weight-relieving caliper, because initially its use results in some obstruction to venous return.

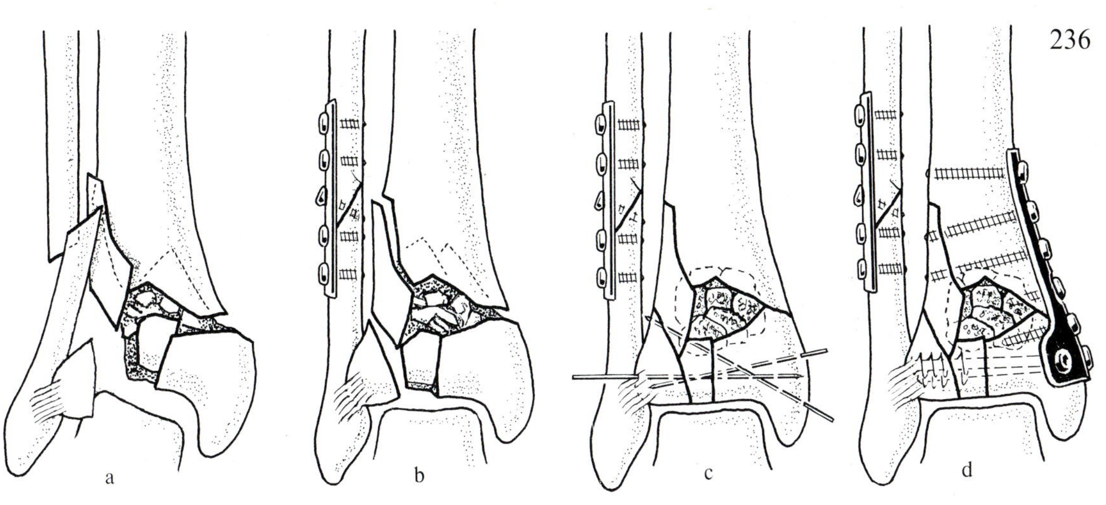
236
a
b
c
d

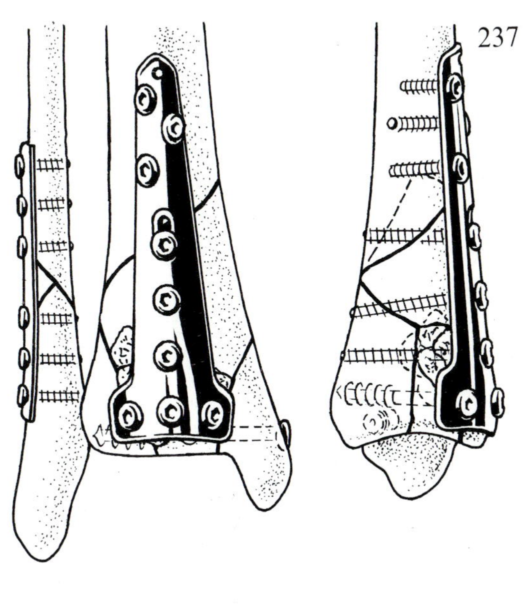
237

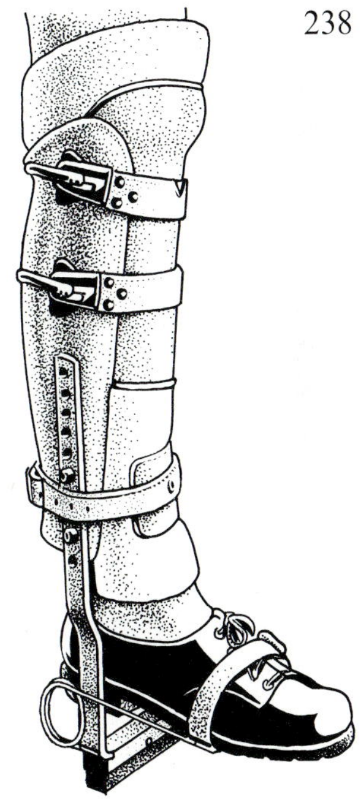
238

1.10 Malleolar Fractures

Most malleolar fractures are indirect injuries resulting from subluxation or dislocation of the talus out of the ankle mortice. Certain fracture patterns are always associated with certain ligamentous injuries. Ligamentous disruptions have, as their equivalent, avulsion fractures of the respective ligamentous insertion.

Exact anatomical reconstruction of the ankle mortice is necessary for perfect guidance of the talus.

The integrity of the ankle mortice depends upon:

a) The correct length of the fibula and its exact position in the fibular notch of the tibia (Fig. 241) and

b) the integrity of the tibiofibular ligaments which consist of three elements: the anterior syndesmotic ligament, the posterior syndesmotic ligament and the interosseous membrane (Fig. 239).

The reconstruction of the lateral complex, the fibula with its tight elastic connection to the tibia, the fibulo-syndesmotic complex, has on biomechanical grounds priority over the reconstruction of the medial malleolus. Even minimal displacement of the lateral malleolus may result in incongruity between the talus and the ankle mortice and may lead as a result to secondary osteoarthritis.

The diagnosis of a malleolar fracture demands the recognition of all the fractures and all ligamentous injuries. The level of the fibular fracture allows one to deduce the injury to the tibiofibular ligamentous connections. If the transverse fracture of the lateral malleolus is at the level of the tibiotalar articulation or distal to it, then the tibiofibular ligamentous connection, the anterior and posterior syndesmotic ligaments and the interosseous membrane are intact. If a spiral fracture of the lateral malleolus begins at the level of the ankle joint, then we have a partial disruption of the tibiofibular ligamentous connection. If the fibular fracture is higher than the ankle joint, then there is always a disruption of the tibiofibular ligamentous complex; that is either a ligamentous disruption or avulsion from bone, and there is as a result always an insufficiency of the ankle mortice. We can therefore classify malleolar fractures according to the level of the fibular fracture. There are three types: type A, type B and type C. For the details of the radiographic diagnosis, see p. 286.

In addition to the malleolar fractures and ligamentous injuries, we have to recognize shear fractures of the medial and lateral edge of the talus. These can be fairly large osteochondral fragments or pure flake fractures. Depending on their size, they can be either recognized radiologically or only when surgery is performed.

In many cases, the perfect reconstruction of the ankle mortice can be obtained only by operative means. These cases therefore comprise a definite indication for surgical intervention as long as there are no general contraindications (arterial damage etc.), or local contraindications such as fracture blisters.

Post-operative Care: Suction drainage is applied in malleolar fractures as in all cases of internal fixation for the first 24–48 hours. In order to prevent an equinus deformity, while the patient is still under anaesthesia the ankle is immobilized at 90° by means of a double U plaster splint (Fig. 102). The patient is encouraged not only to perform isometric exercises of foot and toe muscles but also to begin active dorsiflexion of toes and ankle. Once painless motion can be performed and rest pain has vanished, the plaster splint is removed. The duration depends on the severity of the injury.

The splint is removed usually somewhere between the fourth and tenth post-operative days.

Once the splint is removed, active mobilization is continued until a full range of movement is obtained. We begin with early walking exercises and partial weight bearing, but the amount of weight bearing depends on the severity of the injury and on the stability of the internal fixation. In type C fractures, usually after 1 week, we immobilize the ankle in plaster for 4–5 weeks. No plaster immobilization is carried out as a rule for type A and type B fractures and full weight bearing is usually commenced after a month.

As a rule, all malleolar fractures, including those with ligamentous disruptions, are back to normal in 10–12 weeks.

Metallic implants, since they are in the metaphyseal regions, can be removed after 4 months.

Significant Anatomical and Functional Features

We recognize two groups of ligaments:

A. *The tibiofibular ligamentous connection* which guarantees a tight elastic ankle mortice.

a) At the level of the ankle joint (the distal tibiofibular articulation) the weaker anterior syndesmotic ligament, which joints the anterior tibial tubercle (tubercle of CHAPUT) with the lateral malleolus.

b) The stronger posterior syndesmotic ligament between the lateral malleolus and the tibia in the region of the Volkmann triangle.

c) Proximal to the tibiofibular articulation is the interosseous membrane.

B. *The collateral ligaments* which control the roll of the talus.

a) The lateral collateral ligament: it has three divisions (anterior fibulotalar ligament, the fibulocalcaneal ligament, and the posterior fibulotalar ligament).

b) The medial collateral ligament, the deltoid ligament, which has two layers and three divisions.

The roll of the talus is wider anteriorly. As the talus moves through an arc into dorsiflexion it imparts a slight rotational movement to the lateral malleolus. It also means that the talus is in close contact with its mortice in all attitudes from plantar flexion to dorsiflexion. We would also like to draw the reader's attention to one further most important biomechanical point. On heel strike, the lateral malleolus together with its ligament complex bears a shearing force equal to approximately one-quarter of the joint pressure.

Fig. 239 *The anatomy of the tibiofibular ligaments as seen from the front and the back*. The anterior syndesmotic ligament = Lig. tibiofibulare anterius. The posterior syndesmotic ligament = Lig. tibiofibulare posterius. The interosseous membrane.

Fig. 240 *The collateral ligaments, the "guide ropes" for the talus.* The three divisions of the lateral collateral ligaments (a). The three divisions of the deltoid ligament (b).

Fig. 241 *Cross-section at the level of the syndesmotic ligaments.* The lateral malleolus fits snugly into the fibular notch of the tibia and is held in place by the tight elastic syndesmotic ligaments. Attached to the posterior syndesmosis is the tendon sheath of the peroneal tendons.

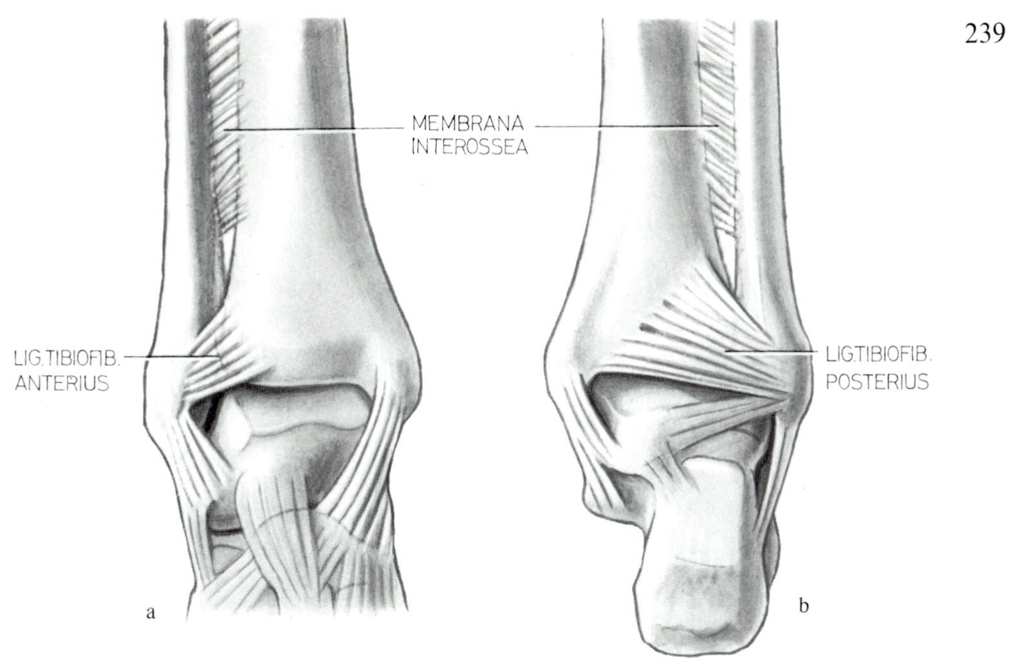
MEMBRANA
INTEROSSEA
LIG.TIBIOFIB.
ANTERIUS
LIG.TIBIOFIB.
POSTERIUS
a
b

239

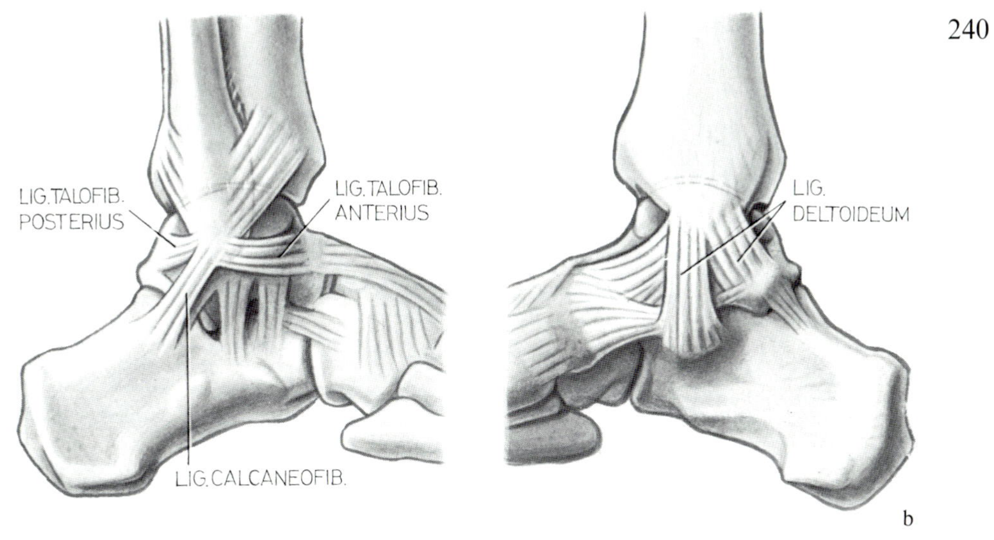
LIG.TALOFIB.
POSTERIUS
LIG.TALOFIB.
ANTERIUS
LIG.CALCANEOFIB.
LIG.
DELTOIDEUM
a
b

240

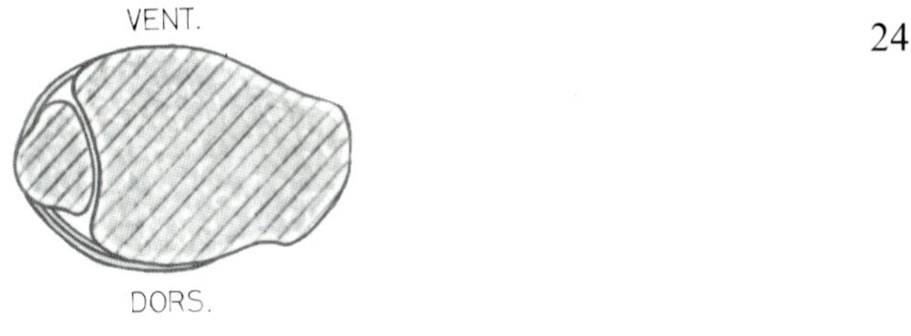
VENT.
DORS.

241

We must have an AP and lateral projection with the X-ray beam centred carefully on the ankle joint. For the AP projection the leg must be in 15°–25° of internal rotation. This position makes the transmalleolar axis lie parallel to the X-ray plate (Fig. 242a).

If an injury to the anterior tibial tubercle (tubercle of TILLAUX-CHAPUT) is suspected, the X-ray must be taken with the leg in 45° of external rotation. Stress AP and stress lateral X-rays demonstrate ligamentous injuries (Fig. 243b). Isolated ruptures of the anterior fibulotalar ligament do not result in varus tilt of the talus on the AP stress view, but there is widening of the joint space between the fibula and the talus. Only with complete disruption of the lateral collateral ligament does the talus tilt into varus.

An occult deficiency of the collateral system can be demonstrated on a lateral stress X-ray with the aid of the apparatus designed by NOESBERGER (Fig. 242c). If a deficiency is present, the talus can be seen to displace forwards.

All stress studies must be done on both sides and the two compared. The oblique fracture of the shaft of the fibula which is often mistakenly judged to be an isolated injury, is almost always a type C malleolar fracture, and in such cases careful radiographic study of the ankle joint must be carried out in order to demonstrate the concomitant injuries to the mortice. If one suspects the complex of type C injury, it is necessary to take an X-ray of the leg which includes the ankle and knee in order to demonstrate the full extent of the injury.

Fig. 242 *The radiological technique.*

a An AP view of the ankle joint: the leg is internally rotated through 20° to bring the transmalleolar axis parallel to the X-ray plate.

b Stress varus AP view of the ankle to demonstrate varus tilting of the talus in the ankle mortice (a tilt greater than 4° is pathological and denotes collateral insufficiency). Always compare with a stress X-ray of the normal side.

c NOESBERGER apparatus designed to demonstrate anterior subluxation of the talus. It denotes insufficiency of the anterior fibulotalar ligament. The leg is internally rotated through an arc of 20° and a strap compresses the tibia posteriorly while the foot is firmly held in position.

Fig. 243 *The evaluation of the X-rays.*

a The ankle articulation with the leg in 20° of internal rotation: the joint space is of equal width throughout. The subchondral bone plate of the talus and tibia are parallel. The line of the subchondral plate of the tibia jumps over the gap and continues along the lateral malleolus without a step.

a′ Even the slightest shortening of the lateral malleolus can be recognized radiologically as a step in the alignment of the subchondral plates of the tibia and of the lateral malleolus.

b An AP stress X-ray of the ankle joint. Note the 10° varus tilt of the talus. This denotes that the important fibulocalcaneal as well as the anterior fibulotalar ligaments are injured.

c An AP stress X-ray as seen from the side, taken with the aid of NOESBERGER's apparatus without subluxation of the talus.

c′ Six-millimetre anterior subluxation of the talus. The distance is measured between the deepest point of the inferior articular surface of the tibia and a tangential point on the dome of the talus. A difference of 3 mm or more in the joint space is pathognomic for an injury to the anterior fibulotalar ligament.

242

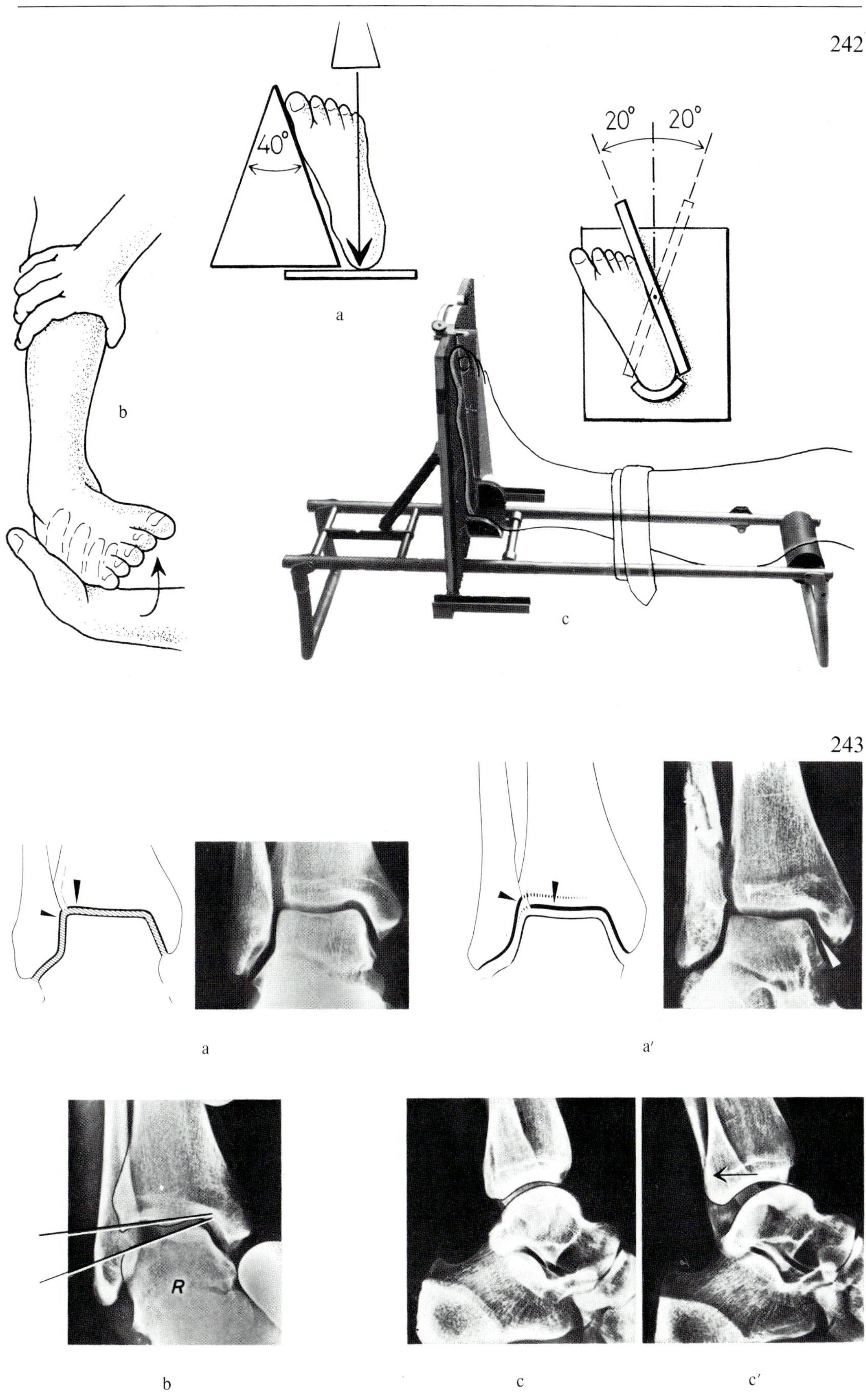

Classification of Malleolar Fractures According to WEBER

The higher the fibular fracture, the more extensive the damage to the tibiofibular ligaments and the greater the danger of an ankle mortice insufficiency. We recognize three types of fractures according to the level of the fibular fracture:

A
Fibula: Transverse avulsion fracture at the level of the ankle joint or below; its equivalent is the rupture of the lateral collateral ligament.
Medial malleolus: Intact or sheared, with the fracture line somewhere between a horizontal and a vertical. Not infrequently there is also a localized compression fracture of the tibial edge.
Posterior edge of the tibia: Usually intact. Rarely there is a medial posterior fragment, which is usually connected with the medial malleolar fragment.
Tibiofibular ligamentous complex: Always intact.

B
Fibula: Spiral fracture beginning at the level of the syndesmosis. The fracture line can be smooth or comminuted depending on the forces involved. Medial malleolus: intact or shows an avulsion fracture of varying size. Its equivalent is the rupture of the deltoid ligament.
Posterior edge of tibia: Intact, or demonstrates a lateral fragment which is an avulsion fracture of the posterior syndesmosis, the so-called Volkmann triangle.
Tibiofibular ligamentous complex: The interosseous membrane is as a rule intact. The posterior syndesmotic ligament is either intact or disrupted as a result of the avulsion fracture of its attachment to the tibia (Volkmann's triangle).

Anterior syndesmosis is intact, if the spiral fracture of the lateral malleolus begins below the level of the ankle joint. If the fracture line begins at the level of the ankle joint, then the anterior syndesmosis is either partially or completely torn. The equivalent of an anterior syndesmotic disruption through its substance is the avulsion of the syndesmotic ligament either from its tibial or fibular insertion.

C
Fibula: Fracture anywhere between the syndesmosis and the head of the fibula. Its rare equivalent is the proximal dislocation of the tibiofibular joint.
Medial malleolus: Transverse avulsion fracture or its equivalent, the rupture of the deltoid ligament.
Posterior edge of the tibia: Commonly a lateral fragment, the avulsion fracture of the posterior syndesmotic ligament (as B).
Tibiofibular ligamentous complex: Always disrupted. Rupture of the interosseous membrane from the ankle joint to the level of the fibular fracture or even more proximally. The syndesmotic ligaments are either ruptured through their substance or avulsed with their bony attachments. The severity of injury to the ligamentous complex and the severity of the malleolar fracture increases progressively from fracture A to B to C.

Fig. 244 *Three types of malleolar fractures.*

244

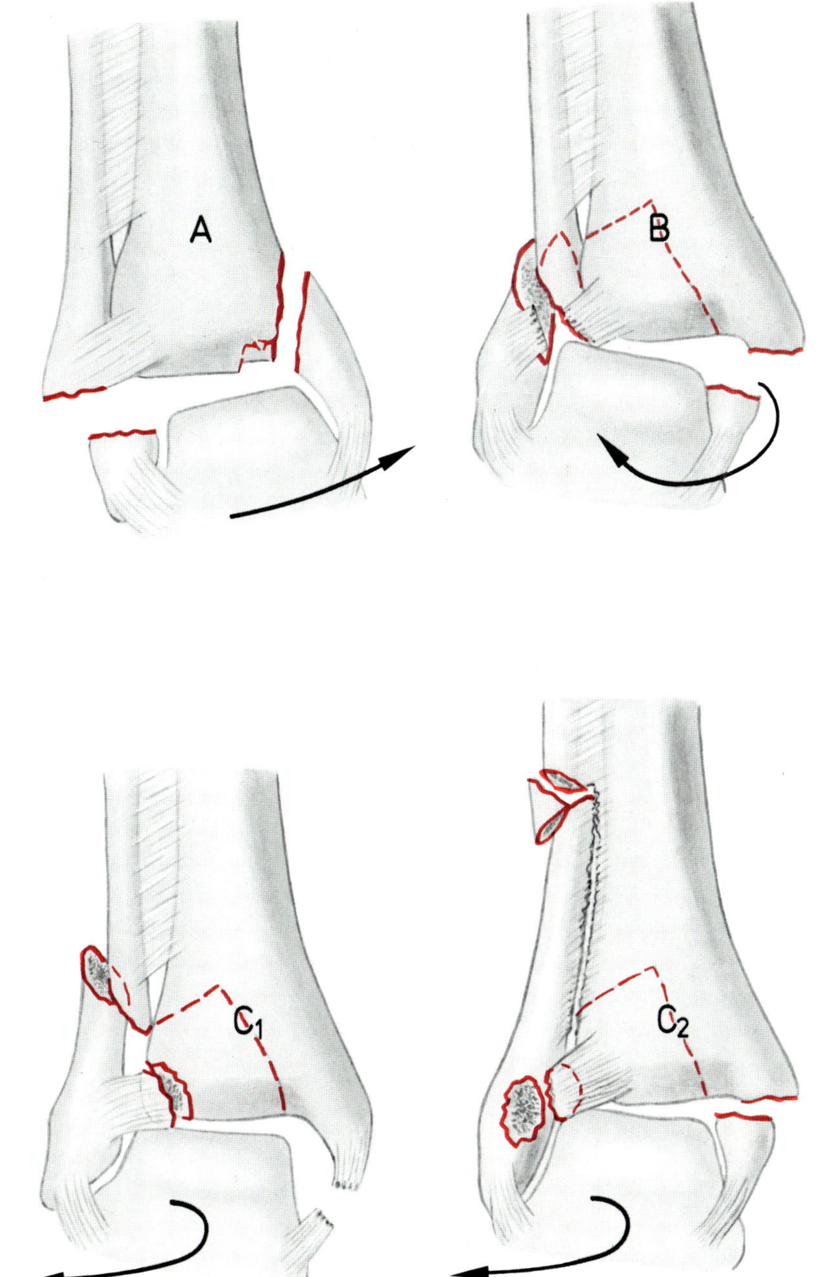

Steps in Open Reduction and Internal Fixation

The reconstruction of the fibula has priority (see p. 282). It is worthwhile therefore to begin the operation with the fixation of the lateral malleolus. Then comes the fixation of the medial malleolus. Occasionally the anatomical reduction of the fibula is blocked by soft-tissue interposition on the medial side such as by the deltoid ligament, or by the tendon of the flexor hallucis longus or by periosteum. In these cases, before the fixation of the fibula can be completed the medial malleolus must be exposed and the block to reduction eliminated. Flake fractures of the talus are removed.

After every anatomical reduction it is necessary to carry out temporary stabilization. For provisional stabilization of the lateral malleolus, use the reduction clamps with tips, or the self-centring bone clamps. For provisional fixation of the medial malleolus, depending on the size of the fragment, use either the reduction clamps with tips or Kirschner wires. The internal fixation is then carried out, either with lag screws or with the one-third tubular plate. All Kirschner wires are removed at the end.

Timing of Surgery: The ideal time for the procedure is within the first 6–8 hours before any true swelling or fracture blisters develop. The initial swelling is due to haematoma formation and not to oedema. In the presence of severe oedema and fracture blisters, the open reduction must be postponed for 4–6 days. The fracture is reduced closed and immobilized in a well-padded plaster. The leg is then kept elevated to reduce the swelling before the open reduction is undertaken.

Fig. 245 *The surgical approaches for malleolar fractures.*

a The exposure of the lateral malleolus and anterior syndesmosis. The skin incision runs more or less parallel to the n. fibularis superficialis. The nerve must not be damaged. The anterior syndesmosis and the anterior edge of the fibula come into view only after transection of the extensor retinaculum. Denude the distal malleolar fragment as little as possible. We prefer either the anterior or posterior curved incision, rarely the incision directly over the lateral malleolus.

b The standard incisions for the exposure of the medial malleolus. Once the joint capsule is opened, one can judge the accuracy of the reduction. This anterior exposure affords an excellent view into the ankle joint.

c Incision for the simultaneous exposure of the medial malleolus and of a large posterior Volkmann's triangle.

d Posterior incision for the posterior exposure and internal fixation of the fibula and of a Volkmann's triangle. The patient is positioned either on his side or prone.

245

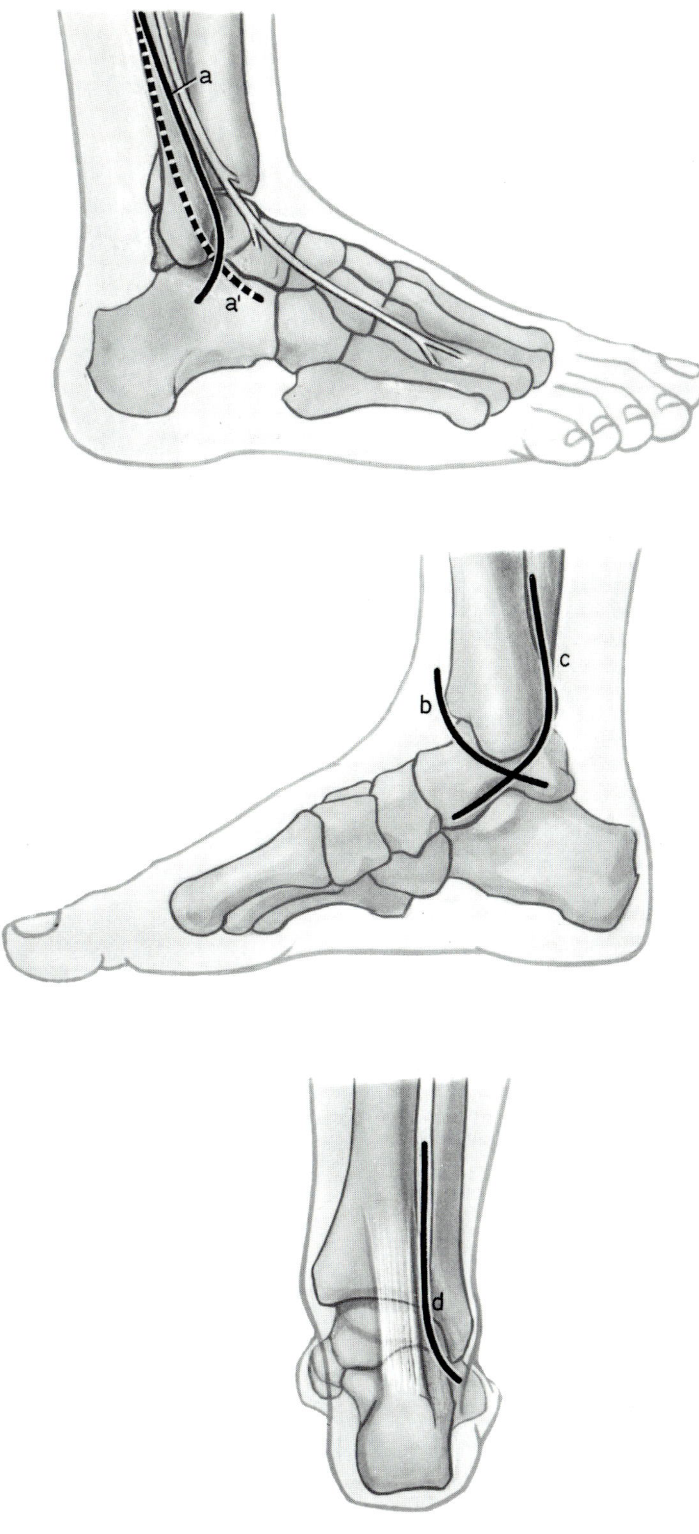

1.10.1 Type A Malleolar Fracture

Type A is the simplest fracture to treat operatively. If the lateral malleolus is avulsed it is fixed as shown in Fig. 246b. Next the fracture of the medial malleolus is exposed and the trapped periosteum is carefully reflected to expose the fracture edges. The anterior capsule is always torn, giving a good view of the intra-articular portion of the fracture. Small bone splinters from the anterior edge of the tibia are removed, but larger pieces should be reduced. Provisional fixation is achieved with Kirschner wires and definitive fixation is carried out according to the principles of interfragmental compression (Fig. 246d). The torn capsule need not be sutured.

The rare posteromedial fragment must be carefully reduced and fixed from the posteromedial aspect (Fig. 246f).

Fig. 246 *Type A.*

a The ruptured lateral collateral ligament and capsule are sutured as one layer.

b A large avulsion fragment of the lateral malleolus is fixed with an obliquely inserted malleolar screw which gains purchase in the cortex above the syndesmosis (the screw lies obliquely in two planes as it passes from anterolateral to posteromedial).

c A small avulsion fragment from the tip of the lateral malleolus is first provisionally fixed with two small Kirschner wires and then fixed under compression by means of a fine tension band wire.

d A large shear fracture of the medial malleolus is fixed with two lag screws, a malleolar screw and a small cancellous screw.

e A small shear fracture of the medial malleolus is fixed with a malleolar screw and a Kirschner wire or with small cancellous screws.

f Posteromedial fragments associated with type A fractures are rare. They always lie next to the medial malleolar fragment. They are therefore easy to expose, reduce and fix from a posteromedial direction. Depending on their size, they are either fixed with malleolar or small cancellous screws.

A 246

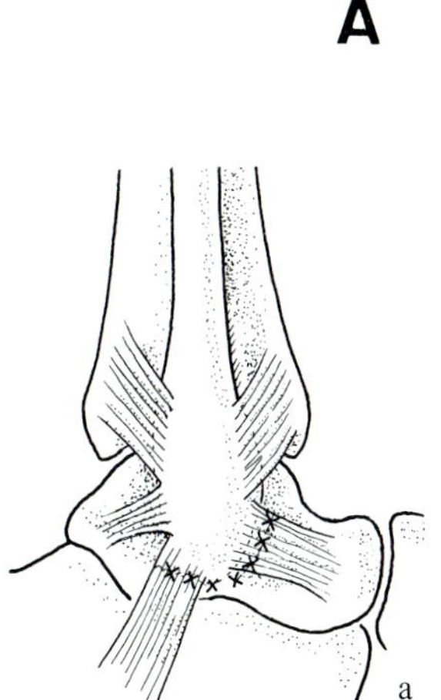

a

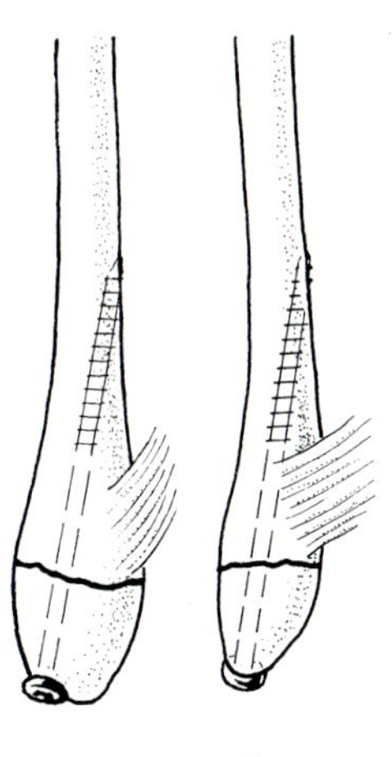

b

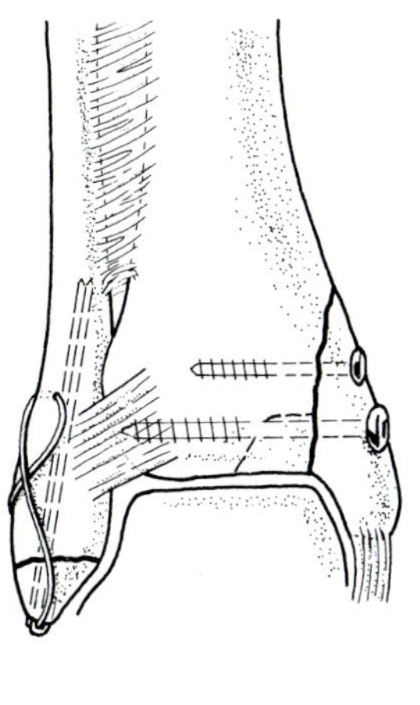

c d

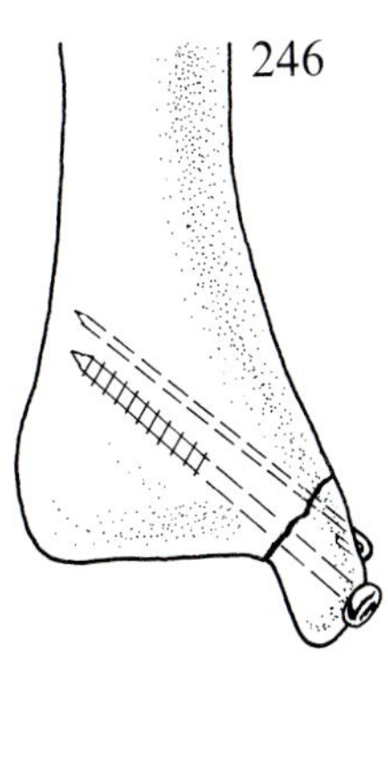

e

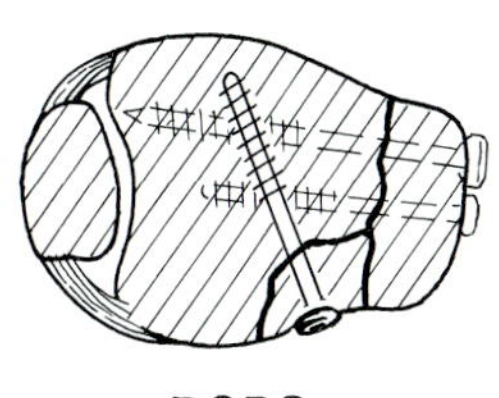

f

Exposure of the lateral malleolus comes first. The anterior capsule is torn. Through careful external rotation, the dome of the talus can be exposed. If flake fractures are present, these are removed. The displacement of the lateral malleolus consists of shortening, posterior displacement, lateral displacement and rotation. The reduction must be anatomical and is best carried out with a reduction clamp with tips, which serves at the same time as a means of provisional fixation. Check the exactness of the reduction along the posterior edge of the fibula. (For the internal fixation see Fig. 247.)

The avulsion fractures of the medial malleolus are fixed, depending on the size of the fragment, either by means of a tension band wire or lag screws. Not all ruptures of the deltoid ligament need be re-sutured. They can be left alone if after reduction of the lateral malleolus the mortice is intact and there is no suspicion of any soft-tissue interposition or medial instability.

Exact reduction of the lateral malleolus results in considerable reduction of the posterolateral fragment because the two are attached by the posterior syndesmotic ligament (Fig. 247a). Thin shell-like avulsion fragments or fragments which comprise less than one-fifth of the joint surface in the sagittal plane need not be reduced and fixed. Larger fragments must be carefully reduced to overcome all steps in the articular surface. The reduction is carried out through a posteromedial exposure. All that is usually required to carry out the reduction is to introduce a narrow instrument and press downwards and forwards with its tip. Temporary fixation is then achieved with a Kirschner wire which is inserted anteromedially. Definitive fixation is carried out with a cancellous lag screw which is inserted anteroposteriorly. In order to do this, it is necessary at times to make a small anterior incision just above the ankle joint.

Fig. 247 *Type B*. Typical internal fixation of a type B fracture.

a The short spiral fracture of the lateral malleolus is fixed with a small cortex lag screw, 2.7 or 3.5 mm. The fixation is supplemented with a one-third tubular neutralization plate. Note the suture of the anterior syndesmotic ligament.

b If exposure of the deltoid ligament becomes necessary because of soft-tissue interposition then it should be sutured (for exposure see Fig. 245b).

c A large posterolateral fragment of the tibia is carefully reduced and then fixed with a cancellous lag screw inserted in an AP direction.

a′, a″, a‴ Different methods of internal fixation of the lateral malleolus in accordance with the different fracture patterns.

b′, b″ The different types of internal fixation of the medial malleolus.

c′, c″, c‴ Lag screw fixation of the posterolateral Volkmann's triangle as seen from the side, from behind, and in cross-section.

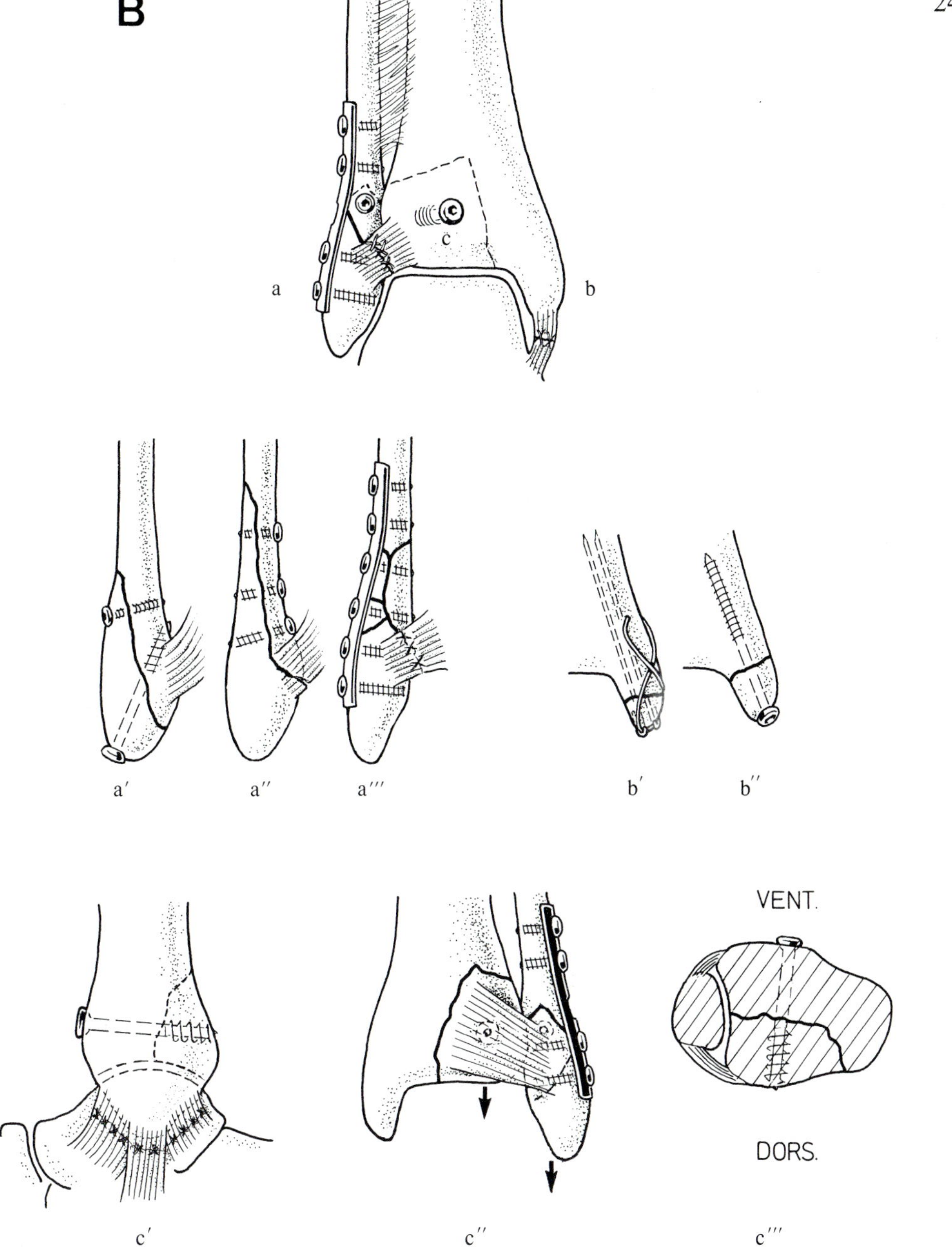
B
a
b
c
a′
a″
a‴
b′
b″
c′
c″
VENT.
DORS.
c‴

1.10.3 Type C Malleolar Fractures

The operation begins with the exposure of the fibular shaft fracture. In badly comminuted fractures, the reconstruction of the exact length of the fibula can pose problems. The shaft fracture is stabilized with lag screws and a one-third tubular plate. Next the anterior syndesmosis is exposed. If avulsed from the tibia (tubercle of TILLAUX-CHAPUT) or from the fibula it is reduced and screwed into place. If the anterior syndesmosis is torn in its substance it is simply re-sutured. The very high fractures of the fibula, through the fibular neck, as a rule need not be exposed. Beware here of the lateral peroneal nerve. If the fibula is shortened, insert a hook or a towel clip into the lateral malleolus and then with strong sudden pulls reduce the fibula into the fibular notch of the tibia. Stabilize it with a Kirschner wire and take an X-ray to check the alignment of the subchondral plate of the distal tibia and that of the lateral malleolus (see Fig. 243a).

Whether any further fibular fixation is necessary depends on the stability of the syndesmosis (this is best tested with a hook – Fig. 248a), on the degree of damage to the interosseous membrane, on the rigidity of the internal fixation of the fibula, and on the repair of the syndesmotic ligament. If upon testing with a hook, residual syndesmosis instability can be demonstrated, then this is stabilized by the insertion of a protection screw 2–3 cm above the ankle joint. This should be a 4.5-mm cortex screw which is inserted through the fibula into the lateral cortex of the tibia. The screw must be inserted obliquely from back to front at an angle of 25°–30° beginning posterolaterally and aiming anteromedially. The screw should also be inserted parallel to the ankle joint. It is enough to get fixation just in the lateral cortex of the tibia. The thread for the screw should be tapped in the fibula as well as in the tibia as this screw must not be inserted as a lag screw but just as a screw to maintain the correct position of the fibula.

We are very much against the fixation of the syndesmosis with a lag screw (Fig. 248e). Such a screw not only blocks all the physiological play of the ankle mortice but may also lead to ossification which would block such movement permanently.

The medial malleolus and the posterolateral fragment are fixed as discussed under Type B Fractures.

Fig. 248 *Type C.*

a The fibular shaft fracture is carefully reduced and stabilized with a one-third tubular plate. The torn anterior syndesmotic ligament is sutured. The stability of the syndesmosis is then tested with a small hook. The small avulsion fracture of the medial malleolus is fixed with two wires and a tension band wire.

b A fracture of the mid-shaft of the fibula is fixed with a plate. The anterior syndesmosis is avulsed from its attachment to the lateral malleolus. It is reduced and fixed either with a small cancellous screw or with a trans-osseous wire suture. The ruptured deltoid ligament is sutured. A large posterolateral fragment which represents the avulsion of the posterior interosseous ligament is carefully reduced and then fixed with a large cancellous screw. This restores stability to the ankle mortice.

c Frequently the subcapital fibular shaft fracture is not shortened and does not require an open reduction. It is most important, however, to check very carefully for any shortening on an AP X-ray of the ankle. Look carefully for any steps in the alignment of the subchondral bone plate of the distal tibia and of the lateral malleolus. Any shortening requires reduction. A small avulsion fracture of the anterior syndesmotic ligament from the tibia is reduced and fixed with a small lag screw. Since this injury involved almost the full extent of the interosseous membrane, the fixation of the anterior syndesmotic ligament does not provide sufficient stability to the ankle mortice and therefore additional fixation of the fibula to the tibia by means of a protection screw is necessary.

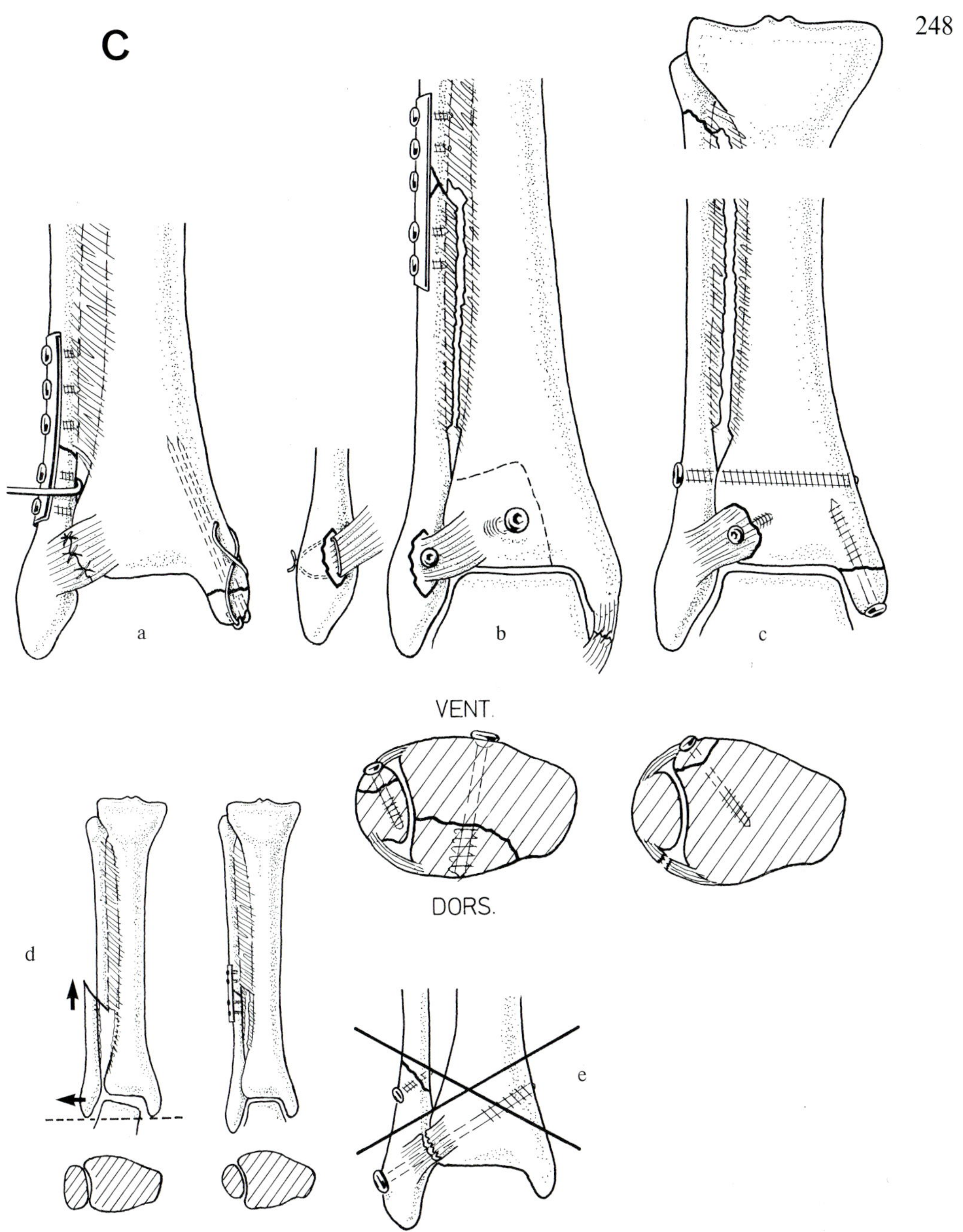

d The avulsion fracture of the medial malleolus is fixed under compression by means of a malleolar screw.

e Exact anatomical reduction of the fibula into the fibular notch in the tibia guarantees a normal ankle mortice. Imperfect reduction with shortening of the fibula leads to a diastasis and a valgus tilt of the talus (see Figs. 241, 243). Even small malreductions and small degrees of valgus tilt of the talus may lead to post-traumatic osteoarthritis.

Fig. 249 *Fracture dislocation of the ankle with splintering of an osteoporotic fibula.* The lateral malleolus was reduced and stabilized with the one-third tubular plate. The avulsed anterior syndesmotic ligament was reduced and fixed with a 4.0-mm cancellous screw. The fracture of the medial malleolus was reduced and fixed with a malleolar screw and a 4.0-mm cancellous screw (type B).

Fig. 250 The internal fixation of the spiral lateral malleolar fracture with two small cortex screws. Suture of the anterior syndesmotic ligament (type B). The Volkmann's trangle was fixed with an AP cancellous screw.

Fig. 251 *A case similar to Fig. 250.* The unusual feature here is the huge posteromedial tibial fragment. Because the posterior syndesmotic ligament is intact, the lateral malleolar fragment follows the posteromedial fragment and is thus pulled and displaced medially. The fracture of the lateral malleolus was fixed with two 3.5-mm cortex screws. The large posteromedial fragment was fixed with a cancellous screw, a malleolar screw and a 3.5-mm cortex lag screw.

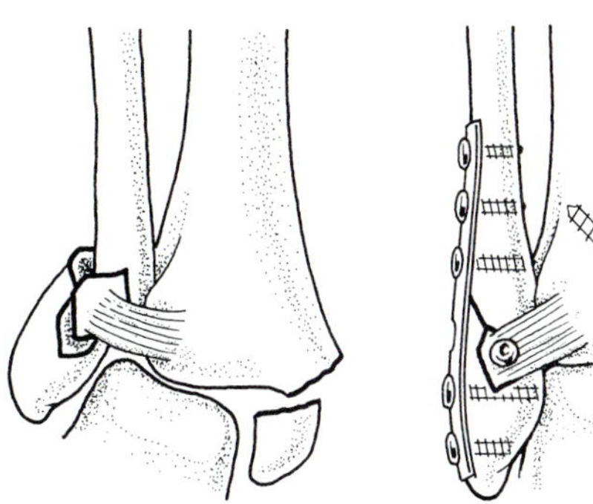

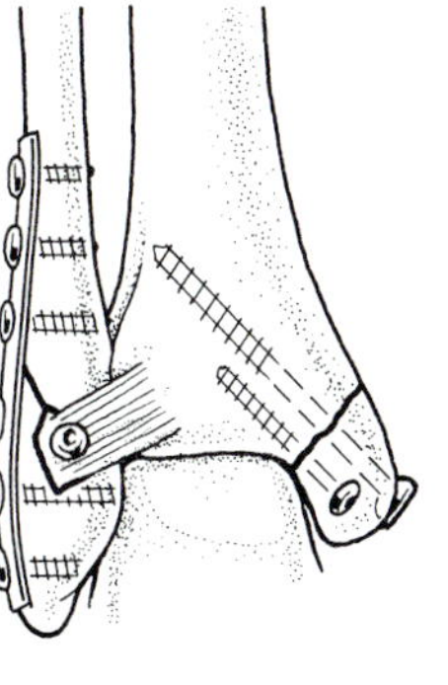

249

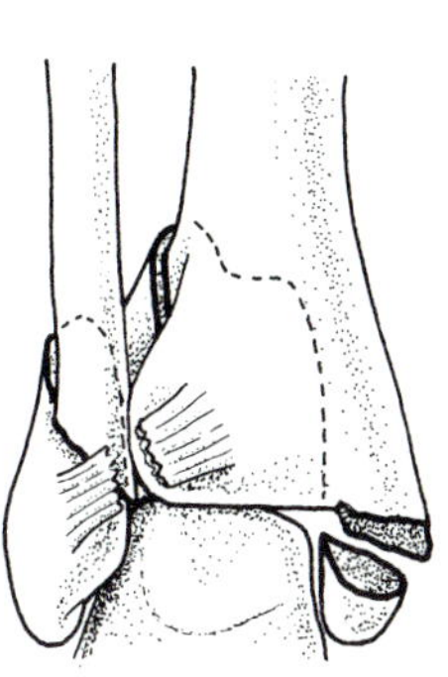

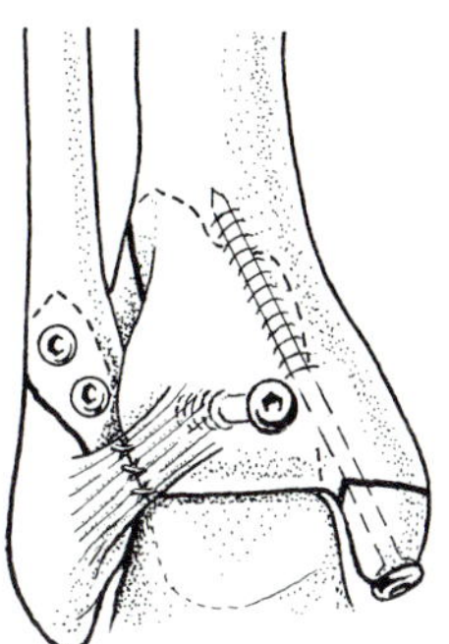

250

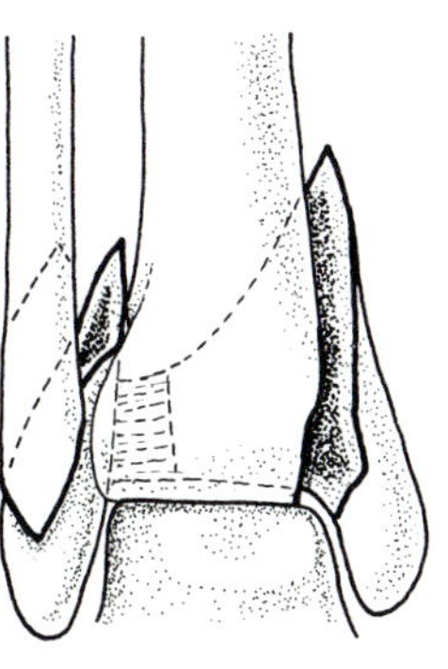

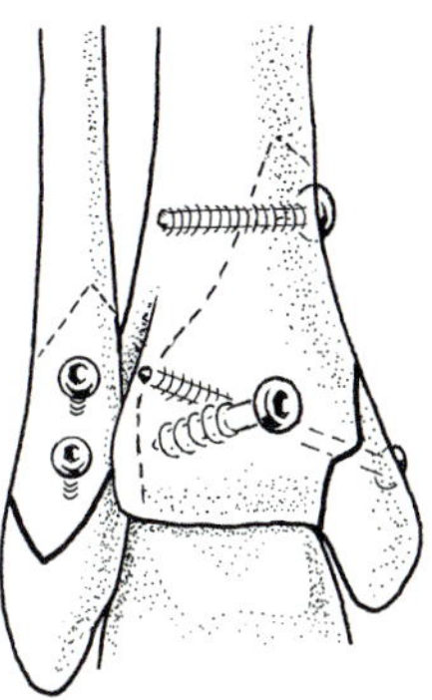

251

1.11 Fractures of the Foot

Fractures of the neck of the talus without displacement are treated conservatively in a non-weight-bearing below-knee plaster. Irreducible fractures are treated operatively. After their reduction, they are stabilized with one or two 4.0-mm cancellous screws.

The mainstay of treatment of most intra-articular fractures of the os calcis depends on elevation of the limb and early mobilization. Displaced tongue-type fractures are reduced and fixed with two lag screws. The procedure is carried out through two small incisions, one lateral and one posterior. In fractures of the os calcis with displacement of the sustentaculum or in displaced fractures in young patients, we have found the single or the double H plate designed for the fixation of the cervical spine most useful. These fractures require cancellous bone grafting at the time of their internal fixation. The same H plate has proven its worth in the arthrodesis of the talonavicular joint which occasionally becomes necessary when the navicular is very badly comminuted.

In the forefoot, internal fixation is required only for unstable fractures of the first and fifth rays (Fig. 255c).

Fig. 252 *Irreducible fracture through the neck of the talus.* Reduced and fixed with one or two 4.0-mm cancellous screws.

Fig. 253 *Displaced fracture of the tuberosity of the os calcis.* Reduced and fixed with cancellous screws.

Fig. 254 *Compression fracture of the os calcis and fracture of the sustentaculum.* Reduced and bone grafted with cancellous bone and fixed with an H plate. In depressed fractures of the articular surface, after elevation a double H plate must be used for the fixation.

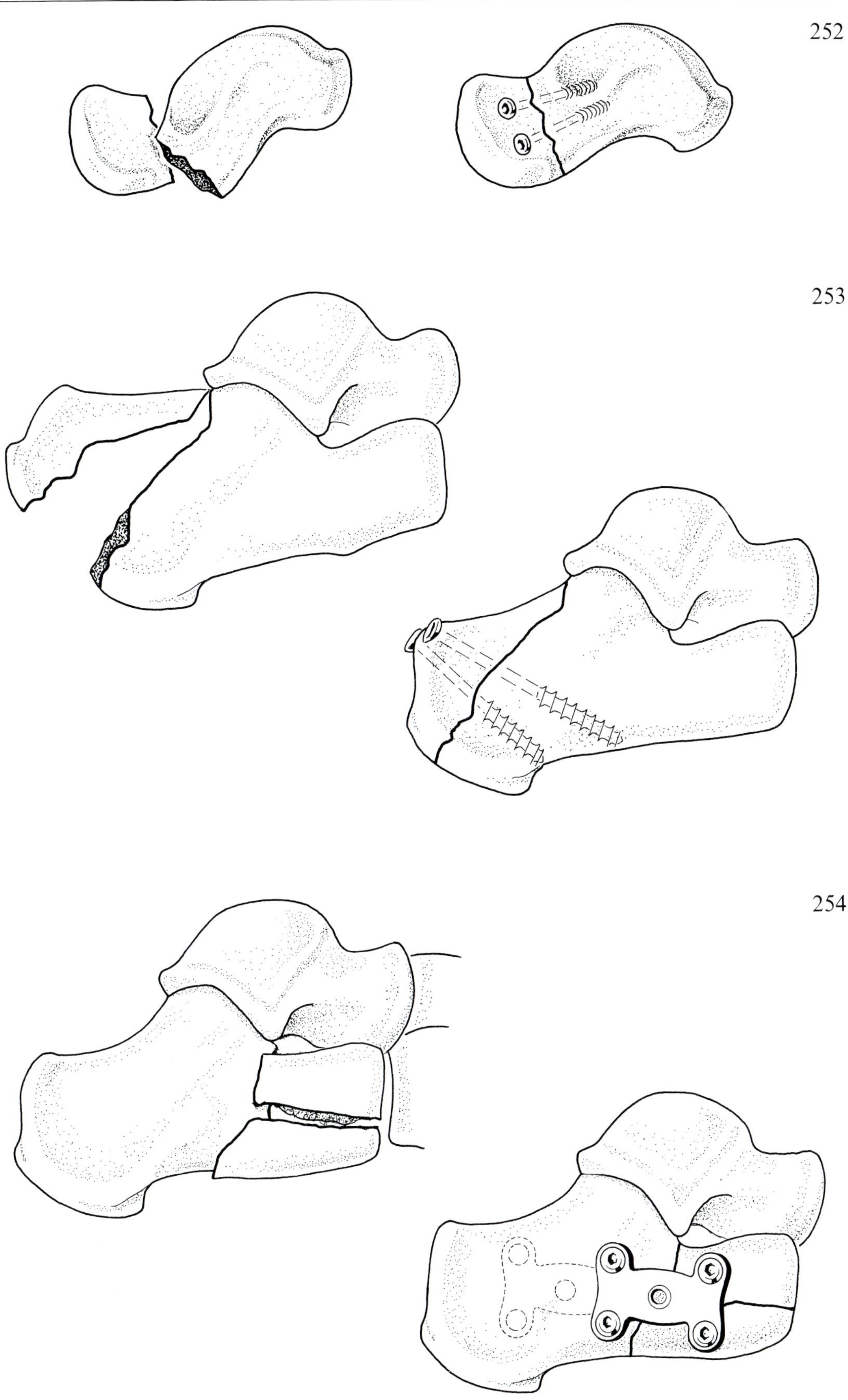
252
253
254

Fig. 255 *Typical examples of internal fixation in the forefoot.*

a Approaches: a straight incision centred on the third ray gives an excellent exposure of the second, third and fourth metatarsals. For the exposure of the first and fifth metatarsals we prefer a dorsomedial and dorsolateral incision, respectively, at the junction between the thick plantar and thin dorsal skin.

b The stabilization of a displaced avulsion fracture of the base of the fifth metatarsal (insertion of the peroneus brevis). If the fragment is large, it is fixed with a small cancellous screw and if it is small with two small Kirschner wires and a tension band wire.

c Typical examples of internal fixation in the forefoot.

(*1*) Displaced, unstable shaft and subcapital fracture of the fifth metatarsal. Reduced and stabilized with a small T plate.

(*2*) Transverse fractures of the second, third and fourth metatarsals are stabilized with intramedullary Kirschner wires.

(*3*) Transverse fractures of the first metatarsal are best stabilized with a one-third tubular plate. This should be inserted as far as possible on the plantar aspect of the metatarsal where it will act as a tension band plate.

(*4*) Displaced unstable transverse fractures to the proximal phalanx of the big toe are reduced and fixed with a small L or T plate.

(*5*) A displaced T or Y fracture through the base of the first metatarsal is reduced and fixed with lag screws and a small T plate.

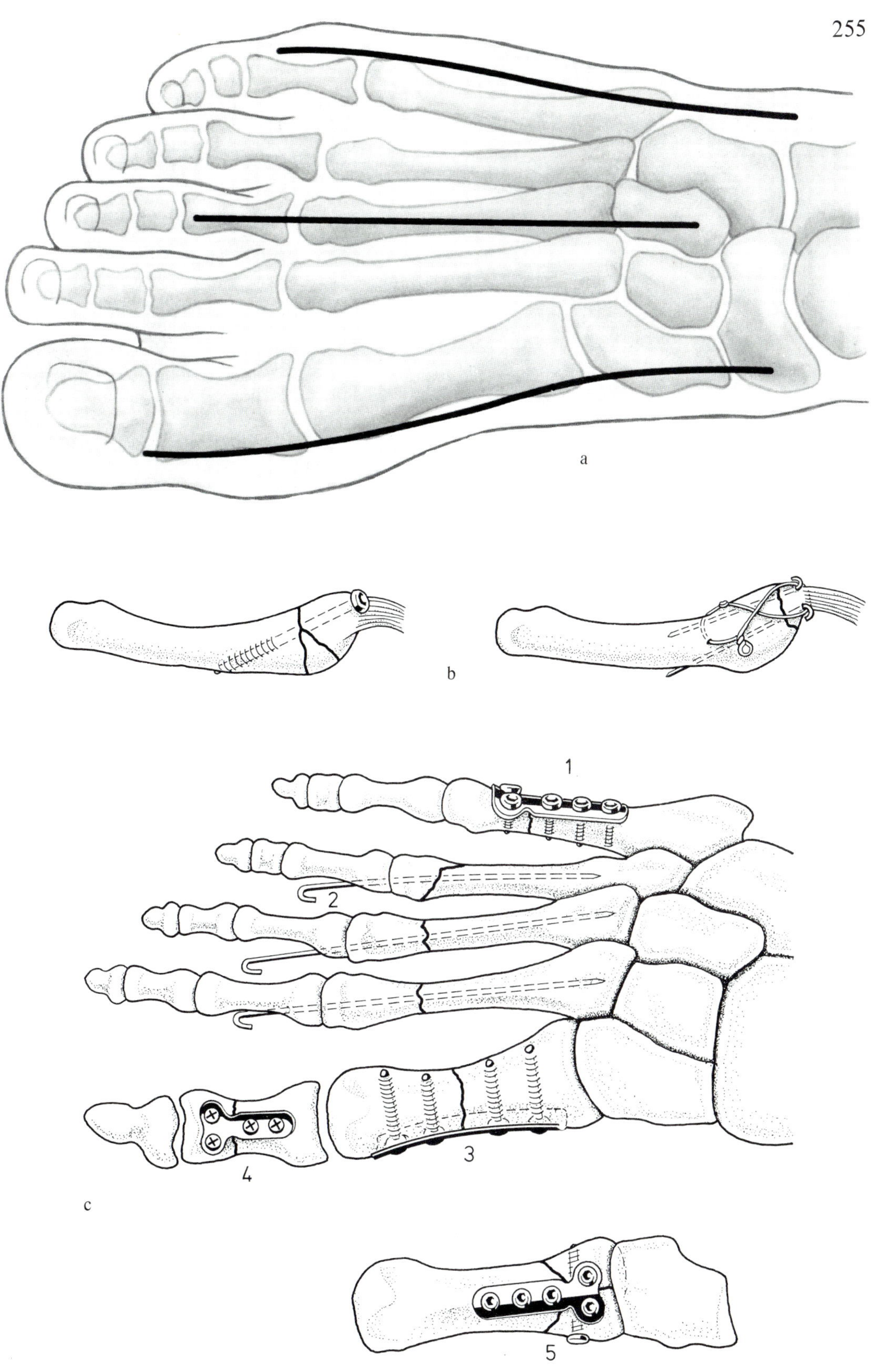
255
a
b
1
2
3
4
c
5

1.12 Fractures of the Vertebral Column

We do not wish to enter into a lengthy discussion on fractures of the vertebral column. The stable fracture should not be immobilized in plaster. Early function and early mobilization is indicated. The problem fractures are the badly displaced ones.

Less than 10% of fractures of the vertebral column require internal fixation. Internal fixation of the vertebral column is, however, particularly difficult because of the necessary transfer and the positioning of the patient. Technical errors can lead to permanent neurological deficits.

Unstable fractures – they are the ones with damage to the posterior complex (spinous process, interspinous ligament, laminae, articular facets). Whether they are with or without neurological deficits they should be stabilized to simplify the care of the patient.

In the cervical region we employ the short or long H plates (OROZCO, SENEGAS, CABOT). They are applied anteriorly after excision of the disc and insertion of an anterior bone graft. Anterior discectomy and fusion is to be carried out only after a pre-operative reduction of the spine has been achieved by means of traction (skull tongs).

In dealing with fractures of the thoracic and particularly of the lumbar spine, we have come to rely on the very difficult posterior procedure of ROY-CAMILLE. We employ the narrow DCP. At the start of the procedure holes are pre-drilled through the joint facets. These are drilled at 90° with a 2-mm drill bit to a depth no greater than 2–3 mm. Two-millimetre Kirschner wires are then inserted into these holes. The fracture is then reduced by the application of posterior and lateral pressure. Once the Kirschner wires come to lie parallel to one another they are marked on one side with a small metal ring and a lateral X-ray of the spine is obtained. Only when the correct direction of the Kirschner wires has been obtained, and checked radiologically, do we slip the chosen DCP over the Kirschner wires. A lateral X-ray will indicate which holes can be used. We then drill with the 3.2-mm drill bit into the pedicles. The anterior defect between T.6 and L.1 is approached and bone grafted through a costo-transversectomy. The defect between L.2 and L.5 is bone grafted through an anterior approach. We carry out at the same time a posterior fusion by placing the bone graft at the base of the spinous processes and over the laminae. This short discourse on fractures of the vertebral column is included merely to illustrate our method of internal fixation of the unstable spine. For further details, we refer the reader to the pertinent literature.

Fig. 256 *The short and the long H plate.*

Fig. 257 *Cervical fusion* from the front with the H plate.

Fig. 258 *Internal fixation of the upper lumbar spine* with DCP; the screws are anchored in the pedicles.

256

257

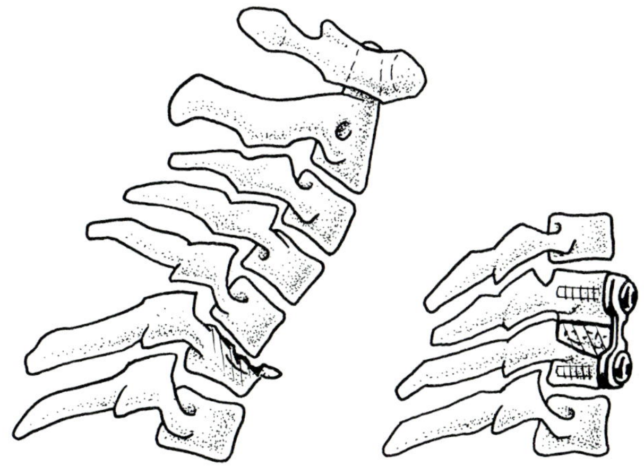

258

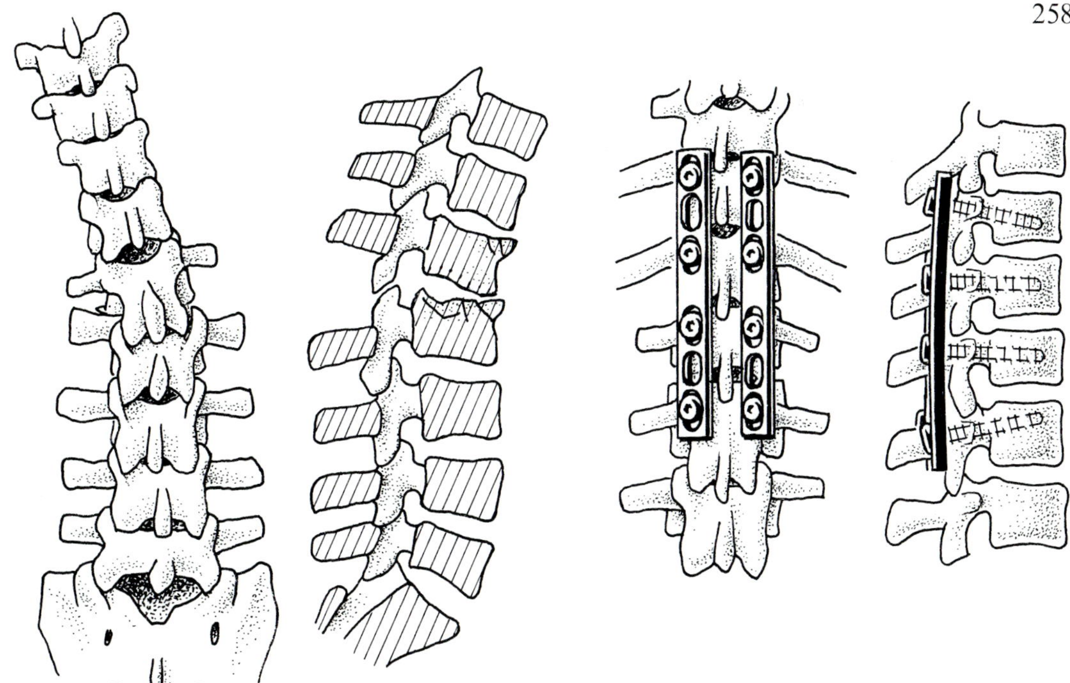

2 Open Fractures in the Adult

The guiding principle in the treatment of open fractures should be the prevention of infection, particularly of osteomyelitis. Two manoeuvres have proven themselves to be particularly important:

1. Debridement
2. Stabilization of the fragments either by means of internal fixation or by means of external skeletal fixators

Bacteria attain after a latent period of only 6–8 hours sufficient numbers and virulence to become pathogenic. If the patient arrives after the initial 6–8 hours, only the external fixator or skeletal traction should be considered as acceptable methods of immobilization, because these methods do not increase the danger of sepsis. If the patient arrives within the first 8 hours, then we feel that we can make an excellent case for primary internal fixation which leads to healing per primum of both the bone and soft tissues. This applies only if the following steps in the prophylaxis of sepsis are carefully followed.

1. Prevention of Further Wound Contamination: Only one-third of open fractures are contaminated with pathogenic material at the time of hospital admission. Secondary contamination with hospital bacteria between hospital admission and operation can be prevented by proper asepsis. A dressing should be left untouched. Only the surgeon responsible for the patient's care can decide whether to remove the dressing and then only if septic precautions are taken (as a minimum – gloves and mask). The dressing is not removed for radiography. The skin preparation is carried out in the operating room when everything is ready for surgery. If a patient with an open fracture arrives in the hospital without a dressing, the wound must be covered immediately with a sterile bandage.

2. Debridement: Extensive and immediate excision of all devitalized soft tissue removes the culture medium for the bacteria. By debridement we mean excision of all the devitalized, contused fat and necrotic muscle. Necrotic muscle is recognized by absence of capillary bleeding, by its failure to react to pressure and by its cyanotic discolouration. Nerve endings are simply approximated. Large vessels are repaired if at all possible. This may require a covering or bridging of defects with grafts. Deep fascia, particularly that covering the tibialis anterior, should be split. The skin edges are either left alone or only the minimum is excised. In de-gloving injuries, the skin flaps should be de-fatted and used as free, full-thickness grafts.

3. The Stabilization of the Fracture: The experience of the AO indicates that once the wound has been debrided, the stabilization of the fracture is the best prophylaxis against sepsis. Therefore if there are no contraindications, all open fractures which come for treatment within 6–8 hours are stabilized either by internal fixation (Fig. 268) or external fixation (Fig. 263) or by a combination of the two methods (Fig. 264).

In carrying out early internal fixation of open fractures one must strive to minimize on additional denuding of bone, and one must use the least amount of metal necessary to achieve a stable internal fixation. The implants should not be left exposed or placed under a damaged soft-tissue envelope. If bone defects are present, we feel it is best to bone graft them at the time of the initial stabilization.

4. Skin Closure Without Tension: If necessary carry out secondary skin closure. Secondary wound-edge necrosis and the resultant infection can only be prevented if skin is closed without tension.

We feel it is best to leave the wound open rather than to close the skin under the least bit of tension. Secondary closure can then be carried out in 5–8 days. A formal secondary closure is rarely necessary because most often it is enough simply to approximate the wound edges with sterile strips.

Skin punctures by bone from within are not excised. They are simply left open as long as they are not part of the surgical approach to the bone. If wound-edge necrosis develops, one should attempt to keep it dry and leave the eschar until the wound heals by granulation tissue and re-epithelialization. It is important to prevent the retention of any secretions under the necrotic skin.

5. It is Necessary to Hold Back with the Functional After-treatment. Post-operatively the limb is splinted in plaster as after internal fixation of closed fractures. One should hold back somewhat with the early post-operative mobilization. The patient is encouraged, however, to begin isometric muscle exercises within the first 24–48 hours after surgery.

6. Antibiotic Prophylaxis has proven itself to be of little value, since it is begun too late to be called and to act as true prophylaxis (see General Principles of Treatment). We feel that in open fractures which have been thoroughly debrided and left open the disadvantages of antibiotic prophylaxis outweigh the advantages. Meticulous debridement and copious irrigation of the wound during surgery are the two most important measures in the prevention of infection.

Classification

First-Degree Open Fracture: The skin is pierced from within by sharp bone fragments. Basically these fractures can be looked upon and treated as closed injuries. If one chooses to carry out an open reduction and internal fixation then it is best to make the incision as far away as possible from the wound. The perforated skin is not excised and is simply left open. If the area of compounding is included in the surgical approach it should be carefully excised with the minimum sacrifice of soft tissue. One should remember as in any internal fixation that the implant should not come to lie directly under the incision.

Second-Degree Open Fracture: The skin is disrupted and crushed from without with moderate damage to the skin, subcutaneous tissue and muscle. After debridement the fracture is stabilized.

Third-Degree Open Fracture: These are compounded from without, with extensive skin, subcutaneous tissue, and muscle necrosis. They are often associated with injury to nerves and vessels (gunshot wounds and explosion injuries, damage to soft tissue after severe motor vehicle accidents, etc.). Such fractures are most often stabilized with the external fixator alone or in combination with lag screws. Here the external fixator takes over the function of the neutralization plate without leading to any added devitalization of bone (Fig. 264). Occasionally such fractures are plated, but the plates must then be covered with healthy soft tissue (Fig. 262). Under exceptional circumstances we use a thin intramedullary nail without reaming. If the skin circulation is precarious we feel that it is best to suspend the limb post-operatively. In open fractures of the tibia of the third-degree, the leg is suspended on two Steinmann pins, one driven through the os calcis and one through the tibial tubercle.

Repeat Internal Fixation: The recommendation to carry out the so-called minimal internal fixation means that quite often after the first 6–8 weeks it is necessary to re-do the internal fixation. This holds particularly true for the cases of intramedullary nailing without reaming. The secondary internal fixation should frequently be combined with a cancellous bone graft either from the pelvis or the greater trochanter.

External Skeletal Fixators: We have discussed in the general section on page 128 the technique of obtaining rotational alignment in fractures of the tibia.

Fig. 259 *The use of the external fixator and the steps necessary to obtain correct rotational alignment* (for text see Fig. 91).

Fig. 260 *The correction of malreduction in the frontal plane.* If at the end of the procedure there is a varus or valgus deformity, this can be easily corrected by swivelling of the double clamp once the fixation screw is loosened. The single clamps can also be exchanged for single swivel clamps if necessary.

Fig. 261 *Correction of malalignment in the sagittal plane.* The long tubes are replaced by four short tubes. Two connecting pieces with their screws completely loosened are slipped over the tubes. As soon as reduction in the sagittal plane is obtained, all screws of the connecting pieces are tightened.

259

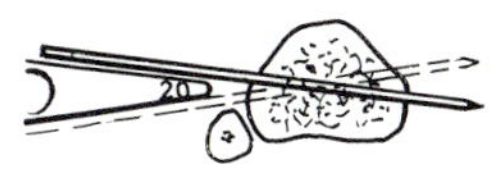

c

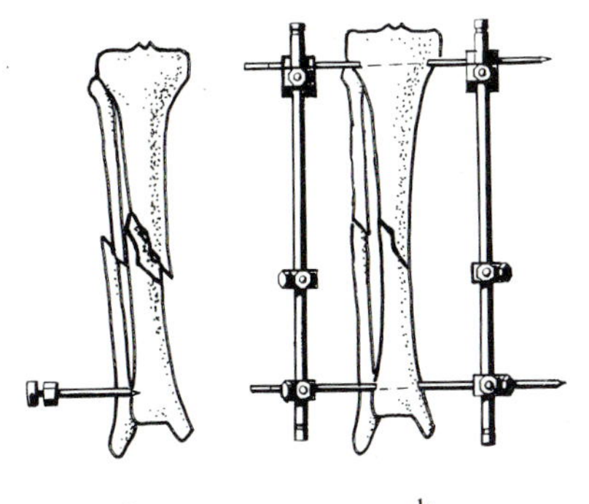

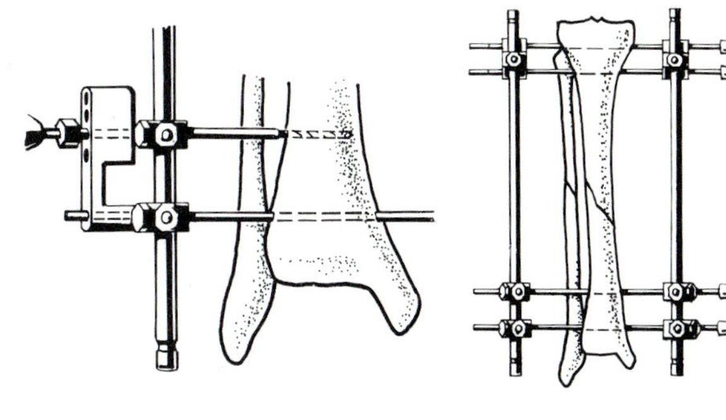

a b d e

260

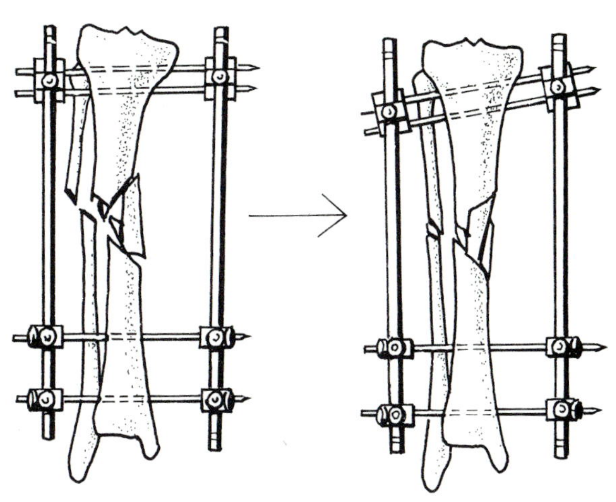

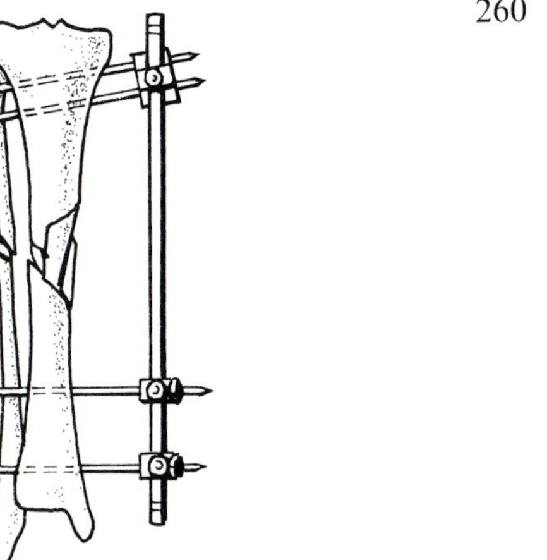

261

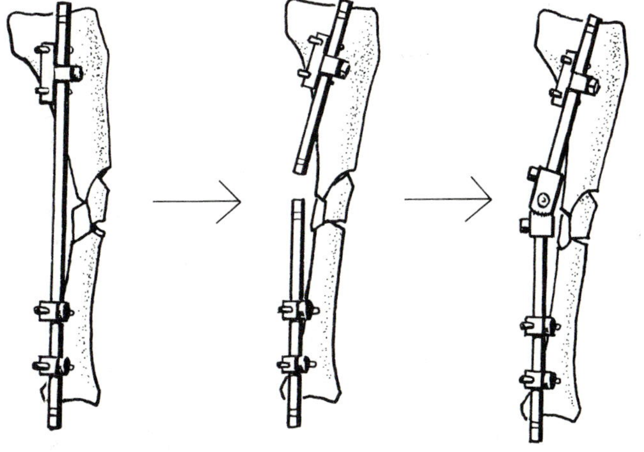

Fig. 262 *The arrangement of the suspension.* The fracture has been stabilized with a narrow plate fixed to the bone with short screws. Note the Steinmann pins, the frame, the wheels, and the mode of suspension. The two Steinmann pins in the middle are not in the bone but are simply passed over the leg and are used here only to strengthen the suspension frame. (*a*) Rope spreader. (*b*) Suspension wheel.

Fig. 263 *Triangulation of the external fixator according to* HIERHOLZER – *used to increase the stability of the fixation.* In addition to the standard frame two Schanz screws are inserted at right angles into the bone and are then joined first to each other by means of a tube, and then to the standard frame. A foot plate is fixed to the frame to prevent equinus deformity. The heel is left exposed. The small bow protects the toes from the overlying sheets.

262

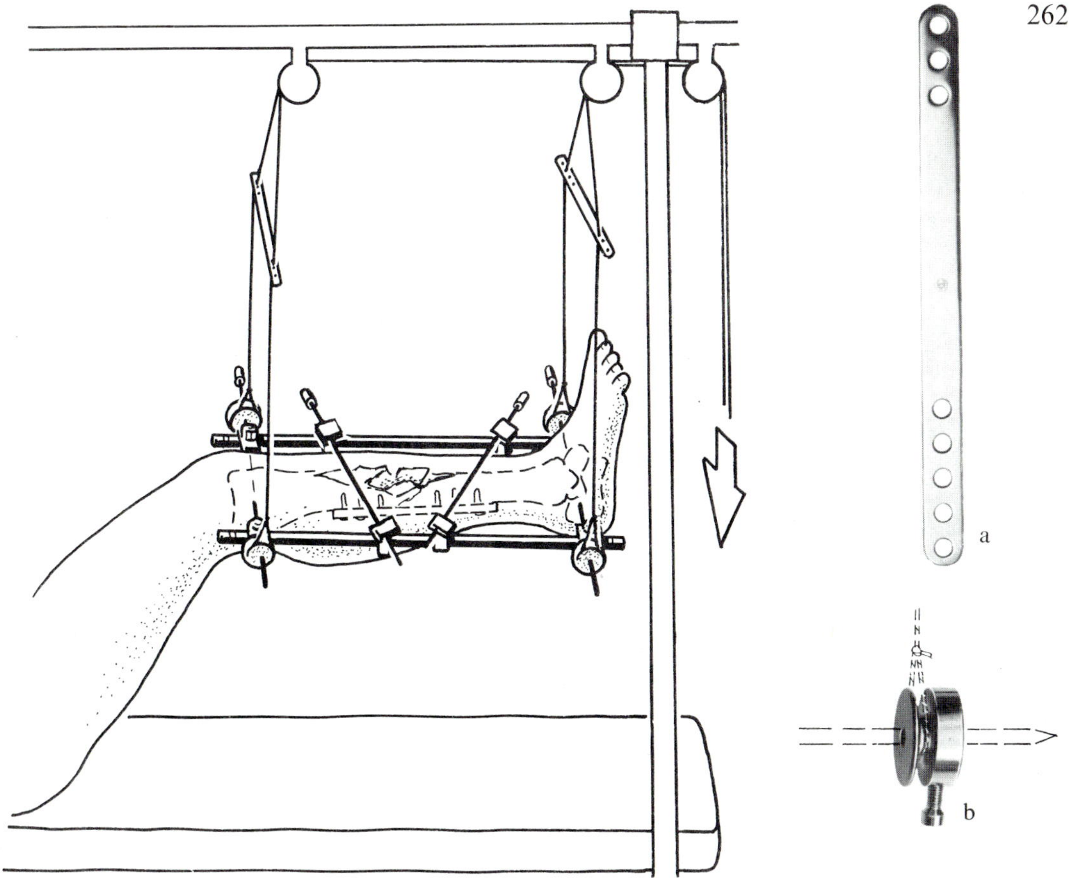

a

b

263

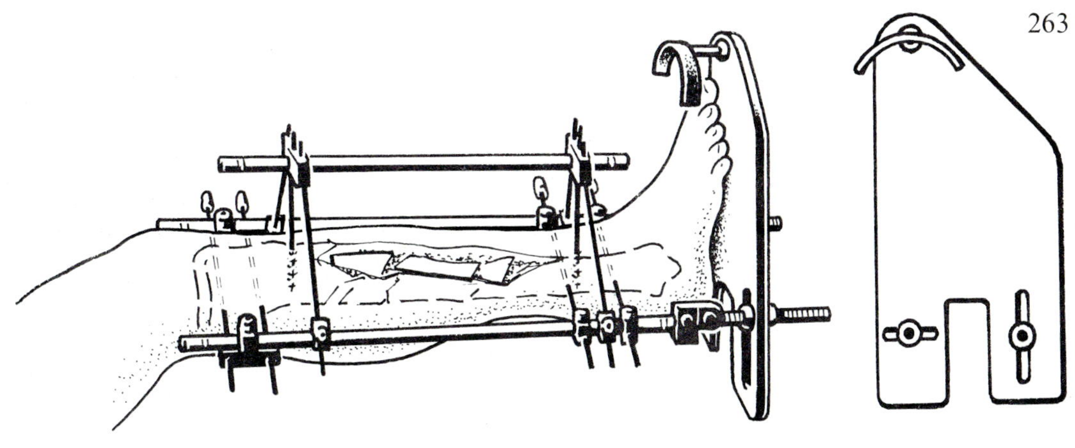

Fig. 264 *Simple fracture lines.* Hold the fracture reduced and stabilize it with two small lag screws. Insert two Steinmann pins proximally and distally and stabilize the tibia with the tubular external fixator.

Fig. 265 *Slight comminution:* Fixed with a thin intramedullary nail without reaming and put in a below-knee plaster.

Fig. 266 *The fracture of the middle third is open. The fracture of the proximal tibia is closed. There is extensive skin contusion on the medial surface.* The proximal fracture was fixed by means of a semi-tubular plate and a tension band wire. The fracture of the middle third was then stabilized with a 9-mm intramedullary nail without reaming. A screw was inserted through the hole in the proximal end of the nail to increase rotational stability.

264

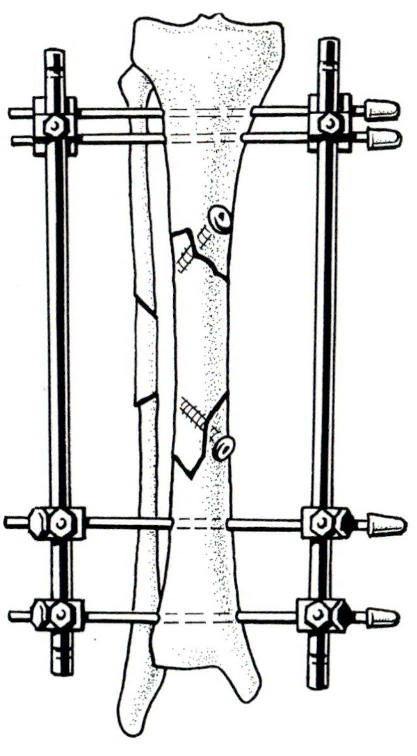

265

266

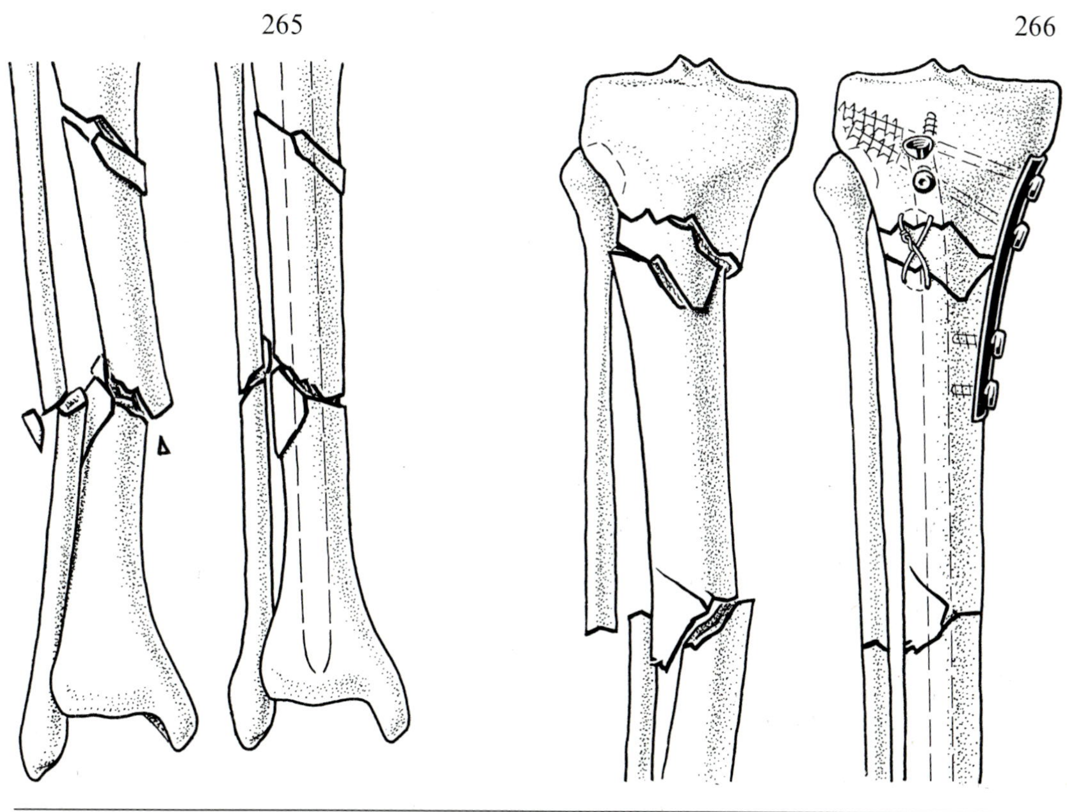

Fig. 267 *Note the posterior releasing incision down the middle of the calf, combined with cruciate incision of the deep fascia.* No stamp graft or splint thickness skin grafting is used. Leave this wound open and cover it with Vaseline gauze. After 1 week, begin to pull the edges together with sterile strips. The compound wound is left open.

Fig. 268 *Soft tissues on the medial side are contused.* The fracture has been fixed with two lag screws inserted anterolaterally running in a posteromedial direction. Subsequently a neutralization plate was placed under the cover of muscle on the lateral side of the tibia.

Fig. 269 *Stabilization of a fracture with a posteriorly applied plate.* All implants are covered by healthy soft tissue. The screws are not allowed to protrude through the open wound.

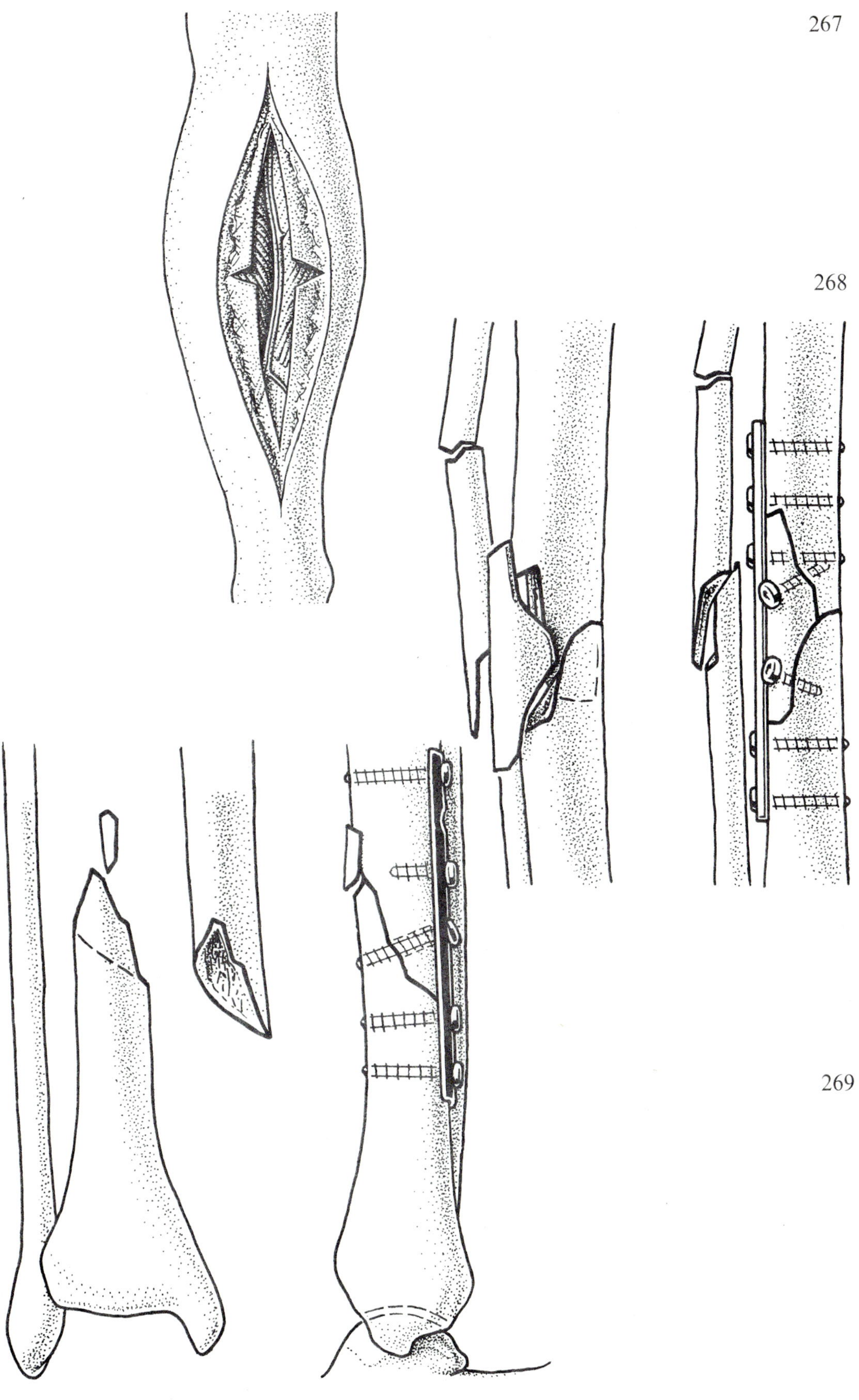
267

268

269

Fig. 270 *Fracture dislocation of the knee.* The shattered patella and necrotic soft tissue were excised. Anatomical reduction and minimal internal fixation of the femoral condyles was performed. Ten percent loss of function occurred because of the patellectomy.

Fig. 271 *Open fracture dislocation of the ankle.* Note the stabilization with minimal internal fixation consisting of intramedullary wiring of the fibula, screwing of the large posterior lip fragment with an AP lag screw, and fixation of the medial malleolus with a lag screw and a Kirschner wire. Double U plaster splint for 1 week was applied to prevent equinus.

Fig. 272 *Fracture dislocation of the distal tibia and avulsion of the medial lip of the talus.* Primary arthrodesis of the ankle was carried out after internal fixation of the fibula and tibia. This was accomplished by two Steinmann pins inserted on each side of the ankle and then joined with two external compressors to bring the surfaces under axial compression as in Fig. 341.

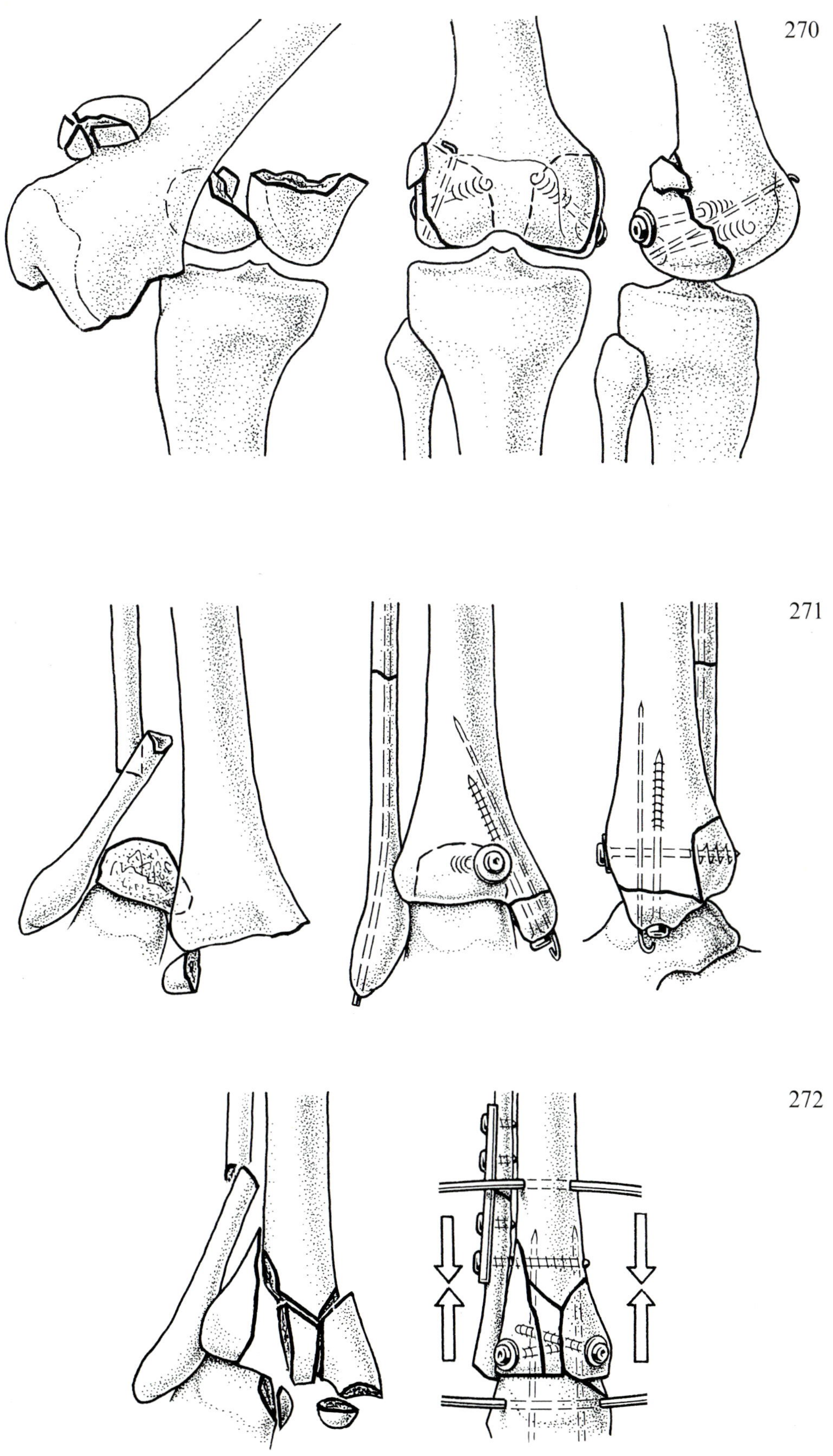
270

271

272

3 Fractures in Children

The treatment of fractures in children is fundamentally different, since only a few fractures require operative intervention. *Diaphyseal* fractures as a rule are treated closed. They heal rapidly and deformities such as varus, lateral displacement and shortening correct spontaneously through remodelling and accelerated growth. It should be noted, however, that rotational deformities never correct spontaneously and are permanent. WEBER'S special traction table has made it possible to control rotational alignment in treating fractures of the femur. Exceptions to the rule of non-operative care is the treatment of long bone fractures such as the femur in teenagers 12 years and over. If opened, these are usually plated.

Articular and periarticular fractures in children are fractures which involve the epiphyseal plate. The fracture lines can either cross the whole epiphyseal plate, which can result in partial closure and resultant growth disturbance, or the fracture line can run through the zone of provisional calcification without involvement of the germinal layer or zone of hypertrophy and pallisading. Compression injuries with involvement of the joint surfaces, the cancellous bone of the epiphyses, and with compression of the epiphyseal plate have the worst prognosis. The damage is permanent and no recovery takes place. Whenever there is danger of growth disturbance of the plate one should strive for accurate reduction and fixation. Certain epiphyseal plate fractures are almost always treated operatively. These do not include the pure epiphyseal plate fracture separations with or without metaphyseal fragments. The open reduction and fixation of the epiphyseal plate fractures is also warranted because the exact reduction and reconstruction of congruent joint surfaces is rarely possible by non-operative means. In epiphyseal plate fracture separations of the distal humerus, one must check carefully the inclination of the epiphyseal plate to the long axis of the humerus. This is a non-weight-bearing extremity and deformities accepted remain permanently without any spontaneous correction (see Fig. 277).

Methods of Internal Fixation: Kirschner wire fixation of fragments is usually all that is necessary because the hard cancellous bone provides excellent fixation for the wires. If anatomically reduced, these fractures heal rapidly and in 2–3 weeks the Kirschner wires can be removed. Epiphyseal plates can be crossed with parallel Kirschner wires. This is one further advantage of this method of internal fixation. We usually leave the Kirschner wires protruding through skin, which makes their removal in 2–3 weeks at the time of plaster change a very simple procedure. Rarely (Figs. 289 and 290) we do use the small cancellous screws. The disadvantage of screw fixation is the second operation required for the removal of the metal. In fractures of the femoral neck, the hard cancellous bone is a definite contraindication to the use of any type of nail. One or two cancellous screws are all that is required to secure excellent fixation of these fractures.

3.1 Epiphyseal Plate Fractures

The treatment and prognosis of epiphyseal fractures depends on the pattern of the injury, i.e. whether the injury involves only the epiphyseal plate, the epiphyseal plate and metaphysis, or the epiphyseal plate and the epiphysis.

3.1 Epiphyseal Plate Fracture Separations

The classification of epiphyseal plate fractures is based on their patho-anatomy (fracture of epiphysis, pure lysis, metaphyseal fracture), their prognosis and their treatment.

A The fracture line does not involve the epiphysis, nor the growth area of the epiphyseal plate. If a proper reduction is carried out, no growth disturbance is to be anticipated.

B The fracture line crosses the epiphysis and the growth zone of the epiphyseal plate. An absolutely accurate reduction must be achieved, or partial closure with resulting eccentric growth disturbance is to be anticipated.

C Compression of epiphyseal plate and epiphysis. If there is severe damage to the growth area a growth disturbance is certain with or without treatment. Partial closure of the epiphyseal plate with growth disturbance is to be anticipated.

Fig. 273 *Classification of articular and periarticular fractures in children with involvement of the epiphyseal plate: three main groups* (M.E. Müller).

A *The fracture line goes through the zone of provisional calcification.* The fracture line does not involve growth zones (germinal zone, cartilaginous columns). There is no growth disturbance even with incomplete reduction. A deformity in the plane of motion will correct spontaneously.

A_1 (Salter I). Simple epiphyseal plate fracture separation – common as a birth injury and during the period of rapid growth spurt prior to epiphyseal plate closure.

A_2 (Aitken I, Salter II). Partial epiphyseal plate fracture, separation with a metaphyseal fragment. Seventy percent of epiphyseal plate injuries.

B *The fracture line crosses the whole epiphyseal plate.* If reduction is not absolutely anatomical, varying degrees of closure of the epiphyseal plate with resultant growth disturbance are certain.

B_1 (Aitken II, Salter III). Partial epiphyseal plate fracture separation with a fracture of the epiphysis. Carry out open reduction of the epiphysis with screw fixation of the epiphyseal fracture. Note that the screw must not cross the epiphyseal plate.

B_2 (Aitken III, Salter IV). The fracture line runs through the epiphysis, through the full thickness of the epiphyseal plate into the metaphysis. After accurate reduction, it is enough to skewer the fragments with two or three Kirschner wires for 2–3 weeks.

B_3 (M.E. Müller). This is an avulsion fracture of the proximal insertion of the medial collateral ligament of the knee which takes with it a portion of the periepiphyseal ring. Accurate reduction and fixation are required. Despite this, one cannot always be certain that partial closure will be prevented.

B_4 (Rang). Damage to the epiphyseal ring, as for instance with a lawn-mower. Partial closure is the usual sequela, because of the partial loss of the periepiphyseal ring.

C (Salter V). *There is compression of the articular surface and of the epiphyseal plate,* with damage to the germinal layer of the epiphyseal plate. Partial closure of the epiphyseal plate with resultant eccentric growth is to be anticipated. An open reduction should be considered only if there is marked widening of the epiphysis, with marked distortion of the articular surface. Normal width of the epiphysis is then maintained with a lag screw. Because of axial deformities, the result of asymmetric growth, subsequent corrective procedures are unavoidable. The commonest example of these injuries is the compression of the medial distal epiphyseal plate of the tibia by the talus in severe supination injuries

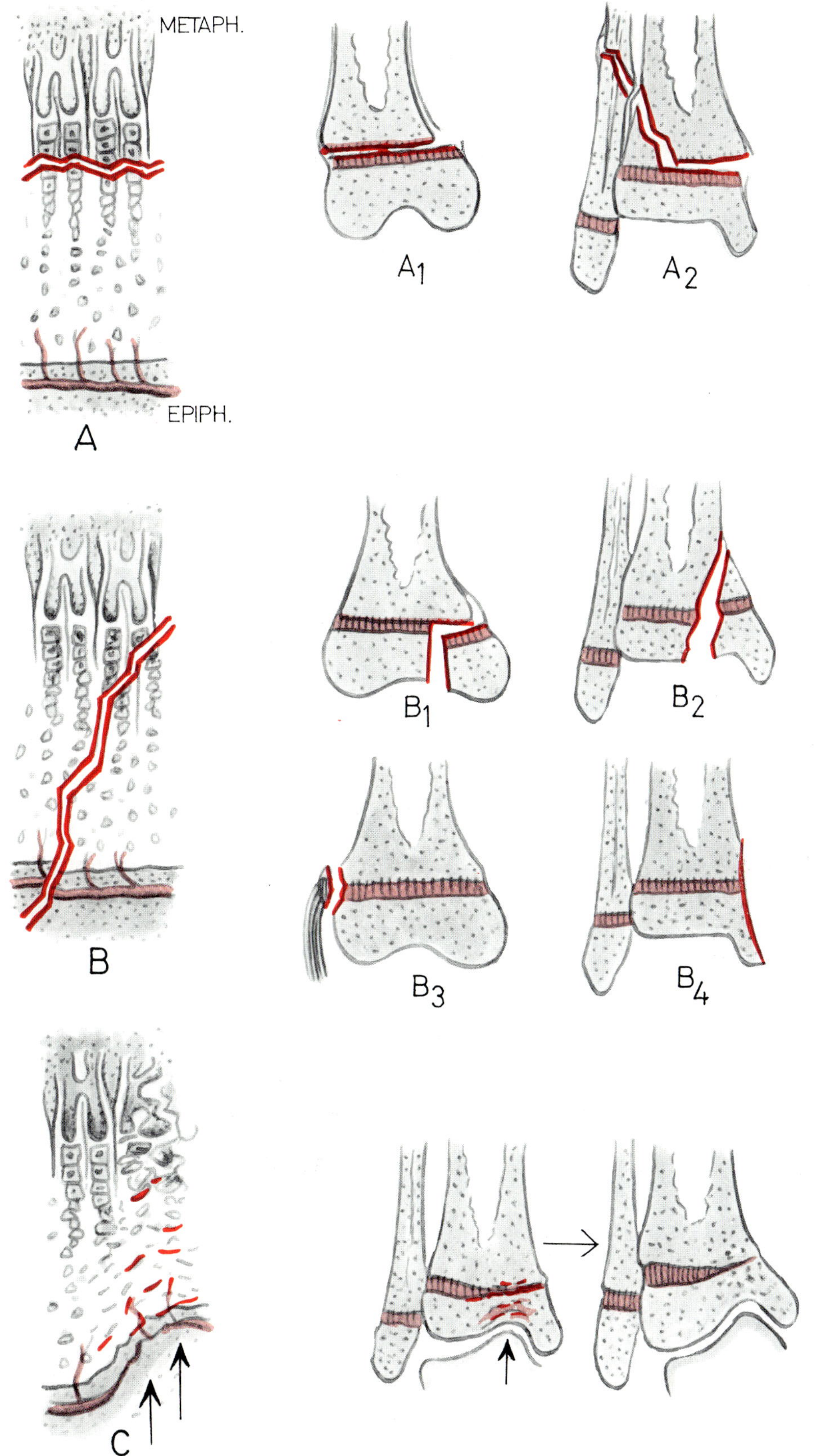
METAPH.
EPIPH.
A
A1
A2
B
B1
B2
B3
B4
C

3.2 Fractures of the Humerus

Most fractures of the proximal humerus and of the humeral shaft can be treated non-operatively either in a Valpeau bandage or in skeletal traction. The immobilization or traction are required usually for 2–3 weeks. For skeletal traction we prefer a cortical screw inserted into the olecranon (Fig. 278) rather than Kirschner wires. The screw has fewer complications associated with it (infection, early and late ulnar palsy, etc.). Occasionally a proximal epiphyseal plate fracture separation has to be operated upon. This is either because it cannot be reduced (interposition of the long tendon of biceps) or because the injury is close to epiphyseal closure. After reduction, all that is necessary is fixation with Kirschner wires for 2–3 weeks. Supracondylar fractures of the humerus are almost always treated non-operatively either by the method of Blount (reduction under general anaesthesia and immobilization in a sling, with the elbow bent to 140°) or in skeletal traction distal to the olecranon (Fig. 278). An open reduction is indicated only if the fracture is irreducible or if it is associated with a vascular or nerve injury. Usually simple Kirschner wire fixation is all that is necessary.

Fractures of the lateral condyle or of the medial epicondyle are usually treated by an open reduction and Kirschner wire fixation for 2–3 weeks. Particular attention must be paid to the restoration of the correct angle between the epiphyseal plate of the lateral condyle and the long axis of the humerus. This angle must subtend 70°–75° (Fig. 277), otherwise one runs the risk of a cubitus valgus with late ulnar nerve palsy or occasionally a cubitus varus.

Fig. 274 *Avulsion of the medial epicondyle* fixed with two Kirschner wires.

Fig. 275 *Fracture of the lateral condyle* fixed with a Kirschner wire. Accurate reduction is obligatory.

Fig. 276 *Supracondylar fracture* (only if closed reduction as described by Blount fails). Stable fixation with two crossed Kirschner wires left in place for 2–3 weeks.

Fig. 277 After a reduction of a *supracondylar fracture*, the angle between the epiphyseal plate of the lateral condyle and the long axis of the humerus must subtend an angle of 70°–75°, otherwise a cubitus valgus or as the case might be a cubitus varus will result.

Fig. 278 *Longitudinal skeletal traction* as described by Baumann. This is accomplished by means of a cortical screw inserted distal to the olecranon. If correctly inserted such a screw provides excellent traction and is not associated with the complications such as sepsis or late nerve paresis, seen with transolecranon Kirschner wires. Occasionally it is necessary to provide longitudinal traction in line with the forearm. This is accomplished by means of skin traction with the forearm pronated.

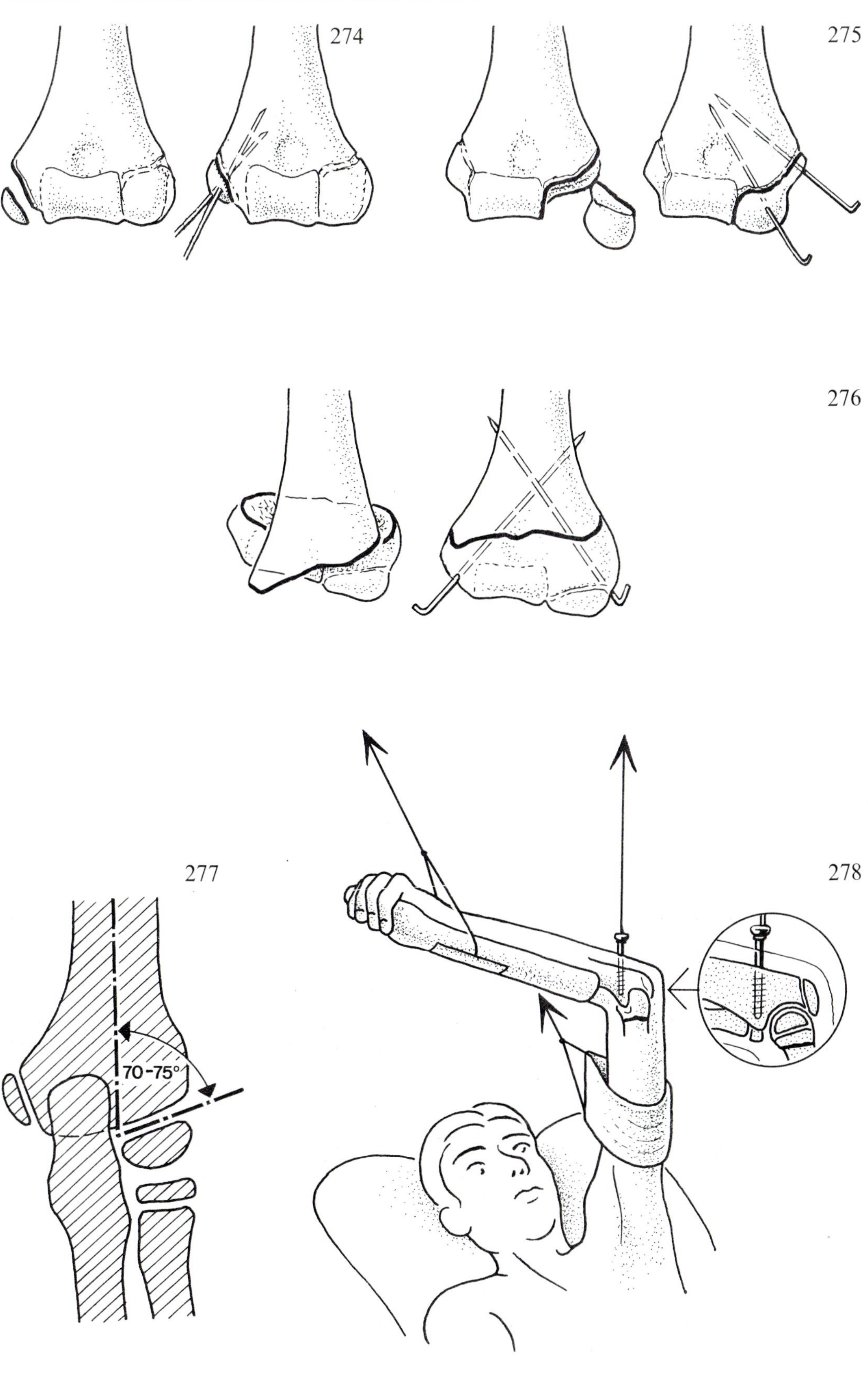
274
275
276
277
70-75°
278

3.3 Fractures of the Forearm

Fractures of the forearm can almost always be treated non-operatively. Because of the intact periosteal hinge, in carrying out a closed reduction, it is usually necessary to increase the deformity. Only in fractures with displacement of the radial head is an open reduction necessary to prevent joint incongruency. Fixation is provided by means of a transhumeral Kirschner wire supplemented by plaster fixation for 3 weeks. In dealing with the Monteggia fracture, if reduction cannot be obtained or maintained, an open reduction and fixation become necessary. The torn annular ligament can only be reconstructed surgically.

Note: There are only two epiphyses which receive their blood supply through synovial retinacular vessels. These are the epiphyses of the proximal femur and of the proximal radius. Thus in epiphyseal plate fractures of the radial head there is danger of avascular necrosis.

Fig. 279 *Fractures of the forearm.*

a The periosteal hinge is intact on one side. The fracture is impacted and cannot be reduced by simple traction.

b Increase the deformity. Simple pressure with the thumb is usually all that is required to hook on the cortices.

c Then apply traction. Use is made of the intact periosteal hinge in completing the reduction. If the periosteal hinge is intact dorsally, plaster immobilization in volar flexion converts the periosteal hinge into a tension band.

Fig. 280 *Epiphyseal plate fracture separation of the radial head.* An open reduction is indicated only if a closed reduction cannot be obtained or maintained. Fix with a Kirschner wire for 2–3 weeks. Inspect the annular ligament.

Fig. 281 *Fracture of the proximal ulna.* Reduction and stabilization with one Kirschner wire.

Fig. 282 *Unstable epiphyseal plate fracture separation of the distal radial epiphyseal plate.* Simple Kirschner wire fixation.

279

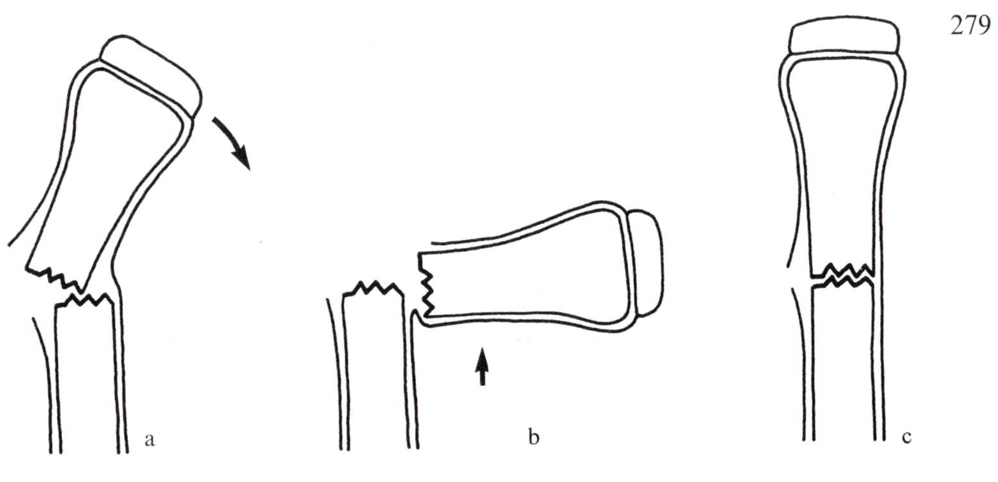

280

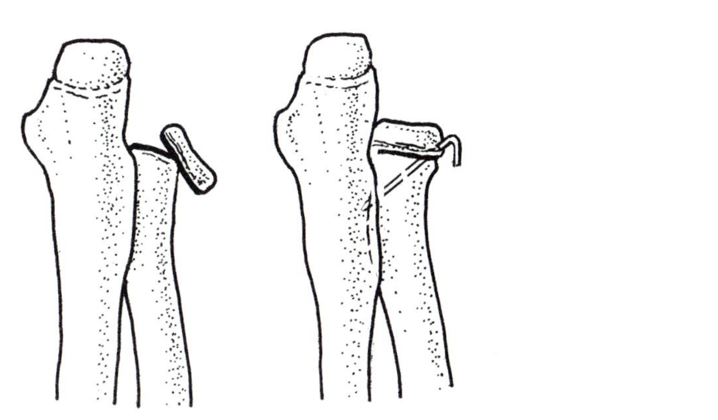

281

282

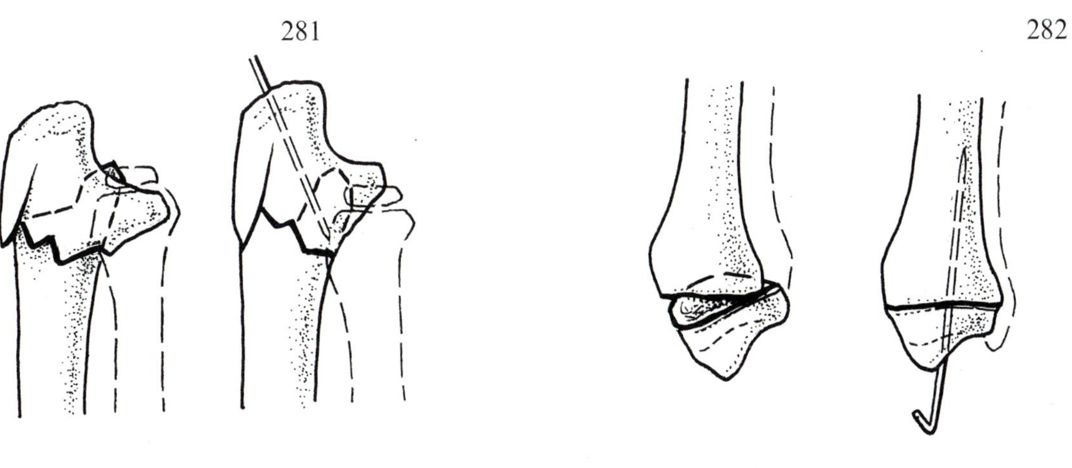

3.4 Fractures of the Femur

Fractures of the femoral neck, whether they be of the lateral type or of the medial subcapital shear type, comprise an absolute indication to open reduction and stable internal fixation.

Stable internal fixation is achieved by the insertion of one or more cancellous screws with care not to cross and thus damage the epiphyseal plate with the screw thread. Nailing of these fractures is definitely contraindicated. The cancellous bone is very hard and in nailing there is great danger of driving the fragments apart and thus tearing the retinacular vessels, which would lead to avascular necrosis of the head.

Immediately after fracture, the vessels are usually intact. Displacement of the fragments leads to kinking of the vessels, which predisposes to their thrombosis. These fractures should be treated as surgical emergencies.

The Technique of Open Reduction and Fixation: The capsule is exposed in the interval between the glutei medius and minimus, and the tensor fascia lata. The capsule is opened and the fracture exposed with the aid of three small retractors. One is inserted over the anterior pelvic brim, one above and one below the femoral neck. Disimpact the fracture and carry out an open reduction. Once reduced, fix the fracture temporarily with Kirschner wires and check the reduction particularly at the level of the calcar by flexing and rotating the hip. Definitive fixation is carried out with two cancellous or malleolar screws. Take care not to damage the epiphyseal plate.

Fractures of the femoral shaft are usually treated non-operatively in skeletal traction on the Weber traction table with the hip and knee in 90° of flexion. The traction table allows one to correct any rotational malalignment.

Fig. 283 *Medial shear subcapital fracture.* The T plate here serves merely as a washer for the lag screws. Note that the screw threads do not cross the fracture line.

Fig. 284 *A lateral base-of-neck (closed) fracture in a child.* Two cancellous screws with long threads were used for the fixation of this fracture in an 11-year-old child. (Malleolar screws would have been just as good, and much easier to remove.)

Fig. 285 *Fracture of the femur in skeletal traction on the Weber traction table.* Both knees and hips are flexed to 90°. The legs are abducted to 20° as for an anteversion X-ray. The uninvolved limb is suspended in skin traction. The side with the fracture is placed in skeletal traction by means of a supracondylar threaded Steinmann pin which is inserted parallel to the knee joint. The weight for the traction is the child's pelvis and trunk. It is wrong to substitute a Kirschner wire for the threaded Steinmann pin. This usually leads to pin tract infection. It is also wrong to substitute pulleys and weights for this system of fixed traction.

Fig. 286 *Supracondylar fracture of the femur.* Stabilize with four Kirschner wires, two inserted on each side of the fracture.

Fig. 287 *Type B_2 epiphyseal plate fracture separation with the fracture line crossing both the distal epiphysis and metaphysis.* After an exact open reduction the fracture was stabilized with two Kirschner wires which were allowed to project through the skin to facilitate their removal.

283 284

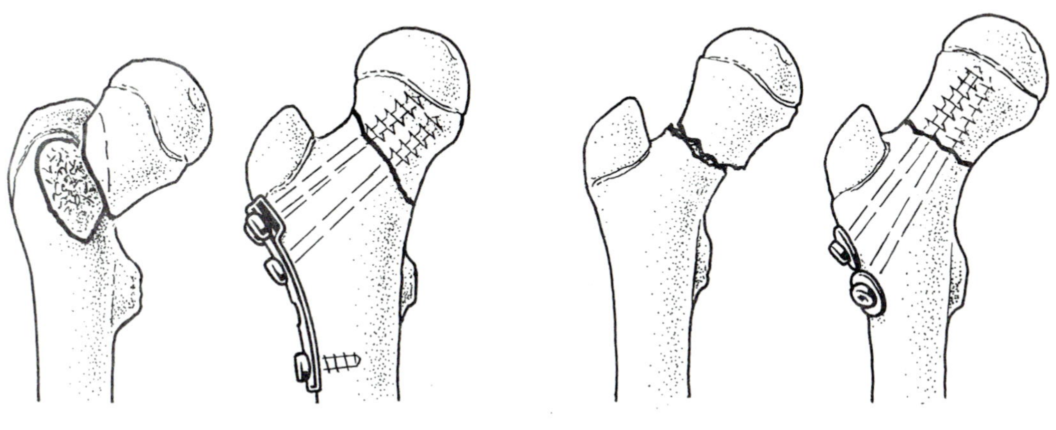

285

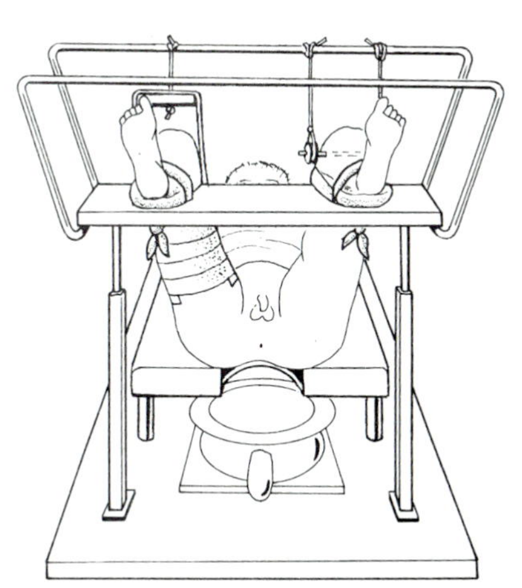

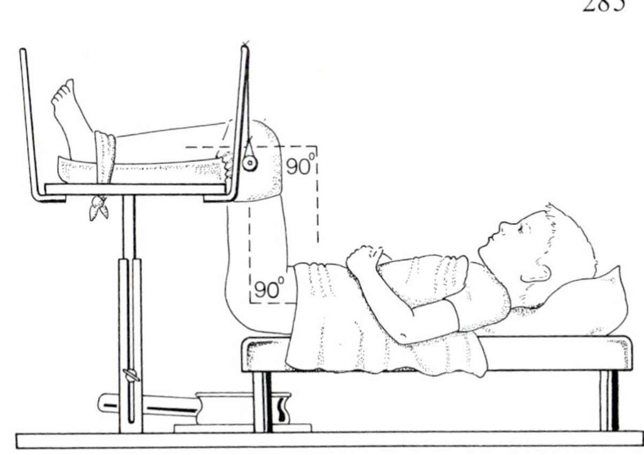

286 287

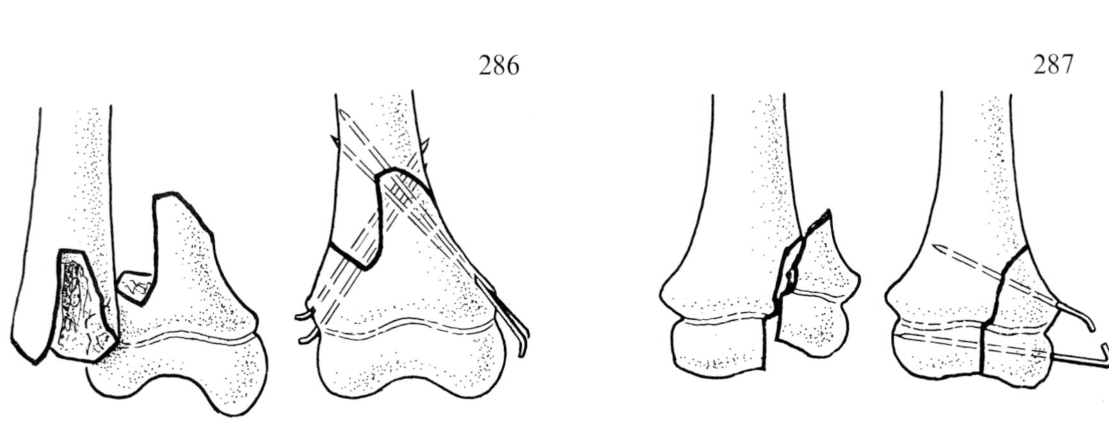

3.5 Fractures of the Tibia

A displaced avulsion fracture of the intercondylar eminence should be treated by means of an open reduction and internal fixation, otherwise there is danger of a permanent flexion deformity. The internal fixation is carried out with a wire loop or even better with a small cancellous screw. Care should be taken not to insert the cancellous screw through the proximal epiphyseal plate.

Technique of Open Reduction and Internal Fixation: Prior to the reduction insert a 1.6-mm-thick Kirschner wire above the tibial tuberosity as far as the fracture, then carry out the reduction. Drill a 2-mm hole through the epiphysis and through the avulsed fragment. Carry out the fixation with a 4.0-mm cancellous screw. Occasionally it is necessary to shorten the screw thread by cutting off the excess threaded portion. If good fixation has been achieved and the screw does not project into the joint, remove the Kirschner wire and immobilize the knee in 30°–40° of flexion in a plaster for the next 4–5 weeks. Remove the screw in 3–6 months.

Fractures of the tibial shaft are treated closed. Only in teenagers 12 years or older in whom the growth period has ended, in order to prevent a varus deformity, one can consider simple screw fixation of an oblique fracture of the tibia with an intact fibula.

Fig. 288 *Avulsion of the intercondylar eminence.* Stabilize with a small cancellous screw, inserted through the epiphysis.

Fig. 289 *Epiphyseal plate fracture separation of the proximal tibial epiphysis.* Anatomical reduction and fixation with two lag screws.

288

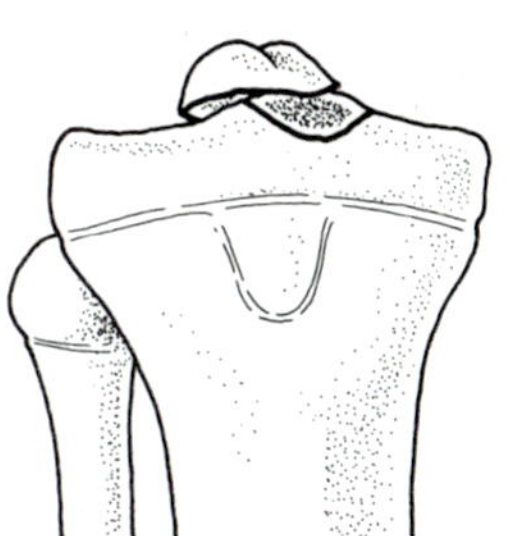

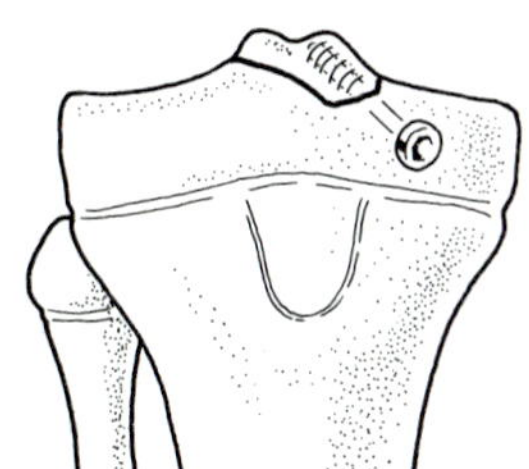

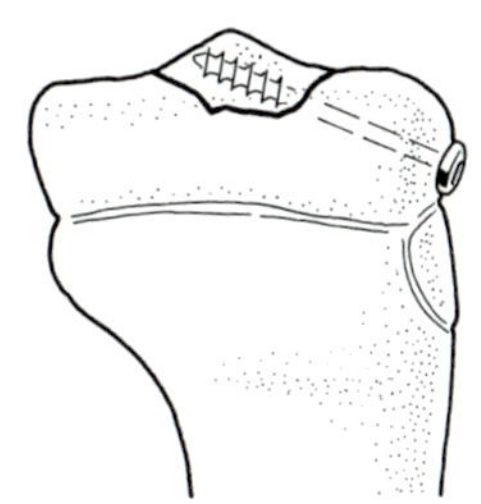

289

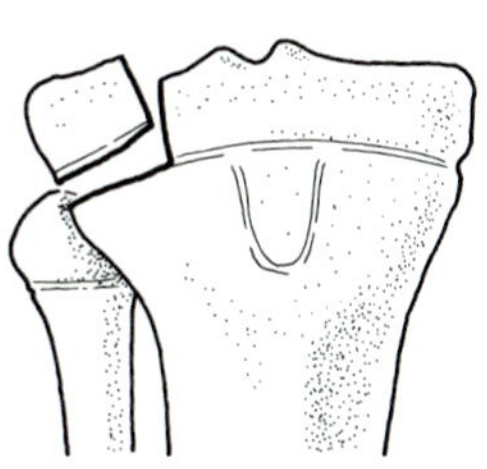

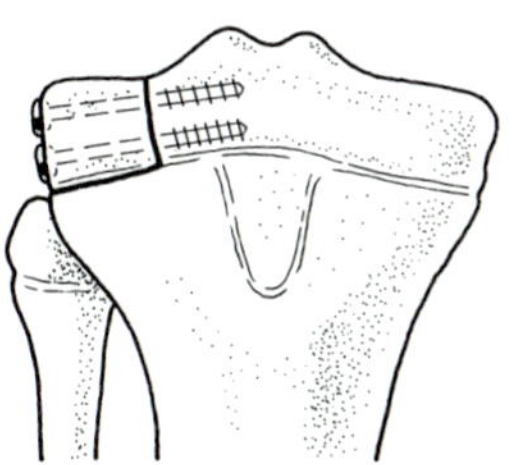

3.6 Malleolar Fractures

An open reduction is indicated when the epiphyseal plate fracture separation is associated with instability of the ankle mortice or when the fracture line crosses both the epiphysis, epiphyseal plate and metaphysis. We prefer lag screw fixation for these rather than simple Kirschner wire fixation, since the lag screw fixation prevents the formation of callus.

Fig. 290 *Type B_1 epiphyseal plate fracture separation with the fracture line running through the epiphysis.* Anatomical reduction and fixation with two 4.0-mm cancellous screws in a 13-year-old boy.

Fig. 291 *Fracture of the tubercle of* CHAPUT. *Type B_2* in a fourteen-year-old girl. Fixation with Kirschner wires. A small cancellous screw could also be used for fixation.

Fig. 292 *Type B_2 epiphyseal plate fracture separation of the medial malleolus.* Anatomical reduction and fixation with two Kirschner wires

Fig. 293 *Fracture dislocation of both the medial and lateral malleolus. Type B_1* – after open reduction, stabilization of both malleoli with Kirschner wires. This perfect anatomical reduction and internal fixation gave this 14-year-old girl an excellent result.

290

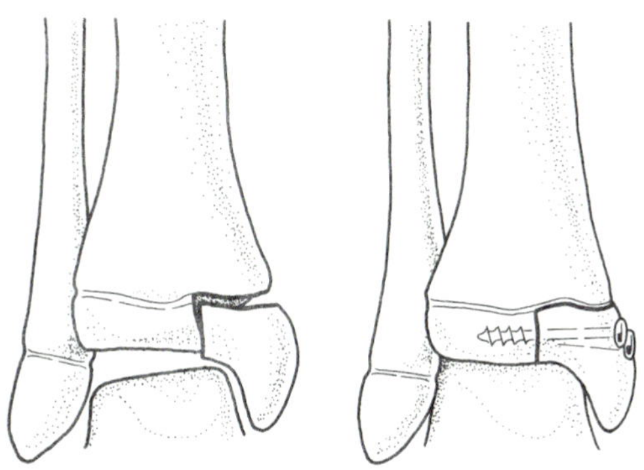

291

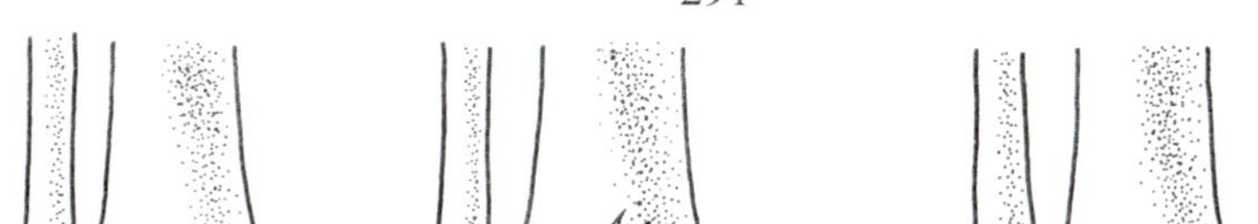

292

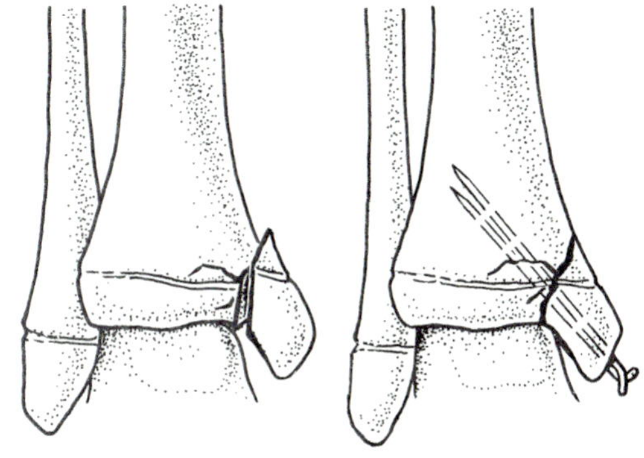

293

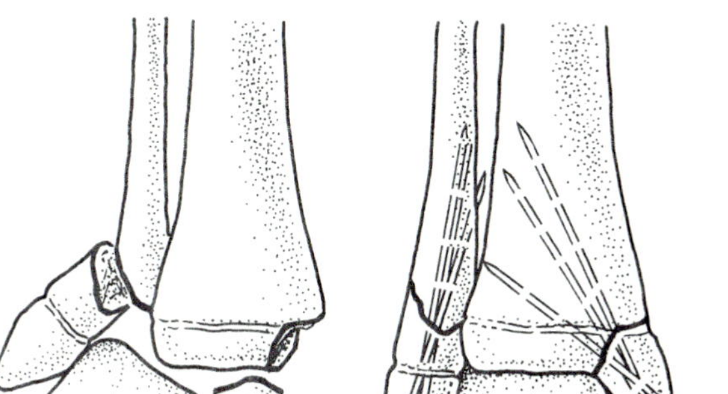

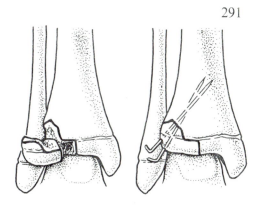

SUPPLEMENT

Reconstructive Bone Surgery

M.E. MÜLLER

Introduction

Stable internal fixation has proven itself not only in the treatment of fresh fractures but also in the treatment of pseudarthroses and reconstructive procedures such as axial corrections or arthrodeses.

We feel that modern reconstructive bone surgery must be accomplished with such stable internal fixation that one can begin with early active painfree mobilization of the involved extremity. Furthermore the surgery must be carefully planned to the very last detail, including all axial corrections and displacements, which then must be carefully noted on the pre-operative drawing of the procedure. The procedure itself should be carried out with great care in accordance with the pre-operative plan, which makes the use of intra-operative radiography superfluous.

In this supplement we shall discuss the procedures which have stood the test of the last 20 years. We shall lay particular emphasis on decortication as described by JUDET (Fig. 296) because it speeds up the rate of consolidation of reconstructive procedures. We shall also lay particular emphasis on the use of autogenous cortico-cancellous bone grafting (Fig. 295) to bridge defects or to bridge de-vitalized bone segments.

1 Delayed Union

A fracture is considered to have gone on to delayed union if it has not healed in 4 months.

Those delayed unions which are not infected and which come about after an accurate reduction, stable fixation, and early mobilization of the limb do not present difficult therapeutic problems. If after 3–4 weeks of unloading and immobilization the fracture still fails to go on to union, then most often, one has to proceed to surgery. In treating delayed unions of shaft fractures of the leg, remove the intramedullary nail or screws and plate, ream the medullary canal of the femur to somewhere between 13–16 mm and of the tibia between 12–14 mm and then insert a suitable intramedullary nail (see p. 116). If there is considerable necrosis of the bone fragments or insufficiency of the medial buttress, then in addition a cancellous bone graft must be carried out. Most of the patients need to remain in hospital only a few days.

In peri-articular fractures of the leg, which go on to delayed union the internal fixation was most often unstable right from the start or the reduction was incomplete. To secure union, the internal fixation must be revised and bone grafting carried out.

The problem of delayed union in fractures of the upper extremity is somewhat more complicated, particularly if the fracture was initially plated. Most of the delayed unions of fractures which were primarily plated are either the results of inadequate internal fixation or of varying degrees of avascular necrosis of the fragments.

The loose plates must be removed, a decortication carried out (Fig. 297b) and after bone grafting the fractures must be stabilized with longer compression plates. Note that the new screw holes must often be drilled at 90° to the old ones. In delayed unions of the arm, consider revision of the internal fixation only in those cases where it is clearly unstable. As long as there is no evidence of screw loosening and as long as the patients are symptom free or become symptom free if the part is put to rest, one can afford to wait a further 1–3 months. Most of these cases, once immobilized, will go on to union despite the usual partial avascular necrosis of the fragments.

In delayed union associated with an implant fracture, the biomechanical construction of the internal fixation must first be revised to prevent further cyclic loading of the new implant. Once a stable internal fixation has been achieved one must make certain that the medial buttress is intact. Most cases require decortication and bone grafting of the medial buttress (Fig. 111b).

2 Pseudarthroses

Introduction

A pseudarthrosis is any fracture which has failed to heal within 8 months. The term is synonymous with non-union. It is no longer reserved for the rare cases of fractures which fail to unite and go on to the formation of a true false joint with sealing of the medullary canal, formation of articular cartilage, joint capsule and synovial fluid.

We classify pseudarthroses as non-infected, previously infected, and infected.

Non-Infected Pseudarthroses

Under the non-infected pseudarthroses we recognize:

a) *The hypertrophic reactive vascular pseudarthroses.* Over 90% of pseudarthroses which come about as a result of conservative fracture treatment are hypertrophic, reactive or vascular. Radiologically they are characterized by the florid bone reaction, which is responsible for the flaring and sclerosis of the bone ends (they are the so-called elephant-foot-type pseudarthroses). The radiologically evident sclerosis does not represent dead bone but rather excessive appositional bone formation with an excellent blood supply. The interposed cartilage or connective tissue once absolutely immobilized rapidly mineralizes and changes into bone. The ends of the bone of such a pseudarthrosis should not be resected and a bone graft is unnecessary. Once immobilized either with an intramedullary nail or one or two compression plates, these pseudarthroses rapidly ossify within a few weeks. An exception to this rule is the hypertrophic reactive pseudarthrosis of the femoral neck which requires for its consolidation a revision of the mechanical configuration, which is achieved by means of the repositioning osteotomy (see Fig. 316).

b) *The atrophic, non-reactive, frequently avascular pseudarthroses.* Radiologically these are characterized by an absence of any bone reaction at the bone ends. From a prognostic and therapeutic point of view they are to be equated with the osteoporotic non-reactive, so-called oligotrophic pseudarthroses of Weber and Čech. The atrophic or oligotrophic pseudarthroses require for their consolidation, in addition to a stable internal fixation (intramedullary nail or compression plate), extensive decortication and bone grafting. With the rise in the number of internal fixations, there has been a marked increase in the incidence of these pseudarthroses.

Note: Epiphyseal and metaphyseal non-reactive pseudarthroses generally indicate necrosis of the bone ends. The prime example of these is the pseudarthrosis of the femoral neck with an avascular head. The congenital pseudarthroses of the tibia are best dealt with by means of a cortical graft from the opposite uninvolved leg (Fig. 301 d).

Previously Infected Pseudarthroses

These are divided into two groups. Those with the fragments *in contact* and those with *a defect.* Where the fragments are in broad contact, as for instance in a pseudarthrosis of the tibia, stability can be achieved either by means of a compression plate with the screws left out of the central four to six holes, or by means of four to six Steinmann pins and an external fixator. Stability is achieved by bringing the pseudarthrosis under axial compression. In addition to achieving stability one must carry out decortication of and add a cancellous graft in those areas which were previously not infected. In the pseudarthroses with a defect where there are no sequestra, stability can be achieved in the upper extremity with one, and in the lower extremity with two semi-tubular plates applied in such a way as to bridge the defect (Fig. 306). The defect is filled with cancellous bone graft (Fig. 98) which is placed between the plates. Here the bone ends must be resected, before the plates are applied. To prevent a flare-up of the infection one should begin with systemic antibiotics (see p. 144), 48 hours prior to the surgical procedure.

Infected Draining Pseudarthroses

These pseudarthroses present two problems: the *consolidation of the pseudarthrosis* and the *clearing of the infection.* The fate of the pseudarthrosis depends on the stability of the internal fixation and on the infection. The infection is propagated by foreign bodies such as the implant or necrotic bone. The bone can lie either as a free sequestrum or be in continuity with normal and healthy bone. The infection is rarely propagated by an inadequate skin cover.

We feel that the prime goal in the treatment of the pseudarthroses is their consolidation. Once the pseudarthrosis is solid, the infection can be eradicated, either by means of a sequestrectomy and cancellous grafting, or in the case of an infected intramedullary nail by further reaming of the medullary canal by 2 mm. Only rarely is skin grafting necessary.

Prior to the definitive surgical procedure we like to obtain cultures. We attempt to decrease the drainage and the severity of the infection by means of elevation and Neomycin/Bacitracin irrigation. Where there is extensive soft-tissue necrosis, we irrigate with Dakin solution. The location, size and number of the sequestra must be determined radiologically, with the help of tomography where necessary. Antibiotics are started 48 hours prior to the procedure.

Metal implants are left in as long as they provide stability. Where stability is still being maintained, the procedure consists of removal of the chief sequestrum, of decortication, and of an extensive posterior cancellous bone graft. Once a sufficiently strong bony bridge develops, the implant and further sequestra can be removed.

Where the internal fixation is no longer providing stability, it is removed. All necrotic tissue is excised until healthy bleeding tissue is encountered.

Note: This applies not only to bone but also to soft tissue. A new internal fixation is carried out to bring about stability and a suction irrigation system is set up. The irrigation is continued with antibiotics such as a Neomycin/Bacitracin solution for the next 5–8 days. To obtain stability in the leg, we have found the following two methods most useful. Stability can be obtained by means of four to six Steinmann pins and the external fixators. This must be combined with extensive decortication and a cancellous bone graft. Where the fibula is still intact, a tibio-pro-fibular bone graft can be carried out, creating a bony bridge between the tibia and fibula above and below the pseudarthrosis. Whether the bone is laid on anteriorly or posteriorly is best determined by means of an arteriogram which will disclose which vessels are still intact, i.e. those anterior or posterior to the interosseous membrane.

In the femur before consolidation is achieved, two or three revisions of the plating often have to be carried out, combined with two or three bone grafts. If the pseudarthrosis is associated with an infected intramedullary nail, a simultaneous re-nailing should be considered (Fig. 307a) combined with a cancellous bone graft. In this way, both the pseudarthrosis and the infection can sometimes be cured on the very first attempt.

The immobilization provided by a stable internal fixation is adequate and supplemental external plaster immobilization is not required. This should not be confused with the elevation of the limb and the splinting of the joints in the position of function. An example of this is the prevention of equinus by the addition of the foot plate to the external fixators. External fixators facilitate the treatment of infected pseudarthrosis, because they allow the suspension of the limb from an overhead beam (Fig. 262).

2.1 Non-Infected Pseudarthroses

Principles of Treatment

a) Hypertrophic reactive pseudarthroses – the so-called elephant-foot-type or horse-hoof-type pseudarthrosis will consolidate rapidly once rigidly immobilized (intramedullary nailing, compression plating) without excision of the interposed pseudarthrosis tissue, bone grafting or post-operative immobilization.

If there is an axial deformity which must also be corrected, then whether the correction is done by means of an osteotomy or by taking the pseudarthrosis apart, additional decortication should be carried out to enhance rapid consolidation of the pseudarthrosis. At times it is also necessary to lengthen the limb. For the technique of lengthening, see Fig. 326.

b) The atrophic or oligotrophic non-reactive pseudarthroses require, in addition to the stable internal fixation, extensive decortication of the viable fragments and an autogenous bone graft.

For stability in the lower extremity we prefer intramedullary nailing with reaming and in the upper extremity, compression plating. Once stabilized the bone should be decorticated through one-half to two-thirds of its circumference and then bone grafted.

Note: If bone grafting is required, we strongly recommend the use of cortico-cancellous bone. As donor site for cortico-cancellous bone we recommend either the inner table of the anterior wing of the ilium or the outer table of the posterior wing of the ilium. The greater trochanter and proximal tibia are good sources of pure cancellous bone which may be required to fill a defect, e.g. the defect of an infected pseudarthrosis.

Fig. 294 *The two types of non-infected pseudarthrosis.*

a The hypertrophic reactive, well-vascularized type (a′) elephant-foot-type (a″) horse-hoof-type

b Non-reactive type, between poorly vascularized or dead bone fragments.

Fig. 295 *To obtain bone from the inner table of the anterior wing of the ilium.* Make your incision 1–2 cm either medially or laterally to the iliac crest. Reflect the skin and keep it retracted with two small Hohmann retractors.

a Use a gouge to obtain 6–10-cm-long cortico-cancellous strips from the inner table.

b Cut these strips into smaller pieces 5–6 mm wide and 15 mm long.

c If you need only cancellous bone, use a broad osteotome. Lift up a large flap of pure cortical bone which you hinge inferiorly. At the end allow it to fall back into place. Once exposed, the cancellous bone is removed with a gouge.

Fig. 296 *The decortication of* Judet. Carry the incision down to bone. Expose the bone by chiselling off small flakes taking care that these flakes retain their soft-tissue attachment and in this way their blood supply. The decortication should extend one-half to three-quarters of the way round the bone and 8–10 cm along its length. This sheet of mobile living cortical flakes of bone usually goes onto rapid consolidation with the formation of a strong fixation callus regardless of whether one is treating a non-infected or an infected pseudarthrosis. In treating pseudarthroses, both fragments are decorticated and in this way pockets are created for the bone graft between the bone and its soft-tissue cover (Fig. 306a). One should also decorticate whenever a corrective osteotomy is made through a fracture callus (Fig. 323).

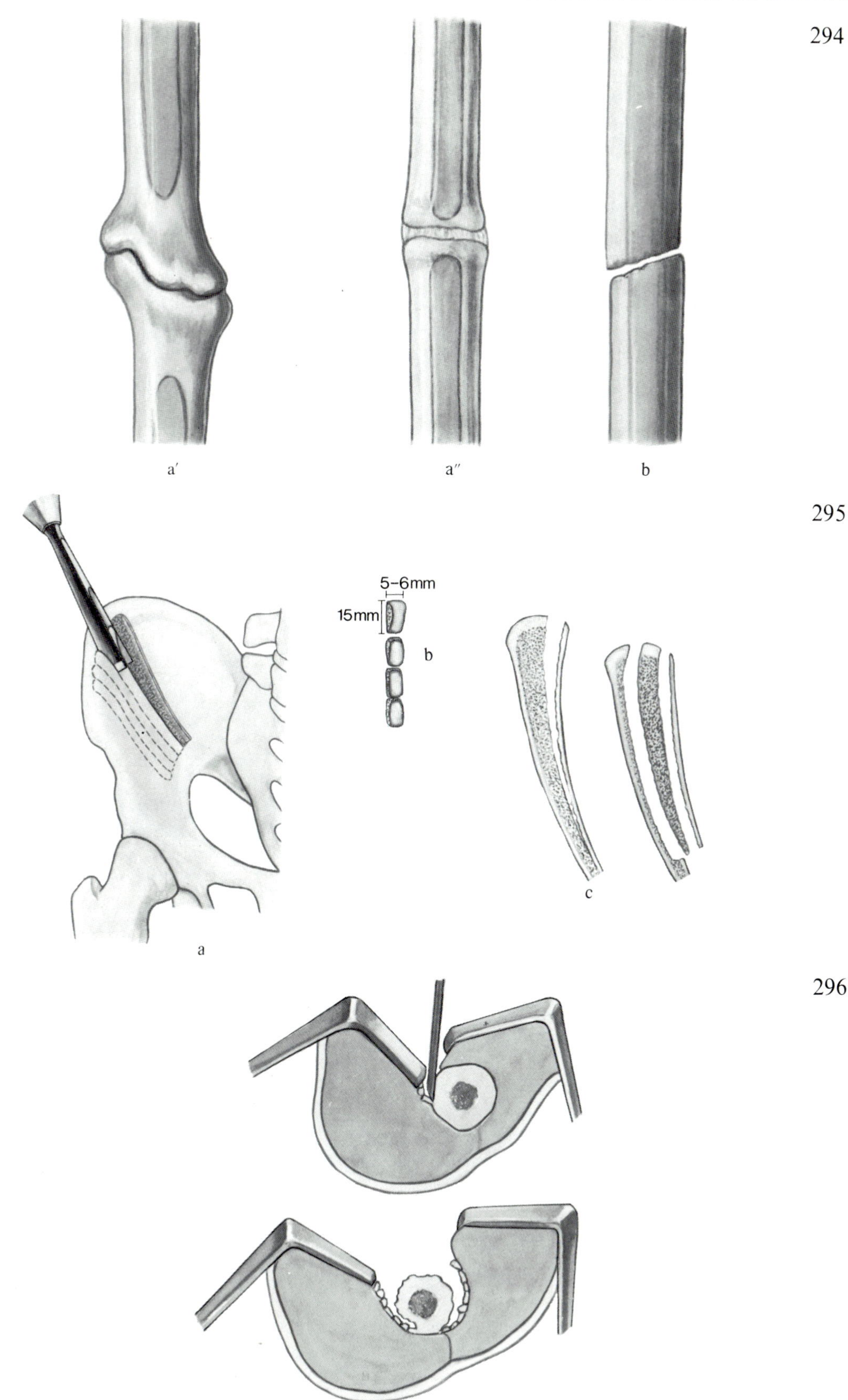
294
a′
a″
b
295
5–6mm
15mm
b
c
a
296

2.1.1 Pseudarthroses of the Upper Extremity

Diaphyseal Pseudarthroses

In the upper extremity the compression plate is our method of choice in securing stable internal fixation. The plates must be carefully contoured to meet the biomechanical requirements of the area. Decortication is almost always necessary and whenever the medial buttress is deficient in the humerus a bone graft must be carried out. Occasionally if the bone is extremely osteoporotic a pure cortical graft can be used, as in a congenital pseudarthrosis of the tibia, to enhance the fixation of the screws.

Fig. 297 *Diaphyseal pseudarthroses of the arm.*

a Pseudarthroses of the clavicle are stabilized with a one-third tubular plate or a narrow six- or seven-hole DCP. Most often it is necessary to resect the ends of the pseudarthrosis and to bone graft. In treating these pseudarthroses an attempt must be made to re-establish the normal length of the clavicle.

b In treating the humerus a wide six- or eight-hole DCP must be used. The screws must grip at least six cortices on each side of the pseudarthrosis. The plate is applied to the lateral cortex. We use the broad plate, not because it is more rigid, but because its holes are staggered. The pseudarthroses of the humerus are most often of the non-reactive, atrophic type. This type of pseudarthrosis appears more often in the humerus than anywhere else. Where the fracture goes on to form a reactive pseudarthrosis, it is often associated with an exceptional degree of osteoporosis of the proximal fragment. We feel that in order to obtain rapid consolidation it is necessary to decorticate as well as to bone graft the pseudarthrosis. *Note:* In the treatment of a pseudarthrosis of the humerus associated with a stiff elbow or where the fixation has not been perfect, the arm should be immobilized in a double plaster splint for 4–6 weeks to protect the internal fixation and prevent the screws from being pulled out.

c For pseudarthrosis of the radius or ulna we use a six-hole tension band plate. For the ulna, we generally prefer the narrow plate. For the radius we prefer the semi-tubular plate or the DCP designed to be used with the 3.5-mm cortex screws.

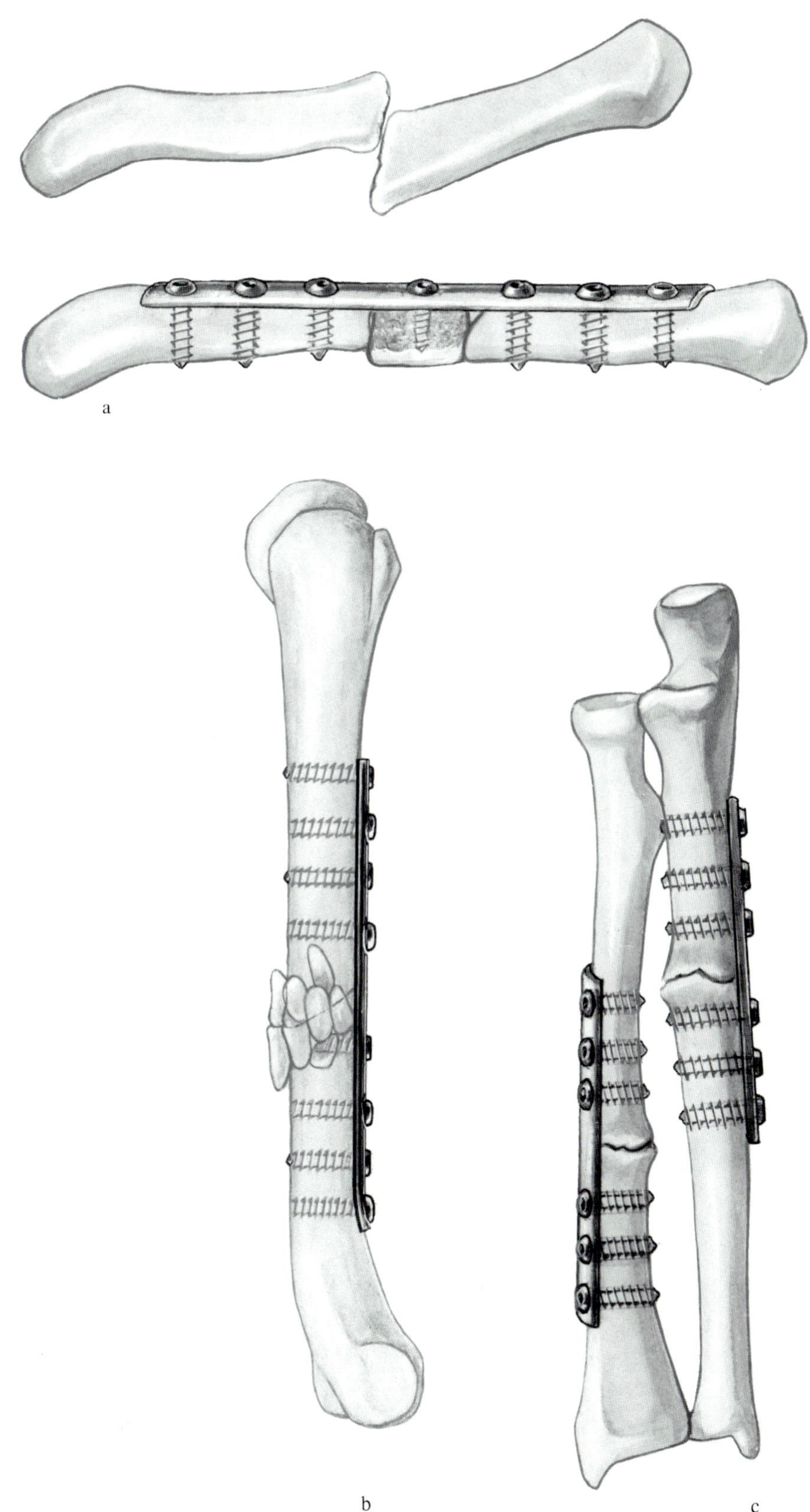
a
b
c

Metaphyseal Pseudarthroses

In order to achieve a stable internal fixation of metaphyseal pseudarthroses, we use the T-plates, angled plates, or even two plates.

Fig. 298 *Metaphyseal pseudarthroses in the arm.*

a Pseudarthrosis through the surgical neck of the humerus fixed with a T-plate.

b An oblique supracondylar pseudarthrosis of the humerus. Note the lag screw across the pseudarthrosis.

c This Y supracondylar pseudarthrosis of the humerus was fixed with two narrow plates. Note also the bone graft. The ulnar nerve was transposed to the front.

d Note the use of the small T-plate in the treatment of the metaphyseal pseudarthrosis of a proximal phalanx.

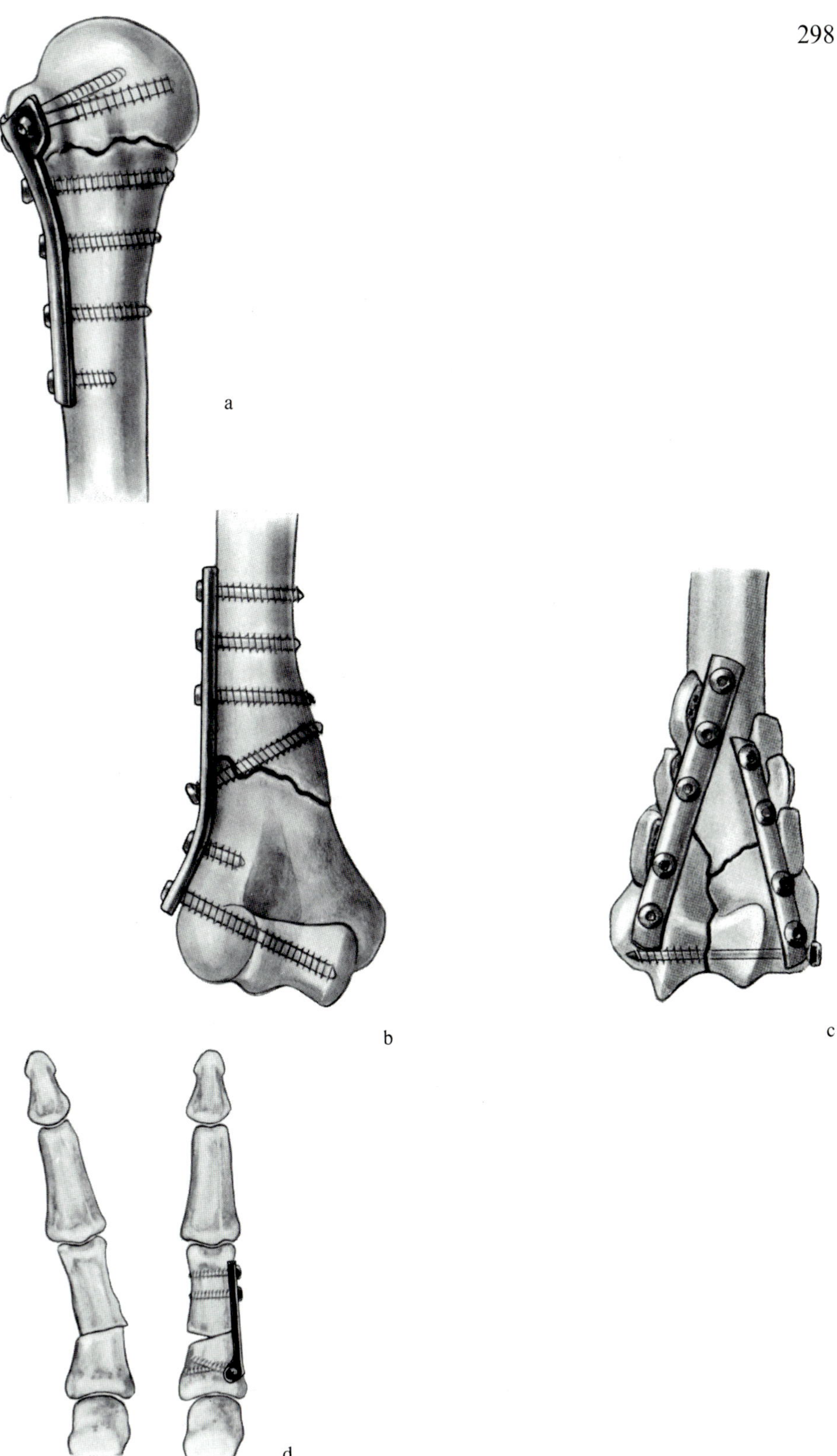
a
b
c
d

2.1.2 Pseudarthroses of the Leg

Diaphyseal Pseudarthroses Without Significant Axial Deformity

These are an ideal indication for intramedullary nailing. The sclerotic seal of the medullary canal is first broached either with a hand reamer (Fig. 300) or with a drill and a long flexible drill bit which is passed through a short curved 150-mm intramedullary nail which serves as a drill guide. Very rarely in a hypertrophic pseudarthrosis of the femur it may be impossible to broach the sclerotic seal as described above. It is then necessary to osteotomize through the pseudarthrosis. Once this is done, one either bends the bone through the pseudarthrosis or simply displaces the fragments the full width of the shaft. The seal is perforated by first drilling through it with a 4.5-mm drill bit and then gradually enlarging the hole with the hand reamer until a 8–9-mm hole is made. If a simultaneous lengthening is undertaken, this is then carried out as described on p. 372.

Re-nailing: One should aim for a new diameter 2–3 mm greater than the previous one. This means a minimum of 15 mm for the femur and 13 mm for the tibia. The new nail must provide the necessary stability, which at times requires a reaming of the medullary canal to 18 mm or more. Furthermore the new nail must extend at least 1 cm further distally than the bony seal which formed around the tip of the old nail.

Exception: If the fracture is too proximal, then the new nail, like the old one, will fail to achieve rotational stability even if one attempts to obtain rotational stability by inserting a screw through the proximal hole of the nail. In these cases, the old nail should be removed and extensive decortication carried out, a bone graft inserted inside the medullary canal and around the pseudarthrosis and the pseudarthrosis fixed with a laterally applied plate.

Fig. 299 *The pseudarthroses of the middle third of the femur and tibia are the ideal indications for an intramedullary nailing*

Fig. 300 *Broaching of the sealed medullary canal*

a With the 6-mm-thick hand reamer.

b With a drill and a drill bit with a long flexible shaft.

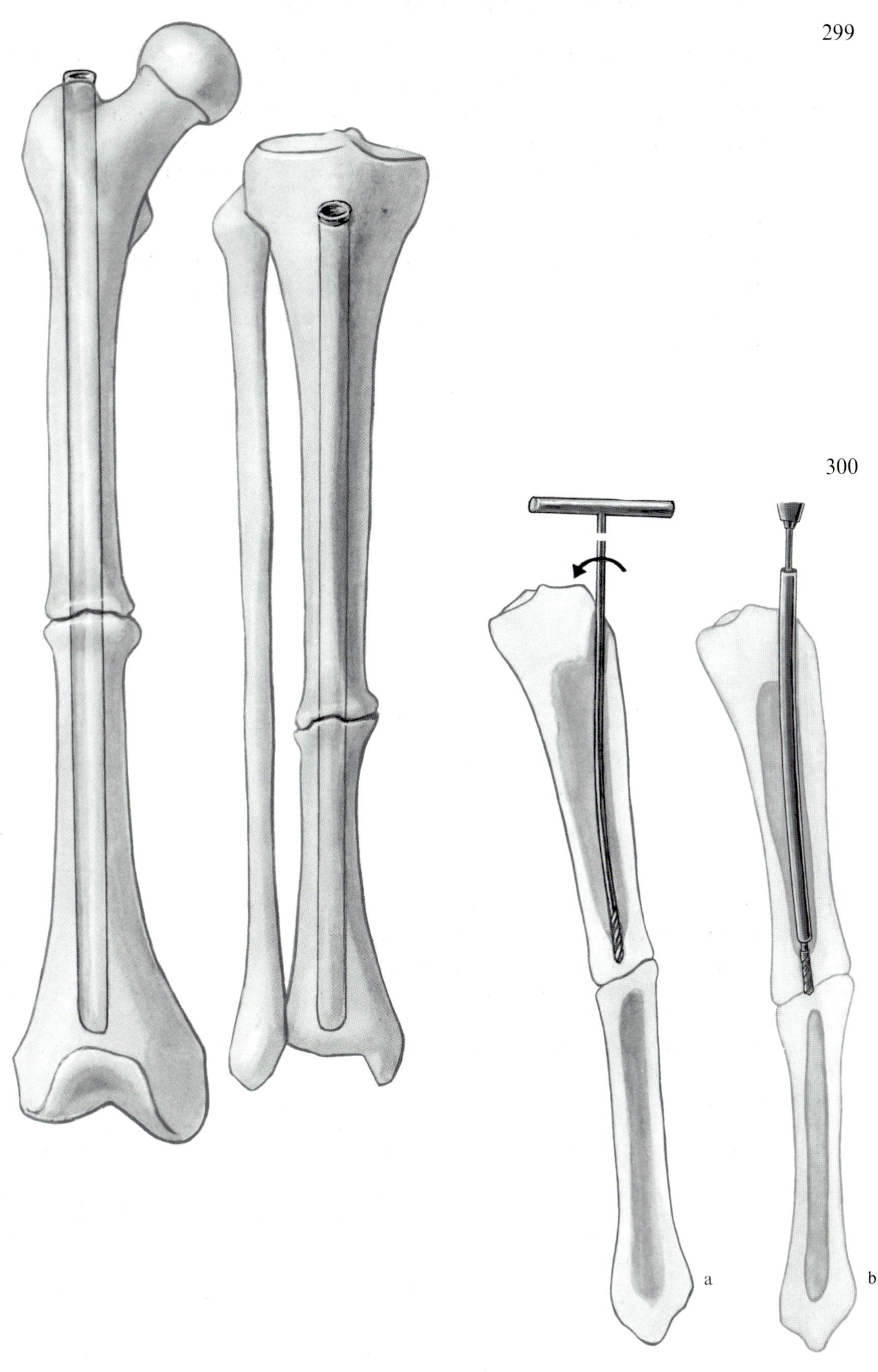
299
300
a
b

Diaphyseal Pseudarthroses with Significant Deformity

Treatment is with the tension band plate. In order to permit a tension band plate to exert its full effect, it must be placed on the convex or tension side of the bone. This means that in a varus deformity it comes to lie laterally, in a valgus deformity, medially, and in a posterior angulation, posteriorly and in an anterior angulation, anteriorly. Excessive callus should be chiselled away to prepare a bed for the plate.

Fibular Osteotomy. An osteotomy of the fibula is necessary only when there is a significant concomitant deformity of the fibula. In a varus deformity of the tibia, despite the application of compression laterally, the correction of the deformity results in an actual lengthening of the tibia, which would result in a gaping of a fibular osteotomy if it were performed.

Fig. 301 *The tension band plate must always be placed on the tension or convex side of the deformity.*

a In the more common varus deformity, the plate is placed on the convex or lateral side.

b The valgus deformity is less common. Here the plate is applied on the medial side. If the fibula is longer, no osteotomy is required. If the fibula has healed with shortening, it must be osteotomized prior to the correction of the tibia.

c In a posterior bowing of the tibia (recurvatum) the plate is placed on the posterior aspect (for approach see Fig. 219c3).

d Congenital pseudarthrosis of the tibia with anterior bowing. According to M.E. MÜLLER, the semi-tubular plate is placed on the anterior crest of the tibia. This follows the first step which is the insertion of a cortical bypass graft posteriorly. The cortical bypass graft is taken from the opposite uninvolved limb.

Fig. 302 *The use of the tension band plate in a significant varus deformity of the tibia.* The plate is first fixed with four screws, usually to the distal fragment. A tension device is then fixed to the proximal fragment with a long cortex screw which is then gradually tightened.
With the development of higher and higher tension on the lateral side, the varus deformity is gradually corrected and the lateral portion of the pseudarthrosis comes under quite marked axial compression.
A fibular osteotomy becomes necessary only if there is a considerable deformity of the fibula.

Fig. 303 *Fibular osteotomy.* This model illustrates how the correction of the varus of the tibia results in an overall lengthening of the bone despite the shortening which takes place as a result of the axial compression.

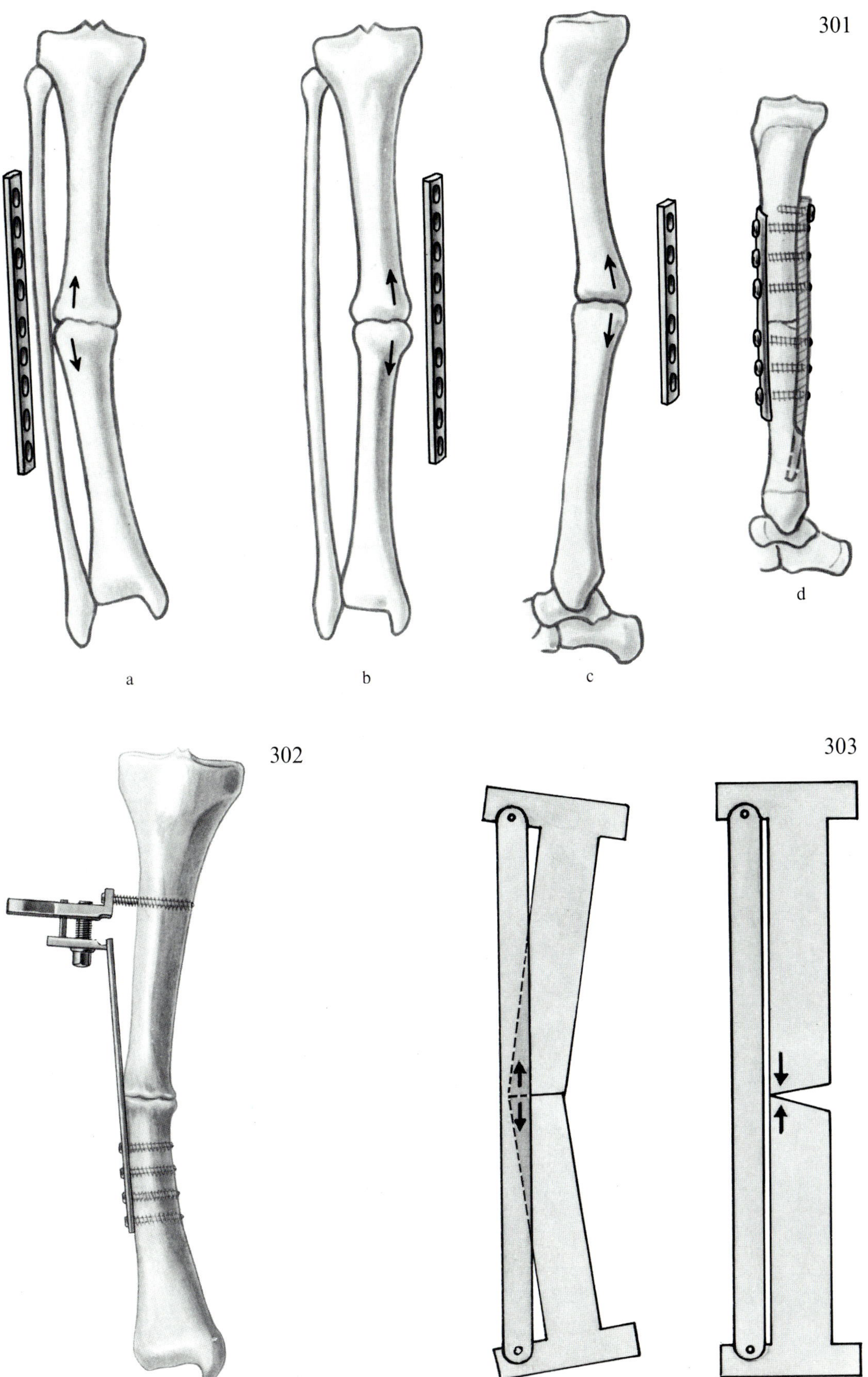
301
a
b
c
d
302
303

Metaphyseal Pseudarthroses in the Lower Extremity

Angled plates and double plates are used in the treatment of metaphyseal pseudarthroses in the lower extremity. An exception is the pseudarthrosis of the malleolus (see p. 360 for pseudarthrosis of the femoral neck).

Fig. 304 *Metaphyseal pseudarthrosis in the leg.*

a Intertrochanteric pseudarthrosis treated with the hook plate.

b Supracondylar pseudarthroses are treated with condylar plates. If a defect is present medially, this is grafted and decortication is carried out.

c A double pseudarthrosis – one on either side of the knee joint. If the joint is destroyed, we recommend two long compression plates at right angles to one another. These not only stabilize the pseudarthroses but enable us at the same time to carry out an arthrodesis of the knee joint in a physiological position (see. p. 390).

d A pseudarthrosis of the proximal tibia stabilized with a T plate laterally and a semi-tubular plate anteriorly.

e In pseudarthrosis of the medial malleolus, it is difficult to achieve any interfragmental compression with a lag screw because of the frequent presence of severe osteoporosis. We therefore insert a graft into a slot cut at right angles to the pseudarthrosis. We adopt this principle of slotting a graft across the pseudarthrosis in all cases where one fragment is very small.

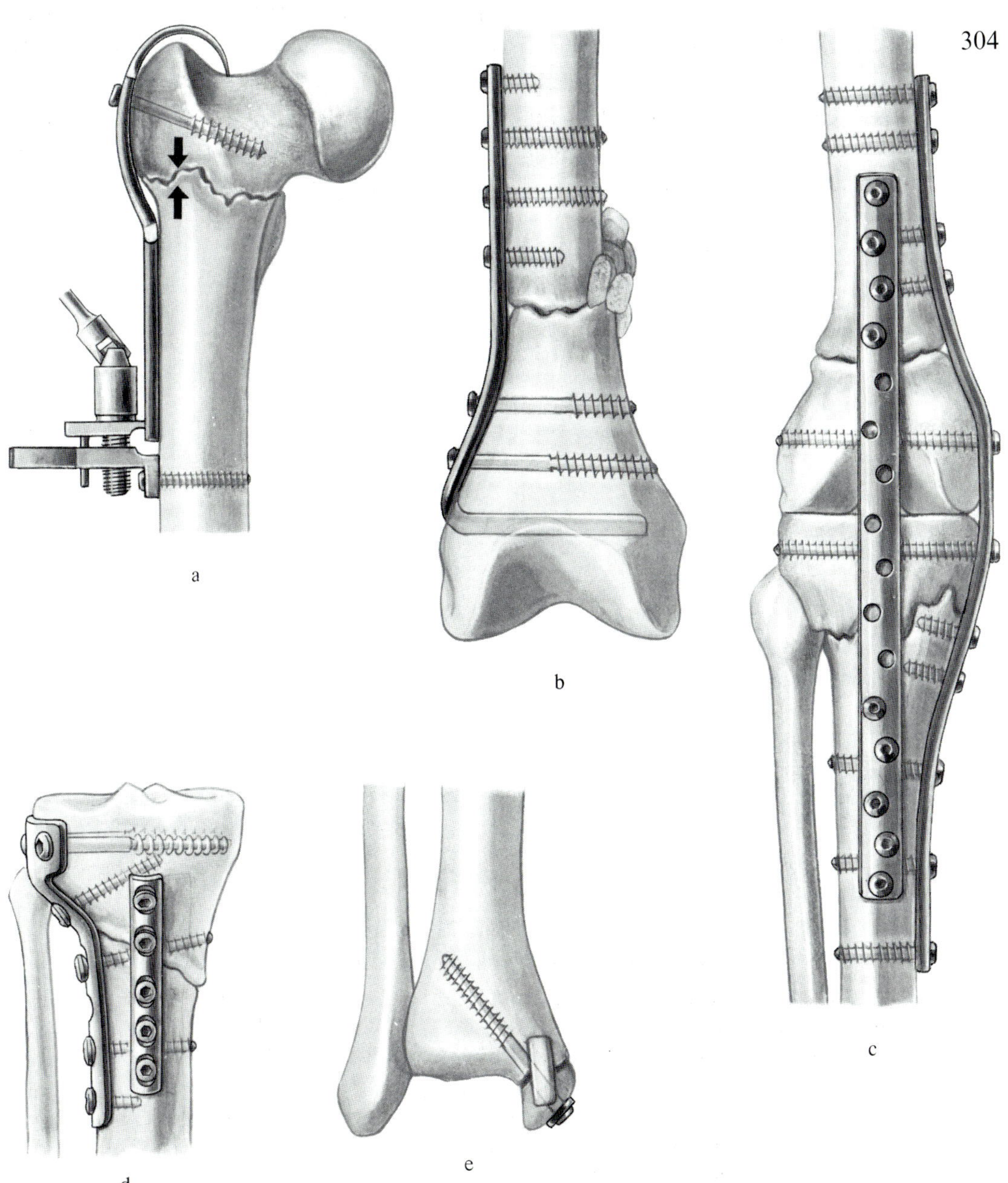
304
a
b
c
d
e

2.2 Previously Infected But Now Dry, Closed Pseudarthroses and Pseudarthroses With Bone Loss

In an area which was previously infected, even after many years of quiescence a new surgical procedure carries with it a great danger of the infection flaring up. We must determine whether there is a broad area of contact between the fragments or whether there is bone loss and how large the defect is that has to be reconstructed.

The principle in the treatment of these pseudarthroses is the combination of stable internal fixation with decortication of the fragments. If there is a defect, a bone graft is required.

Fig. 305 *Previously infected pseudarthroses with a broad area of contact between the ends.*

a In a pseudarthrosis of the tibia with varus deformity and a broad area of contact between the ends, fixation is achieved with a long 11–16-hole plate applied on the lateral side. Screws are inserted only proximally and distally, leaving the pseudarthroses and the surrounding tissue undisturbed.

b Whenever there is a small defect which increases the risk of a flare-up of sepsis, it is better to osteotomize the fibula and fix the pseudarthrosis with the external fixators. Note that the fibular osteotomy is carried out either proximally or distally as in this case, but never at the level of the pseudarthrosis. The external fixation could be made more rigid by triangulating it (see Fig. 263).

Fig. 306 *Pseudarthroses with bone loss.*

a In the humerus it suffices to carry out extensive decortication and a bone graft. A subsequent tension band plating would become necessary if the pseudarthroses were to persist.

b This large defect in the ulna is bridged with a semi-tubular plate which was fixed to the proximal and distal fragment while the hand was maintained in full supination. The trough alongside the plate is filled with small cortico-cancellous bone fragments. Subsequently the arm must be protected in plaster, since a single plate in the presence of a large defect does not provide enough rotational stability.

c A pseudarthrosis of the femur with bone loss is first stabilized using two special long semi-tubular plates which are applied at right angles to one another. These must be left in, even if the infection flares up. The bone graft usually results in bridging of the pseudarthrosis. The same principle of treatment applies to the tibia, particularly in those cases with an intact fibula.

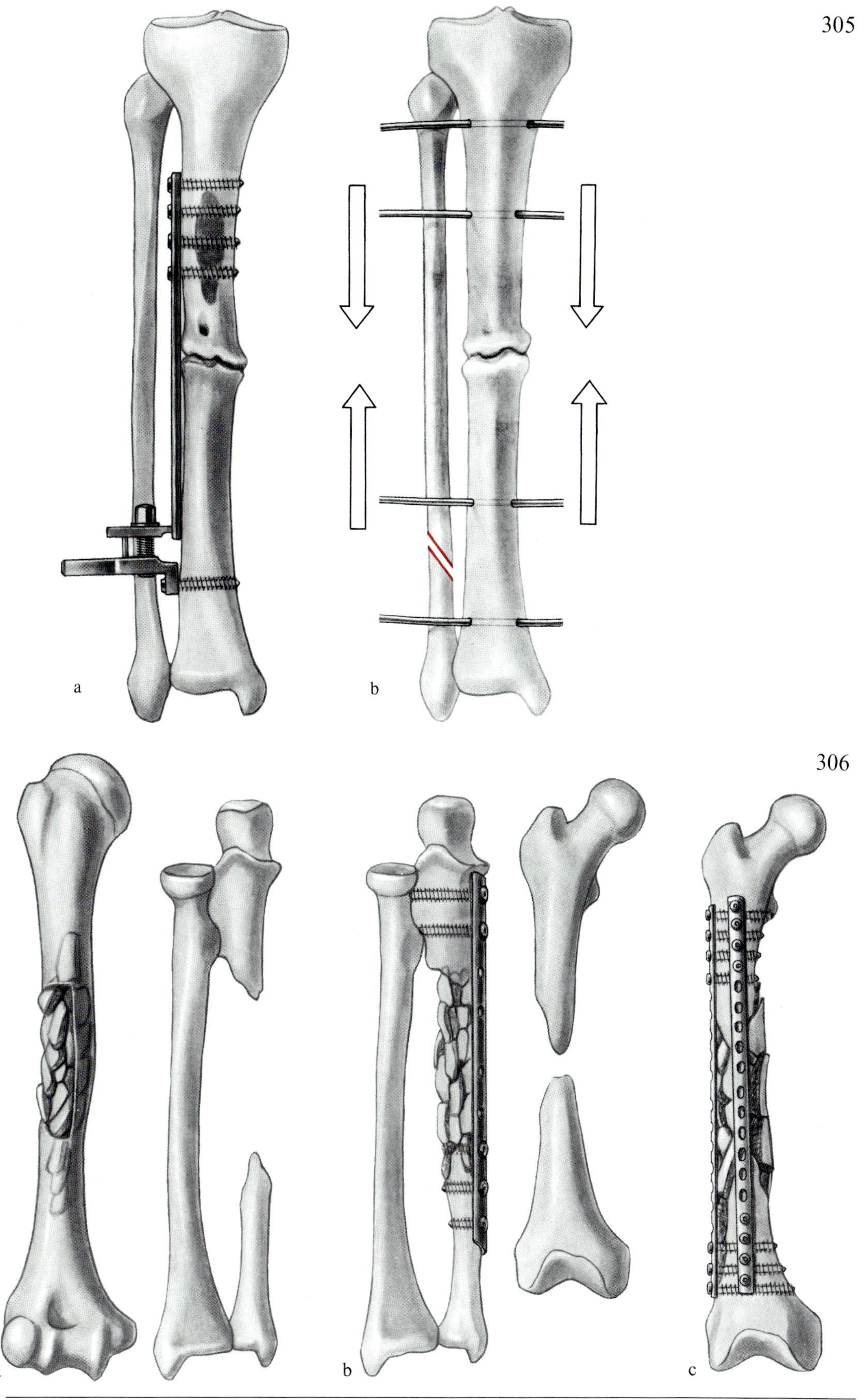
305
a
b
306
a
b
c

2.3 Infected Pseudarthroses

Principle of Treatment: The implant must be left undisturbed as long as it is providing stability. This applies to fresh fractures as well as to pseudarthroses. Where instability exists, stability must be re-established by a revision of the internal fixation. For the principles of treatment see p. 336.

Infected Pseudarthroses of the Femur

The problem with the infected pseudarthroses of the femur is their tendency to shortening. Furthermore the femur is a single bone, and unlike the tibia there is no second bone which can be used to enhance stability. Therefore an existing internal fixation must be left in as long as it provides stability. If there is instability, after decortication and medial bone grafting a new internal fixation must be carried out. We appreciate that the presence of an implant will perpetuate the infection. In most instances, however, the pseudarthrosis will consolidate before the implant has to be removed. The leg-lengthening apparatus is a very useful device when treating infected pseudarthroses of the femur.

Fig. 307 *Infected pseudarthroses of the femur.*

a Note the large resorption cavity in which the distal end of the nail is moving about. If the nail has good purchase above stability can be obtained by simply advancing the nail a further 2–3 cm through the distal end of the resorption cavity. If a small local defect is present, this must be filled with cancellous bone.

b Subtrochanteric pseudarthrosis. We strongly recommend the use of the condylar plate with axial compression as a means of achieving stability. This must be combined with posteromedial bone grafting. A fistula will persist till the plate is removed. In a number of months, however, the pseudarthrosis will be sufficiently consolidated to permit removal of the implant.

c In the middle of the shaft it is possible to use the leg-lengthening device of WAGNER (Fig. 325) as a means of applying axial compression across the pseudarthroses. Note the additional medial bone graft.

d A pseudarthrosis of the distal femur can be stabilized by means of an external fixator. This is accomplished by inserting two Steinmann pins above and two below the pseudarthroses and using the external fixator in such a way that the pseudarthrosis is brought under axial compression.

e An infected pseudarthrosis of the distal femur with a stiff knee joint can at times be salvaged by grafting laterally and then immobilizing the leg in a cast in 60° of valgus through the pseudarthrosis to permit the formation of a lateral bony bridge. Six to seven weeks later, the deformity can be corrected manually. The lateral bony bridge then acts as a tension band and the corrective force results in a strong axial compressive force which will aid in the rapid consolidation of the pseudarthrosis.

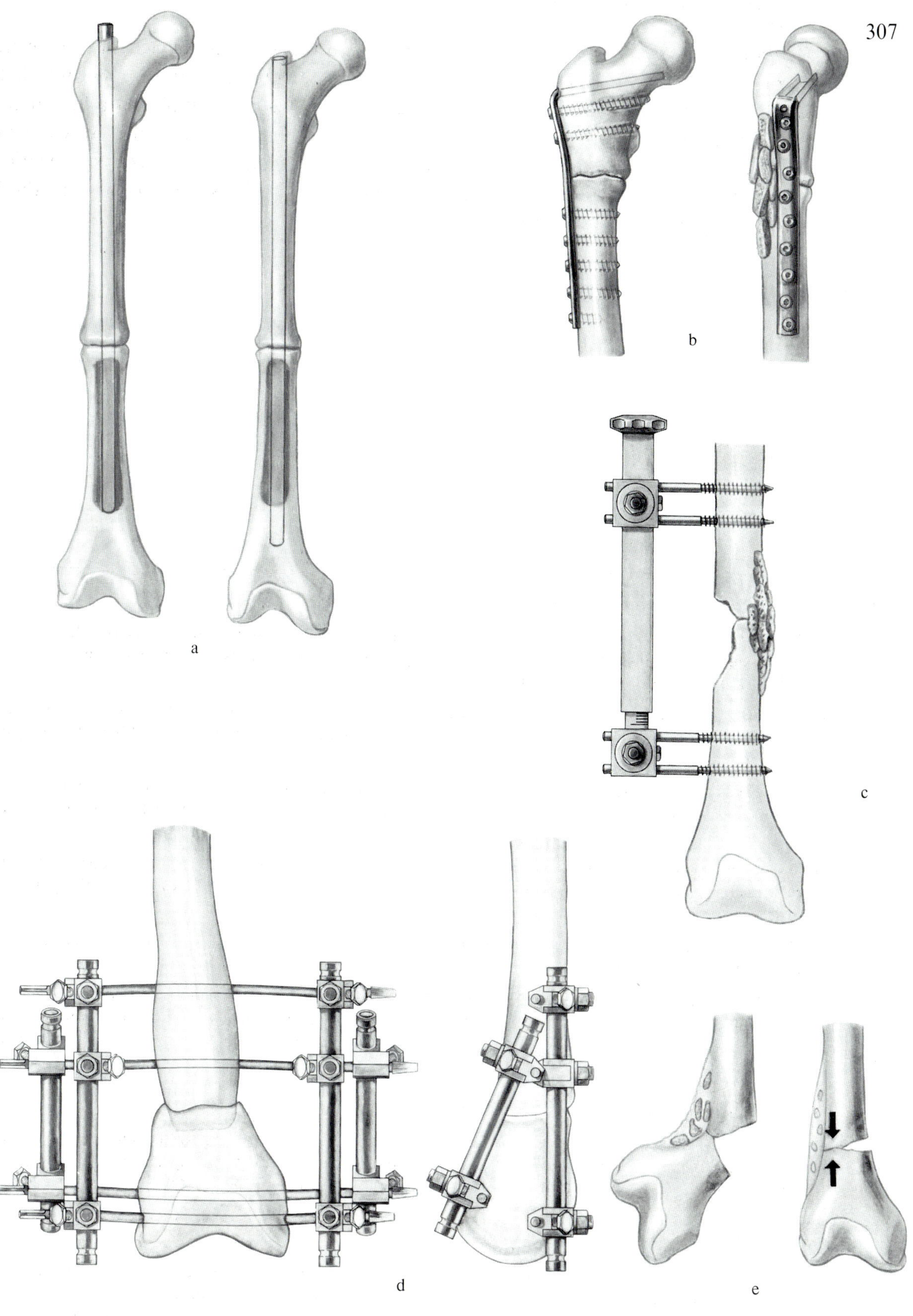
307
a
b
c
d
e

Infected Pseudarthroses of the Tibia

Most infected pseudarthroses of the tibia will go on to healing once the sequestra are removed, the bone is decorticated, and the fragments are stably immobilized. If defects are present then, in addition, a pure cancellous bone graft must be carried out. In the presence of a defect, stability can often be achieved only by carrying out a tibiofibular synostosis. The degree of this will depend on the size, the location and the extent of the defect. This tibiofibular synostosis is an excellent method, particularly if the ankle joint is already stiffened.

Fig. 308 *Infected pseudarthrosis of the tibia.*

a If the intramedullary nail is providing a reasonable degree of fixation, just carry out a posterior decortication and a cancellous graft through a dorsomedial approach. If instability exists, revise the intramedullary nailing. Ream further and insert a larger size nail.

b Where there is good contact between fragments, all that is necessary is a sequestrectomy, a dorsolateral decortication and fixation of the pseudarthrosis under axial compression with the external fixator. Triangulation can increase the degree of stability (Fig. 263).

c This synostosis was carried out in the presence of an intact fibula. There was fairly good bone contact and a medial fistula. The approach was lateral. The tibia and fibula were decorticated and an extensive bone graft was carried out.

d Tibiofibular synostosis in the presence of infection and extensive bone loss: carry out a bone graft as far proximally as possible, but take care not to damage the anterior tibial artery. Carry out the graft also distally to secure union above and below.

e Distal pseudarthrosis with defect: only a tibiofibular synostosis can provide adequate immobilization of the short distal fragment.

f Pseudarthrosis of both the tibia and fibula at the same level with poor contact between the fragments: First immobilize the pseudarthrosis by means of four Steinmann pins and the external fixator. As soon as the infection has abated, carry out a decortication and cancellous bone graft through a posterolateral approach.

Fig. 309 *Treatment of osteomyelitis after consolidation of the pseudarthrosis.*

a Saucerization of the femur. Carry out the saucerization in such a way that the medullary canal is laid widely open, that only good bleeding bone is left behind and that good healthy muscle is laid into the defect. After the saucerization, suction irrigation drainage is carried out for 5–8 days.

b After a radical sequestrectomy the wound is left open but its edges are held approximated with sterile strips which are changed and tightened every day.

c A second sequestrectomy and bone grafting is frequently necessary. This time the wound is closed primarily with the aid of a few retention sutures (Dermalon 0) which must all be removed within 48 hours. The skin tension by then is considerably reduced and the skin edges which were closed by means of the Donati-Allgöwer suture escape damage.

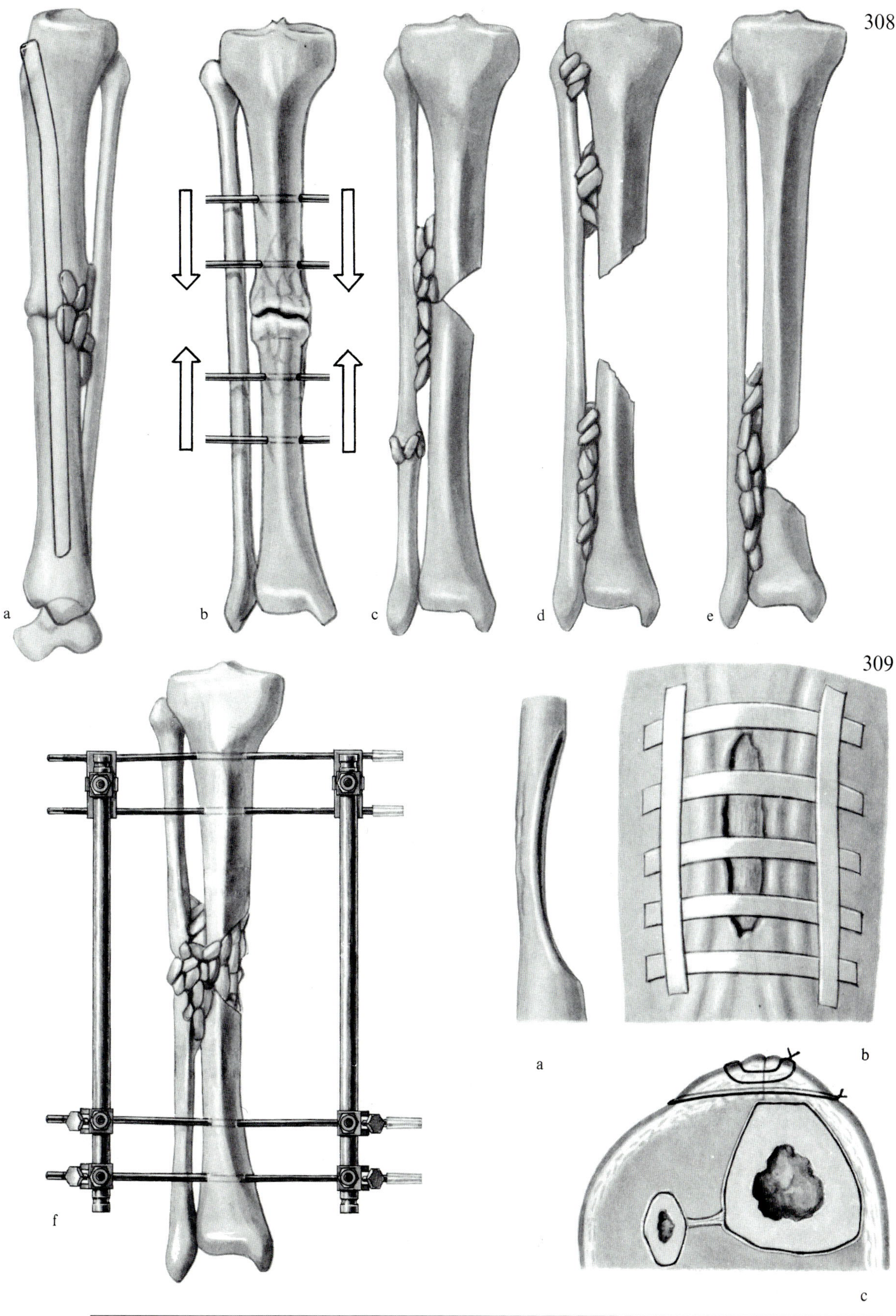
308
a
b
c
d
e
f
309
a
b
c

3 Osteotomies

An osteotomy is the transection of a bone carried out in order to correct axial malalignment or to shorten or lengthen bone. After osteotomy, the fragments must be immobilized by means of stable internal fixation to ensure that no loss of correction occurs while the bone is healing. Most commonly the fragments are immobilized under axial compression either with a plate or the external fixators. Intramedullary nailing is rarely used. The advantages of a stable internal fixation after an osteotomy are great. The limb need not be immobilized in plaster and the patient can begin with early mobilization of the joints. Most patients can be up and about a few days after surgery and can commence with partial weight bearing.

Most osteotomies should be carried out through metaphyseal bone. Union of compressed cancellous bone is rapid and presents few if any biological problems. On the other hand, diaphyseal cortical bone after transsection requires a long time for consolidation. If an osteotomy is carried out through a diaphysis thickened by callus, extensive decortication must always be carried out. After a lengthening osteotomy the resultant defect must always be filled with bone graft.

An osteotomy permits six corrections: valgus-varus, anterior or posterior angulation, internal or external rotation, lengthening or shortening, displacement medially or laterally, or displacement anteriorly or posteriorly.

Prior to carrying out a corrective osteotomy, one must carefully determine all the components of the deformity and what has to be carried out in order to correct them. A pre-operative drawing is essential. It is made on the basis of radiographs of the opposite uninvolved limb.

The beginner always finds it most difficult to diagnose rotational deformities. Here the radiographic projection of Dunn-Rippstein-Müller is most helpful in determining rotational malalignment of the femur. Rotational malalignments of the tibia are diagnosed clinically. As reference points we take the position of the patella while standing and the axis of the foot while sitting. Surgeons frequently neglect lateral or medial displacement, because they do not realize its importance. A mere 5-mm lateral displacement of the tibia can reduce the loading of the talus by more than 50%. Prior to any osteotomy full-length X-rays should be obtained of the limb in two planes.

After the skin incision and exposure of bone, two Kirschner wires at 90° to one another must be introduced into the bone above and below the site of the planned osteotomy. They serve as reference points from which we can determine, at any time, the degree of correction obtained. One of the Kirschner wires above and one below the site of the osteotomy are usually parallel to each other. The other two will indicate the necessary angle of correction in the major plane of the deformity. The anterior Kirschner wires are parallel because there is no anterior or posterior angulation to be corrected. The Kirschner wires inserted from the side on the other hand indicate the desired angle of correction in the frontal plane.

3.1 Osteotomies in the Arm

The osteotomies are carried out with an oscillating saw. The blade of the saw must be cooled with saline while cutting.

All diaphyseal osteotomies are combined with an extensive decortication which aids in their subsequent consolidation.

Fig. 310 *Subcapital osteotomies of the humerus.* This corrective osteotomy became necessary because a subcapital fracture healed in abduction with painful restriction of adduction. Note its fixation with the T plate.

Fig. 311 This deformity in the region of the elbow is usually the result of a malreduced supracondylar fracture in a teenager. The cubitus valgus should be corrected operatively before ulnar palsy develops.

The approach is posterior (Fig. 123). The ulnar nerve is transposed to the front and the osteotomy is fixed with a narrow, carefully contoured plate. Great care must be taken not to cross the olecranon fossa with a screw because this would block extension of the elbow.

Fig. 312 *Shortening osteotomy of the radius for Kienboeck's disease.* Osteochondritis of the lunate has been successfully treated by either lengthening of the ulna or shortening of the radius. It appears that an ulnar plus, i.e. a protruding ulna, results in decompression of the lunate. Whatever the cause of the improvement, the subjective and objective post-operative improvement is always most surprising and gratifying. We prefer the shortening osteotomy of the radius because of the usually slow consolidation of the lengthening osteotomy of the ulna. We excise a 3–4-mm segment of the radius. The osteotomy is oblique to permit its fixation with a lag screw. After the lag screw fixation, we use the narrow five-hole DCP applied dorsally as a neutralization plate.

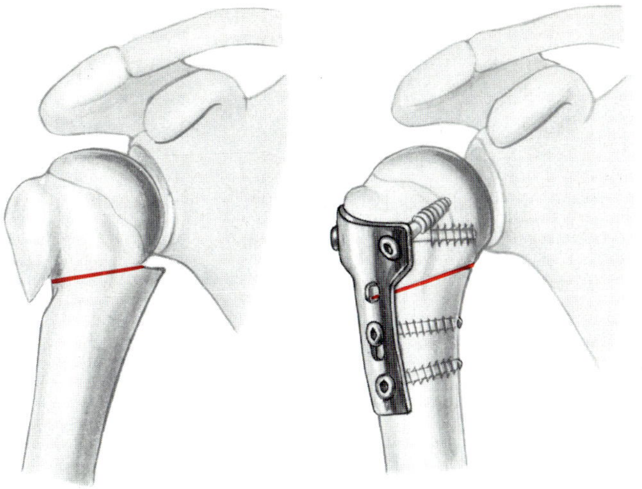

311

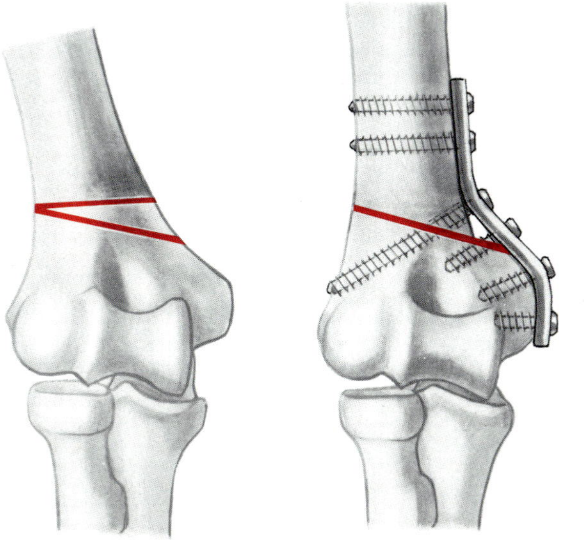

312

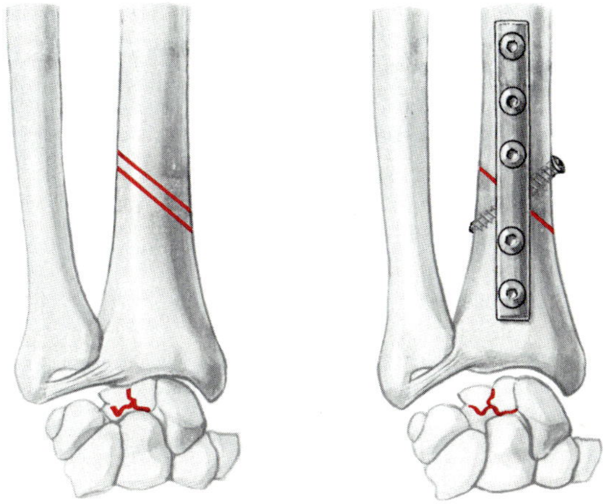

3.2 Osteotomies in the Leg

3.2.1 Intertrochanteric Osteotomies

The repositioning osteotomy is the commonest of the intertrochanteric osteotomies carried out for post-traumatic malalignment. Occasionally it is carried out in the treatment of a fresh subcapital fracture (Fig. 178 b) but most commonly in the treatment of a subcapital pseudarthrosis with a viable head.

The Repositioning Osteotomy in the Treatment of a Pseudarthrosis of the Femoral Neck

The repositioning osteotomy of PAUWELS is indicated in younger patients. We feel that patients over 70, or patients whose health is poor, should be treated with a total hip prosthesis. The treatment of a pseudarthrosis of the femoral neck in patients over 30 years of age is worthwhile only if the head is still alive. We feel that in children and patients under 30, the repositioning osteotomy should be carried out even if there is evidence of avascular necrosis of the femoral head. Avascular necrosis of the femoral head is less likely if on the head side of the pseudarthrosis a sclerotic reaction can be recognized, or if after 8 months the femoral head has retained its sphericity without the slightest evidence of collapse.

In order to stabilize the pseudarthrosis the fragments must be repositioned in such a way that the pseudarthrosis comes to lie at right angles to the compressive forces (PAUWELS). Thereafter the pseudarthrosis heals with surprising speed.

The relationship of the head and neck is best determined from two radiographic projections of the femoral neck. For the AP projection, the leg must be internally rotated through an angle of 20°. For the technique of the axial orthograde projection, see the legend to Fig. 314.

Fig. 313 In order to be able to determine the angle of correction or the repositioning angle the direction of the resultant of forces (*R*) must be known. In a patient with a negative Trendelenburg, its direction is chiefly influenced by the direction of the abductors. According to PAUWELS the resultant (*R*) subtends an angle of 16° with the vertical (*V*) and 25° with the axis of the femur (*A*). The axis of the femur (*A*) subtends an angle of 9° with the vertical (*V*). A pseudarthrosis will be subjected to compressive forces if it subtends an angle of 30° or less with a vertical (*S*) to the shaft axis. Let us suppose that in our example the pseudarthrosis line subtends an angle of 75° with the vertical (*S*) to the shaft axis, then the angle subtended between the pseudarthrosis line and the vertical dropped to the resultant (*R*) would represent the angle of correction, which in this example would be 50°.

Fig. 314 *Axial orthograde projection.* The patient is positioned supine or if a contracture is present in slight rotation. The hips and knees are bent to 90°. If the neck shaft angle is 115°, then the leg is abducted 25° from the vertical.

a As seen from above.

b As seen from the side.

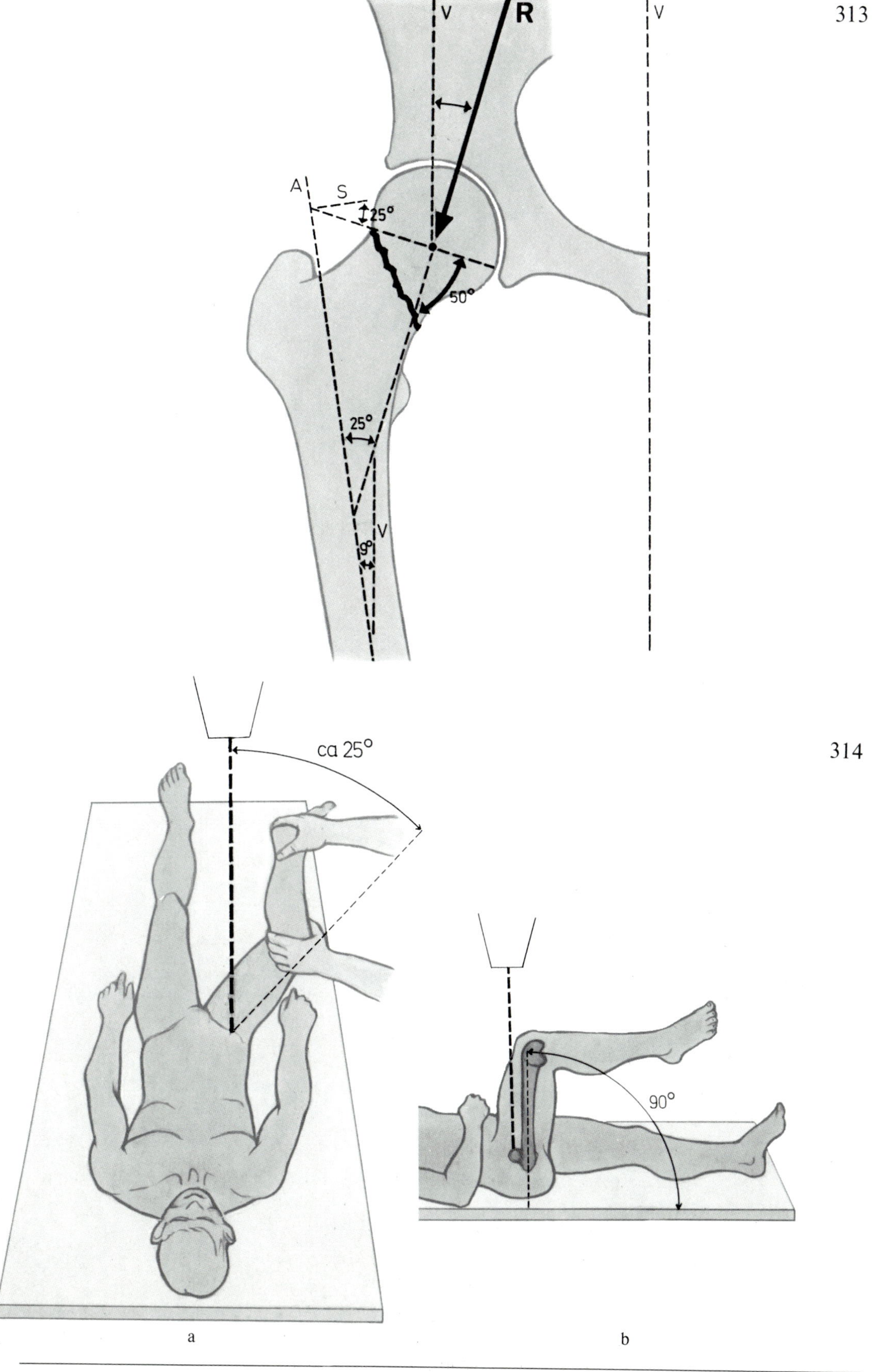
V
R
V
313
A
S
25°
50°
25°
V
9°
ca 25°
314
90°
a
b

The Repositioning Osteotomy in the Treatment of a Non-union of the Femoral Neck

Preparation

Fig. 315 *Pre-operative drawing.* On a copy of the pre-operative orthograde X-ray draw in the following:

a The femoral axis and the perpendicular to the femoral axis.

b Draw a line through the pseudarthrosis to indicate its plane and draw in the angle between this line and a perpendicular to the shaft axis (in our example the angle is 75°). Determine the repositioning angle and the angle of correction (here 50°).

c Draw a perpendicular to the shaft axis so that it will touch the calcar medially.

d Determine the amount of caudal slip of the femoral head. This can be measured either cranially or caudally on the femoral neck. The distance (*d*) is then drawn in on the line (*c*) measuring laterally from the point where (*c*) strikes the calcar. This becomes (d′).

e From this point draw in an angle of 30° below (*c*) as line (*e*) (120° plate less 90° = 30°).

f From the same point but above (*c*) draw in the line (*f*), which here subtends 20° with (*c*). It corresponds to the repositioning angle or angle of correction less 30° (repositioning angle of 50° = 30° + 20°).

g Draw in the guide wire (*g*) as far cranially as possible and parallel to (*f*).

h Draw in the seating chisel, parallel to (*f*) and to (*g*). The tip of the seating chisel should extend as far as possible into the caudal half of the femoral head. Make certain that you retain a bridge of bone of at least 15 mm between the seating chisel and the line (*f*).

i Calculate the angle between the seating chisel, the femoral shaft axis. (60° plus repositioning angle = 110°). (*Note* the now recognizable wedge 50° with its base laterally.)

k Check your drawing with the template of a 120° plate which is placed over the drawing in such a way that the blade lies parallel to the seating chisel. The angle between the femoral shaft and the plate must correspond to 50° or the repositioning angle.

315

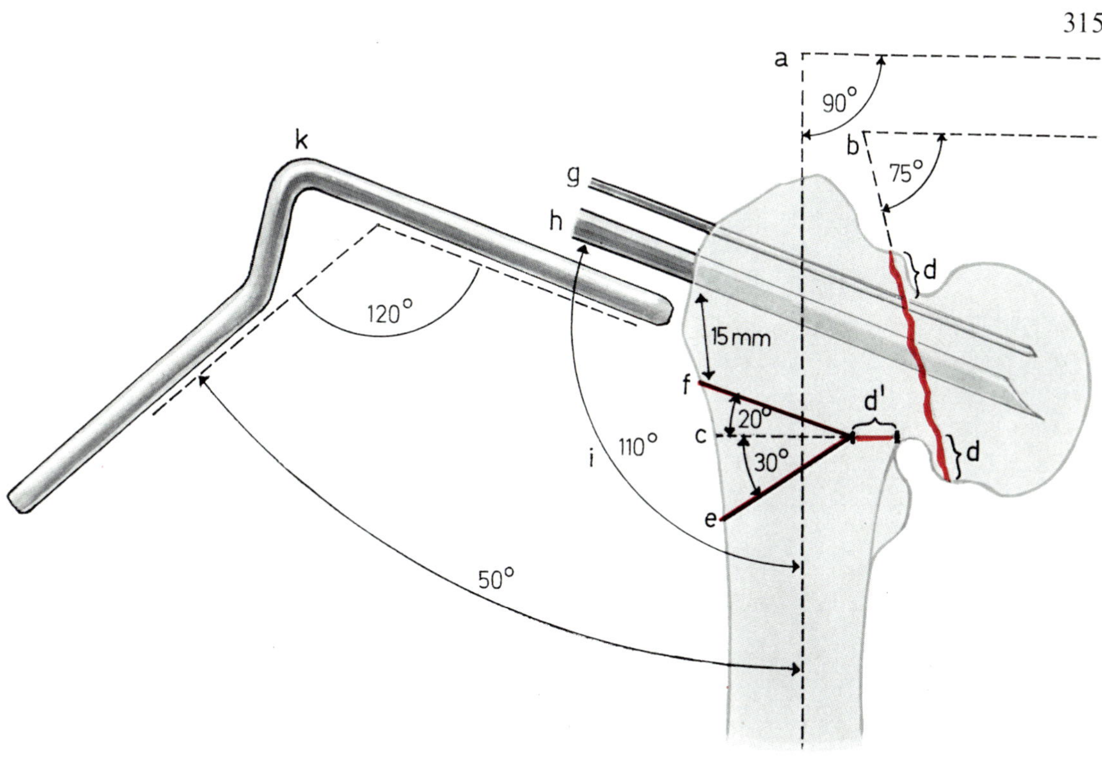

Repositioning Osteotomy in the Treatment of Pseudarthroses of the Neck of the Femur

Fig. 316 *Surgical technique.* The lettering is the same as in Fig. 315.

1 After you have inserted the guide-wire (*g*) and the seating chisel (*h*) with its guide set at 110°, resect the 50° wedge (*l*) with its base laterally.

2 Remove the seating chisel and hammer in the previously selected 120° angled plate. An X-ray may now be taken to check the length of the blade. If all is well proceed now to transsect the femur medially along (d′) at right angles to the femoral shaft.

3 Abduct the leg until the osteotomy is reduced. Begin with fixation of the plate distally with the leg in slightly greater abduction so that when the two screws are tightened the osteotomy surfaces will be brought under inter-fragmental compression.

4 Bone graft medially between calcar and the femoral shaft. Note that the pseudarthrosis is now at a right angle to the resultant of forces (*R*).

5 If the lateral bridge of bone between the blade and osteotomy is less than 15 mm supplement the internal fixation with a tension band wire which is passed around the insertion of the abductors and then through the shaft.

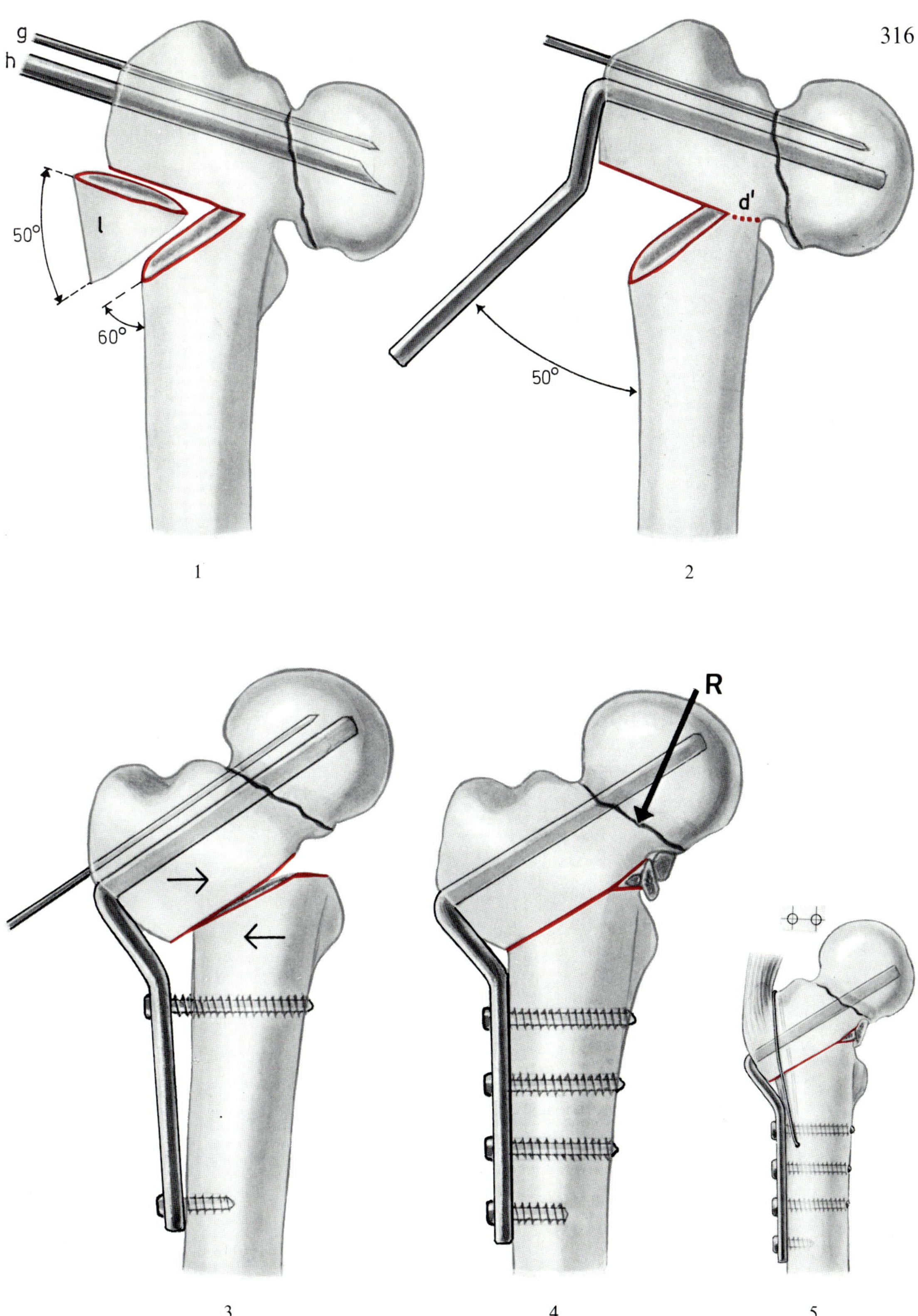
g
h
50°
l
60°
d'
50°
1
2
R
3
4
5

Varus Intertrochanteric Osteotomy

Intertrochanteric osteotomies are frequently carried out to correct a varus or valgus of the femoral neck. The technique of a varus intertrochanteric osteotomy will be illustrated in the following examples.

Fig. 317 *Technique*

1 Start at the tip of the greater trochanter and make a straight lateral incision 20 cm long. Open the fascia in line with the skin incision. Reflect the vastus lateralis forwards and medially. Insert one Hohmann retractor medially around the calcar and one laterally above the neck just medial to the tip of the greater trochanter. Open the joint capsule in line with the neck axis and examine the joint. Insert the Kirschner wire (*a*) which is inserted low along the neck and parallel to the neck. It serves as the guide to the direction of the neck axis or the anteversion. The second Kirschner wire (*b*) is inserted parallel to Kirschner wire (*a*) and parallel to the upper edge of the 60° angled guide (*c*) (Fig. 62). This Kirschner wire is the definitive guide wire for the insertion of the seating chisel. Kirschner wire (*a*) can now be removed.

2 Half a centimetre distal to the site of the planned osteotomy drill a 2-mm hole at right angles to the femoral shaft axis, insert into this hole a Kirschner wire of suitable size.

3 About 2 cm above the site of the planned osteotomy and far to the front, open the cortex of the femur with an osteotome in preparation for the insertion of the seating chisel (*e*). The seating chisel is then inserted parallel to the second Kirschner wire (*b*). The seating chisel is aimed at the centre of the femoral neck and is inserted to a depth of 4.5 cm. The angle which the flap of the seating chisel (*f*) subtends either above or below with the shaft will represent the amount of flexion or extension through the osteotomy.

4 Protect the soft tissues posterior to the osteotomy with a broad Hohmann retractor. With an oscillating saw (*g*) transsect the femur at right angles to its long axis. Use saline to irrigate the blade of the saw to prevent overheating.

5 With the seating chisel as a handle, tilt the proximal fragment into varus. Start in the middle of the osteotomy and cut parallel to the seating chisel. Remove a small wedge (*h*).

6 Remove the seating chisel and insert the blade of the selected right-angle osteotomy plate carefully into the pre-cut channel.

7 Reduce the osteotomy and maintain reduction by clamping the plate to the shaft. Check the rotational alignment of the leg in extension and in flexion. Fix the tension device to the femur and bring the osteotomy under axial compression by tightening the tension device, first with the socket wrench and then with the open spanner. Check the rotational alignment once again. If it is correct fix the plate to the femoral shaft with screws and remove the tension device.

8 The result of the procedure. Note that the last screw is short in order to break up the sudden transition between the rigid plate and the viscoelastic bone.

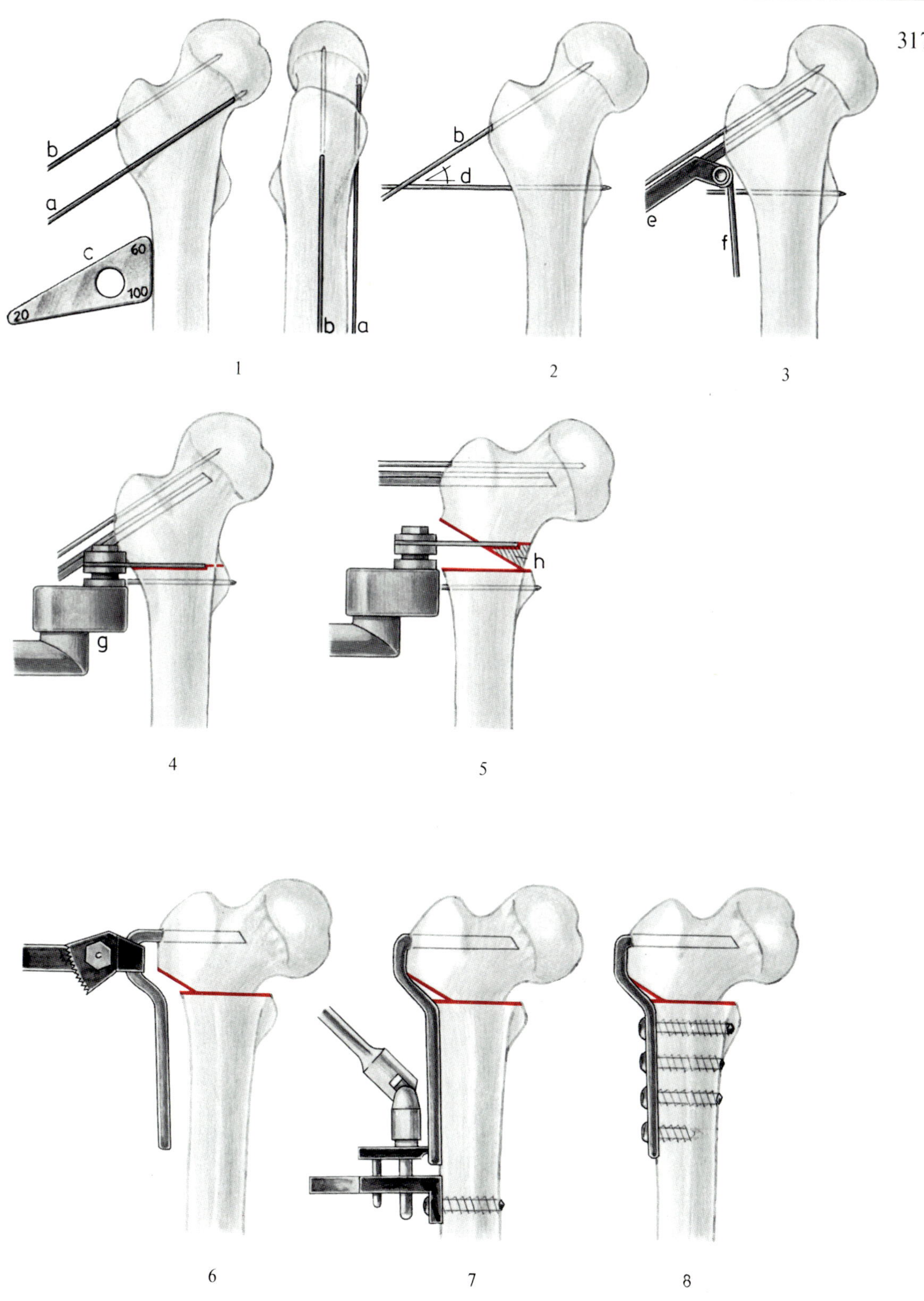

1 2 3

4 5

6 7 8

Fig. 318 *30° varus osteotomy with shortening of the neck and distal transposition of the greater trochanter.*

Fig. 319 *Varus derotation osteotomy in a child.* Note the use of the children's right-angle blade plate. Two further Kirschner wires serve here as guides to the rotational correction.

Fig. 320 *Technique for a 20° varus osteotomy.* Fixation with a right-angle blade plate.

318

319

320

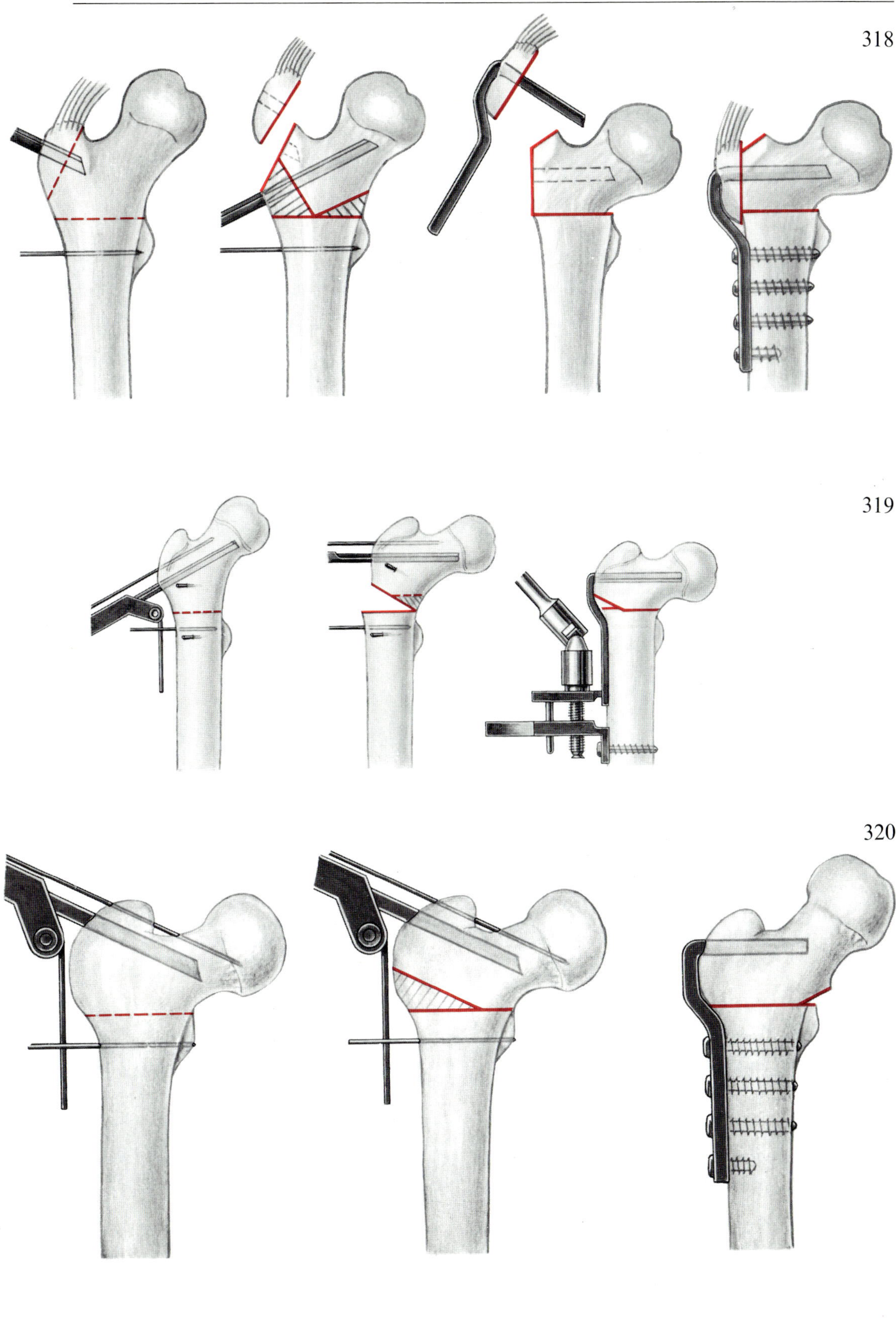

Shortening Osteotomy

Shortening of the femur by 2–3 cm is best carried out through the intertrochanteric area. Shortening of over 3 cm it is best carried out through the middle of the shaft.

Fig. 321 *The technique of an intertrochanteric shortening osteotomy.*

1 Expose the intertrochanteric area and insert three Kirschner wires at right angles to the femoral shaft. The Kirschner wire (*a*) is inserted in the greater trochanter and serves as a guide to the femoral neck axis. The Kirschner wires (*b* and *c*) are inserted more anteriorly at about a 45° angle to the frontal plane. Measure the distance between *b and c*. Note the angle guides (*d*) which assist the insertion of the most proximal Kirschner wire (*a*) at 90° to the shaft axis.

2 Insert the seating chisel at right angles to the shaft. Use the oscillating saw and resect whatever length of bone you wish. Preserve the attachment of the lesser trochanter to the proximal fragment and shape the distal fragment into a slight cone.

3 Remove the seating chisel and insert the right-angle blade. Reduce the osteotomy and rotate the distal fragment until the Kirschner wires *b* and *c* are parallel. Fix the plate to the femur with a bone-holding clamp and insert the tension device.

4 As soon as the fragments are impacted and the desired shortening achieved, fix the plate to the shaft.

Fig. 322 *Shortening osteotomy of over 3 cm through the middle of the femoral shaft.* With an oscillating saw first make a longitudinal cut through the middle of the femur, then make two oblique cuts laterally which represent the amount of shortening desired. Remove this lateral wedge, which consists of one-half of the circumference of the femur.

Insert the reaming guide and ream the medullary canal to a size between 13 and 15 mm and insert the corresponding intramedullary nail just short of the osteotomy.

Complete the osteotomy, taking care not to disturb the soft-tissue attachment of the medial segment. Push out this segment medially and shorten the leg until the osteotomy surfaces are parallel to one another. Insert the nail into the distal fragment. Where the shortening is considerable, further stability may have to be obtained with a four-hole compression plate applied to the posterolateral cortex. In shortening of over 8 cm there is distinct danger of arterial kinking and circulatory obstruction.

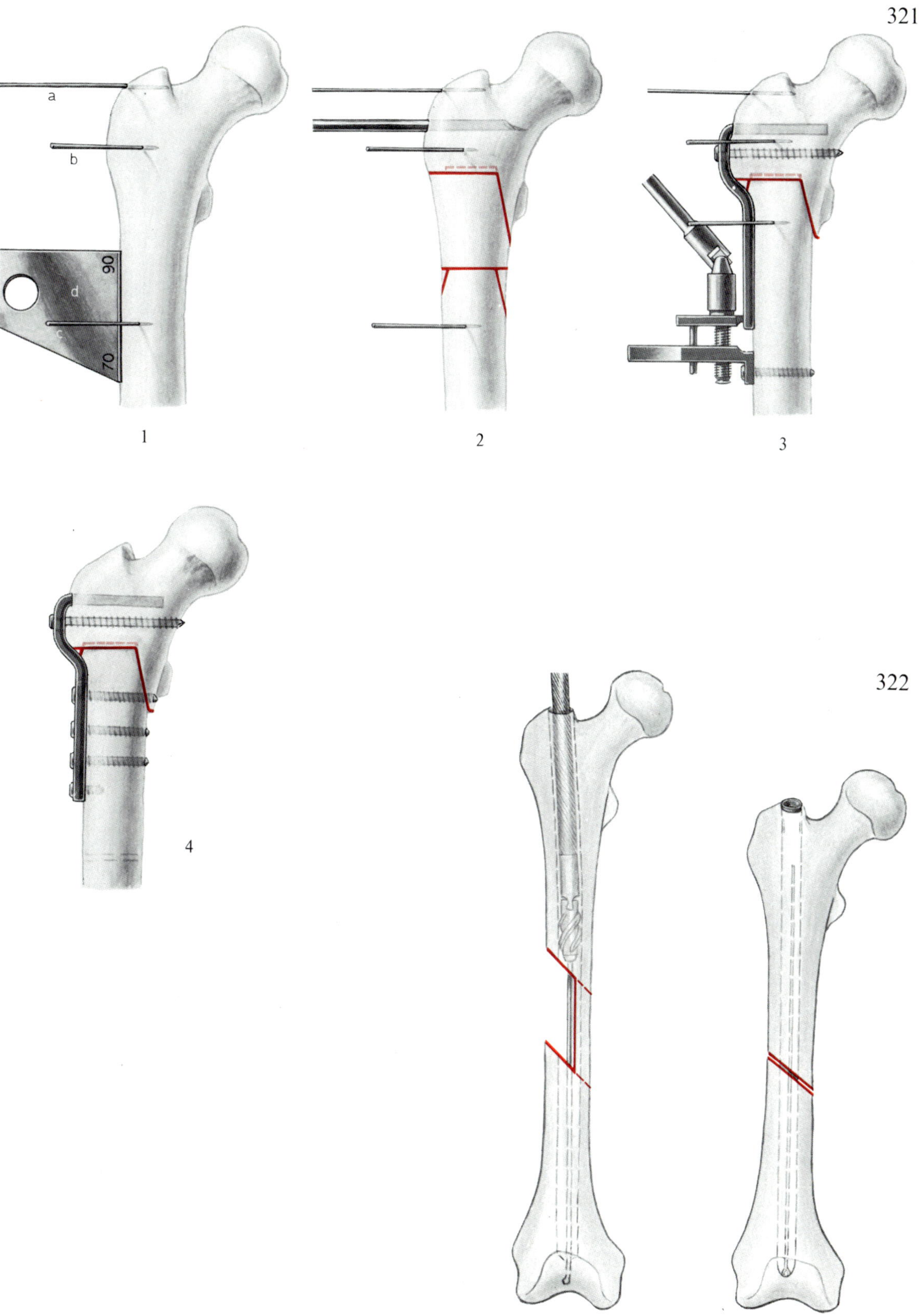
321
a
b
90
d
c
70
1
2
3
4
322

3.2.2 Osteotomies Through the Femoral Shaft

Osteotomies Through Fracture Callus

A post-fracture shortening of the femur can be lengthened 3 cm in one operation without fear of any complications as long as the procedure is combined with an adequate posterior decortication. Shortening in excess of 3 cm must be gradually corrected by continuous distraction which is accomplished by means of the Wagner leg-lengthening apparatus. The lengthening proceeds at the rate of 1.5 mm per day. Once the desired lengthening is achieved an internal fixation should be carried out. Thus the lengthening through continuous distraction requires a two-step procedure.

Fig. 323 *A simple abduction osteotomy of the femur* through the fracture callus. Extensive medial decortication, transverse osteotomy of the femur and if necessary bone grafting of the defect. For fixation in the proximal and distal femur, we recommend the condylar plate. In the middle third, we prefer the intramedullary nail.

Fig. 324 *Lengthening osteotomy of the femur to correct considerable shortening with massive callus formation.*

a A simple transverse osteotomy through the callus (*broken line*) as recommended by CHARNLEY would be the simplest and safest procedure. We prefer, however, the step-wise osteotomy because it allows us to correct as much as possible the concomitant shortening of 2–4 cm. As step number 1, we carry out very extensive posterior decortication over a 10–14-cm-long segment of the femur.

b The reduction of the fragment is carried out with the aid of the distractor.

c Even though the osteotomy surfaces appear to be under high axial compression because of tissue tension, the tension band plate which is applied laterally should still be pre-stressed. If a defect is present medially, it should be copiously bone grafted.

d The internal fixation can also be carried out with an intramedullary nail and reaming of the medullary canal. If some rotational instability results it can be overcome with the addition of a semi-tubular plate laterally.

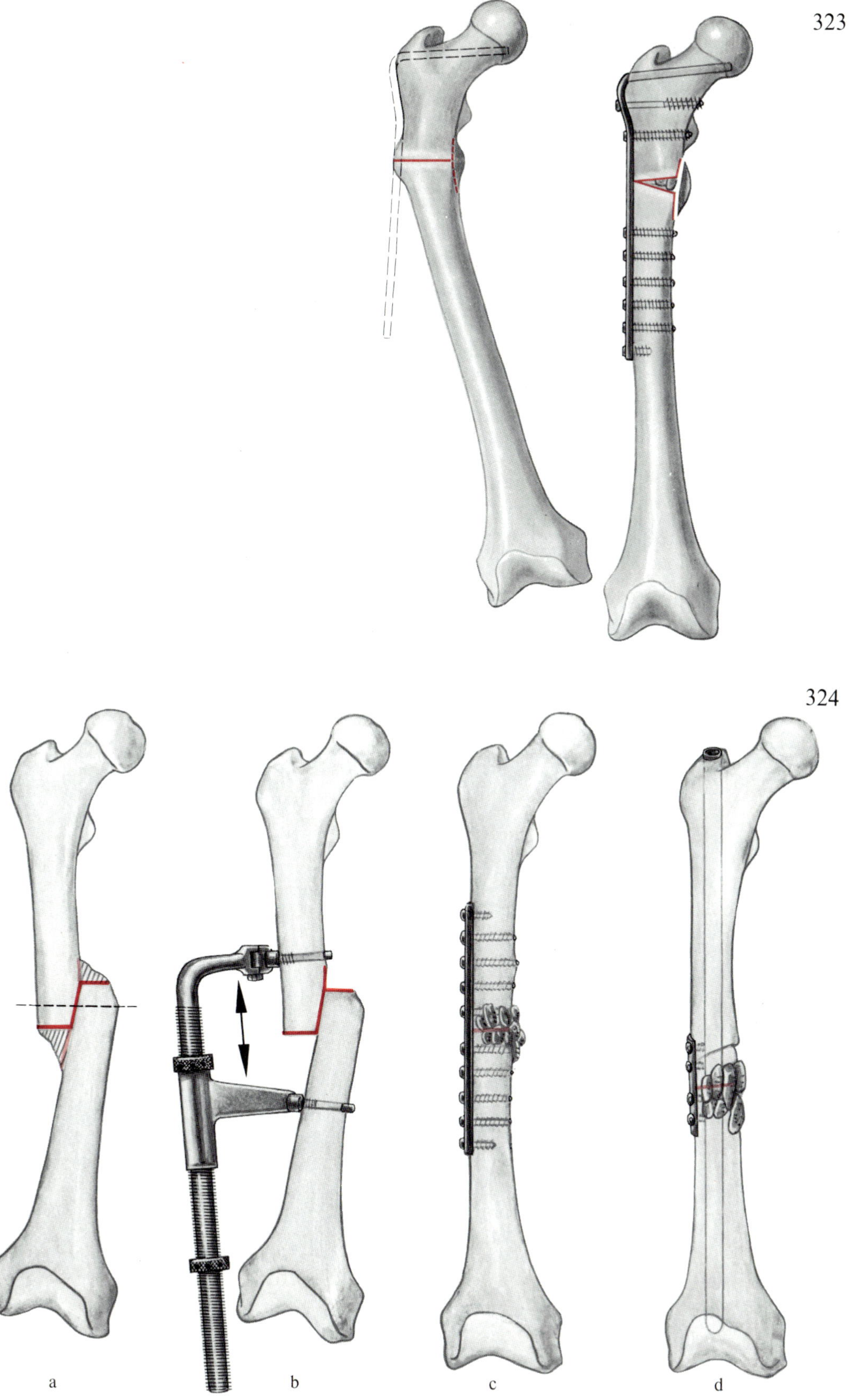
323
324
a
b
c
d

Technique: The patient is supine. Insert two Schanz screws into the proximal and distal metaphysis. Insert them with the aid of the drill sleeve and drill guide parallel to one another and parallel to the knee joint axis. Once the screws are inserted, the lengthening apparatus of WAGNER is passed over them and is fixed to the screws. By turning the black knob the whole system is brought under distraction. The hip and knee are flexed and the leg rotated inwards to expose the posterolateral aspect of the thigh. Four fingerbreadths behind the distraction apparatus and centred between the proximal and distal Schanz screws, make a longitudinal incision and expose the femoral shaft. Go in between the vastus lateralis and biceps femoris. If possible go through the old fracture. Carry out a transverse osteotomy through the femoral shaft. In order to achieve stability with the lengthening device, counterpressure from soft-tissue tension is required, which is obtained by distracting the fragments 1–1.5 cm.

Once the desired lengthening is achieved, the patient is ready for the second step. The patient is positioned prone. The lengthening apparatus is not touched. The femoral shaft is exposed between the vastus lateralis and biceps femoris. The internal fixation is carried out with the special lengthening plate, contoured to correspond to the anterior bow of the femur. The plate is fixed to the posterior surface of the femur. If the callus formation is insufficient, then the lengthened segment is filled with copious quantities of cortico-cancellous bone. The wound is then closed. Only now can the Wagner apparatus be removed together with the Schanz screws.

The lengthening apparatus of WAGNER can also be used to lengthen the tibia. For this, the apparatus is inserted on the medial side. The ankle mortice is protected by the insertion of fixation screws between the fibula and tibia.

Fig. 325 *The lengthening apparatus of* WAGNER.

a The standard model for the femur with 6.0-mm Schanz screws.

b The small model for use in the arm and in children's femora.

c The special plate for internal fixation of the femur after lengthening.

Fig. 326 *Femoral lengthening with the Wagner apparatus.*

a A 5.5-cm lengthening achieved in a span of $3^1/_2$ weeks, about 2 mm per day.

b Fixation with the special lengthening plates after an extensive grafting. Lengthening by 5 cm! At the end the leg-lengthening apparatus is removed.

Fig. 327 *Lengthening of the tibia.*

a, b Insert a threaded Steinmann pin distally through the fibula and tibia. Insert the other three Steinmann pins parallel to the first. One is inserted through the distal fragment, and the other two are inserted through the proximal metaphysis. The Steinmann pins are inserted with the help of a drill sleeve and a drill guide. The skin is then incised one fingerbreadth lateral to the anterior crest of the tibia. (b*2*) Decorticate the anterior crest and the posterolateral aspect of the tibia together with the interosseous membrane. Decorticate over 6–10 cm. Make a number of transverse cuts in the deep fascia. At the end cut the tibia obliquely.
Make the second skin incision (b*1*) one to two fingerbreadths behind the posterior edge of the fibula. Preserve at least a 7-cm skin bridge between the two incisions anteriorly. Make an oblique osteotomy through the fibula. Fix the threaded rods of the old external fixator to the protruding ends of the Steinmann pins and lengthen the tibia 1.5 mm per day.

c Three to four weeks later, make a long posterior incision (b*3*). Lengthen the tendons of tibialis posterior, flexor hallucis longus and tendo-Achilles. Fix the lengthening plate posteromedially and fill the defect with cortico-cancellous bone.

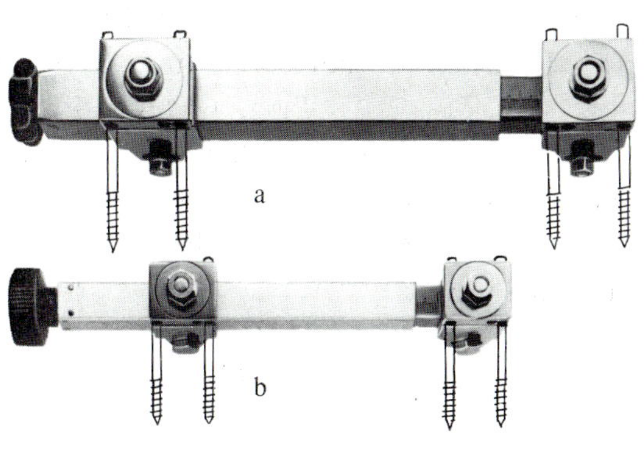

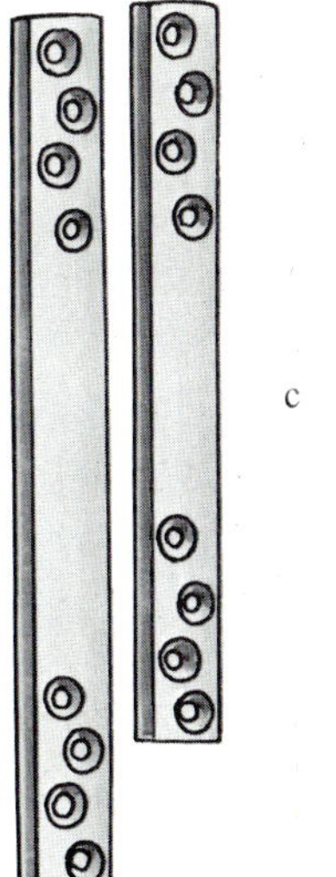

326

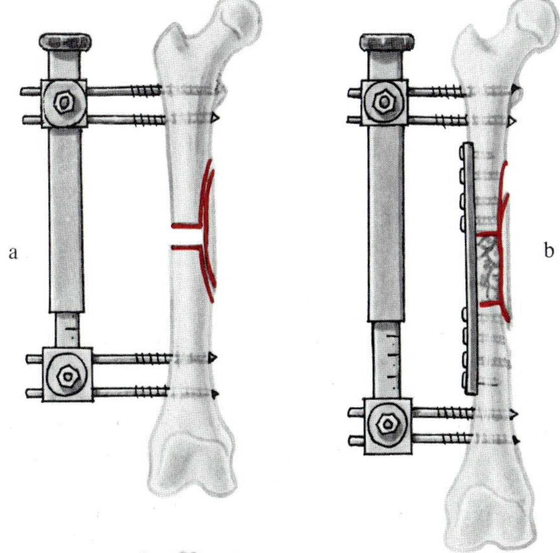

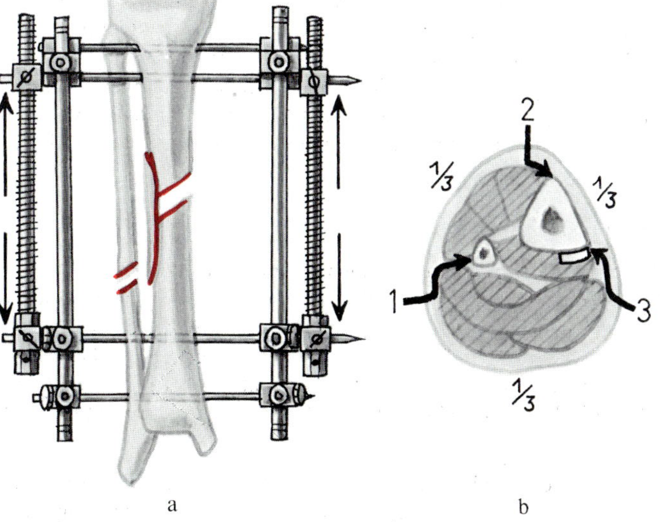

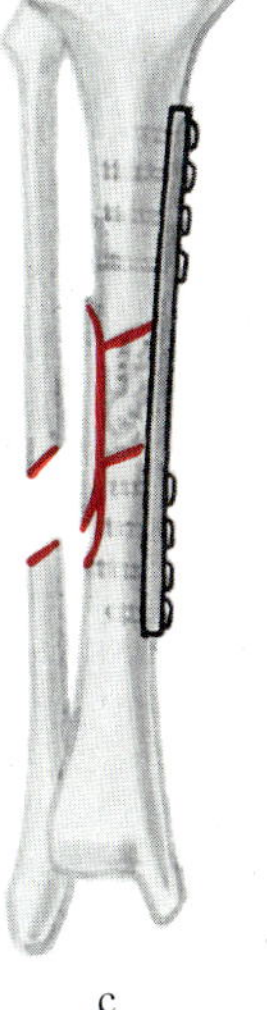

3.2.3 Supracondylar Osteotomies

The laterally inserted condylar plate can be used to correct not only a marked varus but also a valgus deformity. When used to correct a valgus deformity, the condylar plate is used as a buttress plate. In order to correct a marked valgus deformity we prefer to insert the right-angle plate medially. Prior to the procedure, however, we must determine the offset of the plate as well as the length of its blade.

Where the limb is shortened, as for instance in a valgus deformity secondary to a premature post-traumatic lateral closure of the epiphysed plate, we prefer an open wedge osteotomy which is fixed laterally with a condylar plate as shown in Fig. 330b.

Fig. 328 Because of the shape of the distal femur, the valgus osteotomy is best fixed with the condylar plate and the varus osteotomy with the right-angle osteotomy plate.

Fig. 329 *The technique of the valgus osteotomy.* Through a lateral incision, expose the shaft and the lateral condyle. Apply the condylar guide to the lateral surface of the femur. Take the triangular angle guide and apply it to the lower surface of the condylar guide in such a way that the lower edge of the triangular guide makes the desired angle of correction with the lower edge of the condylar guide. Insert a Kirschner wire into the lateral condyle, parallel to the lower edge of the triangular angle guide. The wire should lie parallel to the knee joint axis. Insert the seating chisel parallel to the Kirschner wire and osteotomize the femur. Remove the seating chisel and insert a plate, reduce the osteotomy, and insert the tension device. Insert the first cancellous screw into the supracondylar fragment, tighten the tension device and bring the osteotomy under axial compression. Fix the plate to the shaft and remove the tension device. Post-operatively position the leg with the hip and knee bent to 90°. Begin with active mobilization of the knee the day following surgery and allow the patient out of bed on the fourth or fifth post-operative day.

Fig. 330 *Varus osteotomy to correct genu valgum.*

a Expose the shaft and medial condyle through a medial approach. Apply the varus osteotomy guide to the medial surface of the femur and again with the aid of a triangular angle guide determine the angle of correction and insert a Kirschner wire into the medial condyle, which will serve as a guide for the insertion of the seating chisel and then for the blade of the right-angle plate. Proceed as above. For fixation use the right-angle plate with either the 15- or 20-mm offset. Insert a 60- or 70-mm-long cortex screw through the offset of the plate. This considerably increases the fixation of the angled plate in the distal fragment.

b Varus osteotomy with the condylar plate in a shortened limb. Lateral approach and insertion of the seating chisel, parallel to the knee joint. Carry out a supracondylar osteotomy parallel to the seating chisel. Spread the osteotomy, remove the seating chisel, insert the condylar plate, fill the defect with a bone graft and fix the plate to the femur.

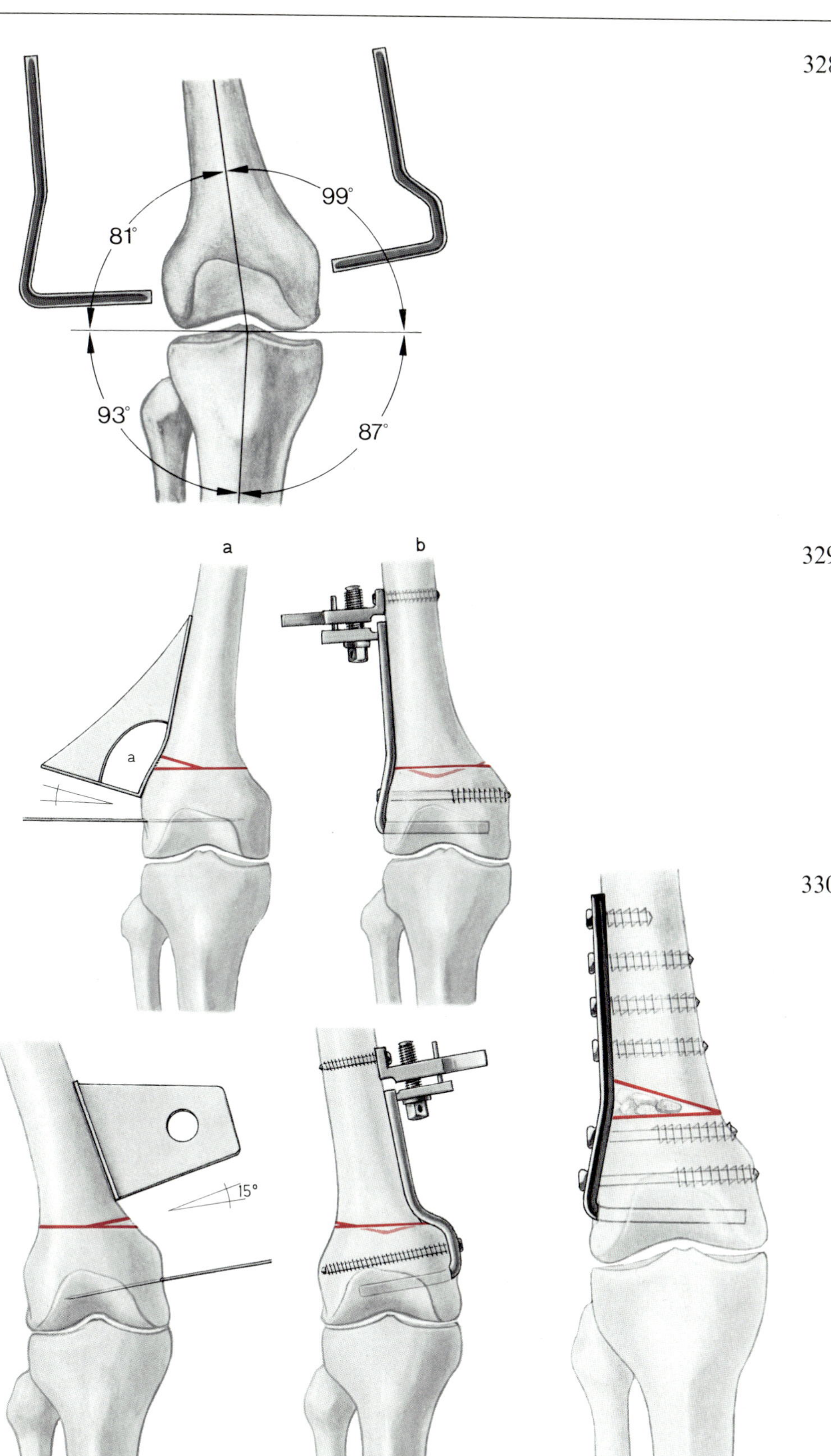

328

329

330

3.2.4 High Tibial Osteotomy

Fix either with two Steinmann pins and external clamps or with a T plate.

Note: We prefer to fix a varus osteotomy with the Steinmann pins and the external fixator because the medial application of a T plate requires not only reflection of muscle but also the reflection of the insertion of the medial collateral ligament.

Fig. 331 *Fixation of a high tibial osteotomy with Steinmann pins and external fixator.*

1 First osteotomize the fibula. Then make your anterior incision and insert the first Steinmann pin about 1–2 cm distal to the knee joint and parallel to it. The Kirschner wire subtends with the Steinmann pin an angle which corresponds to the angle of correction. Two further Kirschner wires (*c and d*) are inserted on each side of the osteotomy at right angles to the Steinmann pin and parallel to one another.

2 Carry out the osteotomy at the level of the tibial tubercle by undermining with an osteotome the distal insertion of the infrapatellar tendon. Remove the wedge either medially or laterally, depending on whether you are carrying out a valgus or a varus osteotomy.

3 Displace the tibia until the Steinmann pin and the Kirschner wire come to lie parallel to one another. Remove the Kirschner wire and replace it with a 4.5-mm Steinmann pin.

4 Compress the osteotomy with two external fixators. This will impact the osteotomy surface. The distance between the osteotomy and the Steinmann pin should be about 1–1.5 cm. If the distance is greater, inadequate fixation of the fragments will result.

Fig. 332 *High tibial osteotomy fixed with a T plate.* First osteotomize the fibula. Undermine the insertion of the infrapatellar tendon from below with an osteotome so that the osteotomy can be carried out through the tibial tubercle. Insert the four Kirschner guide wires as in Fig. 331. Insert the broad-tipped Hohmann retractor to protect the posterior soft structures and osteotomize the tibia at the level of the tibial tubercle. Use the undermined anterior portion of the tibial tubercle as a living graft. Correct the deformity and fix the osteotomy with a T plate.

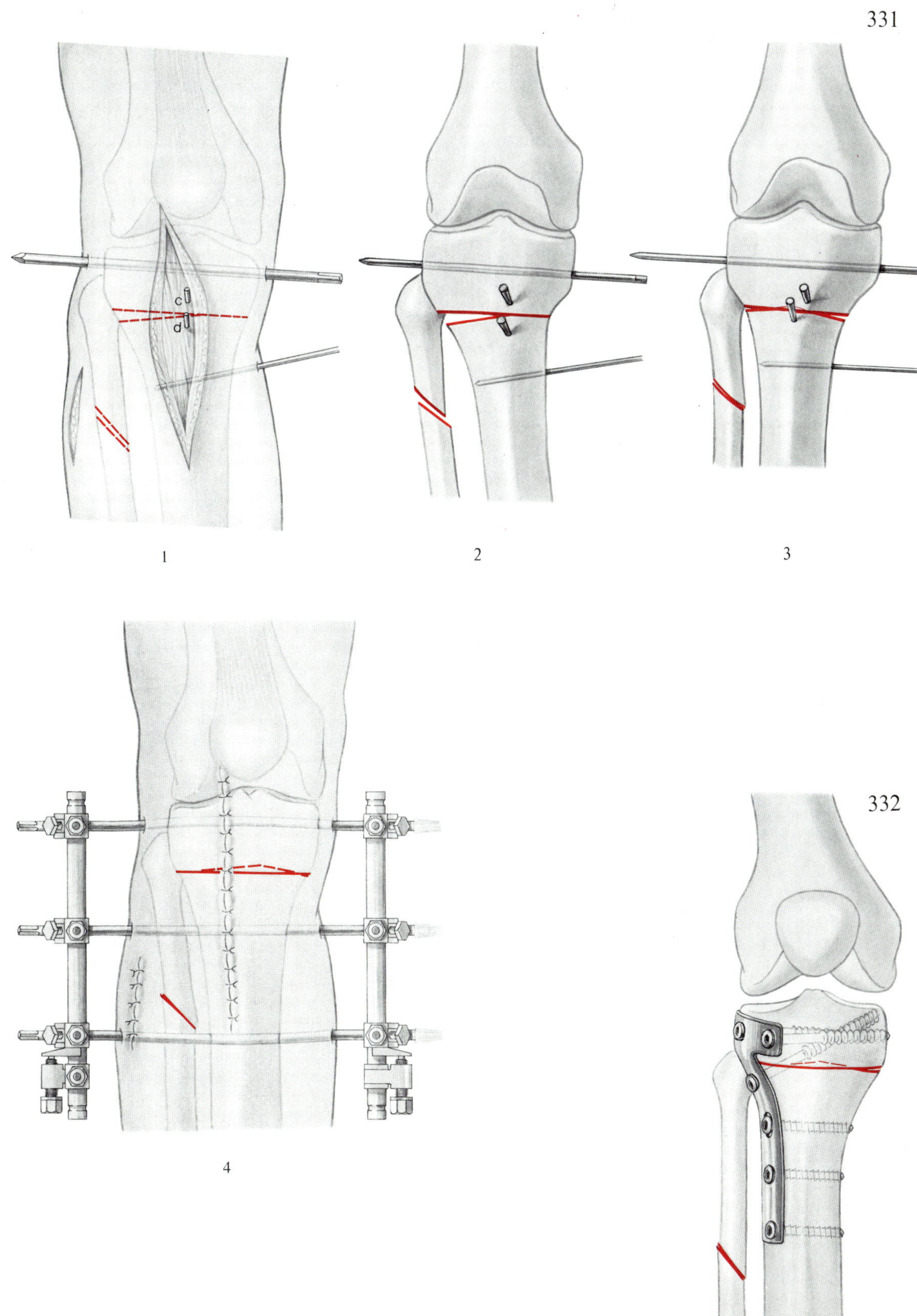
331
c
d
1
2
3
4
332

3.2.5 Osteotomies of the Tibial Shaft

Osteotomies through the shaft of the tibia should be considered only if the soft-tissue envelope is perfect. Otherwise there is a very high risk of a wound-edge necrosis.

A long straight lateral skin incision should be made and an extensive decortication carried out. Frequently the skin incision and relaxing incision must be left open.

In lengthening osteotomies the bone defect must be filled with bone graft. If the necessary lengthening is greater than 1.5 cm then we employ three or four Steinmann pins and insert them transversely through the fragments and then by means of the external fixator we carry out a gradual distraction of about 2 mm per day. After 2–3 weeks the lengthening osteotomy is stabilized with a plate which we apply posteriorly or laterally. As a general rule, in the adult, a lengthening of 3 cm or more is very difficult to obtain and requires lengthening of the tibialis posterior, flexor hallucis longus and gastrocnemius tendons (see Fig. 327).

Fig. 333 *Technique of a corrective and lengthening osteotomy of the tibia through the fracture callus.*

a Insert two Kirschner wires into the proximal and two into the distal metaphysis, above and below the area of callus formation. These will serve as your guide wires. Carry out an extensive decortication and then an oblique osteotomy through the callus.

b In a valgus osteotomy apply the tension band plate laterally and fill the medial defect with bone graft.

c If the osteotomy is slightly more distal, then after correction of the osteotomy the medullary canal can be reamed out and the osteotomy fixed with an intramedullary nail. Rotational instability can be controlled with a lag screw or a plate.

The plate should be removed after 1–2 months as soon as there is some rotational stability.

333

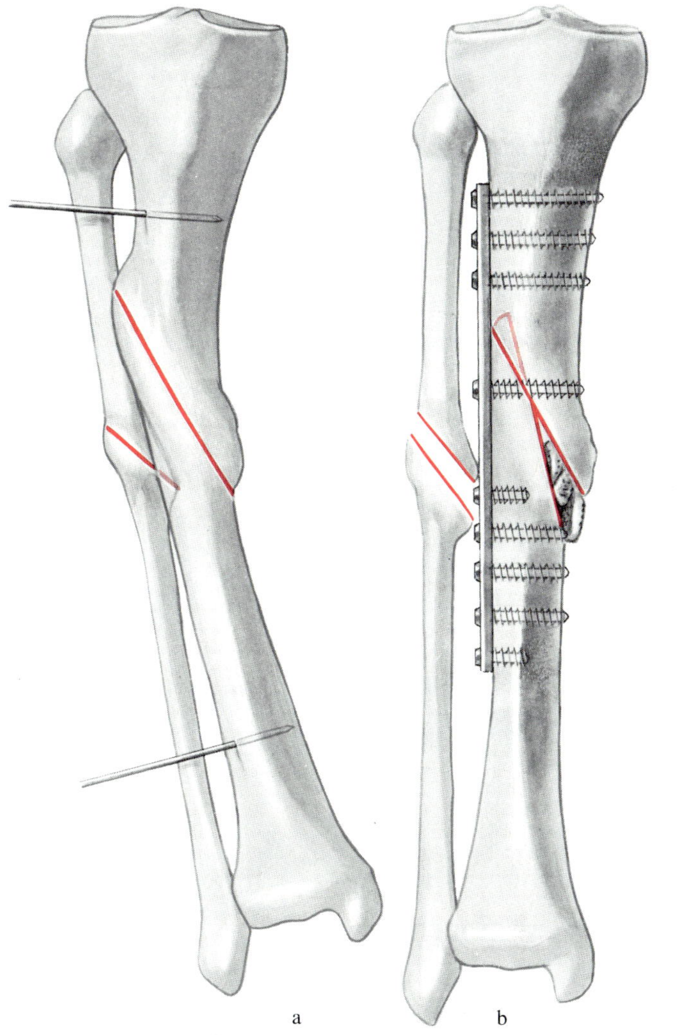

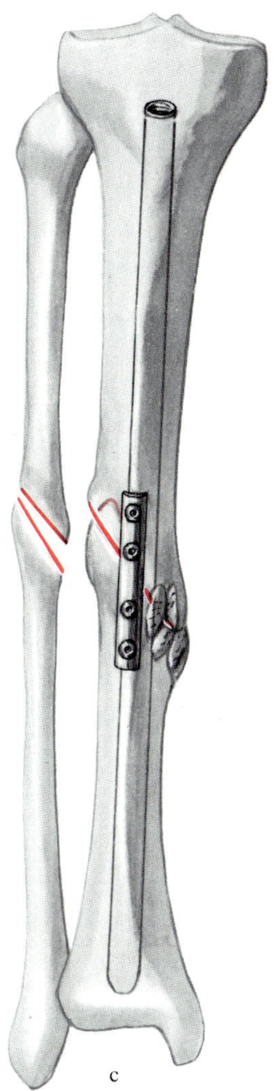

3.2.6 Supramalleolar Osteotomies of the Tibia

Fig. 334 *Stable fixation with two Steinmann pins and external fixators.*

a Resect a small segment of the fibula and insert a Steinmann pin parallel to the distal articular surface of the tibia. From the medial side and proximal to the planned osteotomy insert a 3.2 mm drill bit into the tibia. This drill bit should subtend the desired angle of correction with the Steinmann pin. Into the anterior crest insert two Kirschner wires, one on each side of the osteotomy. They will serve to control the rotational alignment as well as the angulation of the osteotomy.

b Carry out the osteotomy between the Kirschner wires.

c Correct the deformity.

d Remove the drill bit and replace it with a Steinmann pin. Reduce the osteotomy and compress it with two external fixators.

Fig. 335

a Two semi-tubular plates used for the fixation of a supramalleolar osteotomy in an adult. Most of the screws are short.

b In children where the deformity results from a premature partial exposure of the epiphyseal plate we prefer an open wedge osteotomy, which restores some length and corrects the deformity. The defect is bone grafted and the osteotomy fixed with a small T plate.

334

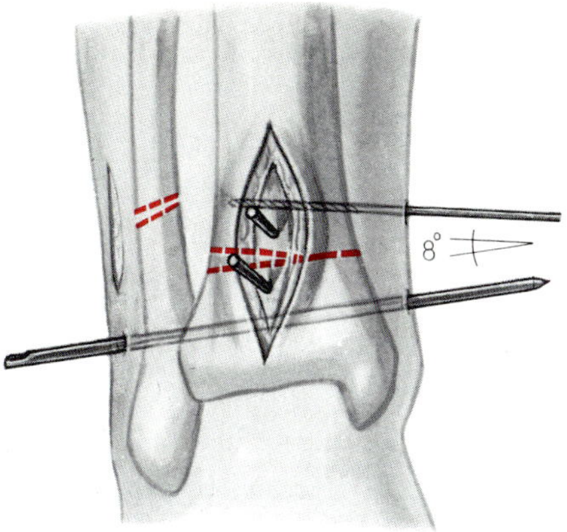

8°

a

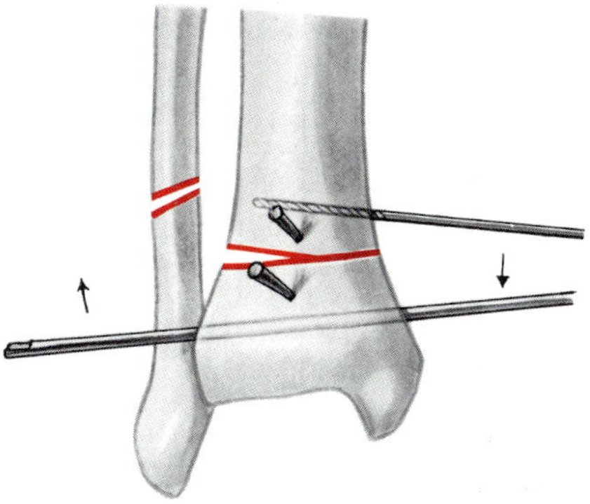

b

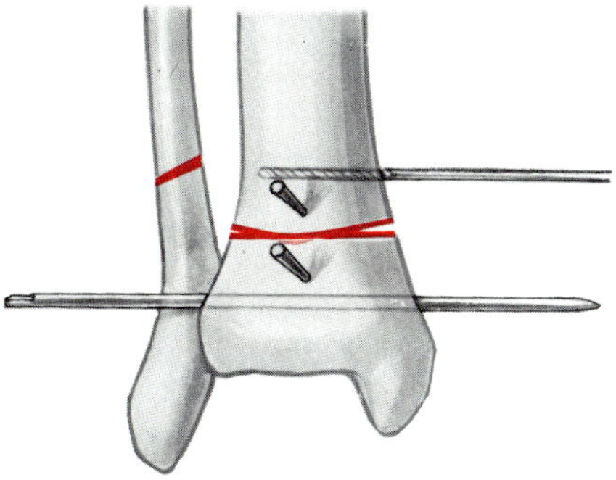

c

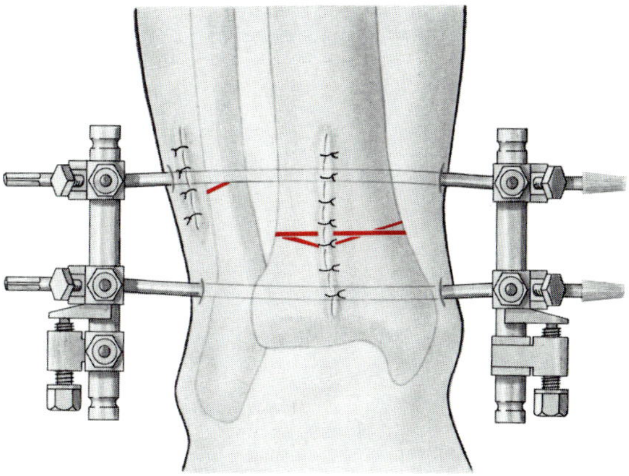

d

335

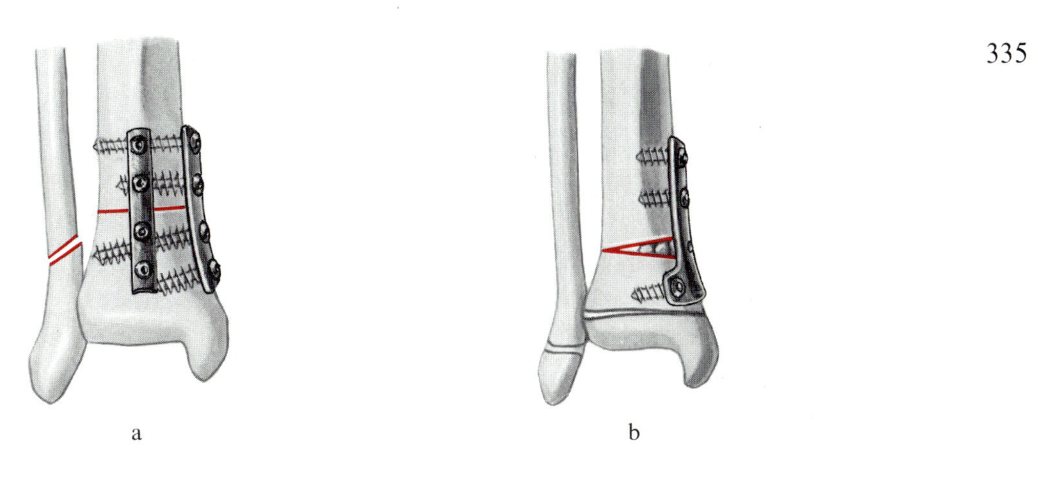

4 Arthrodeses

4.1 Arthrodesis of the Shoulder

Compression arthrodesis of the shoulder is a spectacular procedure which one would not have thought possible a few years ago. A patient with a compression arthrodesis of the hip or knee joint does not require supplementary plaster immobilization. These arthrodeses are so stable that these patients are able to be ambulant and begin partial weight bearing within a few days after surgery. Compression arthrodesis of the shoulder is just as stable. These patients, too, can and may begin with active mobilization and use of their arm just a few days after surgery.

Fig. 336 First of all determine the desired angle between the vertebral border of the scapula and the humerus. In most cases the position of choise for an arthrodesis is in 50° of abduction, 40° of internal rotation and 25° of flexion. A humerus abducted to 50° subtends an angle of 60°–70° with the axillary border of the scapula.

The operation is carried out with the patient lying on his side. A straight skin incision is made just below the spine of the scapula, down to the deltoid insertion. The lateral portion of the spine of the scapula is exposed and the acromion is transsected 1 cm from its tip, leaving the deltoid insertion undisturbed. The deltoid, together with the transsected tip of the acromion is then reflected distally and laterally. Carry out a wide capsulotomy. Resect all articular cartilage, and decorticate not only the joint surfaces but also the under-surface of the acromion. In preparing the glenoid one must make sure to resect its cartilaginous rim. The humerus is then placed in the desired position and all the surfaces are trimmed to fit.

The first hole is drilled through the acromion, approximately 1 cm medial to the resected surface of the glenoid through the neck of the scapula. This is a very important anchor hole for the plate medially, through which a 50- or 60-mm cortex screw (*a*) will be inserted.

The next step is the careful contouring of the seven- or eight-hole plate with the plate press. The plate is then fixed medially with a screw (*a*). Two further screws are inserted into the spine of the scapula. The arm is then placed in the desired position and is pushed upwards to make good contact with the decorticated undersurface of the acromion. Compression is applied with the tension device and the plate is fixed to the humerus. Fixation can be improved by inserting two long cortex screws through the plate, through the head, into the neck of the scapula. Care should be taken that these do not come in contact with the long anchor screw (*a*).

If necessary the fixation can be supplemented by buttressing it posteriorly with a second six-hole plate which is fixed to the scapula and humerus. This is particularly necessary if the bone is osteoporotic. The upper limb is a long lever arm and there is danger of fracture of the humerus just distal to the two plates. To prevent this complication, we advocate prophylactic bone grafting of the humerus just below the plates.

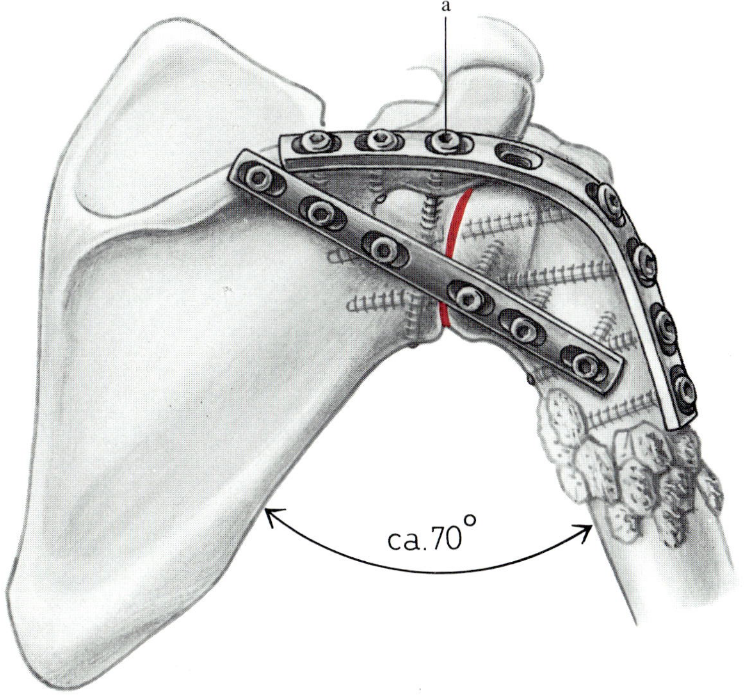
a
ca. 70°

4.2 Arthrodesis of the Elbow and of the Wrist

Fig. 337 *Arthrodesis of the elbow*. This is one of the most difficult arthrodeses to perform. The position of the nerves and vessels make a tension band fixation impossible. Most often for fixation we combine axial compression, obtained by means of external compression clamps, with a cancellous lag screw inserted in the axis of the humerus. The surfaces of the trochlea of the humerus and of the olecranon fossa are resected to fit. The osteotomy is reduced and a thick Kirschner wire is driven through the olecranon into the medullary canal of the humerus. The head and neck of the radius are then resected as far as the insertion of the biceps tendon. A Steinmann pin is then inserted transversely through the olecranon in line with the anterior cortex of the humerus. The Kirschner wire is removed and is replaced with a long cancellous screw. This cancellous screw should be as long as possible and should be supported with a washer to prevent its head from sinking into the olecranon.

The second Steinmann pin should be inserted and the compression of the arthrodesis increased by means of the external fixators.

Fig. 338 *Arthrodesis of the wrist*. Resect cartilage from all joint surfaces. Obtain a broad cortico-cancellous slab of bone from the wing of the ilium. Cut it to shape. This bone graft is slotted into the distal end of the radius, and is laid across the freshened carpal bones as far as the base of the second and third metacarpal. The arthrodesis is bridged with a seven-, eight- or nine-hole plate which is fixed with three screws to the shaft of the second metacarpal and with three screws to the distal radius. The wrist is fused, depending on the age of the patient, in 10° to 30° of dorsiflexion. If you wish to obtain axial compression with the tension device, you must resect the distal end of the ulna.

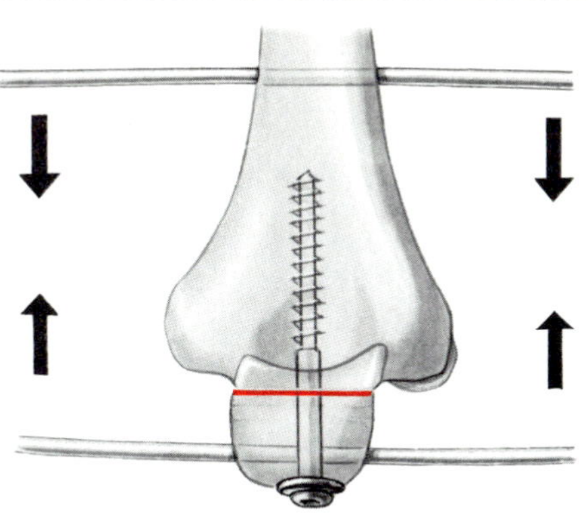

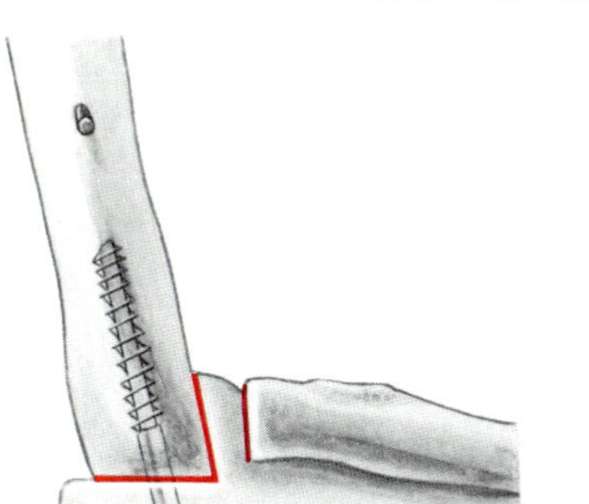

337

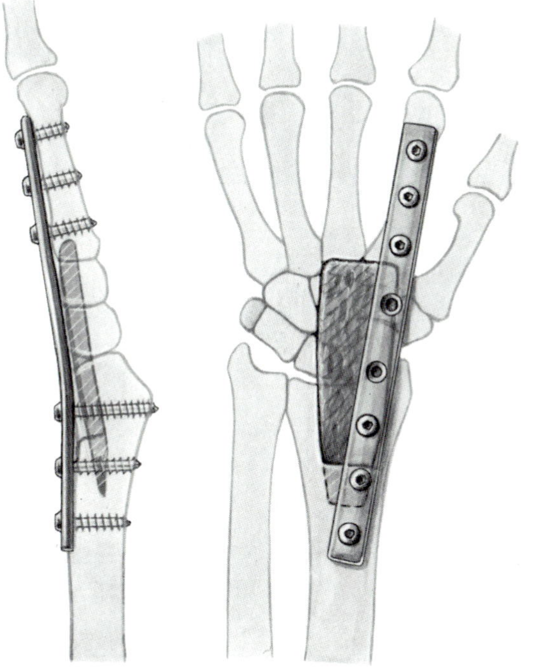

338

4.3 Arthrodesis of the Hip with the Cobra Plate

Technique of R. SCHNEIDER

Fig. 339 The patient is positioned supine. The pelvis and both legs are prepared and draped in such a way that both iliac crests and both legs are draped free. The free draping of the uninvolved leg is necessary to permit determination of leg length during the procedure.

a The approach is carried out through a 30-cm-long, straight lateral incision.

b A superficial osteotomy of the greater trochanter is carried out. This frees the gluteus medius and minimus at their insertion. The abductor muscle mass is reflected upwards till the joint and the wing of the ilium above the joint are exposed. A broad-tipped Hohmann retractor is inserted along the bone into the greater sciatic notch to protect the sciatic nerve. Next an osteotomy of the pelvis is carried out with the oscillating saw and osteotomes. During the osteotomy, use is made of bone spreaders. These make the osteotomy easier and safer because the surgeon can visualize the depth of the cut and thus he will not plunge medially. The hip is not dislocated. Little bone is removed from the superior surface of the femoral head. Two broad slices of bone are resected with the saw from the greater trochanter. One of these will be inserted between the plate and the head and the second will be slotted as a graft anteriorly.
Note: The osteotomy must be either horizontal or even sloped slightly medially upwards, otherwise an abduction deformity will result once compression is applied.

c The femur is exposed by reflecting the vastus lateralis forwards. The leg is adducted 10°, kept in neutral rotation and flexed somewhere between 10° and 25° depending on the age of the patient. The cobra plate is laid across the joint and is tested for fit. Further contouring of the plate may be necessary and further bone may have to be resected from the greater trochanter to make it fit the contour of the plate.

d The plate is then fixed to the pelvis with one screw which is inserted 1 cm above the osteotomy. This screw must be inserted carefully, because it is an important anchor screw. The large tension device is fixed distally to the femur and is then slightly tightened. This brings the surfaces under slight compression.

e The position of the leg must now be carefully checked. Two Kirschner wires are driven into the anterior superior spines. Over these two Kirschner wires is slipped one arm of a special right-angle guide. At the start before compression is applied, the long arm of this right-angle guide should lie three to four fingerbreadths lateral to the patella. At the end it should touch the lateral edge of the patella.

f, g The remaining six screws are inserted proximally into the pelvis. The tension device is tightened to achieve maximal axial compression. The position of the leg must be carefully checked and if necessary adjustments must be made. Once the position of the arthrodesis is perfect, the plate is fixed distally to the femur. The slice of bone from the greater trochanter is now used as a bone graft and is slotted in front of the plate into the ilium and the femoral head. The second piece of bone is inserted between the plate and femoral head. Leg length difference should also be checked during the procedure. Ideally the arthrodesed leg should be 0.5–1 cm shorter. If excessive shortening results at a future date, the uninvolved leg should be shortened by 2–3 cm as required through the intertrochanteric area.

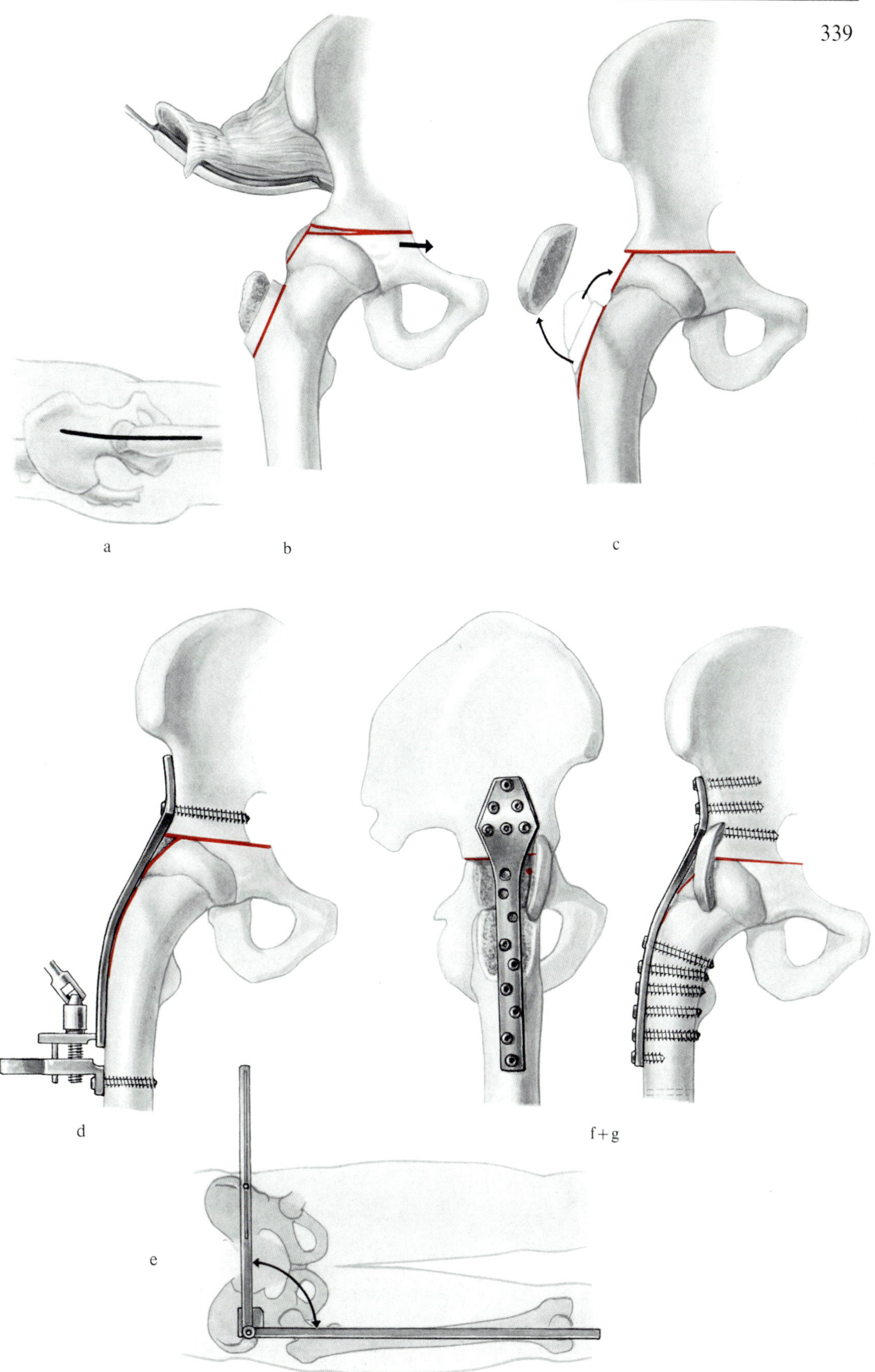
a
b
c
d
f+g
e

4.4 Arthrodesis of the Knee Joint

Fig. 340 *Technique.*

a Make a straight longitudinal incision directly over the middle of the patella. With a broad chisel, resect from the patella as large a cube of bone as possible.

b Open the joint widely so that you can flex the knee to 90°. Keeping the knee flexed to 90°, resect the distal articular surface of the femur in line with the long axis of the tibia.

This means that the cuts are at 90° to the long axis of the femur. Next the proximal surface of the tibia is cut, sloping the cut slightly to the back, so that the arthrodesis, when reduced, will be in a slight degree of flexion.

c The position of the arthrodesis once reduced is checked with a long string which, when stretched from the anterior superior spine to the web space between the first and second toe, should cross the middle of the knee joint.

d The flexion and the valgus should be about 10° each.

e Once the surfaces have been suitably trimmed to obtain correct alignment of the leg, four Steinmann pins are inserted. The two posterior pins are inserted as close to the arthrodesis as possible. The two anterior ones are inserted about 4 cm away from the arthrodesis. The external fixator is inserted and axial compression is applied. Check once again the alignment of the arthrodesis. At the end slot in the cube of bone which you removed from the patella as a bridge across the arthrodesis.

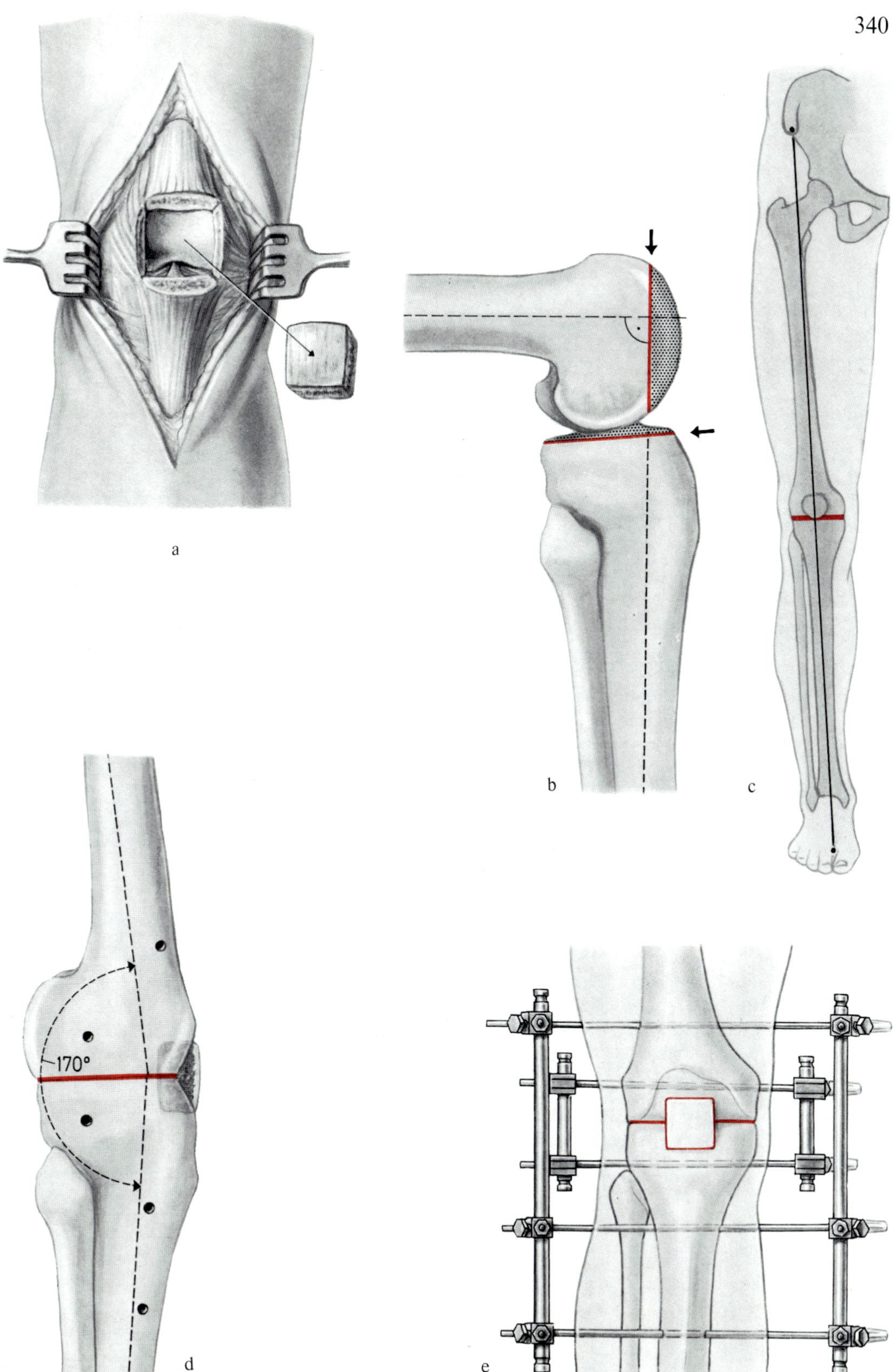
a
b
c
170°
d
e

4.5 Arthrodesis of the Ankle Joint

Fig. 341 *Technique with the use of the external fixator.*

a The procedure is carried out under pneumatic tourniquet control. The whole limb is prepared and draped to expose the knee and ankle. The first Steinmann pin is then inserted 6–7 cm above the ankle joint. It is inserted in a lateral to medial direction and in 20° of external rotation with respect to the knee joint axis. To ease the insertion of the Steinmann pin, pre-drill the bone with a 3.2-mm drill bit.

Make an 8-cm incision over the lateral malleolus and a 5-cm-long medial incision over the medial malleolus. If no great deformity exists, particularly as far as rotation is concerned, it is best to insert a 2.5-mm Kirschner wire through the neck of the talus, through the place where the second Steinmann pin will subsequently be inserted. This point lies directly in the neck of the talus, directly in line with the anterior edge of the tibia.

b The Steinmann pin and Kirschner wire are now removed so that they are not in the way of the surgeon during the procedure. The lateral malleolus is now osteotomized obliquely about 3 cm from its tip. The distal articular surface of the tibia is now resected either with a chisel or an oscillating saw. Great care must be taken to resect the posterior lip of the tibia. The resection is completed through the medial incision through which also the medial malleolus is resected and the distal medial flare of the tibia is flattened.

c The knee is flexed to 90° and the foot externally rotated so that its external rotation corresponds to the external rotation of the uninvolved leg. Two Kirschner wires are now inserted through the previous drill holes through the tibia and neck of the talus respectively. In a woman we aim for a right angle between the long axis of the tibia and the foot. In a man with a mobile midtarsal joint we prefer 10° of dorsiflexion.

d While the foot is held in the desired position, the articular surface of the dome of the talus is resected parallel to the distal osteotomy through the tibia.

e The proximal Steinmann pin is now inserted into the tibia. The position of the foot is checked once again and if all is well the second Steinmann pin is inserted into the hole previously drilled through the neck of the talus. The foot is now displaced posteriorly, which improves the posterior lever arm as well as the forwards roll of the foot from heel to toe.

f Fix the external fixators to the two Steinmann pins and place the arthrodesis under maximal axial compression. The lateral malleolus is now shaped to fit and is used as a bridge across the arthrodesis. It is fixed with a malleolar screw and a washer.

Fig. 342 *The fixation of an ankle arthrodesis with a spoon plate.* The joint surfaces are resected and the arthrodesis is fixed with a spoon plate. Note that the medial and lateral malleoli are preserved and that at the end the fixation of the arthrodesis is supplemented with two lag screws, one inserted through the medial and one through the lateral malleolus. This technique makes it impossible to displace the foot posteriorly.

341

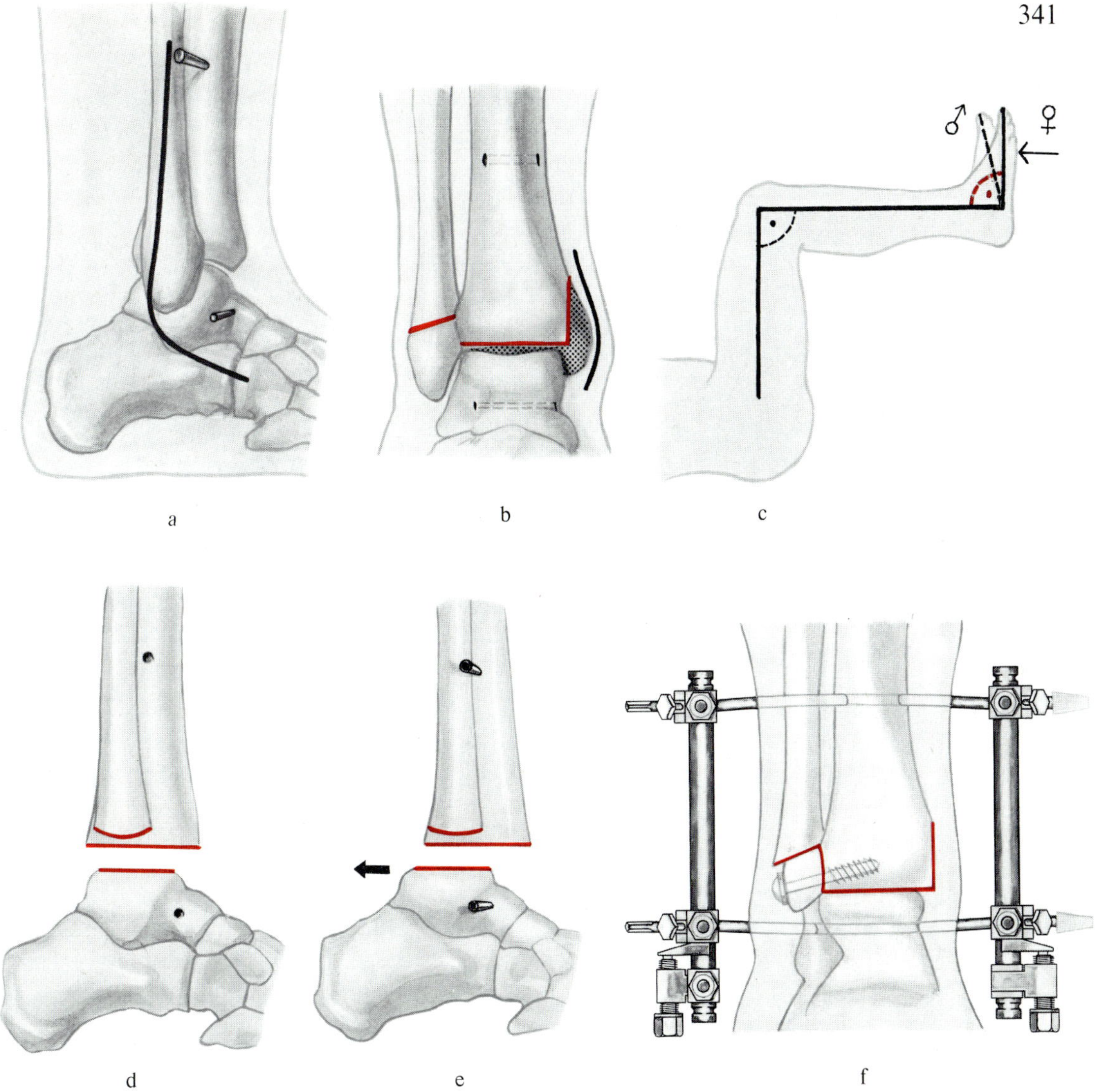

♂ ♀

a b c

d e f

342

Fig. 343 *Primary arthrodesis of the ankle following trauma.* If in distal fractures of the distal tibia the dome of the talus is badly damaged or the fracture cannot be reconstructed, one should consider an ankle arthrodesis. Under such circumstances we prefer to carry out a primary arthrodesis of the ankle. Reconstruction of the distal tibia with lag screws and a buttress plate is carried out. The articular cartilage must be resected and the joint surfaces cut to shape. At the end the Steinmann pins are inserted and axial compression is achieved by means of the external fixators (see also Fig. 270).

Fig. 344 *Pantalar arthrodesis.* Resect the whole talus. Resect the articular surface of the navicular, the cuboid and the articular surfaces of the os calcis. Reduce the foot and shape the talus as a free bone graft in such a way that it fits accurately between all the resected surfaces. At the end a Steinmann pin is inserted into the tibia, one into the os calcis and one across the cuboid and navicular, taking great care not to damage the neurovascular bundle. Axial compression is then obtained between the different segments by using four or six external compression clamps.

Fig. 345 *Interphalangeal arthrodesis of the big toe.* Such an arthrodesis may become necessary in post-traumatic arthritis of the interphalangeal joint or in paresis of the extensor hallucis longus.

a Make two incisions, one transversely across the joint and the second as a 10–15-mm-long fish-mouth incision through the pulp.

b Resect the articular surfaces with a small saw. Insert a 2-mm drill bit through the distal phalanx in line with its long axis.

c Remove the 2-mm drill bit and without using the drill insert the bit through the tip of the toe into the pre-drilled hole in the distal phalanx. Reduce the fragments and check their position. Occasionally slight adjustments are necessary. Advance the drill into the proximal phalanx, remove it and fix fragments with a long 4.0-mm cancellous screw.

343

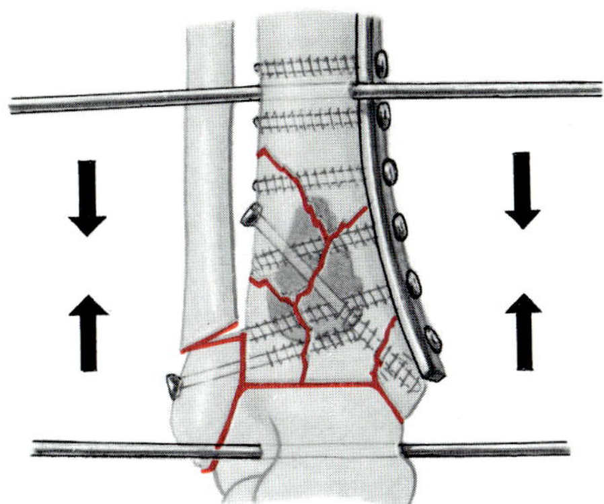

344

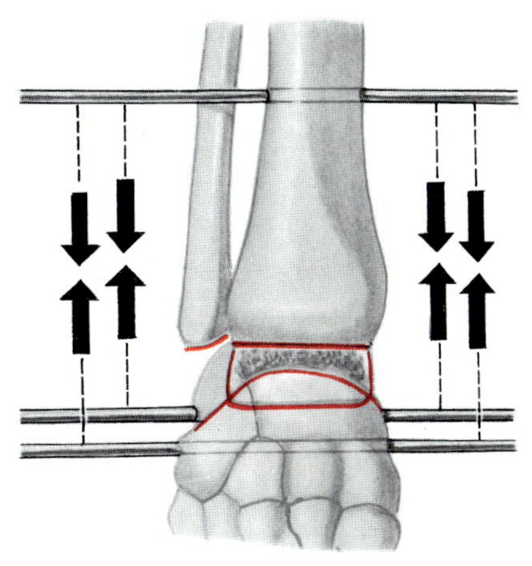

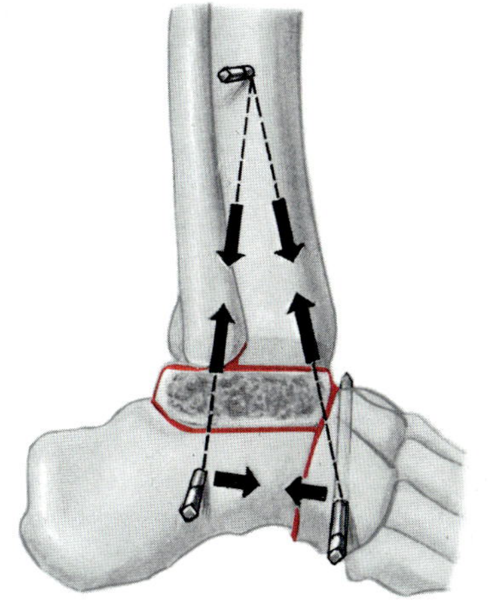

345

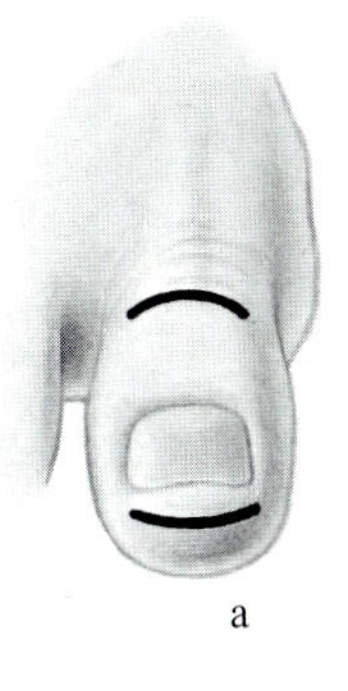

a

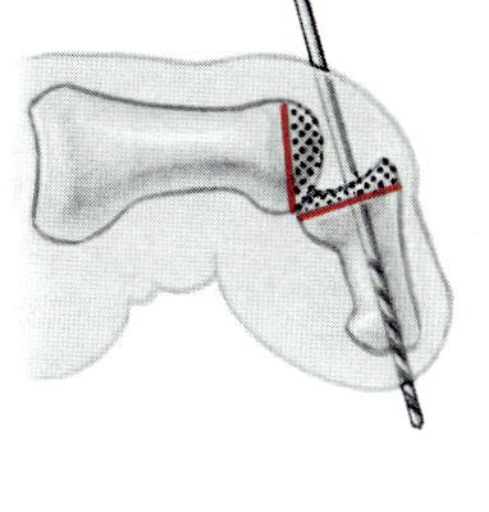

b

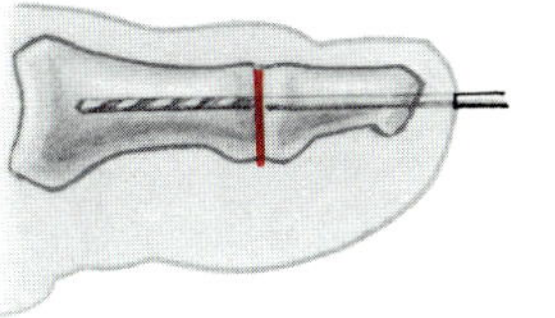

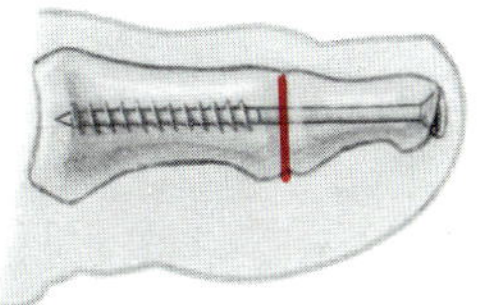

c

Bibliography

Allgöwer, M.: Funktionelle Anpassung des Knochens auf physiologische und unphysiologische Beanspruchung. Langenbecks Arch. klin. Chir. **319**, 384–391 (1967)

– A healing of clinical fractures of the tibia and rigid internal fixation. Reprint from "The healing of osseous tissue". Nat. Acad. Sciences – Nat. Res. Council pp. 81–89 (1967)

– Die intraartikulären Frakturen des distalen Unterschenkelendes. Helv. chir. Acta **35**, 556–582 (1968)

– Luxationsfrakturen im Ellbogenbereich. (Auszug aus dem Hauptreferat, vorgetragen an der Deutsch-Oesterreichisch-Schweizerischen Unfalltagung vom 26.–28.10.1972 in Bern.). Z. Unfallmed. u. Berufskr. **2**, 71–74 (1973)

– Kinzl, L., Matter, P., Perren, S.M., Rüedi, T.: The Dynamic Compression Plate. Berlin-Heidelberg-New York: Springer 1977

– Perren, S.M., –, Cordey, J., Russenberger, M.: Developments of compression plate techniques for internal fixation of fractures. Progr. Surg. (Basel) **12**, 152–179 (1973)

– Cueni, Th.A., –: Knochenszintigraphische und röntgenologische Untersuchungen während der Heilung der Talusfraktur. Helv. chir. Acta **41**, 459–468 (1974)

– Rüedi, Th., Kolbow, H., –: Erfahrungen mit der dynamischen Kompressionsplatte (DCP) bei 418 frischen Unterschenkelschaftbrüchen. Arch. orthop. Unfall-Chir. **82**, 247–256 (1975)

– Gruber, U.F., Pestalozzi, A., Rüedi, Th., –: Bilateral fractures of the lower leg due to skiing accidents. Orthop. Clin. N. Am. **7**, No. 1, 215–222 (1976)

– Scharplatz, D., –: Fracture-dislocations of the elbow. Injury **7**, 143–159 (1976)

Bandi, W.: Zur Problematik der Korrektur posttraumatischer Achsenfehlstellungen der kindlichen Tibia. Z. Unfallmed. Berufskr. **4**, 289–294 (1966)

– Indikation und Technik der Osteosynthese am Humerus. Helv. chir. Acta **31**, 89–100 (1964)

– Die gelenknahen Frakturen des Oberarms. Chirurg **40**, 193–198 (1969)

– Die distalen, intraartikulären Schienbeinbrüche des Skifahrers. Actuelle Traumatologie **4**, 1–6 (1974)

– Allgöwer, M.: Zur Therapie der Osteochondritis dissecans. Helv. chir. Acta **26**, 552 (1959)

Blount, W.P.: Knochenbrüche bei Kindern. Stuttgart: Thieme 1957

– Fractures in children. Baltimore: Williams & Wilkins 1955

Böhler, J.: Gelenknahe Frakturen des Unterarmes. Chirug **40**, 198–203 (1969)

Böhler, L.: Die Technik der Knochenbruchbehandlung. Wien: Maudrich 1953, 1954, 1957 und 1963

– Neues zur Behandlung der Fersenbeinbrüche. Langenbecks Arch. klin. Chir. **297**, 698 (1957)

Boyd, H.B., see Crenshaw, A.H.: Campbell's operative orthopaedics. Saint Louis: Mosby 1963

Boitzy, A.: La fracture du col du fémur chez l'enfant et l'adolescent. Paris: Masson 1971

Burke, J.F.: The effective period of preventive antibiotic action in experimental incisions and dermal lesions. Surgery **50**, 161–168 (1961)

– Preoperative antibiotics. Surg. Clin. N. Amer. **43**, 665–676 (1963)

Burri, C., Ecke, H., Kuner, E.H., Pannike, A., Schweiberer, L., Schweikert, C.H., Spier, W., Tscherne, H.: Unfallchirurgie. Heidelberger Taschenbücher. Berlin-Heidelberg-New York: Springer 1974

– (Herausgeber): Posttraumatische Osteitis. Band 18: Aktuelle Probleme in der Chirurgie. Bern-Stuttgart-Wien: Huber 1974

Charnley, J.: Compression arthrodesis. Including central dislocation as a principle in hip surgery. Edinburgh: Livingstone 1953

– The closed treatment of common fractures. Third edit. Edinburgh and London: Livingstone 1961

Danis, R.: Théorie et pratique de l'ostéosynthèse. Paris: Masson 1949

– Le vrai but et les dangers de l'ostéosynthèse. Lyon chir. **51**, 740 (1956)

Dehne, E.: Die Osteosynthese der Unterschenkelbrüche. Arch. orthop. Unfall-Chir. **39**, 328 (1938)

– Treatment of fractures of the tibial shaft. Clin. Orthop. **66**, 159 (1969)

– et al.: The natural history of the fractured tibia. Surg. Clin. N. Amer. **41**, 1495 (1961)

Dunn, D.M.: Anteversion of the neck of the femur. A method of management. J. Bone Jt Surg. B **34**, 181 (1952)

– The treatment of adolescent slipping of the upper femoral epiphysis. J. Bone Jt Surg. B **46**, 621 (1964)

Eggers, G.W.N.: The contact splint. Rep. Biol. Med. **4**, 42 (1946)

– Internal contact splint. J. Bone Jt Surg. A **30**, 40 (1948)

– Schindler, Th.O., Pomerat, Ch.M.: The influence of the contact-compression factor on osteogenesis in surgical fractures. J. Bone Jt Surg. A **31**, 693 (1949)

ENDER, J.: Per- und subtrochantere Oberschenkelbrüche. Probleme beim frischen per- und subtrochanteren Oberschenkelbruch. Hefte Unfallheilk. **106**, 2 (1970)

— SIMON-WEIDNER, R.: Die Fixierung der Trochanterbrüche mit runden elastischen Condylennägeln. Acta chir. Austriaca 1970, **1**, 40–42

GALEAZZI, R.: Über ein besonderes Syndrom bei Verletzungen im Bereich der Unterarmknochen. Arch. Orthop. Unfall-Chir. **35**, 557 (1935)

GALLINARO, P., PERREN, S., CROVA, M., RAHN, B.: La osteosintesi con placca a compressione. Bologna: Aulo Gaggi 1969

— RAHN, B.A., —, BALTENSPERGER, A., PERREN, S.M.: Primary bone healing. An experimental study in the rabbit. J. Bone Jt Surg. **A53**, 783–786 (1971)

— RAHN, B.A., —, SCHENK, R.K., BALTENSPERGER, A., PERREN, S.M.: Compression interfragmentaire et surcharge locale des os. Voir BOITZY, A.: Ostéogénèse et compression. Bern-Wien-Stuttgart: Huber 1972

GANZ, R., ALLGÖWER, M., EHRSAM, R., —, MATTER, P., PERREN, S.M.: Clinical experience with a new compression plate DCP. Acta orthop scand., Suppl. 125 (1969)

— BRENNWALD, J., HUNTER, W., PERREN, S.M.: The recovery of medullary circulation after osteotomy and internal fixation of the rabbit tibia. Europ. Surg. Res. **2**, 106 (1970)

— Isolierte traumatische Knorpelläsion am Kniegelenk. Hefte Unfallheilk. **110**, 146 (1971)

— BRENNWALD, J.: L'ostéosynthèse à compression du tibia du lapin. Etude de la revascularisation du canal médullaire et de la corticale sous fixation stable. In: Périarthrite de l'épaule. Ostéogénèse et compression, Travaux divers. Bern: Hans Huber 1971

MÜLLER, M.E., —: Luxationen und Frakturen: Untere Gliedmassen und Becken. In: REHN, J., Unfallverletzungen bei Kindern. Berlin-Heidelberg-New York: Springer 1974

— FREIBURGHAUS, P.: Operativer Beinlängenausgleich im Bereich der Hüfte. Ther. Umsch. **32**, 329 (1975)

— RIESEN, H.: Primary arthrodesis of ankle and subtalar joints. See: CHAPCHAL, G. (editor): The Arthrodesis. Stuttgart: Thieme 1975

— Ioslierte Knorpelabscherungen am Kniegelenk. Hefte Unfallheilk. **127**, 79 (1976)

STOLTZ, M.R., —: Fracture after arthrodesis of the hip and knee. Clin. Orthop. related Res. **115**, 177 (1976)

HACKETHAL, K.H.: Die Bündelnagelung. Wien: Springer 1961

HARRINGTON, P.R.: Spine instrumentation. Amer. J. Orthop. **8**, 228–231 (1964)

HEIM, U.: Die Technik der operativen Behandlung der Metacarpalfrakturen. Helv. chir. Acta **36**, 619 (1969)

— Behandlung der Frakturen der Metatarsalia und Zehen. Z. Unfallmed. Berufskr. 1970, 305

— PFEIFFER, K.M. und unter Mitarbeit von MEULI, H.CH.: Periphere Osteosynthesen, unter Verwendung des Kleinfragment-Instrumentariums der AO. Berlin-Heidelberg-New York: Springer 1972

— PFEIFFER, K.M.: Small Fragment Set Manual. New York: Springer 1974

— PFEIFFER, K.M.: Ostéosynthèses périphériques. Paris: Masson 1975

— PFEIFFER, K.M.: Osteosintesis periferica. Barcelona: Editorial Cientifico-Médica 1975

— Indications et techniques de l'ostéosynthèse AO dans le traitement des fractures de la main. Acta orthop. belg. **39**, 957 (1973)

— L'ostéosynthèse rigide dans le traitement des fractures de la base du premier métacarpien. Acta orthop belg. **39**, 1073 (1973)

— PFEIFFER, K.M., MEULI, H.CH.: Resultate von 332 AO-Osteosynthesen des Handskelettes. Handchirurgie **5**, 71 (1973)

— Periphere Osteosynthesen, Indikation und Technik. Zbl. Chir. **99**, 1319 (1974)

HERZOG, K.: Nagelung der Tibiaschaftbrüche mit einem starren Nagel. Dtsch. Z. Chir. **276**, 227 (1953)

— Die Technik der geschlossenen Marknagelung des Oberschenkels mit dem Rohrschlitznagel. Chirurg **31**, 465 (1960)

HIERHOLZER, G.: Arthrodese nach Schienbeinkopfbrüchen. Hefte Unfallheilk. **126**, 283–289 (1976)

HUTZSCHENREUTER, P.: Beschleunigte Einheilung von allogenen Knochentransplantaten durch Präsensibilisierung des Empfängers und stabile Osteosynthese. Langenbecks Arch. Chir. **331**, 321 (1972)

— et al.: Some effects of rigidity of internal fixation on the healing pattern of osteotomies. Injury **1**, No. 1 (1969)

— ALLGÖWER, M., BOREL, J.F., PERREN, S.M.: Second-set reaction favouring incorporation of bone allografts. Experientia (Basel) **29**, 103 (1973)

— CLAES, L.: Struktur und Festigkeit neu aufgebauter Knochenanteile im Corticalisgleitloch bei liegender Zugschraube (histologische und mechanische Befunde). Arch. Orthop. Unfall-Chir. **85**, 161 (1976)

JUDET, R.: Luxation congénitale de la hanche. Fractures du cou-de-pied, rachis cervical. Actualités de chirurgie orthopédique de l'Hôpital Raymond-Poincaré. Paris: Masson 1964

— The use of an artificial femoral head for arthroplasty of the hip joint. J. Bone Jt Surg. B**32**, 166 (1950)

— LETOURNEL, E.: Les fractures du cotyle. Paris: Masson 1974

KROMPECHER, S.: Die Knochenbildung. Jena: Fischer 1937

KÜNTSCHER, G.: Die Marknagelung. Berlin-Göttingen-Heidelberg: Springer 1962

— Praxis der Marknagelung. Stuttgart: Schattauer 1962

– Das Kallus-Problem. Stuttgart: Enke 1970
– MAATZ, R.: Technik der Marknagelung. Leipzig: Thieme 1945

LAMBOTTE, A.: Le traitement des fractures. Paris: Masson 1907
– Chirurgie opératoire des fractures. Paris: Masson 1913

LANE, W.A.: The operative treatment of fractures. London: Medical Publishing Co., 1914

Lanz, T., Wachsmuth, W.: Praktische Anatomie. Bd. I, vierter Teil: Bein und Statik. 2. Aufl. Berlin-Heidelberg-New York: Springer 1972

LETOURNEL, E.: Les fractures du cotyle, étude d'une série de 75 cas. J. Chir. (Paris) **82**, 47 (1961)

MATTER, P., BRENNWALD, J., RÜTER, A., PERREN, S.M.: Die knöcherne Heilung von Schraubenlöchern nach Metallentfernung. Z. Orthop. **110**, 920 (1972)
– BRENNWALD, J., PERREN, S.M.: Biologische Reaktion des Knochens auf Osteosyntheseplatten. Helv. chir. Acta, Suppl. 12 (1974)
– – – : Die Heilung der Knochendefekte nach Entfernung von Osteosyntheseschrauben. Z. Unfallmed. Berufskr. **68**, 104 (1975)
– Skitraumatologie und Unfallprophylaxe. Schweiz. Z. Sozialvers. **20**, H. 1 (1976)

MÜLLER, J., SCHENK, R., WILLENEGGER, H.: Experimentelle Untersuchungen über die Entstehung reaktiver Pseudarthrosen am Hunderadius. Helv. chir. Acta **35**, 301–308 (1968)
SCHENK, R., –, WILLENEGGER, H.: Experimentell-histologischer Beitrag zur Entstehung und Behandlung von Pseudarthrosen. Hefte Unfallheilk. **94**, 15 (1968)
– PLAASS, U., WILLENEGGER, H.: Spätergebnisse nach operativ behandelten Malleolarfrakturen. Helv. chir. Acta **38**, 329–337 (1971)
– SCHENK, R.: Zuggurtungsplattenosteosynthese zur Behandlung von Pseudarthrosen der langen Röhrenknochen. Mschr. Unfallheilk. **74**, 253–271 (1971)

MÜLLER, M.E.: Die hüftnahen Femurosteotomien. 1. Aufl. 1957, 2. Aufl. mit Anhang: 12 Hüfteingriffe 1971. Stuttgart: Thieme

MÜLLER, M.E., DEBRUNNER, H.: Sulla diagnosi delle lesioni con sublussazione dell'articolazione tibioastragalica. Reforma méd. **72**, 961 (1958)
– Internal fixation for fresh fractures and for nonunion. Proc. roy. Soc. Med. **56**, 6, 455–460 (1963)
– ALLGÖWER, M., WILLENEGGER, H.: Technik der operativen Frakturenbehandlung. Berlin-Göttingen-Heidelberg: Springer 1963
– A propos de la guérison per primam des fractures. Rev. Orthop. **50**, 5, 697 (1964)
– ALLGÖWER, M., WILLENEGGER, H.: Technique of internal fixation of fractures. Berlin-Heidelberg-New York: Springer 1965
– Treatment of Non-Unions by compression. Clin. Orthop. **43**, 83–92 (1966)
– Intertrochanteric osteotomy in arthrosis of the hip joint. Proc. Sect. Meeting ACS in coop. with Germ. Surg. Soc., Munich June 26–29, 1968. Berlin-Heidelberg-New York: Springer 1969
– Compression as an aid in orthopaedic surgery. From "Recent advances in orthopaedics" by APLEY, A.G. (editor), pp. 79–89. London: Churchill 1969
– Fractures basses du fémur. Acta orthop. belg. **36**, 566–575 (1970)
– Biomechanische Fehlleistungen. Langenbecks Arch. Chir. **329**, 1144–1151 (1971)
– Femoral shaft surgery. See HALL Int. Inc.: Air instrument surgery, orthopaedics, Volume 2 (1972)
– Intertrochanteric osteotomy in the treatment of the arthritic hip joint. See TRONZO, R.G.: Surgery of the hip joint. Philadelphia: Lea & Febiger 1973
– Schenkelhalsfraktur beim Kind. Orthop. Prax. **2**, 65–67 (1974)
– GANZ, R.: Luxation und Frakturen: Untere Gliedmaßen und Becken. Beitrag zu REHN, J.: Unfallverletzungen bei Kindern. Prophylaxe, Diagnostik, Therapie, Rehabilitation. Berlin-Heidelberg-New York: Springer 1974
– Schrauben- und Plattenosteosynthese. Beitrag zu BIER, A., BRAUN, H., KÜMMELL, H.: Chirurgische Operationslehre. Achte Aufl., herausgeg. von DERRA, E., HUBER, P., SCHMITT, W., Bd. 6: Operationen an Extremitäten, Becken und Haut. Leipzig: Barth 1975
– Intertrochanteric Osteotomies in Adults: Planning and Operating Technique. Chapter 6 in CRUESS, R.L., MITCHELL, N.S.: Surgical management of degenerative arthritis of the lower limb. Philadelphia: Lea & Febiger 1975
– Zur Einteilung und Reposition der Kinderfrakturen. Unfallheilkunde **80**, 187–190 (1977)

NOESBERGER, B.: Ein Halteapparat zum differenzierten Nachweis der fibularen Bandläsion. Helv. chir. Acta **43**, 195–203 (1976)
HACKENBRUCH, W., –: Die Kapselbandläsion am Sprunggelenk. Ther. Umsch. **33**, 1976, Heft 6, 433–439

OROZCO, R.: Fractura muy conminuta de cubito y radio, abierta. Rev. Ortop. Traum. (Iberila) Abril 1970
– Osteosintesis en las fracturas de raquis cervical. Rev. Ortop. Traum. (Iberila) Julio 1970
– LLOVET TAPIES, J.: Osteosintesis en las lesiones traumaticas y degenerativas de la columna cervical. Trauma. cirug. Rehab. **1**, No. 1 (1971)
– Fracturas de la diafisis femoral. Memories de 1er. Congreso nacional de la Asoc. Mex. de Ortop. y Traumat. en Febrero de 1972, p. 96
– Valor de la compresion en cirugia osea, p. 87
– Osteosintesis diafisaria. Técnica AO. Barcelona: Editorial Cientifico-Médica 1974
– HAMOUI, N.: Tratamiento de las fracturas condileas de femur. La Paz-Madrid Fines de Semanes Traumatologicos 1974
– VALLHONRAT, F.: Fracturas conminutas y polifragmentarias de tercio medio de femur. Rev. Ortop. Traum. (Madrid) **19**, 4, 887–894 (1975)

— — —: Resección-reconstrucción del tercio superior de tibia. Rev. Ortop. Traum. **19**, 3, 673–682 (1975)
— El principio de neutralización. Aplicación a las fracturas de tibia. Técnica AO. Tarragona: Ediciones Tarraco 1976
PAUWELS, F.: Der Schenkelhalsbruch, ein mechanisches Problem. Stuttgart: Enke 1935
— Gesammelte Abhandlungen zur funktionellen Anatomie des Bewegungsapparates. Berlin-Heidelberg-New York: Springer 1965
PERREN, S.M.: Naht- und Implantatmaterialien in der Extremitätenchirurgie. Chirurg **46**, 10, S. 447–453 (1975)
— ALLGÖWER, M., EHRSAM, R., GANZ, R., MATTER, P.: Clinical experience with a new compression plate "DCP". Acta orthop. scand., Suppl. **125** (1969)
— HUGGLER, A., RUSSENBERGER, M., ALLGÖWER, M., MATHYS, R., SCHENK, R.K., WILLENEGGER, H., MÜLLER, M.E.: The reaction of cortical bone to compression. Acta orthop scand., Suppl. **125** (1969)
— HUGGLER, A., RUSSENBERGER, M., STRAUMANN, F., MÜLLER, M.E., ALLGÖWER, M.: A method of measuring the change in compression applied to living cortical bone. Acta orthop. scand., Suppl. **125** (1969)
— HUTZSCHENREUTER, P., STEINEMANN, S.: Some effects of rigidity of internal fixation on the healing pattern of osteotomies. Z. Surg. **1**, 77 (1969)
— RUSSENBERGER, M., STEINEMANN, S., MÜLLER, M.E., ALLGÖWER, M.: A dynamic compression plate. Acta orthop scand., Suppl. **125** (1969)
— ALLGÖWER, M.: Biomechanik der Frakturheilung nach Osteosynthese. Nova Acta Leopoldina **44**, 223, 61–84 (1976)
— GANZ, R., RÜTER, A.: Mechanical induction of bone resorption. 4th Int. Osteol. Symp., Prag 1972
— ALLGÖWER, M., CORDEY, J., RUSSENBERGER, M.: Developments of compression plate techniques for internal fixation of fractures. Progr. Surg. (Basel) **12**, 152 (1973)
— GANZ, R., RÜTER, A.: Oberflächliche Knochenresorption um Implantate. Med. Orthop. Tech. **95**, 6 (1975)
— MATTER, P., RÜEDI, T., ALLGÖWER, M.: Biomechanics of fracture healing after internal fixation. Surg. Ann. 361–390 (1975)
POHLER, O., STRAUMANN, F.: Charakteristik der AO-Implantate aus rostfreiem Stahl. AO-Bulletin Sept. 1975
PUDDU, G.C.: Unveröffentlichte Arbeit aus dem Laboratorium für experimentelle Chirurgie, CH-7270 Davos.
RAHN, B.A., PERREN, S.M.: Xylenol orange, a fluorochrome useful in polychrome sequential labeling of calcifying tissues. Stain Technol. **46**, 125 (1971)
— GALLINARO, P., BALTENSPERGER, A., PERREN, S.M.: Primary bone healing. An experimental study in the rabbit. J. Bone Jt Surg. A **53**, 4, 783–786 (1971)
— PERREN, S.M.: Alizarinkomplexon-Fluorochrom zur Markierung von Knochen- und Dentinanbau. Experientia (Basel) **28**, 180 (1972)
RITTMANN, W.W., PUSTERLA, C., MATTER, P.: Früh- und Spätinfektionen bei offenen Frakturen. Helv. chir. Acta **36**, 537–540 (1969)
— MATTER, P., ALLGÖWER, M.: Behandlung offener Frakturen und Infekthäufigkeit. Acta chir. austr. **2**, 18–21 (1970)
— Antibiotika bei der posttraumatischen Osteitis. Beitrag zu BURRI, C., Posttraumatische Osteitis, pp. 98–107, Bern-Stuttgart-Wien: Huber 1973
— MATTER, P., BRENNWALD, J., KAYSER, F.H., PERREN, S.M.: Biologie und Biomechanik infizierter Osteosynthesen. Fortschr. Kiefer- u. Gesichtschir., Vol. XIX, pp. 48–50 (1975)
— PERREN, S.M.: Corticale Knochenheilung nach Osteosynthese und Infektion. Berlin-Heidelberg-New York: Springer 1974
— PERREN, S.M.: Cortical bone healing after internal fixation and infection; biomechanics and biology. Berlin-Heidelberg-New York: Springer 1974
ROY-CAMILLE, R., BARCAT, E., DEMEULENCARE, C., SAILLAUT, G.: Chirurgie par abord postérieur du rachis dorsal et lombaire. Encylcopédie Méd.-Chir. 1–44178
RÜEDI, TH., MATTER, P., ALLGÖWER, M.: Die intraartikulären Frakturen des distalen Unterschenkelendes. Helv. chir. Acta **35**, 5, 556–582 (1968)
— Frakturen des Pilon Tibial: Ergebnisse nach 9 Jahren. Arch. orthop. Unfall-Chir. **76**, 248–254 (1973)
— MOSHFEGH, A., PFEIFFER, K.M., ALLGÖWER, M.: Fresh fractures of the shaft of the humerus. Conservative or operative treatment? Reconstr. Surg. Traumat. **14**, 65–74 (1974)
— ALLGÖWER, M.: Die Frakturheilung nach Osteosynthese im Röntgenbild. Helv. chir. Acta **41**, 213–216 (1974)
— WOLFF, G.: Vermeidung posttraumatischer Komplikationen durch frühe definitive Versorgung von Polytraumatisierten mit Frakturen des Bewegungsapparates. Helv. chir. Acta **42**, 507–512 (1975)
— Titan und Stahl in der Knochenchirurgie. Hefte zur Unfallheilkunde 123. Berlin-Heidelberg-New York: Springer 1975
— KOLBOW, H., ALLGÖWER, M.: Erfahrungen mit der dynamischen Kompressionsplatte (DCP) bei 418 frischen Unterschenkelschaftbrüchen. Arch. orthop. Unfall-Chir. **82**, 247–256 (1975)
— ALLGÖWER, M.: Richtlinien der schweizerischen AO für die Nachbehandlung operativ versorgter Frakturen. AO-Bulletin Frühjahr 1975
— WEBB, J.K., ALLGÖWER, M.: Experience with the dynamic compression plate (DCP) in 418 recent fractures of the tibial shaft. Injury **7**, 4, 252–257 (1976)

STADLER, J., GAUER, E. —: Operativ versorgte Malleolarfrakturen — Ergebnisse nach 3–4 Jahren mit besonderer Berücksichtigung des Talusprofils. Arch. Orthop. Unfall-Chir. **82**, 311–323 (1975)

HELL, K., MÜLLER, C., —: Nachkontrolle von 50 operativ behandelten Tibiakopffrakturen. Helv. chir. Acta **42**, 27–29 (1975)

SARMIENTO, A.: A functional below-the-knee brace for tibial fractures. A report on its use in one hundred thirty-five cases. J. Bone Jt Surg. A **52**, 295 (1970)

— Functional bracing of tibial and femoral shaft fractures. Clin. Orthop. **82**, Jan./Febr. (1972)

See also: WIEDMER, U., GMÜR, D., STÜHMER, G., DOERIG, M., BIANCHINI, D.: Die Behandlung der Unterschenkelfraktur mit der funktionellen konservativen Methode (DEHNE-SARMIENTO). Unfallheilkunde **80**, 303–311, 1977

SIMON-WEIDNER, R.: Die Fixierung trochanterer Brüche mit multiplen elastischen Rundnägeln nach SIMON-WEIDNER, Hefte Unfallheilk. **106**, 60 (1970)

SMITH, R.W.: A treatise on fractures in the vicinity of joints and of certain forms of accidental and congenital dislocations. Dublin: Hodges and Smith (1847)

SMITH-PETERSEN, M.N.: Treatment of fractures of the neck of the femur by internal fixation. Surg. Gynec. Obstet. **64**, 287 (1937)

— Approach to and exposure of the hip joint for mold arthroplasty. J. Bone Jt surg. A **31**, 40 (1949)

— CAVE, E.F., VAN GORDER, G.W.: Intracapsular fractures of the neck of femur. Arch. Surg. **23**, 715 (1931)

SCHATZKER, J.: Compression as an aid in the surgical management of fracture of the tibia. Clin. Orthop. & Related Research 1975

— WADDELL, J., HORNE, G.: The Toronto experience with the supracondylar fractures of the femur. Injury 1975

— SANDERSON, R., MURNAGHAN, P.: The holding power of orthopaedic screws in vivo. Clin. Orthop. & Related Research 1975

— HORNE, G., SUMMER-SMITH, G.: The effect of compression of cortical bone by screw threads and the effect of movement of screws in bone. Clin. Orthop. & Related Research 1975

SCHENK, R.K., MÜLLER, J., WILLENEGGER, H.: Experimentell-histologischer Beitrag zur Entstehung und Behandlung von Pseudarthrosen. Hefte Unfallheilk. **94**, 15–24 (1968); 31. Tagg. Berlin 1967

— WILLENEGGER, H.: Zum histologischen Bild der sogenannten Primärheilung der Knochenkompakta nach experimentellen Osteotomien am Hund. Experientia (Basel) **19**, 593–595 (1963)

— WILLENEGGER, H.: Histologie der primären Knochenheilung. Langenbecks Arch. klin. Chir. **308**, 440–452 (1964)

— WILLENEGGER, H.: Morphological findings in primary fracture healing. Symp. Biol. Hung. **7**, 75–86 (1967)

— Fracture repair—Overview. Ninth European Symp. on Calcified Tissues, Baden near Vienna, 1972, XIII–XXII, Facta-Publication, 1973

SCHNEIDER, R.: Die Marknagelung der Tibia. Helv. chir. Acta **28**, 207 (1961)

— Komplikationen bei der Marknagelung der Tibia. Helv. chir. Acta **30**, fasc. 1–2, 95–97 (1963)

— Mehrjahresresultate eines Kollektivs von 100 intertrochanteren Osteotomien bei Coxarthrose. Helv. chir. Acta **35**, fasc. 1–2, 185–205 (1966)

— Die Gefahren der Osteosynthese. Aktuelle Chirurgie **6**, 2, 89–102 (1971)

— Mechanische Fehlleistungen bei der Druckosteosynthese. Zbl. Chir. **100**, 201–209 (1975)

— Indikation zur konservativen oder operativen Frakturbehandlung. Sonderfall: Der alte Mensch. Langenbecks Arch. Chir. **337** (Kongreßbericht 1974)

SCHWEIBERER, L., HOFMEIER, G.: Die Bündelnagelung bei Unterschenkel- und Oberarmfrakturen. Zbl. Chir. **92**, 48, 2903 (1967)

— Experimentelle Untersuchungen von Knochentransplantaten mit unveränderter und mit denaturierter Knochengrundsubstanz. (Ein Beitrag zur causalen Osteogenese). Hefte Unfallheilk. **103** (1970)

— VAN DE BERG, A., DAMBE, L.T.: Das Verhalten der intraossären Gefäße nach Osteosynthese der frakturierten Tibia des Hundes. Therapiewoche **20**, 1330 (1970)

— Neuere Ergebnisse zur Knochenregeneration und ihre klinische Bedeutung. Langenbecks Arch. Chir. **329** (1971)

— Der heutige Stand der Knochentransplantation. Chirurg **42**, 252 (1971)

— LINDEMANN, M.: Infektion nach Marknagelung. Chirurg **44**, 542 (1973)

— DAMBE, L.T., EITEL, F., KLAPP, F.: Revaskularisation der Tibia nach konservativer und operativer Frakturbehandlung. Hefte Unfallheilk. **119**, 18 (1974)

— BÖS, T., POEPLAU, P.: Verzögerte Bruchheilung. Akt. Traumatologie 4, 163 (1974)

HERTEL, P., —: Die Ergebnisse nach operativer Behandlung von 48 frischen Monteggia-Verletzungen. Akt. Traumatologie **4**, 147 (1974)

— KLAPP, F., CHEVALIER, H.: Platten- und Schraubenosteosynthese bei Frakturen und Pseudarthrosen des Ober- und Unterschenkels. Chirurg **46**, 155 (1975)

— Weichteilschaden beim Knochenbruch. Langenbecks Arch. Chir. **339** (1975)

— Bedeutung der autologen Spongiosatransplantation sowie Fragen der Vaskularisation von Transplantaten. In: Callus—Nova acta Leopoldina **223**, 44, 371 (1976)

— HERTEL, P.: Unfallchirurgische Eingriffe im Kindesalter. In: Breitner Operationslehre, Bd. 6. München-Berlin-Wien: Urban & Schwarzenberg 1976

— Theoretisch-experimentelle Grundlagen der autologen Spongiosatransplantation im Infekt. Unfallheilk. **79**, 151 (1976)

EITEL, F., DAMBE, L.T., KLAPP, F. —: Vaskularisation der Diaphyse langer Röhrenknochen unter Cerclagen. Unfallheilk. **79**, 41 (1976)
EITEL, F., KLAPP, F., DAMBE, L.T., —: Revaskularisierung hypertrophischer Pseudarthrosen nach Druckplattenosteosynthese. Chir. Forum 76, Suppl. zu Langenb. Arch. klin. Chir., Berlin-Heidelberg-New York: Springer 1976
KLAPP, F., EITEL, F., DAMBE, L.T., —: Revaskularisation devitalisierter Corticalissegmente unter stabilisierenden Cerclagen. Chir. Forum 76, Suppl. zu Langenb. Arch. klin. Chir., Berlin-Heidelberg-New York: Springer 1976

WAGNER, H.: Die Einbettung von Metallschrauben im Knochen und die Heilungsvorgänge des Knochengewebes unter dem Einfluß der stabilen Osteosynthese. Langenbecks Arch. klin. Chir. **305**, 28 (1963)
— Technik und Indikation der operativen Verkürzung und Verlängerung von Ober- und Unterschenkel. Orthopädie **1**, 59–74 (1972)

WATSON-JONES, R.: Fractures and joint injuries. Edinburgh: Livingstone 1955

WEBER, B.G.: Die Verletzungen des oberen Sprunggelenkes. Aktuelle Probleme in der Chirurgie, Bd. 3. Bern u. Stuttgart: Huber 1966 u. 1972
— Fractures of the femoral shaft in childhood. Injury **1**, 1, July (1969)
SÜSSENBACH, F., —: Epiphysenfugenverletzungen am distalen Unterschenkel. Bern-Stuttgart-Wien: Huber 1970
— Die Verletzungen des oberen Sprunggelenks. Bern-Stuttgart-Wien: Huber 1966 und 1972
— ČECH, O.: Pseudarthrosis. Pathophysiology, Biomechanics, Therapy, Results. Bern-Stuttgart-Wien: Huber 1976

WELLER, S.: Die Marknagelung von Ober- und Unterschenkelbrüchen. Dtsch. med. Wschr. **16**, 681–684 (1965)
— Grenzen der konservativen und operativen Fraktur-Behandlung. Hefte Unfallheilk. **87**, 138–140 (1966)
— Behandlungsprinzipien von Pseudarthrosen. Chirurg **10**, 445–448 (1967)
— Gedanken zur konservativen und operativen Knochenbruchbehandlung. Mschr. Unfallheilk. **6**, 233–242 (1967)
— Therapeutische Gesichtspunkte und Behandlungsergebnisse bei Mehrfachfrakturen. Langenbecks Arch. klin. Chir. **322**, 1073 (1968)
— Der Oberschenkel-Mehrfragmentenbruch im Schaftbereich. Dtsch. med. Wschr. **13**, 645–652 (1969)
— Allgemeine Indikationsfehler bei der Knochenbruchbehandlung. 87. Tg. Dtsch. Ges. Chir., 1.–4. Apr. 1970, Bd. 327, 813 (1970)
— Zur Behandlung offener Gelenkverletzungen. Therapiewoche **20**, 27, 1320 (1970)
— Für und Wider die Indikation zur Drahtumschlingung bei Schaftbrüchen. Act. Traumatol. **1**, 1, (1971)
— Vermeidung technischer Fehler bei der operativen Behandlung von Frakturen. Chirurg **43**, 100–104 (1972)
— Distale Femurfrakturen im Wachstumsalter. Act. Traumatol. **2**, 2 (1972)
— Komplikationen bei der Marknagelung. Therapiewoche **22**, 47, 4178 (1972)
— Grundsätzliche Fehler und Komplikationsmöglichkeiten der Marknagelung. Chirurg **44**, 533–539 (1973)
— Die Marknagelung von Unterschenkelschaftbrüchen. Heft 117, 98–102, 37. Jahrestag. Dtsch. Ges. Unfallheilk., Berlin 1973 (1974)
— Indikationen zum chirurgischen Eingriff — Wandlungen und Entwicklungen in der Unfallchirgie. Langenbecks Arch. Chir. **337**, 57–63 (1974)
— Die Marknagelung — Gute und relative Indikationen, Ergebnisse. Chirurg **46**, 152–154 (1975)
— Die Indikation zur Versorgung proximaler Femurfrakturen mit elastischen Rundnägeln. Akt. Traumatolog. **6**, 167–170 (1976)

WILLENEGGER, H., GUGGENBÜHL, A.: Zur operativen Behandlung bestimmter Fälle von distalen Radiusfrakturen. Helv. chir. Acta **26**, 81 (1959)
— Die Behandlung der Luxationsfrakturen des oberen Sprunggelenks nach biomechanischen Gesichtspunkten. Helv. chir. Acta **28**, 225 (1961)
SCHENK, R., —: Zum histologischen Bild der sogenannten Primärheilung der Knochenkompakta nach experimentellen Osteotomien am Hund. Experientia (Basel) **19**, 593 (1963)
SCHENK, R., MÜLLER, J. —: Experimentell-histologischer Beitrag zur Entstehung und Behandlung von Pseudarthrosen. Hefte Unfallheilk. **94**, 15 (1968)
RIEDE, U., —, SCHENK, R.: Experimenteller Beitrag zur Erklärung der sekundären Arthrose bei Frakturen des oberen Sprunggelenks. Helv. chir. Acta **36**, 343 (1969)
— Problems and results in the treatment of comminuted fractures of the elbow. Reconstr. Surg. Traumat. **11**, 118 (1969)
— SCHNEIDER, R., BANDI, W.: Irrungen und Wirrungen in der Frakturenbehandlung. Acta Chir. Austriaca **2**, 6 (1970)
— Klinik und Therapie der pyogenen Knocheninfektion. Chirurg **41**, 215 (1970)
— Spätergebnisse nach konservativ und operativ behandelten Malleolarfrakturen. Helv. chir. Acta **38**, 321 (1971)
— PERREN, S.M., SCHENK, R.: Primäre und sekundäre Knochenbruchheilung. Chirurg **42**, 241 (1971)
— BIRCHER, J.: Sudecksche Dystrophie des Fußes. Z. Unfallmed. Berufskrh. **64**, 109 (1971)
RIEDE, U.H., SCHENK, R., —: Gelenkmechanische Untersuchungen zum Problem der posttraumatischen Arthrosen im oberen Sprunggelenk. 1. Die intraartikuläre Modellfraktur. Langenbecks Arch. Chir. **328**, 258 (1971)

— Licht und Schatten über Indikation zur Knochenbruchbehandlung. Mschr. Unfallheilk. **75**, 455 (1972)
— Präliminäre Überbrückungs-Osteosynthese bei der Resektion von Knochentumoren. Helv. chir. Acta **40**, 185 (1973)
Terbrüggen, D., —: Die operative Versorgung des Außenknöchels bei Malleolarfrakturen. Med. Techn. (Berlin) **93**, 80 (1973)
Riede, U.N., Schweizer, G., Marti, J., —: Gelenkmechanische Untersuchungen zum Problem der posttraumatischen Arthrosen im oberen Sprunggelenk. III. Funktionell-morphometrische Analyse des Gelenkknorpels. Langenbecks Arch. Chir. **333**, 91 (1973)
— Therapie der traumatischen Osteomyelitis. Langenbecks Arch. Chir. **334**, 529 (1973)
Ledermann, M., Burckhardt, A., —: Zur Indikation der Navikulareverschraubung an Hand von Nachkontrollen. Helv. chir. Acta **41**, 239 (1974)
Rehn, J., —: Indikationen zur konservativen und operativen Knochenbruchbehandlung. Langenbecks Arch. Chir. **337**, 389 (1974)
— Die überregionale Zusammenarbeit in der klinischen Forschung. (Am Beispiel der Internationalen Arbeitsgemeinschaft für Osteosynthesefragen). Chirurg **45**, 494 (1974)
— Verplattung und Marknagelung bei Femur- und Tibiaschaftfrakturen: Pathophysiologische Grundlagen. Leitthema: Nagelung oder Plattenosteosynthese? Chirurg **46**, 145 (1975)
Meyer, St., Weiland, A., —: The treatment of infected non-union of fractures of long bones. J Bone Jt Surg. **57**, 836 (1975)

Subject Index

R. Bombelli

Osteoarthritis of the Hip

Pathogenesis and Consequent Therapy
With a Foreword by M. E. Müller

1976. 160 figures (70 in color). X, 136 pages
ISBN 3-540-07842-8

J. Charnley

Low Friction Arthroplasty of the Hip

Theory and Practice

1979. 440 figures (205 in colour).
XII, 376 pages
ISBN 3-540-08893-8

The Dynamic Compression Plate (DCP)

By M. Allgöwer, P. Matter, S. M. Perren, T. Rüedi

Revised printing. 1978. 29 figures. IV, 48 pages
ISBN 3-540-06466-4

U. Helm, K. M. Pfeiffer

Small Fragment Set Manual

Technique Recommended by the ASIF-Group
In collaboration with H. C. Meuli
Translators: R. Kirschbaum, R. L. Batten

1974. 157 figures (414 separate illustrations).
IX, 299 pages
ISBN 3-540-06904-6

Late Reconstructions of Injured Ligaments of the Knee

Editors: K.-P. Schulitz, H. Krahl, W. H. Stein
With contributions by M. E. Blazina, D. H. O'Donoghue, S. L. James, J. C. Kennedy, A. Trillat

1978. 42 figures, 21 tables. V, 120 pages
ISBN 3-540-08720-6

R. Liechti

Hip Arthrodesis and Associated Problems

Foreword by M. E. Müller, B. G. Weber
Translated from the German edition "Die Arthrodese des Hüftgelenks und ihre Problematik" by P. A. Casey

1978. 266 figures, 35 tables. XII, 269 pages
ISBN 3-540-08614-5

P. G. J. Maquet

Biomechanics of the Knee

With Applications to the Pathogenesis and the Surgical Treatment of Osteoarthritis

1976. 184 figures. XIII, 230 pages
ISBN 3-540-07882-7

M. E. Müller, M. Allgöwer, H. Willenegger

Technique of Internal Fixation of Fractures

With contributions by W. Bandi, H. R. Bloch, A. Mumenthaler, R. Schneider, S. Steinemann, F. Straumann, B. G. Weber
Revised for the english edition by G. Segmüller

1965. 244 figures. XII, 272 pages
ISBN 3-540-03374-2

New Concepts in Maxillofacial Bone Surgery

Editor: B. Spiessl
With contributions by numerous experts

1976. 183 figures, 36 tables. XIII, 194 pages
ISBN 3-540-07929-7

Springer AV Instruction Program

Films:

Theoretical and practical bases of internal fixation, results of experimental research:

Internal Fixation – Basic Principles and Modern Means

The Biomechanics of Internal Fixation

The Ligaments of the Knee Joint Pathophysiology

Internal fixation of fractures and in corrective surgery:

Internal Fixation of Forearm Fractures

Internal Fixation of Noninfected Diaphyseal Pseudarthroses

Internal Fixation of Malleolar Fractures

Internal Fixation of Patella Fractures

Medullary Nailing

Internal Fixation of the Distal End of the Humerus

Internal Fixation of Mandibular Fractures

Corrective Osteotomy of the Distal Tibia

Internal Fixation of Tibial Head Fractures
(available in German only)

Joint replacement:

Total Hip Prostheses (3 parts)

Part 1: Instruments. Operation on Model

Part 2: Operative Technique

Part 3: Complications. Special Cases

Elbow-Arthroplasty with the New GSB-Prosthesis

Total Wrist Joint Replacement

Slide Series:

ASIF-Technique for Internal Fixation of Fractures

Manual of Internal Fixation
(in preparation)

The Ligaments of Knee Joint – Pathophysiology
(in preparation)

Small Fragment Set Manual

Internal Fixation of Patella and Malleolar Fractures

Total Hip Prostheses
Operation on Model and in vivo, Complications and Special Cases

Asepsis in Surgery

- Further films and slide series in preparation
- Technical data: 16 mm and super-8 (color, magnetic sound, optical sound), videocassettes. Slide series in ringbinders
- Films in English or German, several in French; slide series with multilingual legends
- Please ask for information material

Sales:
Springer-Verlag,
Heidelberger Platz 3,
D-1000 Berlin 33,

or

Springer-Verlag New York Inc
175 Fifth Avenue,
New York, NY 10010

Springer-Verlag
Berlin
Heidelberg
New York

RL Jacobs